International Conference on Materials, Mechanical and Energy Engineering (ICMMEE 2021)

IOP Conference Series: Materials Science and Engineering
Volume 1185

Bapatla, India
7 – 8 May 2021

Editor:

B. Ratna Sunil

ISBN: 978-1-7138-4645-1
ISSN: 1757-8981

Printed from e-media with permission by:

Curran Associates, Inc.
57 Morehouse Lane
Red Hook, NY 12571

Some format issues inherent in the e-media version may also appear in this print version.

Institute of Physics
Dirac House, Temple Back
Bristol BS1 6BE UK

Phone: 44 1 17 929 7481
Fax: 44 1 17 920 0979

techtracking@iop.org

Additional copies of this publication are available from:

Curran Associates, Inc.
57 Morehouse Lane
Red Hook, NY 12571 USA
Phone: 845-758-0400
Fax: 845-758-2633
Email: curran@proceedings.com
Web: www.proceedings.com

TABLE OF CONTENTS

IOP Conf. Series: Materials Science and Engineering 1185 (2021) 011001 doi:10.1088/1757-899X/1185/1/011001

International Conference on Materials, Mechanical and Energy Engineering (ICMMEE 2021)
07th – 08th May 2021

Preface

Recent developments in the field of materials engineering revolutionized the engineering applications in different strategic sectors including automobile, aerospace, marine, biomedical, agriculture, energy engineering and structural and construction engineering. Developing new material systems and novel manufacturing processes enhanced the ability of engineers to address several technological challenges in various industrial applications. Understating the behavior of the materials helps to develop potential engineering applications to improve the quality of the human life. Utilizing computational tools to analyze the materials and processes brings additional advantage to understand the complex issues and accelerates the development of science and technology. Interdisciplinary research couples the expertise of several professionals belong to diversified fields and promotes rapid growth in achieve a common goal in addressing the research challenges. In this context, we believe that the International Conference on Materials, Mechanical and Energy Engineering (ICMMEE 202) organized on 07th & 08Th May 2021 has become a viable platform for knowledge shearing among the experts from various research fields in materials and mechanical engineering. Dr. M. Ananda Rao, Principal Scientist, CSIR – NML Madras Complex, Chennai, India delivered inaugural address on "*Metallurgical failure investigations of industrial components*". Dr. Martin Sebesta, Institute of Laboratory Research on Geomaterials, Comenius University in Bratislava, Slovakia, delivered a key-note address on "*Transformation of ZnO and* TiO_2 *nanoparticles in the soil environment (influence of soil microorganisms and plants)*". Several post graduate and doctoral students have presented their research work at ICMMEE 2021 in the areas of materials engineering, thermal engineering, mechanical processing, physics, materials in civil engineering, design and analysis of structures etc. We believe that the knowledge exchanged at ICMMEE 2021 immensely helps to the researchers working on relevant fields and contributes to the growth of science and technology. Due to the COVID-19 pandemic and imposed lock down, the conference was virtually conducted through Google meet. All the keynote addresses and the contributory talks were presented through online. We thank the Bapatla Education Society, Bapatla for supporting the event. We also sincerely thank the Head of the Institute, Bapatla Engineering College, Head of the Mechanical Engineering Department, Bapatla Engineering College for their encouragement in successfully organizing the conference. Our sincere thanks to the keynote speakers, reviewers, technical and advisory committee members, organizing committee, and all authors for making ICMMEE 2021 a grand success.

Dr. B. Ratna Sunil
Editor/Convener
ICMMEE 2021

IOP Conf. Series: Materials Science and Engineering 1185 (2021) 011001 doi:10.1088/1757-899X/1185/1/011001

ICMMEE 2021 committees

Chairman

Dr. T. Nancharaiah, Professor & Head of the department, Department of Mechanical Engineering, Bapatla Engineering College, Bapatla 522101, India.

Conveners

Dr. B. Ratna Sunil, Associate Professor, Department of Mechanical Engineering, Bapatla Engineering College, Bapatla 522101, India.

Dr. D. Sameer Kumar, Associate Professor, Department of Mechanical Engineering, Bapatla Engineering College, Bapatla 522101, India.

Co-Convener

Dr. V.V. Kondaiah, Associate Professor, Department of Mechanical Engineering, Bapatla Engineering College, Bapatla 522101, India.

Technical & Advisory Committee

Dr. Martin Urik, Faculty of Natural Sciences, Comenius University in Bratislava, Slovak republic.
Dr. Marek Colencik, Slovenská poľnohospodárska univerzita v Nitre, Slovak republic.
Dr. M. Paparao, Senior Scientist, Jawaharlal Nehru Aluminium Research Development and Design Centre, Nagpur, India.
Dr. B. Balakrishna, Professor & Principal, JNTU College of Engineering, Kakinda, A.P. India.
Dr. M. Jagannatham, Manager, R&D, TVS Wheels India, Chennai, India
Dr. N. Rameshbabu, Associate Prof., Dept. of MME, NIT Tiruchirapalli, India
Dr. Ravikumar Dumpala, Asst. Prof. Department of Mechanical Engineering, VNIT Nagpur, India.
Dr. S. Anandkumar, Asst. Prof., Department of Mechanical Engineering, IIT Jammu, India.
Dr. T. Hanas, Asst. Prof., School of Materials Science and Engineering, NIT Calicut, India.
Dr. K. N. S. Suman, Associate Prof., Department of Mechanical Engineering, Andhra Univeristy, India
Dr. P. Poddar, Principal Scientist, CSIR-NML Jamshedpur, India
Dr. T. K. Mahanta, Asst. Prof., Department of Mechanical Engineering, VIT Chennai, India
Dr. M. Ravikumar, Asst. Prof., Department of Mechanical Engineering, MNIT Bhopal, India

Organizing committee

Dr. M. Venkateswara Rao, Dean, R &D, Depart. of Mechanical Engineering, Bapatla Engineering College.
B. Ravi Sankar, Professor, Department of Mechanical Engineering, Bapatla Engineering College.
Ch. Lakshmi Srinivas, Professor, Department of Mechanical Engineering, Bapatla Engineering College.
Sri. T. Siva Sankar, Associate Professor, Department of Mechanical Engineering, Bapatla Engineering College.
Sri. G. Kantha Prasad, Associate Professor, Dept. of Mechanical Engineering, Bapatla Engineering College.
Sri. K. Prasada Rao, Associate Professor, Dept. of Mechanical Engineering, Bapatla Engineering College.

Sri. D Narayana Chowdary, Assoc. Prof., Dept. of Mechanical Engineering, Bapatla Engineering College,
Dr. B. Ifthekar Hussain, Associate Professor, Dept. of Mechanical Engineering, Bapatla Engineering College.
Dr. J. Laxmi Lalitha, Associate Professor, Dept. of Mechanical Engineering, Bapatla Engineering College.
Dr. P. Ravikumar, Associate Professor, Department of Mechanical Engineering, Bapatla Engineering College.
Sri. K. Srinivasa Rao, Assistant Professor, Dept. of Mechanical Engineering, Bapatla Engineering College.
Sri. P. Uma Maheswara Rao, Asst. Prof., Dept. of Mechanical Engineering, Bapatla Engineering College.
Sri. S. Vijay, Assistant Professor, Department of Mechanical Engineering, Bapatla Engineering College.
Sri. Y. Narendra Babu, Assistant Professor, Dept. of Mechanical Engineering, Bapatla Engineering College.
Dr. S. Krugon, Assistant Professor, Dept. of Mechanical Engineering, Bapatla Engineering College.
Sri. S. Ravi Kumar, Assistant Professor, Dept. of Mechanical Engineering, Bapatla Engineering College.
Sri. D. Vijay Praveen, Assistant Professor, Dep. of Mechanical Engineering, Bapatla Engineering College.
Sri. K. Rajasekar Babu, Assistant Professor, Dept. of Mechanical Engineering, Bapatla Engineering College.
Sri. M. Pardha Saradhi, Assistant Professor, Dept. of Mechanical Engineering, Bapatla Engineering College.
Dr. A. Praveen, Assistant Professor, Department of Mechanical Engineering, Bapatla Engineering College.
Sri. V. Srinivasulu, Assistant Professor, Department of Mechanical Engineering, Bapatla Engineering College.
Sri. M. Subramanyam, Assistant Professor, Dept. of Mechanical Engineering, Bapatla Engineering College.
Sri. A Suresh, Assistant Professor, Department of Mechanical Engineering, Bapatla Engineering College.
Sri. Ch. L. Prudhvi Raz, Assistant Professor, Dept. of Mechanical Engineering, Bapatla Engineering College.
Sri. G. Moshe Babu, Assistant Professor, Department of Mechanical Engineering, Bapatla Engineering College.
Sri. V. Krishna Koundinya, Asst. Professor, Dept. of Mechanical Engineering, Bapatla Engineering College.

Session chairs

Dr. N. Raghu Kiran, Assoc. Professor, Department of Mechanical Engineering, VIT Chennai, India.
Dr. J. Lakshmi Lalitha, Assoc. Professor, Mechanical Engineering Department, Bapatla Engineering College, Bapatla, India
Dr. Anilkumar Birru, Asst. Professor, Mechanical Engineering Department, NIT Manipur, India.
Dr. B. I. Hussain, Assoc. Professor, Mechanical Engineering Department, Bapatla Engineering College, Bapatla, India
Dr. S. Krugon, Asst. professor, Mechanical Engineering Department, Bapatla Engineering College, Bapatla, India.
Dr. B. Sarath Chandra Kumar, Assoc. Professor, Civil Engineering Department, Kallam Haranadhareddy Institute of Technology, Guntur, India.
Dr. P. Sundara Kumar, Professor, Department of Civil Engineering, Bapatla Engineering College, Bapatla, India.
Dr. V. V. Kondaiah, Assoc. Professor, Department of Mechanical Engineering, Bapatla Engineering College, Bapatla, India.
Dr. J. S. Rao , Asst. Professor, Mechanical Engineering Department, Rajiv Gandhi University of Knowledge Technologies (RGUKT), Nuzvid, India.
Dr. L. Ratna Raju, Asst. Professor, Mechanical Engineering Department, VFSTR University,Vadlamudi, Guntur, India
Dr. B. Ravisankar, Professor, Mechanical Engineering Department, Bapatla Engineering College, Bapatla, India.
Dr. I Ramakanth, Asst. Professor, Department of Chemistry, Rajiv Gandhi University of Knowledge Technologies (RGUKT), Nuzvid, India.
Dr. K. Prasada Rao, Sr. Asst. Professor, Chemistry Department, Bapatla Engineering College, Bapatla, India.

Editorial board

Reviewers

IOP Conf. Series: Materials Science and Engineering 1185 (2021) 011001 doi:10.1088/1757-899X/1185/1/011001

IOP Conf. Series: Materials Science and Engineering 1185 (2021) 011002 doi:10.1088/1757-899X/1185/1/011002

Peer review declaration _ ICMMEE 2021

All papers published in this volume of *IOP Conference Series: Materials Science and Engineering* have been peer reviewed through processes administered by the Editors. Reviews were conducted by expert referees to the professional and scientific standards expected of a proceedings journal published by IOP Publishing.

- **Type of peer review:**

  We have adopted **double – blind** peer review process for ICMMEE. Initially, the editorial team has gone through the submitted papers and based on the research area, experts from relevant field were identified and sent for review. Before, sending the full papers to reviewing, detailed guidelines were provided to the reviewers.

- **Conference submission management system:** Google forms & icmmee@gmail.com

  We have created Google forms for every step right from the registration to submission of final version of the full paper. We have also provided a dedicated e-mail for communication

- **Number of submissions received:** 190
  We have received a total of 190 abstracts. Among the submitted abstracts, 124 were accepted for presentation and among them, 55 authors have submitted full papers and were sent for the peer review process. We completed review process and accepted 39 papers to publish as proceedings in IOP: Materials Science and Engineering.

- **Number of submissions sent for review: 55**

- **Number of submissions accepted: 39**

IOP Conf. Series: Materials Science and Engineering 1185 (2021) 011002 doi:10.1088/1757-899X/1185/1/011002

- **Acceptance Rate (Number of Submissions Accepted / Number of Submissions Received X 100): 70 %**

- **Average number of reviews per paper: 3**

- **Total number of reviewers involved: 41**

- **Any additional info on review process:**

  In addition to the reviewer's assessment, all the papers were also reviewed by at least one of the editors form the editorial board. The editor comments were also provided to the authors to revise the papers.

  All the papers were also verified for English corrections and language quality with the help of faculty from Department of English, Bapatla Engineering College. The authors have provided suggestions to improve the language quality of the papers and requested to incorporate the corrections while submitting the revision.

  The form provided to the reviewers to complete the review process of ICMMEE is herewith attached.

- **Contact person for queries:**

Dr. B. Ratna Sunil
Convener, ICMMEE 2021
Editor, Proceedings of ICMMEE 2021
Associate Professor, Department of Mechanical Engineering,
Bapatla Engineering College, Bapatla 522101, Andhra Pradesh, India
e-mail: bratnasunil@becbapatla.ac.in, phone: +91 9677119819

Application of MADM Methods for the Synthesis of Alumina/Lanthanum Hexa-aluminate Powders by Chemical Precipitation and Filtration Method

Y N V Sai Ram [1, 2*], C Tara Sasanka[3] and J Prabakaran[4]

[1] Research Scholar, Annamalai University, India
[2, 3] Assistant Professor, R.V.R. & J.C. College of Engineering, India
[4] Associate Professor, Annamalai University, India

* Corresponding author e-mail: sairamynv@gmail.com

Abstract. In product design, the selection of apt material plays a vital role as each material has individual uniqueness that add numerous aspects to suit the specific application. Picking of wrong material selection promotes product failure and also high-cost contribution. Consequently, it becomes hard to pick a specific one among the different materials available. For the choice of content, a better methodology is increasingly appreciated. MADM methods are the most looming concepts in today's world which facilitate the decision maker to select best strategies among the different variable alternatives. Multi-Attribute Decision Making (MADM) strategies based on mathematical models include a ranking of the various alternatives, making it easier to make decisions. As a thermal barrier coating (TBC), Lanthanum Hexa-aluminate (LHA) with a magneto plumbite structure is a promising competitor to yitria partially stabilized zirconia (Y-PSZ) and is synthesized via a chemical precipitation and filtration process.LHA is proved to be an efficient material however the correct composition reflecting better mechanical properties is a challenging task. Therefore, this paper presents the selection of best composition of Lanthanum Hexa-aluminate ceramic powder using AHP and TOPSIS methods and also the impact of weighting factors has been addressed as well.

Keywords: Lanthanum Hexa-aluminate (LHA),Multi-Attribute Decision Making (MADM), Analytic Hierarchy Process (AHP), Technique for Order Preference by Similarity to Ideal Solution (TOPSIS),Entropy based weighting(EBW),measure of output (MOP),

1. Introduction

Materials with elevated temperature capability are imperative in many sectors like manufacturing including processing of materials, chemical engineering, power generation, transport and aerospace [1-3]. The work capacity at elevated temperatures is therefore a vital factor for industrial efficiency. The material from which the part is produced basically meets the mechanical requirements of the process and provides a ounce of environmental resistance. Nevertheless, to provide resistance to environment a coating that is physically and chemically friendly with the matrix material may be desired. However,

Selection of material is characterized as a factor that influences the choice of a material for a specific application, and it is typically done by many young engineers [4-5]. It should be mentioned that there may not be always a single objective to select the elite material, so designers and engineers must weigh a diverse range of materials selection criteria. Initial screening, production and comparison of alternatives and determination of the best solution are the three stages of material selection [6-7]. Many aspects must be considered when selecting a material for agreeable economic output in a specific application. Multi-Criteria Decisions Making (MCDM) [8-10] divides decision making in the presence of multiple competing criteria into two categories: Multi-Attribute Decision Making (MADM) approaches and Multi-Objective Decision Making (MODM) [11]. Recently, MADM methods have become more commonly available to assist researchers and practitioners in assessing and picking the best compromise choices. These MADM models are capable of classifying materials based on a range of properties (attributes) [12]. Analytical Hierarchy Process (AHP) [13–16] and Technique for Order Preference by Similarity to Ideal Solution (TOPSIS) [17–19] are two common MADM methods that have earned a lot of attention for dealing with serious problems with friction factors. TOPSIS and AHP methods are used in this analysis to rate competing materials for their applications.

Lanthanum Hexa-aluminate ($LaAl_{11}O_{18}$) is truncated as LHA which belongs to the rare-earth aluminate family with magneto plumbite structure with a unique stoichiometry and it is the key compound formed in La_2O_3–Al_2O_3 system and it is used as a material in thermal barrier coating [20-21]. This paper focuses on application of MADM methods to opt best condition for synthesis of LHA. Problem formulation and methodology details are mentioned in section 2 & 3 while the results are discussed in the subsequent sections.

2. Problem formulation

Three different compositions of Alumina/LHA powder ranging from 20 to 60 vol% at 20 vol% LHA interval were synthesized using chemical precipitation and filtration technique [22]. But the economical production of Lanthanum Hexa-aluminate nano particles is a major challenge. Experiments are designed based on the Taguchi L9 orthogonal array by considering the three parameters at three levels as in Table 1. From Table 2, the decision making to pick the specific combination is complex as it is based on the application perspective. The present problem involves identification of best combination of parameters among the nine runs available trying to get the preeminent combination fulfilling the requirements thereby decision maker has to judge the best one.

3. Methodology

3.1. *Determining the weightage of each attribute* A subsection

The problem proposed here is the implementation entropy based weighting factors for each attribute and to take an accurate decision.

3.2. Employment of different techniques

When the decision maker relies on the weightages, a variety of different methodologies may be used. The MADM methods are discussed in simple terms below:

- Note down the attributes, alternatives.
- Pick out relative value weightages for each attribute.
- The table should be normalized based on beneficiary and non-beneficiary variables.
- Using various MADM methodologies, the measure of output (MOP) of each alternate is determined.
- On the basis of measure of output (MOP), ranking is allotted to various alternates.

Despite the fact that there are more MADM approaches, only a few have a greater potential for solving problems in the world of manufacturing. Among all MADM methods, the Simple Additive

Weighing Method (SAW method) and the Weighted Product Method (WPM) are the most common. To observe the preference, AHP and TOPSIS procedure and its variant modifications are used.

Table 1. Investigated parameters and their levels.

Parameters	Level I	Level II	Level III
% of Composition	1:20	1:40	1:60
Calcination Temperature (°C)	750	900	1000
Time (Hr)	1	2	3

Table 2. Experimental measured values for Hardness and Ultimate compressive strength (Taguchi Orthogonal array L9).

Runs	% of Composition	Calcination Temperature (°C)	Time (Hr)	Hardness (VHR)	Ultimate Compressive strength (N/mm^2)
1	1:20	750	1	151	12.0
2	1:20	900	2	164	5.5
3	1:20	1000	3	152	5.7
4	1:40	750	2	147	12.0
5	1:40	900	3	158	12.0
6	1:40	1000	1	142	12.0
7	1:60	750	3	145	12.0
8	1:60	900	1	159	12.0
9	1:60	1000	2	148	5.7

3.3. *AHP Method*

In multiple domains of science and technology decision making plays a vital role to hash out that Satty [23] developed the AHP (Analytic Hierarchy Process) in 1981. AHP is mostly used to solve multi-criteria and uncertain decision problems [24-25]. Because of its flexibility, ease of use, and simplicity this tool has a wide range of applications.

3.4. *TOPSIS Method*

To spot out solutions from a finite set of available alternatives Hwang et.al developed the TOPSIS method. The selection criteria for the best possible alternative using TOPSIS method is as follows.

a) By considering both beneficial and non-beneficial variables a standardized matrix is prepared that meets the objective on the basis of decision table.

b) Obtain normalized decision matrix,

$$R_{pq} = \frac{m_{pq}}{\left[\sum_{q=1}^{M} m_{pq}^{2}\right]^{1/2}} \quad (1)$$

Where m_{pq} is the element on decision table of pth alternative and q^{th} attribute

c) Find the weighted Normalized matrix,

$$U_{pq} = w_q\, R_{pq} \quad (2)$$

d) Obtain Ideal and Negative Ideal solutions by the expressions given below

$$U^{+} = \{(\sum_{p}^{max} U_{pq}/q \in Q),(\sum_{p}^{min} U_{pq}/q \in Q)/p \quad \forall p = 1,2,3,\dots M\} \\ = \{U_1^{+}, U_2^{+}, U_3^{+}, \dots\dots\dots, U_N^{+}\} \quad (3)$$

$$U^{-} = \{(\sum_{p}^{min} U_{pq}/q \in Q),(\sum_{p}^{max} U_{pq}/q \in Q)/p \ \forall p = 1,2,3,\dots N\} \\ = \{U_1^{-}, U_2^{-}, U_3^{-}, \dots\dots\dots, U_M^{-}\} \quad (4)$$

The positive ideal best and negative ideal worst values of the considered attribute among the available values of the same attribute in various alternatives are U^+ and U^-. In case of non-beneficiary variables $U^+$is the lower and U^- is the higher value.

e) For each alternative, Euclidian distance is calculated.

$$S_p^{+} = \left\{\sum_{q=1}^{M}\left(U_{pq} - U_q^{+}\right)^2\right\}^{0.5} \ \forall \ p = 1, 2, \dots..N \quad (5)$$

$$S_p^{-} = \left\{\sum_{q=1}^{M}\left(U_{pq} - U_q^{-}\right)^2\right\}^{0.5} \ \forall \ p = 1, 2, \dots..N \quad (6)$$

f) For each alternative X_p relative closeness is calculated by using the formula

$$X_p = S_p^{-}/(S_p^{+} + S_p^{-}) \quad (7)$$

g) For each alternative, ranking is allotted based on S_{p+} and S_p^{-}.

4. Results and Discussion

4.1. Entropy based weighting (EBW) method

It's not always appropriate to use randomized weights and replay the experiment until the decision maker is pleased. The EBW method is a very much accurate information calculation method for enacting accurate weights [12]. When subjective weights are unreliable, objective weighting is especially useful. Reduction in decision makers experiments as much as possible for obtaining the weights in EBW method to solve a MADM problem. These additionally supportive in distinguishing the weights to attributes in a less amount of time. In EBW method, the attributes with performance rating vary with each other has privileged significance in the problem. For decision making process, bigger the entropy value corresponding to the particular attribute signifies the smaller attribute's weight and the less power of that attribute in decision making process. The bigger the entropy values referring to a specific attribute, the lower the attribute's weight and the less influence it has in the decision-making process. Table 3 shows the weighting given to each attribute using EWB method.

Table 3. The weightings given to each attribute using EBW method

Hardness	**Ultimate compression strength**
0.9806	0.0194

4.2. AHP method

Since the weights have already been determined, the alternatives are equated one by one to gain a better understanding of how each attribute compares to the others. The Relative mode is used to rank alternatives in order of choice. Table 4 shows the AHP score and ranking structure. As a result, the AHP method selection priority is S2 – S3 – S9 – S8 – S5 – S1 – S4 – S7 – S6.

4.3. TOPSIS method

Using MATLAB programme formulae of the method have been developed to calculate the attributes (U+, U-), alternatives (S+, S-) which are based on entropy weights and tabulated in Table 5. For each alternative performance score is evaluated using TOPSIS method and tabulated in Table 6.

Table 4. Performance score and ranking based on AHP method with EBW method

Sample	% of Composition	Calcination Temperature (°C)	Time (Hr)	AHP	Ranking Structure for AHP
S_1	1:20	750	1	0.0813	6
S_2	1:20	900	2	0.1750	1
S_3	1:20	1000	3	0.1687	2
S_4	1:40	750	2	0.0812	7
S_5	1:40	900	3	0.0814	5
S_6	1:40	1000	1	0.0811	9
S_7	1:60	750	3	0.0812	8
S_8	1:60	900	1	0.0814	4
S_9	1:60	1000	2	0.1687	3

Table 5. Calculated Values of Attributes and Alternatives

Sample	ATTRIBUTES		ALTERNATIVES	
	U^+	U^-	S^+	S^-
S_1	0.3799	0.0064	0.0019	0.0004
S_2	0.1741	0.0070	1.0000	0.2058
S_3	0.1805	0.0065	0.9691	0.1995
S_4	0.3799	0.0063	0.0010	0.0002
S_5	0.3799	0.0067	0.0033	0.0007
S_6	0.3799	0.0060	0	0
S_7	0.3799	0.0062	0.0006	0.0001
S_8	0.3799	0.0068	0.0035	0.0007
S_9	0.1805	0.0063	0.9691	0.1995

Table 6. Performance score and ranking based on TOPSIS method with EBW method

Sample	% of Composition	Calcination Temperature (°C)	Time (Hr)	TOPSIS	Ranking Structure for TOPSIS
S_1	1:20	750	1	0.0019	6
S_2	1:20	900	2	1	1
S_3	1:20	1000	3	0.9691	3
S_4	1:40	750	2	0.001	7
S_5	1:40	900	3	0.0033	5
S_6	1:40	1000	1	0	9
S_7	1:60	750	3	0.0006	8
S_8	1:60	900	1	0.0035	4
S_9	1:60	1000	2	0.9691	2

As a result, the TOPSIS method's selection priority is S2 – S9 – S3 – S8 – S5 – S1 – S4 – S7 – S6.

5. Conclusions

This paper successfully implemented the AHP and TOPSIS methods for opting the best combination for the synthesis of Lanthanum Hexa-Aluminate Powders (LHA). These methods provided simple and powerful ranking criteria for decision making. Out of all highest preference is for 1:20, 900^{0}C, 2hrs (Sample 2) and least preference is for 1:40, 1000^{0}C, 1hr in both the methods. It is detected that the powder prepared with 1:20, 900^{0}C, 2hrs has yielded good mechanical properties among the conditions compared in this work. The same problem can be extended further to measure the mechanical behaviour usingTaguchi method and GRA analysis as well.

References

[1] Smialek, J.L. and Jacobson, N.S., 2014. Oxidation of high-temperature aerospace materials. *High temperature materials and mechanisms*, pp.95-162.

[2] Bredesen, R., Jordal, K. and Bolland, O., 2004. High-temperature membranes in power generation with CO2 capture. *Chemical Engineering and Processing: Process Intensification*, *43*(9), pp.1129-1158.

[3] Meetham, G.W. and Van de Voorde, M.H., 2000. *Materials for high temperature engineering applications*. Springer Science & Business Media.

[4] Sapuan, S.M., 2001. A knowledge-based system for materials selection in mechanical engineering design. *Materials & Design*, *22*(8), pp.687-695.

[5] Rao, R.V., 2006. A material selection model using graph theory and matrix approach. *Materials Science and Engineering: A*, *431*(1-2), pp.248-255.

[6] Khorshidi, R., Hassani, A., Rauof, A.H. and Emamy, M., 2013. Selection of an optimal refinement condition to achieve maximum tensile properties of Al–15% Mg2Si composite based on TOPSIS method. *Materials & Design*, *46*, pp.442-450.

[7] Farag, M.M., 2002. Quantitative methods of materials selection. *Handbook of materials selection*, pp.1-24.

[8] Ashby, M.F. and CEBON, D., 1993. Materials selection in mechanical design. *Le Journal de Physique IV*, *3*(C7), pp.C7-1.

[9] Babanli, M.B., Prima, F., Vermaut, P., Demchenko, L.D., Titenko, A.N., Huseynov, S.S., Hajiyev, R.J. and Huseynov, V.M., 2018, August. Material selection methods: A review. In *International Conference on Theory and Applications of Fuzzy Systems and Soft Computing* (pp. 929-936). Springer, Cham.

[10] Shanian, A.A. and Savadogo, O., 2006. A material selection model based on the concept of multiple attribute decision making. *Materials & Design*, *27*(4), pp.329-337.

[11] Jahan, A., Ismail, M.Y., Sapuan, S.M. and Mustapha, F., 2010. Material screening and choosing methods–a review. *Materials & Design*, *31*(2), pp.696-705.

[12] Kumar, D.S. and Suman, K.N.S., 2014. Selection of magnesium alloy by MADM methods for automobile wheels. *International Journal of Engineering and Manufacturing*, *2*, pp.31-41.

[13] Önüt, S. and Soner, S., 2008. Transshipment site selection using the AHP and TOPSIS approaches under fuzzy environment. *Waste Management*, *28*(9), pp.1552-1559.

[14] Dağdeviren, M., Yavuz, S. and Kılınç, N., 2009. Weapon selection using the AHP and TOPSIS methods under fuzzy environment. *Expert systems with applications*, *36*(4), pp.8143-8151.

[15] Albayrak, E. and Erensal, Y.C., 2004. Using analytic hierarchy process (AHP) to improve human performance: An application of multiple criteria decision making problem. *Journal of intelligent manufacturing*, *15*(4), pp.491-503.

[16] Gupta, N., 2011. Material selection for thin-film solar cells using multiple attribute decision making approach. *Materials & Design*, *32*(3), pp.1667-1671.

[17] Milani, A.S., Shanian, A., Madoliat, R. and Nemes, J.A., 2005. The effect of normalization norms in multiple attribute decision making models: a case study in gear material selection. *Structural and multidisciplinary optimization*, *29*(4), pp.312-318.

[18] Shanian, A.A. and Savadogo, O., 2006. A material selection model based on the concept of multiple attribute decision making. *Materials & Design*, *27*(4), pp.329-337.

[19] Kumar, D.S., Radhika, S. and Suman, K.N.S., 2013. MADM methods for finding the right personnel in academic institutions. *International Journal of u-and e-Service, Science and*

Technology, *6*(5), pp.133-144.
[20] Djambazov, S. and Yoleva, A., 2001. Synthesis and properties of lanthanum hexaaluminate ceramic La1− xCaxAl11− y− zMgyTizO18. *Ceramics international*, *27*(8), pp.899-901..
[21] Cinibulk, M.K., 1995. Synthesis and characterization of sol-gel derived lanthanum hexaluminate powders and films. *Journal of materials research*, *10*(1), pp.71-76.
[22] Ram, Y.S., Tarasasanka, C. and Prabakaran, J., 2021. Preparation and characterization of lanthanum hexa aluminate powders for high temperature applications. *Materials Today: Proceedings*, *39*, pp.1472-1475.
[23] Rao, R.V., 2004. Evaluation of metal stamping layouts using an analytic hierarchy process method. *Journal of Materials Processing Technology*, *152*(1), pp.71-76.
[24] Saaty, T.L., 1988. What is the analytic hierarchy process?. In *Mathematical models for decision support* (pp. 109-121). Springer, Berlin, Heidelberg.
[25] Saaty, R.W., 1987. The analytic hierarchy process—what it is and how it is used. *Mathematical modelling*, *9*(3-5), pp.161-176.

Experimental study on fault detection in roller bearing used in rolling mill industry

Nilesh Nirwan[1*], Hardik Ramani[1]

[1]Department of Mechanical Engineering, Dr. A.P.J. Abdul Kalam University, Indore-452016, Madhya Pradesh, India

* Corresponding author e-mail: nilesh.nirwan@gmail.com

Abstract. Rolling element bearings are extremely important components of rotating machines, and bearing defects can cause machines to fail. As a result, early detection of such defects, as well as the severity of damage while the bearing is in operation, may help to prevent machine malfunction and breakdown. Vibration is produced by defective bearings when no. of passes are increases to reduce down the thickness of raw materials and these vibration signals can be used to diagnose the problem. This article provides a summary of recent developments in bearing defect studies, as well as sources of vibration and vibration measurement methods in the time, frequency, and time frequency domains. An experimental setup has been established in this paper to determine roller bearing faults. The fault evaluation progress is extracted using the signal processing method. The results are presented as time series and FFT plots. Misalignment fault, Indentation event, Smoothened dent, Defects, and Damage growth state are the five-wear fault assessment events examined. The test lasts 480 hours and takes place at temperatures ranging from 68 to 78 degrees Celsius. After 480 hours of testing, it was discovered that the roller bearing is damaged due to a high wear rate, with amplitude of acceleration in the frequency domain of 32.51 m/s^2 at fifth event. The amplitude of acceleration in the frequency domain is also found to be greater than the critical value of 30 m/s^2. As a result, the state of the roller bearing after 480 hours is the damage growth state.

Keywords: bearing defects; misalignment fault; Indentation event; Smoothened dent; roller bearing

1. Introduction

The bearing is an important component of all rotating machines, as it serves to stabilize the machine and allow the shaft to rotate in relation to a fixed structure. Rolling element bearings are used in more than 90% of machines, and their failure will cause the machine to malfunction. Therefore, they are the most essential components in industrial applications. As a result, bearing reliability and robustness are critical for machine efficiency[1]. Bearings are subjected to heavy and dynamic loads produced by machines and transmitted via rolling element-bearing components during operation. As a result, bearing condition is critical in high-volume production systems with a large number of rotating machines contributing to the system. Any bearing defect must be detected rapidly to prevent increased downtime, manufacturing time, and catastrophic machinery failure. As a result, detecting these flaws is critical for bearing condition control and quality inspection. Acoustic measurements, temperature control, wear debris analysis and vibration measurements are some of the techniques used to diagnose these defects in bearings. Vibration tracking is a commonly used and cost-effective method of

IOP Conf. Series: Materials Science and Engineering 1185 (2021) 012002 doi:10.1088/1757-899X/1185/1/012002

identifying, finding, and separating defects in rolling element bearings[2]. Bearings create noise and vibration because of varying compliance or the presence of a defect. Even when these bearings are geometrically fine, they produce vibration when filled radially. The most critical parameter for bearing working life is the roller bearings used in rotating systems. As a result, prior to the design and manufacture of the bearing, a fault analysis is required[3]. Since a long time, scientists and engineers have been working both experimentally and theoretically to determine the various forms of bearing faults. When there are faults in them, the vibration level rises dramatically[4], [5]. Vibration and acoustic measurement have been used to diagnose bearing defects using a variety of techniques, including vibration measurement in the time domain, frequency domain, time-frequency domain, shock pulse system, and acoustic emission technique. Several articles on vibration signal analysis techniques have been performed, as well as various reviews focused on defect and fault diagnosis techniques[6], [7].

Our study provides a first-of-its-kind experimental review of roller bearing wear evaluation using time series signals and FFT plots. It is important to measure the bearing's working life before designing and manufacturing it, so that rotary machineries experience minimal wear and repair costs. During action, a fluid film lubrication between the roller and the inside of the outer race of the bearing should be maintained for low wear in rotary machineries.The wear rate is calculated for five events: Misalignment fault, Indentation event, Smoothened dent, Defects and Damage growth state at different times using time series signals and FFT plots, and wear rate is calculated for five events: Misalignment fault, Indentation event, Smoothened dent, Defects and Damage growth state at different times using time series signals and FFT plots. The amplitude of a time series signal of acceleration is calculated at different points in this evaluation.

2. Defects in Bearings

The mechanism of bearing vibration has been clarified by several researchers. Even a safe bearing produces vibration, but the presence of defects greatly raises vibration levels. Premature bearing failures can be caused by a variety of causes, the most common of which are fatigue, wear, plastic deformation, corrosion, brinelling, inadequate lubrication, defective installation, and incorrect design. Identification of these defects, as well as the vibration they cause, is critical for bearing condition monitoring. Bearing defects can be divided into two categories: dispersed defects and localized defects [8].

2.1. Distributed Defects

Surface roughness, waviness, misaligned races, and off-size rolling components are examples of distributed defects. Manufacturing mistakes, abrasive wear, and incorrect installation are the most common causes of distributed defects[9]. The contact force between rolling elements and raceways differs in distributed defects, resulting in vibration. The vibration response caused by a distributed defect is primarily used in bearing quality inspection and condition monitoring.

2.2. Localized Defects

Pits, cracks, and spalls that form on the rolling surfaces fall into this category of defects. Spalling is the most common mode of failure among these. Fatigue cracks originate underneath the surface and extend to the surface until the material fails and leaves localized defects[10]. According to Bentley's article, localized defects cause damage to the inner ring, outer ring, and rolling elements in 90% of all bearing faults [11].

3. Considerations from a theoretical stand point

A complex of friction appearance ways produces the resistant torque that occurs in roller bearings. The rolling friction phenomenon is complicated by a variety of variables that arise at the same time.

Friction caused by contact deformations, rolling friction on contact surfaces, friction created by lubricants, slipping of bearing elements, and friction from seals are the most common sources of bearing friction. Friction torque can be calculated with reasonable precision using relationships derived from experimental findings in standard calculations. The gross friction torque (M_{gc}) is calculated using equation (1).

$$M_{gc} = M_f + M_l \quad \text{....(1)}$$

Where, Mf is the bearing load produces a resistant torque and M_l is the lubricant-induced resistant torque created by bearing elements in contact with the lubricant.

For bearings that work at a moderate speed and load, equation (1) is used. When using roller bearings for high-speed applications, the friction torque generated by the spin and gyroscopic motions should be considered.The bearing load produces a resistant torque (M_f), which is calculated using the equation (2).

$$M_f = f_1.F.d_m \quad \text{....(2)}$$

Where, F is the bearing radial load, in [N]; d_m is the mean bearing diameter, in [m]; Ball bearings factor f_l is established with the following relation.

$$f_1 = 0.0009\left(\frac{F_r}{C_0}\right)^{0.55} \quad \text{....(3)}$$

Where, F_r is the radial force acting on roller bearings, in [N]; C_0 is the dynamic load capacity, in [N]. To calculate the resistant torque (M_l) provided by the fluid friction of bearing elements in contact with the lubricant, use the formula given below.

$$M_l = f_0.(v.N)^{2/3}.d_m.10^3 \quad \text{for } (v.N) \geq 20\text{x}10^{-3} \quad \text{....(4)}$$

$$M_l = 16.f_0.(d_m)^3 \quad \text{for } (v.N) < 20\text{x}10^{-3} \quad \text{....(5)}$$

Where, v is the kinematic viscosity of the lubricants which would be taken $(25 \sim 33)\text{x}10^{-6}$, in [$m^2$/s]; N is the rotational speed of operated bearing, in [rpm] and the bearing factor f_0 should be taken 1.5 ~ 2.

4. Setup and procedure for the experiment

Experiments were carried out on a test rig in which the roller bearing (Bearing No. 202512) was filled with 100 N using the arrangement depicted in Figure 1. The roller bearing specifications are described in Table 1. Via a drive system consisting of a single phase induction motor of 1 HP, the shaft is rotated by a V-Belt and single stage cone pulley arrangement. The shaft's speed is maintained at 900 RPM using the arrangement shown in Figure 1.With the aid of a vibration metre and sensing element, the signals of acceleration and corresponding frequency are registered. Table 2 onward shows an example of a reading. Time series data and FFT plots are produced using those signals, which are then used to evaluate roller bearings.

Table 1. Specification of Test Bearing

Sr. No.	Parameter	Value
1	Bearing Number	202512
2	Category	Roller Bearing
3	Inner Race Diameter (mm)	20.04
4	Outer Race Diameter (mm)	26.06
5	Width (mm)	11.96
6	Speed (rpm)	900

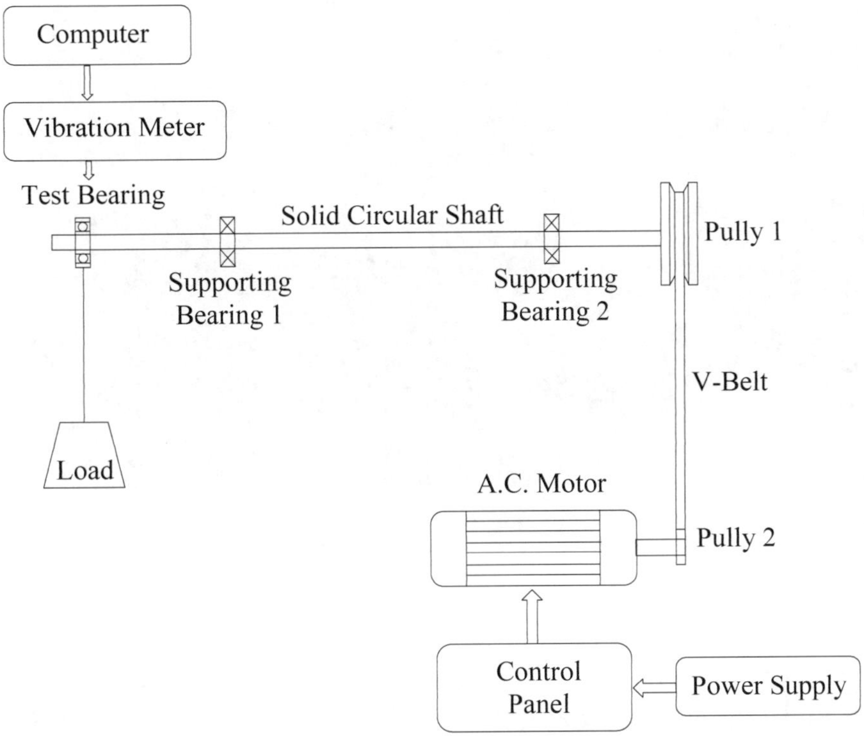

Figure 1. Block Diagram of Experimental Setup

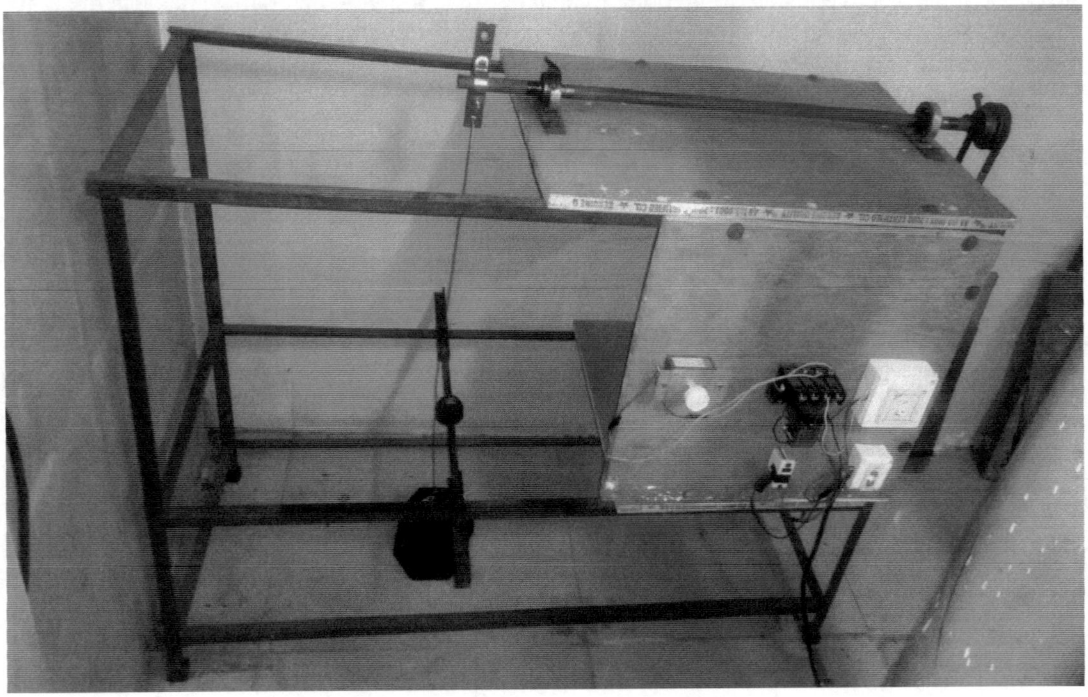

Figure 2. Practical Experimental Setup

Figure 3. Practical Bearing photos outside the machine (Left) and inside attached to a machine (right)

5. Results and Discussion

Figure 3 shows the photographs of bearing outside and inside of the machine. For the wear test, a roller bearing (202512) is used. The fault induced signal is detected using frequency domain methods. The test bearing is run for 480 hours in the experimental test. After 430 hours of testing, a misalignment fault occurred due to alubrication film that is not properly lubricated being formed between the roller and the inside of the outer race. After that, a high temperature was generated during the test run, which was registered as 68 °C. After 450 hours of work of operating the test bearing, an indentation event occurred due to unbalanced rolling contact between the roller and the inside of the outer race, resulting in a high temperature of 70 °C during the test cycle. After 460 hours of work of running the test bearing, a smoothened dent appeared due to an unbalanced spinning speed of the roller, and a high temperature was created during the test run, which was registered as 73 °C. After 470 hours of work of test bearing operation, a defect appears due to wear between the roller and the inside of the outer race, and then a high temperature is created during the test run, which is registered as 75 °C. Damage growth state occurred at the end of 480 hours of test bearing run due to high wear rate generated between roller and inside of outer race, followed by high temperature caused during the test run of 78 °C. The tests are carried out five times to investigate the five wear assessment events of misalignment fault, indentation, smoothened dent, defects and damage growth condition. When the amplitude of acceleration in the frequency domain exceeds 30 m/s^2, the roller bearing is found to be affected. As a result, in the frequency domain, the critical value for harm is 30 m/s^2.

5.1. Misalignment fault

The frequency range is 200 Hz to 1000 Hz. The data in Table 2 are used to produce the time series signal of acceleration. In Figure 4, the amplitude of the time series signal of acceleration ranges from -121.6 to 121.6 m/s^2. As shown in Figure 5, the amplitude of the FFT of acceleration reaches a maximum of 15.80 m/s^2 at 880 Hz and a minimum of 8.10 m/s^2 at 402 Hz. The amplitude of acceleration in the frequency domain is found to be far below the critical value. As a consequence, the roller bearing has a misalignment fault.

IOP Conf. Series: Materials Science and Engineering 1185 (2021) 012002 doi:10.1088/1757-899X/1185/1/012002

Table 2. Acceleration and Frequency using roller bearing

Reading No.	Acceleration (in m/sec^2)	Frequency in Hz
1	8.1	402
2	8.8	430
3	12.7	710
4	10.5	462
5	11.6	620
6	12.7	687
7	15.7	855
8	14.5	820
9	10.7	468
10	11.3	670
11	9.7	435
12	9.8	443
13	10.1	450
14	11.3	614
15	12.8	700
16	13.5	765
17	14.3	770
18	15.8	880
19	11.8	680
20	10.2	455

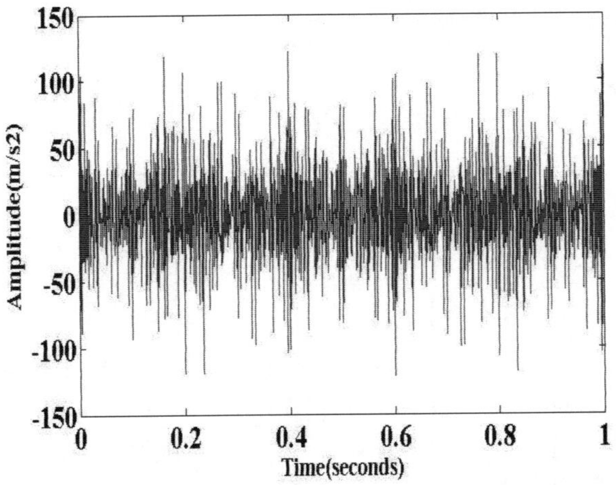

Figure 4. Time series Signal

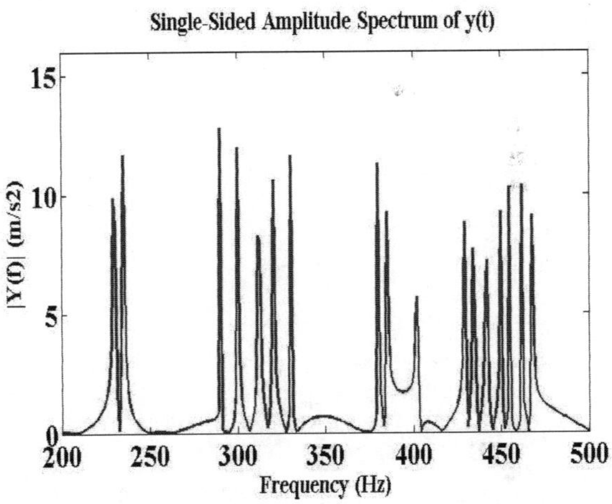

Figure 5. Fast Fourier Transform (FFT)

5.2 Indentation event

The frequency range is 0 Hz to 1000 Hz. The data in Table 3 are used to produce the time series signal of acceleration. In Fig. 6, the amplitude of the time series signal of acceleration ranges from -225.9 to 225.9 m/s^2. As shown in Fig. 7, the amplitude of the FFT of acceleration reaches a maximum of 20.7 m/s^2 at 944 Hz and a minimum of 11.4 m/s^2 at 480 Hz. The amplitude of acceleration in the frequency domain is found to be far below the critical value. As a result, this is a roller bearing indentation case.

Table 3. Acceleration and Frequency using roller bearing

Reading No.	Acceleration (in m/sec^2)	Frequency in Hz
1	12.7	492
2	11.4	480
3	14.5	610
4	12.8	522
5	13.6	530
6	15.6	640
7	14.8	624
8	16.2	725
9	17.3	790
10	17.7	793
11	18.4	815
12	15.8	704
13	16.2	710
14	16.7	734
15	19.2	885
16	19.7	902
17	17.6	790
18	18.7	840
19	20.7	944
20	19.2	860

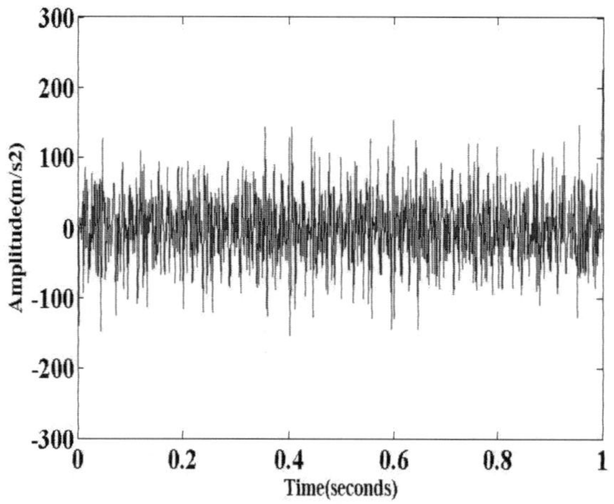

Figure 6. Time series Signal

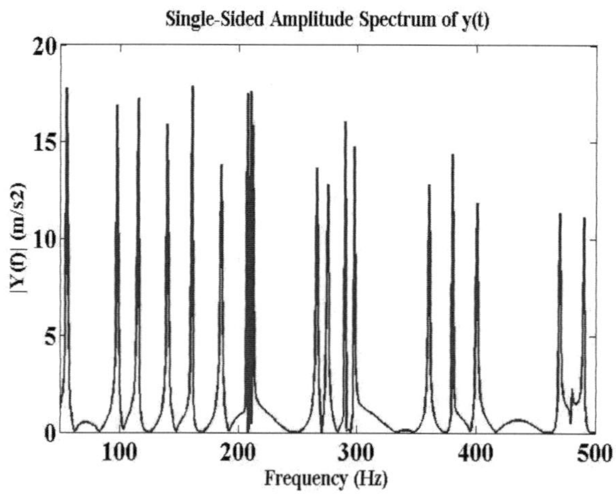

Figure 7. Fast Fourier Transform (FFT)

5.3 Smoothened dent

The frequency range is 0 Hz to 1000 Hz. The data in Table 4 is used to produce the time series signal of acceleration. In Fig.8, the amplitude of the time series signal of acceleration ranges from -292.5 to 292.5 m/s^2. As shown in Fig.9, the amplitude of the FFT of acceleration reaches a maximum of 28.70 m/s^2 at 943 Hz and a minimum of 14.70 m/s^2 at 477 Hz. The amplitude of acceleration in the frequency domain is found to be less than the critical value. As a consequence, the roller bearing's dent has been smoothed out.

Table 4. Acceleration and Frequency using roller bearing

Reading No.	Acceleration (in m/sec^2)	Frequency in Hz
1	15.3	482
2	16.2	498
3	18.4	621
4	14.7	477
5	17.8	517
6	22.5	724
7	21.5	688
8	24.4	762
9	25.8	812
10	24.2	753
11	26.7	821
12	28.7	943
13	17.8	520
14	27.2	910
15	26.9	902
16	20.9	649
17	21.8	713
18	22.7	732
19	20.6	643
20	24.2	720

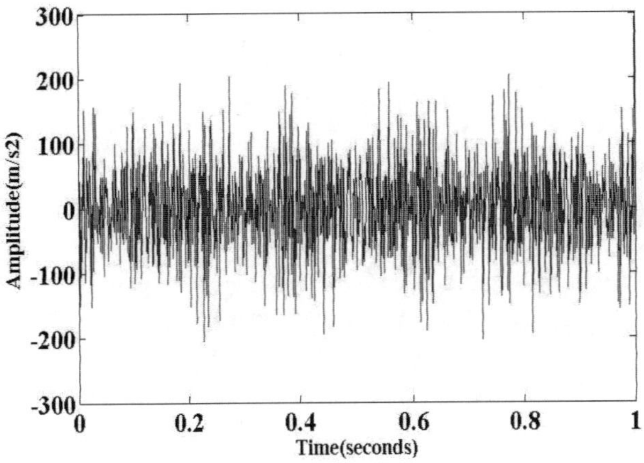

Figure 8. Time series Signal

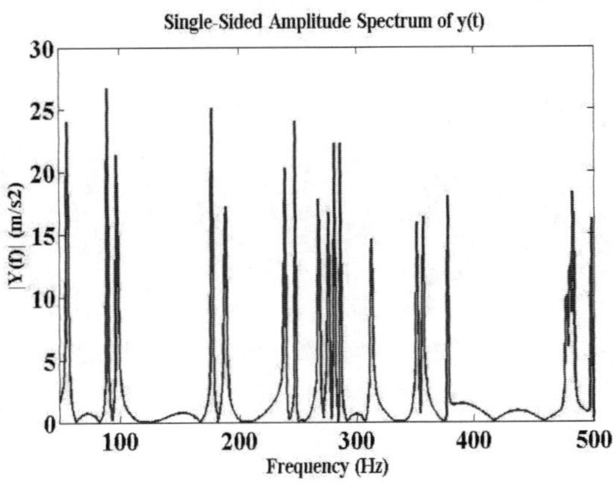

Figure 9. Fast Fourier Transform (FFT)

5.4 Defects

The frequency range used is 0 to 1000 Hz. The data in Table 6 are used to produce the time series signal of acceleration. In Fig.10, the amplitude of the time series signal of acceleration varies from -363.1 to 363.1 m/s^2. As shown in Fig.11, the amplitude of FFT of acceleration has a maximum of 32.51 m/s^2 at 86.91 Hz and a minimum of 15.26 m/s^2 at 352.5 Hz. The amplitude of acceleration in the frequency domain is found to be greater than the critical value. As a result, the roller bearing is in a weakened growth state.

Table 5. Acceleration and Frequency using roller bearing

Reading No.	Acceleration (in m/sec^2)	Frequency in Hz
1	16.6	612
2	17.7	628
3	14.3	588
4	21.2	704
5	20.3	688
6	22.2	712
7	24.7	755
8	25.1	806
9	27.8	850
10	26.8	832
11	26.3	816
12	27.6	842
13	28.6	956
14	28.2	941
15	18.9	657
16	19.8	671
17	20.7	693
18	29.5	971
19	29.2	968
20	28.1	872

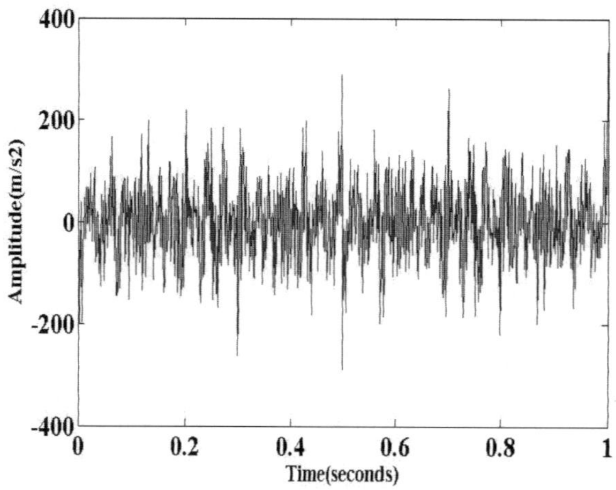

Figure 10. Time series Signal

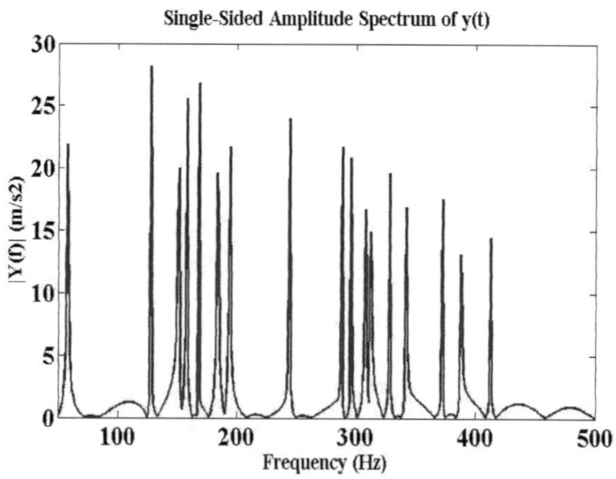

Figure 11. Fast Fourier Transform (FFT)

5.5 Damage growth state

Frequency range is taken from 0 to 1000 Hz. The time series signal of acceleration is generated using data given in Table 6. The amplitude of time series signal of acceleration is observed from -363.1 to 363.1 m/s^2 in Fig.12. The amplitude of FFT of acceleration is observed maximum of 34.6m/s^2 at 998 Hz and minimum 22.5m/s^2 at 613 Hz as shown in Fig.13. It is found that the amplitude of acceleration in frequency domain is more than the criticalvalue. Therefore, this is damage growth state of the roller bearing.

Table 6. Acceleration and Frequency using roller bearing

Reading No.	Acceleration (in m/sec^2)	Frequency in Hz
1	24.5	654
2	25.7	671
3	27.5	723
4	28.8	797
5	22.5	613
6	30.5	845
7	34.6	998
8	31.4	872
9	32.8	957
10	33.7	984
11	27.9	744
12	32.4	912
13	33.2	978
14	34.1	989
15	30.2	813
16	27.3	711
17	23.5	648
18	28.7	768
19	31.8	898
20	23.2	642

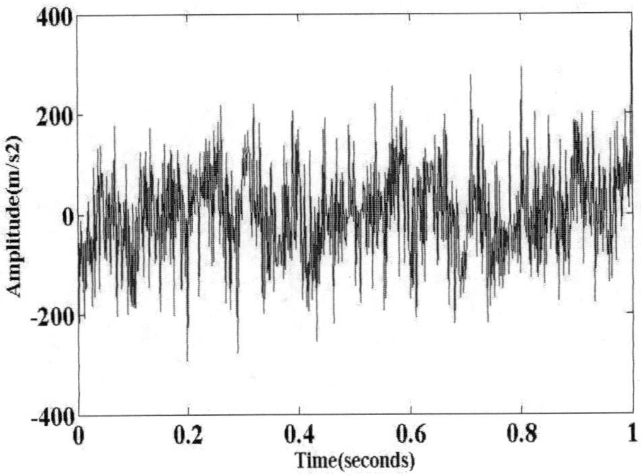

Figure 12. Time series Signal

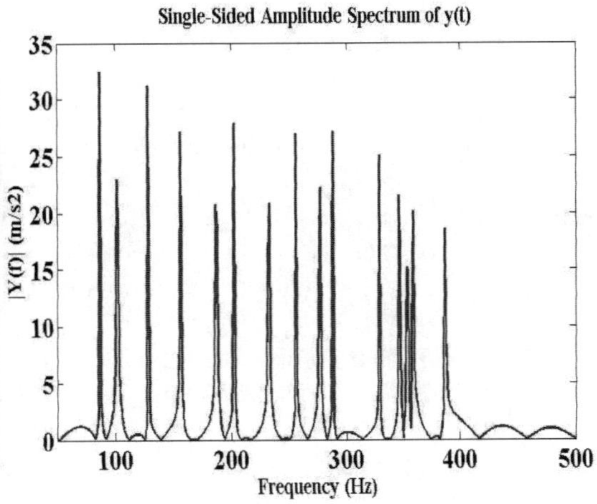

Figure 13. Fast Fourier Transform (FFT)

6. Conclusion

The research describes an experimental model for roller bearing wear evaluation. With the assistance of signal processing techniques, the wear assessment was observed. The wear assessment of roller bearing is evaluated using five events at different times. After 430 hours of work of testing, a misalignment fault occurred in the first case. In the second event, an indentation event occurred after 450 hours of work of test bearing activity. Smoothened dent occurred after 460 hours of work of test bearing operation in the third case. After 470 hours of work of test bearing defects appears in the fourth event. Damage growth state occurred at the end of the 480-hour test bearing run in the fifth event. The temperatures measured during all five events were 68°C, 70°C, 73°C, 75°C and 78°C, respectively, with the fifth event having the highest temperature. The magnitude of acceleration in the frequency domain is also observed to be greater than the critical value at the fifth event as compared to the other four events. In this regard, the research demonstrates the maximum wear rate in roller bearings at the fifth event of wear assessment.

Acknowledgement

The authors appreciatively acknowledge Dynamics of Machine Lab, Department of Mechanical Engineering, Dr. A.P.J. Abdul Kalam University, Indore-452016, (M.P.), Indiafor providing laboratory equipment. Also, we are thankful to"MetalFab Hightech Pvt. Limited" E-21, 25 & 33 B/2, MIDC, Hingna Road, Nagpur for their facilities and giving supports in this study.

References

[1] Al-Ghamd, A.M. and Mba, D., 2006. A comparative experimental study on the use of acoustic emission and vibration analysis for bearing defect identification and estimation of defect size. *Mechanical systems and signal processing*, *20*(7), pp.1537-1571.

[2] Ebert, F.J., 2010. Fundamentals of design and technology of rolling element bearings. *Chinese Journal of Aeronautics*, *23*(1), pp.123-136.

[3] El-Thalji, I. and Jantunen, E., 2015. Fault analysis of the wear fault development in rolling bearings. *Engineering Failure Analysis*, *57*, pp.470-482.

[4] Harvey, T.J., Wood, R.J.K. and Powrie, H.E.G., 2007. Electrostatic wear monitoring of rolling element bearings. *Wear*, *263*(7-12), pp.1492-1501.

[5] Rayo Ramírez, S.M., 2020. Análisis de la distribución del peso de la cúpula sobre los mecanismos de rotación del Very Large Telescope ubicado en el Observatorio Paranal, y desarrollo de un método de monitoreo de condición= Analysis of the load distribution of the dome weight on the rotation mechanisms of the Very Large Telescope located at the Paranal Observatory, and development of a condition monitoring method.

[6] Jagdale, M.N. and Diwakar, G., 2018, December. A critical review of condition monitoring parameters for fault diagnosis of rolling element bearing. In *IOP Conference Series: Materials Science and Engineering* (Vol. 455, No. 1, p. 012090). IOP Publishing.

[7] Liu, J., Wu, H. and Shao, Y., 2018. A theoretical study on vibrations of a ball bearing caused by a dent on the races. *Engineering Failure Analysis*, *83*, pp.220-229.

[8] Matsunaga, H., Komata, H., Yamabe, J., Fukushima, Y. and Matsuoka, S., 2014. Effect of size and depth of small defect on the rolling contact fatigue strength of bearing steel JIS-SUJ2. *Procedia materials science*, *3*, pp.1663-1668.

[9] Mishra, C., Samantaray, A.K. and Chakraborty, G., 2016. Bond graph modeling and experimental verification of a novel scheme for fault diagnosis of rolling element bearings in special operating conditions. *Journal of Sound and Vibration*, *377*, pp.302-330.

[10] Espejel, G.E.M. and Gabelli, A., 2016. A model for rolling bearing life with surface and subsurface survival: sporadic surface damage from deterministic indentations. *Tribology International*, *96*, pp.279-288.

[11] Warda, B. and Chudzik, A., 2014. Fatigue life prediction of the radial roller bearing with the correction of roller generators. *International Journal of Mechanical Sciences*, *89*, pp.299-310.

IOP Conf. Series: Materials Science and Engineering 1185 (2021) 012003 doi:10.1088/1757-899X/1185/1/012003

Optimization of dry-sliding wear parameters on carbon fiber reinforced polyester composites using taguchi based greyrelation analysis

K Maneesh Geeth [1], M Chandra Sekhar Reddy[2] and MV Sai Kumar[3]

[1,2] Department of Mechanical Engineering, University College of Engineering, Osmania University, Hyderabad, India
[3]Department of Mechanical Engineering, R. V. R. & J.C. College of Engineering, Guntur, India

* Corresponding author e-mail: geethmaneesh22@gmail.com

Abstract. Fiber reinforced polymer composite materials gradually replace traditional metallic materials due to their high specific strength. The cost of composites, however, is still higher than the traditional materials. This paper examined the dry-sliding wear behaviour using a pin-on-disc type of wear arrangement on carbon fiber reinforced polyester (CFRP) composites, which were fabricated with the help of die-casting technique. The carbon fiber reinforced polyester specimens have been subjected to dry-sliding wear tests against a EN32 steel disc. The rate of wear and frictional force were selected as responses variables to the varying percentages of reinforcement, loads, sliding velocity and sliding distance were opted as control variables. The experimental design was carried out by using the Taguchi L_{16} OA (Orthogonal Array). The approach used in the initial step of optimization is the dry-sliding performance of carbon fiber reinforced polyester composites was GRA (Grey Relational Analysis). Based on the GRA, the rate of wear and frictional force were found to be the parameters at optimum level for the whole GRG (Grey Relational Grade).According to the ANOVA (Analysis of Variance) result based on GRG, the most important process parameters influencing the wear performance of the composites were found to be load and percentage of reinforcement, which influenced the wear performance of the composites by 60% and 28% respectively, followed by sliding distance and sliding velocity.

Keywords. Carbon fiber reinforced composite (CFRP), Grey relational analysis (GRA),Orthogonal Array (OA), Taguchi method, Grey Relational Grade (GRG).

1. Introduction

Carbon fibre is an excellent choice for achieving a good balance of weight, mechanical properties, performance, and aesthetic. Its use enables innovative shapes to be combined with superior technical efficiency, allowing for greater design freedom. The key problems with carbon fibre processing are the manufacturing costs and the recyclability. As a result, its key uses remain in the racing industry and niche markets [1]. Enhance fibres and matrix structures (such as thermosets & thermoplastics) led to an increase use and practise, resulting in CFRP (Carbon Fibre Reinforced Plastics) composites with enhance mechanical properties, enabling them to displace more traditional materials for primary structures such as titanium and aluminium alloys [2]. In jute fibre and CFRP composites, a glucose-based epoxy resin component was tested. The novel glucose-based epoxy monomers with complex mechanical properties, tensile strength, glass transition temperature and tensile modulus were calculated, indicating that it could be a feasible alternative to the widely used mineral oil-based diglycidyl-ether of bisphenol-A (DGEBA), also at high temperatures up to 160°C by using an amine-type curing agent certified for aircraft composites [3]. Practical use of PEEK (Poly-Ether-Ether-Ketone) and 30% carbon fibre reinforced PEEK (PEEK -CF30) for orthopaedic applications led to the decision to investigate the effect of carbon fibre reinforcement and the meaning of various parameter's, such as contact stress, sliding velocity and counter-face roughness on frictional behaviour. Abrasion and adhesion wear processes are clearly caused by complex conditions applied to all components. Iron transferred films from the counter-face to the surface of pin, as well as PEEK content transferred from the surface of pin to the surface of disc, showed a significant adhesion wear [4].

Because of their major advantages in dielectric properties, cost, resistance to corrosion properties and processability, UPR (Unsaturated Polyester Resin) is commonly used in the fields of industry as a thermosetting content. TMSAP (phosphorus and nitrogen-containing silane coupling agent) and ammonium polyphosphate (APP) combined action had a multifunctional effect on mechanical properties of UPR composition, toxic pyrolysis risks and flame retardancy effectiveness [5]. The carbon composite filled with polyester resin has a electrical conductivity with low threshold value of 0.7-0.8 vol percent of a filler [6]. The effect of fibre glass wastes reinforced with polyester matrix composites boosted mechanical properties like impact and tensile strength significantly [7]. Under mechanical agitation, add the necessary amount of GNS (Graphene Nano Sheets) to unsaturated polyester resin. As a result, flexural and tensile performance have also been improved. Also, GNS increased the high voltage and resistance properties greatly [8].

Numerous decision-making methods, such as AHP (Analytic Hierarchy method), TOPSIS (Technique for Order of Preference by Similarity to Ideal Solution), SMART (Simple Multi Attribute Rating Technique) and GRA (Grey Relational Analysis) have been proposed in the literature to investigate the various attributes correlated with wear resistance of composites. GRA is one of them, Deng discussed it in 1989 is one of the most commonly used strategy when the meaning of the knowledge is incomplete and unclear [9]. Basavarajappa et al. studied the impact of filler materials on the dry sliding wear behaviour of polymer matrix, and found that adding SiC & graphite to glass epoxy composite significantly improved wear resistance. Using Taguchi's orthogonal series, the dry-sliding wear behaviour of SiC & graphite particles reinforced with aluminium matrix was tested, and the findings demonstrated that graphite particles were efficacious for continuing to increase the wear resistance of Al/SiCp composites [10]. Rajmohan et al. used Taguchi based GRA on to refine the drilling parameters of fly ash packed with CFRP composites. According to GRA, the weight percent of fly ash & drill diameter are the most critical variables affecting drilling efficiency [11].

The current research is to improve the dry sliding wear factors in order to establish a desirable CFRP composite with potential applications in breaking and friction. Initially, a CFRP composite was fabricated using the die-casting technique. An experimental design using a Taguchi L_{16} OA with four varying parameters, namely the load applied (L), sliding speed (S), sliding distance (D) &% of reinforcement (R). For each of these four factors, four levels were defined. From the L_{16} OA, sixteen experimental runs were carried out in order to acquire data for the various responses of the wear and frictional force. Followed up by S/N ratios and GRA, the response to the optimum values was assessed. ANOVA was then used to identify the most important factors influencing CFRP. Finally, validation tests were performed using the optimum levels of the design criteria to validate the enhancement of the consistency attribute, namely the GRG (Grey Relational Grade). Figure 1 depicts the entire methodology used for the successful completion of current research.

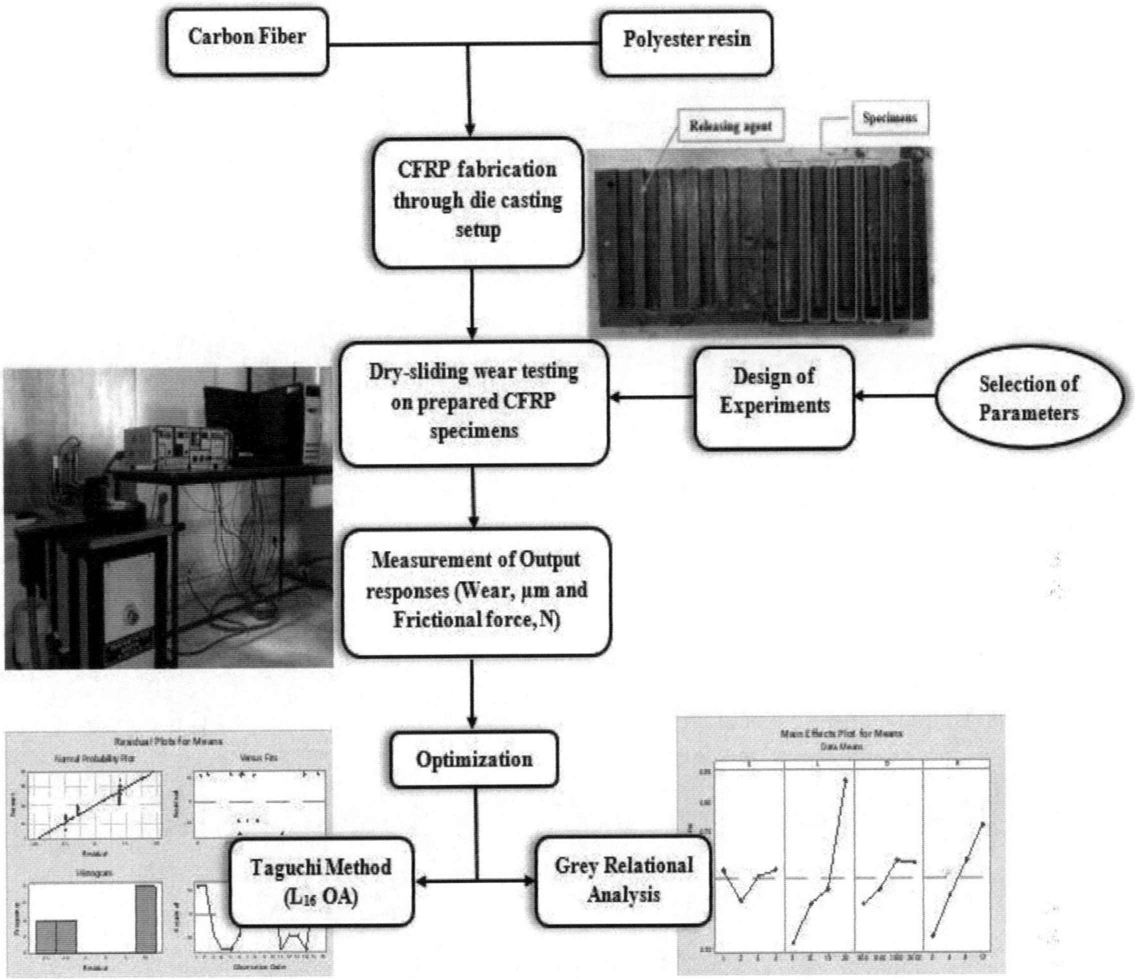

Figure 1. Systematic procedure for the current research work.

2. Experimental details

2.1. Materials

Figure 1 shows the procedure for the current research work. In this study, the materials used for composite preparation were the carbon fibers of 85 µm, added as reinforcement material and matrix material as polyester resin added to it. A mild steel mould of Ø15mm×150mm in height was manufactured for preparation of samples used in current study.

2.2. Composite Fabrication

Polyester resin was blended with hardener [Methyl Ethyl Ketone Peroxide MEKP (MEKP), CO (Cobalt octoate)] and then carbon fiber of 85 µm was added to the matrix at a ratio of 0, 4, 8 and 12Wt%. The weighted quantity is mixed in a beaker by mechanical stirring at a suitable speed of 110 rpm. It is mixed up to when it comes to the semi solid state. For the easy removal of composites from the die a release agent (wax) coat was applied to the mould before pouring the mixture. The semi solid state mixture is now poured into the mild steel mould and left it for 24 hours at room temperature. After 24 hours, mild steel mould is removed and the specimens were taken out and then it is left for post cured. The samples in the mould had a diameter of 15 mm. They were machined to an 8 mm diameter after curing in order to conduct wear tests. The composition of fabricated specimens is illustrated in Table 1.

Table 1. Composition of fabricated specimens.

Sl. no	Composite Identification	Wt% of Carbon fiber	Wt% of Polyester resin	Wt% of Hardener (MEKP+CO)
1.	Pure Polyester	0	200	4
2.	CF 4Wt% + Polyester	4	200	8
3.	CF 8Wt% + Polyester	8	200	10
4.	CF 12Wt% + Polyester	12	200	16

2.3. *Dry-sliding Wear test*

Dry-sliding wear characteristics of prepared CFRP composites, according to ASTM G99-95 standards [12] were investigated using pin-on-disc wear measuring machine (Figure 2) with counter face of a hardened & polished disc made of EN32 steel disc (58-60 HRC). Figure 3 shows the wear specimens (pins) that were Ø8 mm×50 mm in height. The specimen's initial wt. was determined using an electronic balancing system with a precision of 0.0001 gm.

During the test, the load was applied to the pin, which was pressed against the counterpart on an EN32 disc made of steel. The specimens were removed after running through a sliding distance, cleanse with acetone, dried & weighted to assess the loss of weight due to wear. The wear loss of the composite was determined by the variation in weight measured before and after each examination. Normal loads (L), % of reinforcement (R), sliding velocity (S) and sliding distance (D) were examined as a function of the wear (W, μm) and frictional force (FF, N) of CFRP composites.

Figure 2. Pin-on-disc wear test setup.

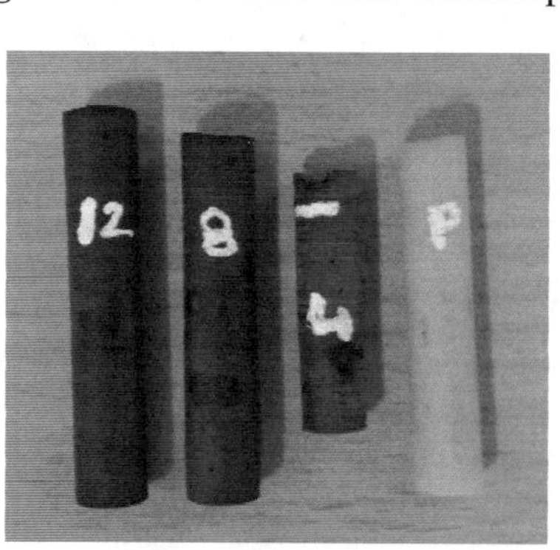

Figure 3. Wear pin samples (P- Pure Polyester, 4- CF 4Wt%+Polyester, 8- CF 8Wt%+Polyester, 12- CF 12Wt%+Polyester).

IOP Conf. Series: Materials Science and Engineering 1185 (2021) 012003 doi:10.1088/1757-899X/1185/1/012003

2.4. Design of Experiments using Taguchi orthogonal array

Experimental design using Taguchi is a very powerful tool for determining the effect of various parameter degrees on output responses. The most crucial stage in the design of experiments is selecting the control factors influencing performance of output. Several factors are considered at first, and then the less important factors are eliminated, leaving only the more important factors. As shown in table 2, four major factors were considered in the current dry-sliding wear test: load (L), sliding velocity (S), sliding distance (D), and percent of fiber reinforcement (R), each at four levels according to L_{16} OA. At room temperature the experiments are conducted. Using Taguchi L_{16} OA, thereby saving much experimental time and money, reduced the number of Experiments from 4^4=256 traditional runs to just 16 runs (Table 3). Again, these experiments are transformed into signal-to-noise ratios (S/N).S/Nratio for minimal wear and frictional force is determined as a logarithmic transformation of the loss function [13-14] as in equation (1).

"Smaller-the-Better", $$S/N_{SB} = -10\log_{10}\left[\frac{1}{n}\sum_{i=1}^{n} y_i^2\right] \quad (1)$$

where n equals to number of observations and y equals to the response variables.
For the present study $n = 16$ and $y = 2$

Table 2. Dry-sliding wear test levels and controlling factors.

Control Factors	Notation	Levels				Units
		1	2	3	4	
Sliding velocity	S	1	2	3	4	m/s
Load	L	5	10	15	20	N
Sliding distance	D	500	1000	1500	2000	m
% of Reinforcement	R	0	4	8	12	%

Table 3. Experimental design of L_{16} OA.

Runs	Independent Parameters			
	S	L	D	R
1	1	1	1	1
2	1	2	2	2
3	1	3	3	3
4	1	4	4	4
5	2	1	2	3
6	2	2	1	4
7	2	3	4	1
8	2	4	3	2
9	3	1	3	4
10	3	2	4	3

11	3	3	1	2
12	3	4	2	1
13	4	1	4	2
14	4	2	3	1
15	4	3	2	4
16	4	4	1	3

2.5. Grey Relational Analysis

Taguchi's experimental approach is sufficient for evaluating the optimum process parameter setting for a particular response characteristic. Multi-response optimization with GRA is the optimal approach where there are two or more responses with dissimilar output characteristics. Grey relation analysis can also be used to evaluate the resemblance of unusual finite data [15-16]. As a result, multi-response optimization for wear parameters is done in this analysis using the following procedure in GRA, as seen in figure 4. The ANOVA approach describes the most critical factor impacting wear behaviour.

Taguchi's Grey Relational Analysis is calculated using following steps are as follows [17-19]:
Step-I: Normalized data for wear and frictional force equivalent to the "Lower-the-Better" criteria, represented in grey relational generation by equation (2).

"Lower-the-Better", $$y_i(k) = \frac{\max x_i(k) - x_i(k)}{\max x_i(k) - min x_i(k)} \quad (2)$$

Where $i = 1, 2, 3 \ldots 16$ (no. of runs) & $k = 1, 2$ (no. of responses), $x_i(k)$ is obtained value, $max\ x_i(k)$equals to maximum value of $x_i(k)$ and $min\ x_i(k)$equals to minimum value of $x_i(k)$.
Step-II: Equation (3) determines the deviation sequence for all corresponding process variables.

$$\Delta_{oi}(k) = |x_o(k) - x_i(k)| \quad (3)$$

Where $\Delta_{oi}(k)$is the deviation sequence for the reference sequence $x_o(k)$& the comparability sequence $x_i(k)$.
Step-III: Following formulae [Equation(4)] are used to calculate the GRC (Grey Relational Coefficients) from normalized values in GRA method.

GRC, $$\xi_i(k) \quad (4)$$

Where ζ is distinguishing coefficient (i.e., ~0.5).
Step-IV: The GRG (Grey Relational Grade), which is predicted by averaging the number of GRC, is the key step using equation (5).

GRG, $$\gamma_i \quad (5)$$

Whereγ_i ranges in between 0 - 1 and *'n'* equals to the number of output responses.

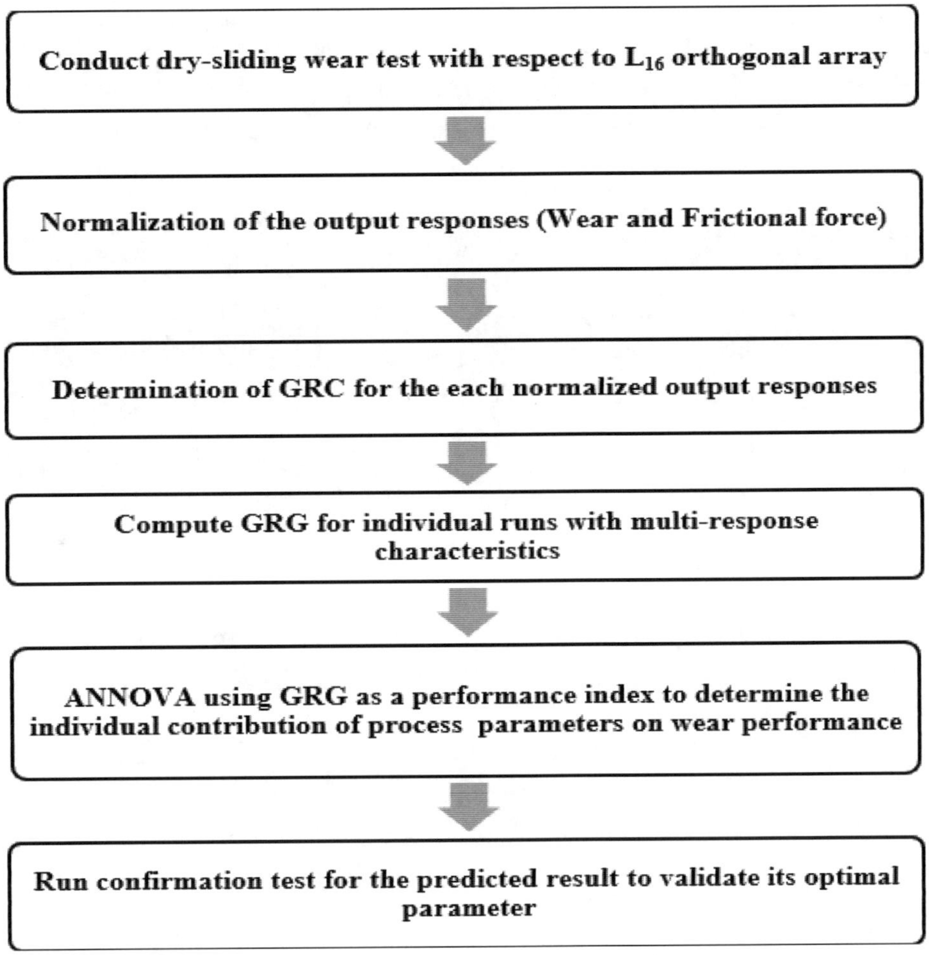

Figure 4. Steps followed in GRA technique.

3. Results and Discussion

3.1. Taguchi Approach and Optimization

3.1.1. Analysis of S/N Ratio. Minitab 18 was used as software for statistical analysis of the experimental findings made. Table 4 predicts the experimental results of wear and frictional force along with the associated S/N ratios, while tables 5 & 6 shows S/N ratio responses for wear (W) and frictional force (FF).

Figures 5& 6 illustrate how control factors influence the S/N ratio for wear and frictional force, respectively. By considering the lowest S/N ratios values, an optimal control factor setting for a better performance can be achieved. The bold values in tables 5 and 6 denote the lowest S/N ratios.From figure 5 &table 5, % of reinforcement (R) is an influencing parameter on the wear (W) and from figure 6 &table 6,load (L) is an influencing parameter on the frictional force (FF). It can be perceive from figure 5 &table 5, that the combination of factors; Sliding velocity: S (Level 3 i.e., 3 m/s), Load: L (Level 4 i.e., 20 N), Sliding distance: D (Level 4 i.e., 2000 m) and % of Reinforcement: R (Level 4 i.e., 12%) gives minimum wear. Furthermore, as illustrated in figure 6 and table 6, the combination of factors; Sliding velocity: S (Level 1 i.e., 1 m/s), Load: L(Level 4 i.e., 20 N), Sliding distance: D (Level 3 i.e., 1500 m) and % of Reinforcement: R (Level 1 i.e., 0%) gives minimum frictional force.

Table 4. Taguchi L_{16} OA & multi-response findings with S/N ratio.

Runs	Control Factors				Response variables		S/NRatios	
	S (m/s)	L (N)	D (m)	R (%)	W (μm)	FF (N)	W (dB)	FF (dB)
1	1	5	500	0	254	10.50	-48.0967	-20.4238
2	1	10	1000	4	124	9.00	-41.8684	-19.0849
3	1	15	1500	8	88	6.30	-38.8897	-15.9868
4	1	20	2000	12	24	4.40	-27.6042	-12.8691
5	2	5	1000	8	52	19.30	-34.3201	-25.7111
6	2	10	500	12	45	15.10	-33.0643	-23.5795
7	2	15	2000	0	213	8.00	-46.5676	-18.0618
8	2	20	1500	4	96	4.40	-39.6454	-12.8691
9	3	5	1500	12	37	13.80	-31.3640	-22.7976
10	3	10	2000	8	40	11.30	-32.0412	-21.0616
11	3	15	500	4	83	9.90	-38.3816	-19.9127
12	3	20	1000	0	122	5.30	-41.7272	-14.4855
13	4	5	2000	4	79	13.30	-37.9525	-22.4770
14	4	10	1500	0	94	9.55	-39.4626	-19.6001
15	4	15	1000	12	41	8.80	-32.2557	-18.8897
16	4	20	500	8	73	4.90	-37.2665	-13.8039

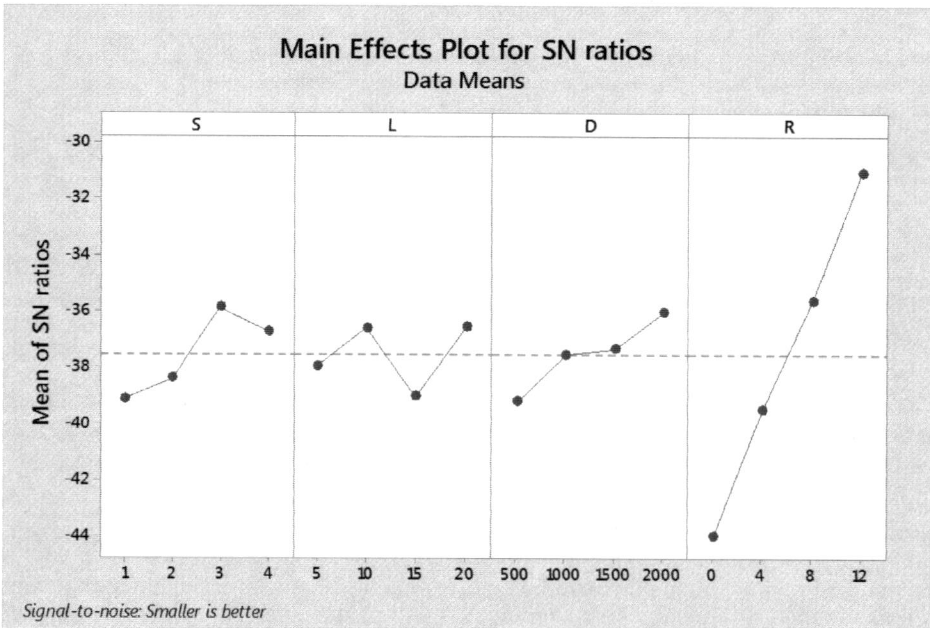

Figure 5. Main effect of wear for the S/N ratio plots.

Table 5. S/N ratio for Wear response table.

Level	S	L	D	R
1	-39.11	-37.93	-39.20	-43.96
2	-38.40	-36.61	-37.54	-39.46
3	**-35.88**	-39.02	-37.34	-35.63
4	-36.73	**-36.56**	**-36.04**	**-31.07**
Delta	3.24	2.46	3.16	12.89
Rank	2	4	3	1

Table 6. S/N ratio for Frictional force response table.

Level	S	L	D	R
1	**-17.09**	-22.85	-19.43	**-18.14**
2	-20.06	-20.83	-19.54	-18.59
3	-19.56	-18.21	**-17.81**	-19.14
4	-18.69	**-13.51**	-18.62	-19.53
Delta	2.96	9.35	1.73	1.39
Rank	2	1	3	4

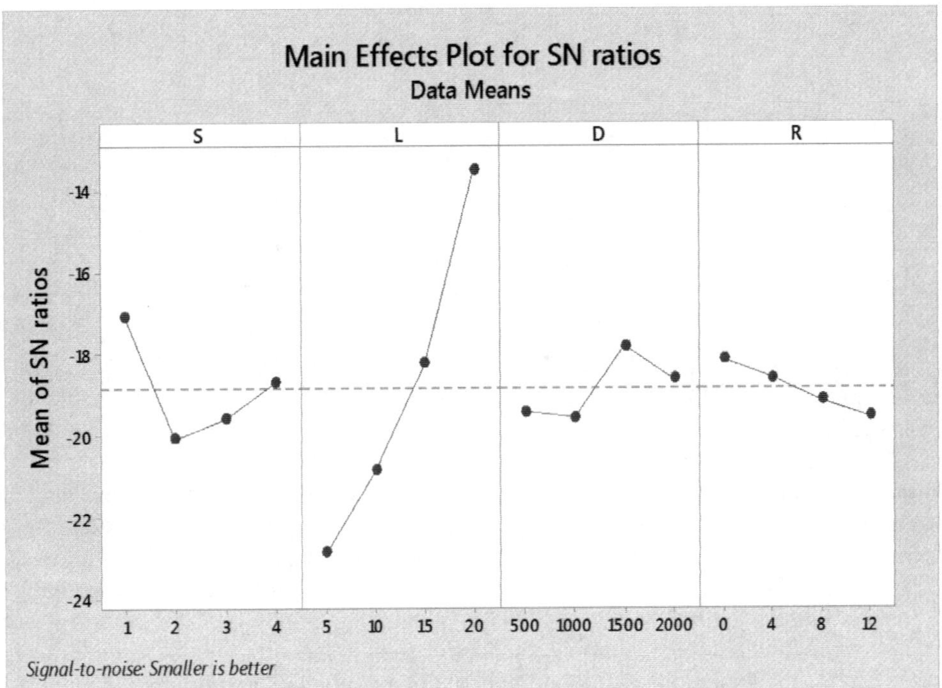

Figure 6. Main effect of frictional force for the S/N ratio plots.

3.1.2. Analysis of Variance. ANOVA (Analysis of Variance) was used to identify the relevant control factors that influence wear and frictional force response values. The overall sum of squared value is used to calculate the relevant importance of factors. The greater the sum of squared value, is the most influencing factor is to control the response values. The percentage contribution factors are calculated using these values [20]. At a confidence level of 95% & a significance level of 5%, the study is carried out. Tables 7 and 8 illustrates the ANOVA results for wear and frictional force, respectively.

As per table 7, the most significant factor influencing wear is % of reinforcement (R), with a contribution percentage of 77.71%, followed by sliding velocity (S) with 5.69%, sliding distance (D) with 4.33%, and load (L) with 3.59%. As per table 8, the most significant factor influencing frictional force is load (L), with a contribution percentage of 84.25 %, followed by sliding velocity (S), with 8.75 %, sliding distance (D), with 3.35 %, and % of reinforcement (R), with 1.93 %.

Table 7. ANOVA table for the S/Nratio of Wear.

Source	SS[a]	DF[b]	MS[c]	F-value	P-value	Contribution %
S	26.51	3	8.837	0.66	0.631	5.69
L	16.72	3	5.575	0.41	0.756	3.59
D	20.19	3	6.731	0.50	0.708	4.33
R	361.76	3	120.587	8.97	0.052	77.71
Error	40.35	3	13.449			8.66
Total	465.53	15				

$S = 3.66$, $R^2 = 91.33\%$, $R^2_{adj} = 56.67\%$

Table 8. ANOVA table for the S/Nratio of Frictional force.

Source	SS[a]	DF[b]	MS[c]	F-value	P-value	Contribution %
S	20.327	3	6.776	5.11	0.107	8.75
L	195.602	3	65.201	49.20	0.005	84.25
D	7.780	3	2.593	1.96	0.298	3.35
R	4.489	3	1.496	1.13	0.461	1.93
Error	3.976	3	1.325			1.71
Total	232.173	15				

$S = 1.15$, $R^2 = 98.29\%$, $R^2_{adj} =91.44\%$

[a] Sum of Squares.
[b]Degrees of Freedom.
[c]Mean squares.

The value of 'Prob > F', also known as the 'P-value'< 0.05 at a 95% confidence level, the interactions and factorsare considered relevant. Furthermore, a high F-value indicates that a process parameter has a characteristic impact on thesignificant performance [21-22]. Table 7 shows that the P-value of % of reinforcement is <0.05, indicating that it is a contributing factor to the change in wear. Similarly, the P-value of load in table 8 is <0.05, indicating that it is a contributing factor to the change in frictional force.Tables 7 and 8 also show that R^2 (coefficient of correlation) and R^2_{adj} (adjusted coefficient of correlation) values are high & comparable. This predicts the model's goodness of fit [23-24].

3.2. Multi-Response Optimization Using GRA

Wear and frictional force are process parameters that happen at the same time. As a result, they must be optimised in combination. GRA was selected for this role because it can transform a multi-objective to a single-objective problem, which it then optimises [25].

3.2.1. Calculation of GRG. Table 4 displays the responses for various parameter combinations depending on the L_{16} OA. Since minimum wear and frictional force was acceptable, equation (2) was used to normalize wear and frictional force. The deviation sequences were calculated using equation (3) based on the normalized value. Once the deviation sequences were ascertained, GRC (Grey Relational Coefficients) are determined for all responses using equation (4). Finally, using equation (5) the GRG (Grey Relational Grade) was calculated by taking the mean of all GRC.

Table 9 displays the GRA results for all run's, beginning with the normalisation process and ending with the calculation of GRG. The bold values in table 9 indicates that experiment number (4) has the maximum GRG value of 1.00, suggesting that the outcomes of the subsequent experiments are similar to the optimal value. The greater the GRG rating, the best the efficient characteristics, and the particular experimental run has better multiple response characteristics.

Table 9. Normalized, Deviation sequence, GRC and GRG for Wear and Frictional force.

Runs	Normalized data $[y_i(k)]$		Deviation sequence $[\Delta_{oi}(k)]$		GRC $[\xi_i(k)]$		GRG (γ_i)	Rank
	W	FF	W	FF	W	FF		
1	0.000	0.591	0.909	0.409	0.333	0.550	0.442	16
2	0.514	0.691	0.395	0.309	0.535	0.618	0.577	12
3	0.656	0.872	0.253	0.128	0.642	0.797	0.720	5
4	**0.909**	**1.000**	**0.000**	**0.000**	**1.000**	**1.000**	**1.000**	**1**
5	0.798	0.000	0.111	1.000	0.804	0.333	0.569	13
6	0.826	0.282	0.083	0.718	0.846	0.410	0.628	9
7	0.162	0.758	0.747	0.242	0.378	0.674	0.526	15
8	0.624	1.000	0.284	0.000	0.615	1.000	0.807	3
9	0.857	0.369	0.051	0.631	0.898	0.442	0.670	8
10	0.845	0.537	0.063	0.463	0.878	0.519	0.699	7
11	0.676	0.631	0.233	0.369	0.661	0.575	0.618	10
12	0.521	0.940	0.387	0.060	0.540	0.892	0.716	6
13	0.691	0.403	0.217	0.597	0.676	0.456	0.566	14
14	0.632	0.654	0.277	0.346	0.622	0.591	0.606	11
15	0.841	0.705	0.067	0.295	0.871	0.629	0.750	4
16	0.715	0.966	0.194	0.034	0.701	0.937	0.819	2

Table 10 shows the GRG responses calculated at their respective levels for each parameter and figure 7 shows the GRG's main effect plot on control factor derived from Minitab 18 application. Based on the GRG, S_4 L_4 D_3 R_4 (i.e., sliding velocity = 4 m/s, load = 20 N, sliding distance = 1500m and 12Wt% carbon fiber) is used for the optimum wear and friction force obtained from CFRP composite. Higher is the GRG, closer the quality of the product to ideal value. As a result, for optimal performance, a higher GRG is desired.

Table 10. GRG Response table.

Level	S	L	D	R
1	0.6844	0.5617	0.6267	0.5726
2	0.6326	0.6274	0.6528	0.6421
3	0.6757	0.6535	**0.7010**	0.7015
4	**0.6854**	**0.8357**	0.6977	**0.7621**
Delta	0.0528	0.2740	0.0742	0.1895
Rank	4	1	3	2

Average value of the GRG = 0.670

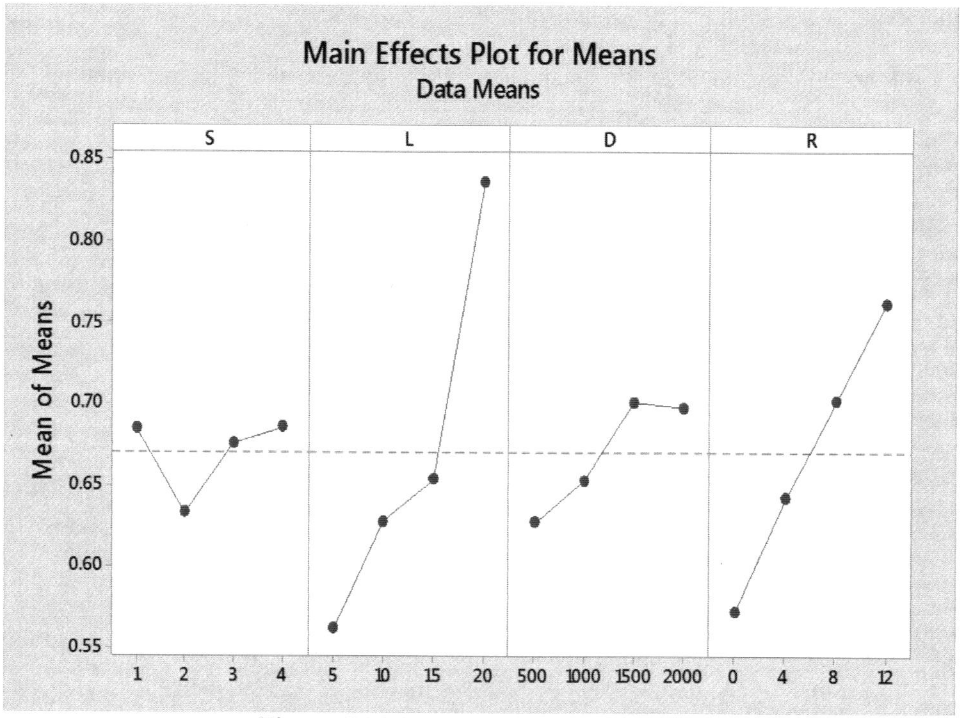

Figure.7 Main effect plots of GRG

3.2.2. ANOVA for GRG. Essence of ANOVA (Analysis of Variance) is to look into factors that have a significant impact on characteristic performance. This is accomplished by splitting overall variability of the GRG, as determined by the sum of squared deviations from the averages value of the GRG, in contributions from each wear parameter and listing them in table 11. According to ANOVA, a factor's relative ability to minimise variance is indicated by the percent of contributions. The factor with a high contribution % has a significant impact on performance. The percent of contribution of load, L (60.48 %) was known to be the most important factor influencing performance to wear. The % reinforcement, R (28.93%) was found to be the 2nd most influencing factor, followed by sliding distance, D (5.71%) and sliding velocity, S (2.74%).

Table 11. ANOVA table for GRG.

Source	SS [a]	DF [b]	MS [c]	F-value	P-value	Contribution %
S	0.007498	3	0.002499	1.28	1.28	2.74
L	0.165087	3	0.055029	28.28	28.28	60.48
D	0.015574	3	0.005191	2.67	2.67	5.71
R	0.078959	3	0.026320	13.52	13.52	28.93
Error	0.005838	3	0.001946			2.14
Total	0.272956	15				
			S = 0.044, R^2 = 97.86%, R^2_{adj} =89.31%			

[a] Sum of Squares.
[b]Degrees of Freedom
[c]Mean squares.

*3.2.3. Confirmation test.*Based on identification of the optimal parameter's impacting the multiple responses, a confirmation test was conducted to validateexperimental results. Equation (6) is used to calculate the predicted GRG. As per table 12, the predicted & experimental findings are in close agreement, indicating that the study was implemented successfully. For the optimal combination level,experimental GRG is 0.9532, while the predicted GRG is 0.9742. The experimental and predicted results agreed well. This means that the grey relation method is useful in order to optimize the wear parameter, where several characteristics must be examined simultaneously [26].

$$\hat{y} = y_m + \sum_{i=1}^{q} (\bar{y}_i - y_m) \quad (6)$$

Where $\hat{y}$ equals to the predicted grey relation grade, y_m equals to the mean value of GRG, y_i GRG at optimum levels and q equals to number of factors.

Table 12. Confirmation test findings.

	Optimal process parameters	
	Predicted	Experimental
Combination level	S_4 L_4 D_3 R_4	S_4 L_4 D_3 R_4
GRG	0.9742	0.9532
	% of improvement in GRG = 2.20%	

4. Conclusion

The fabrication of carbon fiber reinforced polyester (CFRP) composites using die-casting technique with varying concentrations of 0, 4, 8, and 12Wt percent and the predicted results show that sliding wear decreases significantly with increase in the content of carbon fiber.

- The Grey Relational Analysis technique can be used to integrate multi-objective optimization into single-objective optimization,& the Taguchi orthogonal array can be used to lower the number of experiments required to obtain an optimised result.
- Using Taguchi method and GRA, the dry-sliding wear parameters with multi-response characteristics was effectively optimized. The S_4 L_4 D_3 R_4 combination (i.e., sliding velocity(S= 4 m/s), load (L = 20 N), sliding distance (D = 1500 m),& 12Wt percent carbon

fiber) was found to have the optimal value dry sliding wear behaviour accountable for low wear and frictional force. The confirmation test demonstrated that the optimal parameters formulti responses were found to be effective.

- The significant and influencing factors from ANOVA (Analysis of Variance) revealed that load (L) and percent of reinforcement (R) were significantly contributed to performance of wear, but other parameters were non-significant.
- The carbon fiber reinforced polyester composites with 12Wt%reported excellent wear performance for the optimized wear variables combination.
- Finally, due to their superior mechanical &wear properties despite their low density, fabricated CFRP composites may be contemplateas a propitious material in the field of aerospace, automotive & marine applications.
- The Taguchi optimization approach may be used in the future to determine the effect of parameters in certain forms of wear testing methods.

Acknowledgement
The authors would like to thank Y.N.V. Sai Ram, Assistant Professor, Departmentof Mechanical Engineeringfor their assistance in the preparation of composites, planning of experiments as well as the Research Centre of R.V.R. & J.C College of Engineering, Guntur for thewear and friction measurement facilities.

References

[1] Messana, A., Sisca, L., Ferraris, A., Airale, A.G., de Carvalho Pinheiro, H., Sanfilippo, P. and Carello, M., 2019. From design to manufacture of a carbon fiber monocoque for a three-wheeler vehicle prototype. *Materials*, *12*(3), p.332.

[2] Soutis, C., 2005. Fibre reinforced composites in aircraft construction. *Progress in aerospace sciences*, *41*(2), pp.143-151.

[3] Niedermann, P., Szebényi, G. and Toldy, A., 2015. Characterization of high glass transition temperature sugar-based epoxy resin composites with jute and carbon fibre reinforcement. *Composites Science and Technology*, *117*, pp.62-68.

[4] Davim, J.P., Marques, N. and Baptista, A.M., 2001. Effect of carbon fibre reinforcement in the frictional behaviour of Peek in a water lubricated environment. *Wear*, *251*(1-12), pp.1100-1104.

[5] Chu, F., Yu, X., Hou, Y., Mu, X., Song, L. and Hu, W., 2018. A facile strategy to simultaneously improve the mechanical and fire safety properties of ramie fabric-reinforced unsaturated polyester resin composites. *Composites Part A: Applied Science and Manufacturing*, *115*, pp.264-273.

[6] Vilčáková, J., Sáha, P. and Quadrat, O., 2002. Electrical conductivity of carbon fibres/polyester resin composites in the percolation threshold region. *European Polymer Journal*, *38*(12), pp.2343-2347.

[7] Araújo, E.M., Araújo, K.D., Pereira, O.D., Ribeiro, P.C. and de Melo, T.J., 2006. Fiberglass wastes/polyester resin composites: mechanical properties and water sorption. *Polímeros*, *16*, pp.332-335.

[8] Swain, S., 2013. Synthesis and characterization of graphene based unsaturated polyester resin composites. *Transactions on Electrical and Electronic Materials*, *14*(2), pp.53-58.

[9] Pervez, M., Shafiq, F., Sarwar, Z., Jilani, M.M. and Cai, Y., 2018. Multi-response optimization of resin finishing by using a taguchi-based grey relational analysis. *Materials*, *11*(3), p.426.

[10] Basavarajappa, S., Chandramohan, G. and Davim, J.P., 2007. Application of Taguchi techniques to study dry sliding wear behaviour of metal matrix composites. *Materials & design*, *28*(4), pp.1393-1398.

[11] Rajmohan, T., 2019. Experimental investigation and optimization of machining parameters in drilling of fly ash-filled carbon fiber reinforced composites. *Particulate Science and Technology*, *37*(1), pp.21-30.

[12] Gee, M.G. and Owen-Jones, S., 1998. Wear testing methods and their relevance to industrial wear problems.
[13] Tarasasanka, C. and Ravindra, K., 2017. Application of Taguchi techniques to study dry sliding wear behaviour of magnesium matrix composites reinforced with alumina nano particles. *Journal of Engineering Science and Technology*, *12*(11), pp.2855-2865.
[14] TR, H.K., Swamy, R.P. and Chandrashekar, T.K., 2011. Taguchi technique for the simultaneous optimization of tribological parameters in metal matrix composite. *Journal of Minerals & Materials Characterization & Engineering*, *10*(12), pp.1179-1188.
[15] Achuthamenon Sylajakumari, P., Ramakrishnasamy, R. and Palaniappan, G., 2018. Taguchi grey relational analysis for multi-response optimization of wear in co-continuous composite. *Materials*, *11*(9), p.1743.
[16] Antil, P., Singh, S. and Manna, A., 2018. Electrochemical discharge drilling of SiC reinforced polymer matrix composite using Taguchi's grey relational analysis. *Arabian journal for Science and Engineering*, *43*(3), pp.1257-1266.
[17] Tarasasanka, C., Snehita, K., Ravindra, K. and Sameerkumar, D., 2019. Optimization of dry sliding wear properties ofAZ91E/nano Al 2 O 3 reinforced metal matrix composite with grey relational analysis. *International Journal of Engineering, Science and Technology*, *11*(4), pp.41-48.
[18] Raju, S.S., Senapathi, A.K. and Rao, G.S., 2017. Estimation of tribological performance of Al-MMC reinforced with a novel in-situ ternary mixture using grey relational analysis. *Indian J. Sci. Techno*, *10*(15), pp.1-9.
[19] Parihar, R. and Jathar, S., 2015. GREY RELATIONAL ANALYSIS TO OPTIMIZE WELDING PARAMETERS FOR DISSIMILAR SHEETS OF MATERIAL IN RESISTANCE SPOT WELDING. *Technology*, *6*(11), pp.23-32.
[20] Ozsoy, N., Ozsoy, M. and Mimaroglu, A., 2017. Taguchi approach to tribological behaviour of chopped carbon fiber-reinforced epoxy composite materials. *Delta*, *2*, p.3.
[21] Çiçek, A., Kıvak, T. and Ekici, E., 2015. Optimization of drilling parameters using Taguchi technique and response surface methodology (RSM) in drilling of AISI 304 steel with cryogenically treated HSS drills. *Journal of Intelligent Manufacturing*, *26*(2), pp.295-305.
[22] Sudheer, M., Prabhu, R., Raju, K. and Bhat, T., 2013. Modeling and analysis for wear performance in dry sliding of Epoxy/Glass/PTW composites using full factorial techniques. *International Scholarly Research Notices*, *2013*.
[23] Kilickap, E., Yardimeden, A. and Çelik, Y.H., 2017. Mathematical modelling and optimization of cutting force, tool wear and surface roughness by using artificial neural network and response surface methodology in milling of Ti-6242S. *Applied Sciences*, *7*(10), p.1064.
[24] Rakić, T., Kasagić-Vujanović, I., Jovanović, M., Jančić-Stojanović, B. and Ivanović, D., 2014. Comparison of full factorial design, central composite design, and box-behnken design in chromatographic method development for the determination of fluconazole and its impurities. *Analytical Letters*, *47*(8), pp.1334-1347.
[25] Wojciechowski, S., Maruda, R.W., Krolczyk, G.M. and Niesłony, P., 2018. Application of signal to noise ratio and grey relational analysis to minimize forces and vibrations during precise ball end milling. *Precision Engineering*, *51*, pp.582-596.
[26] Siriyala, R., Alluru, G.K., Penmetsa, R.M.R. and Duraiselvam, M., 2012. Application of grey-taguchi method for optimization of dry sliding wear properties of aluminum MMCs. *Frontiers of Mechanical Engineering*, *7*(3), pp.279-287.
[27] Laxminarayana, P., 2018. Optimization of electrode tool wear in micro holes machining by die sinker EDM using Taguchi approach. *Materials today: proceedings*, *5*(1), pp.1824-1831.
[28] Kumar, U.A., Saidulu, G. and Laxminaryana, P., 2020. Experimental investigation of process parameters for machining of Nimonic alloy 75 using wire-cut EDM. *Materials Today: Proceedings*, *27*, pp.1362-1368.

Developing hybrid composites of Al5083-carbon nanotubes and fly ash by friction stir processing: machining studies

G.V.N.B. Prabhakar [1], *, Lingaraju Dumpala [2], N. Ravikumar [3]

[1] Department of Mechanical Engineering,
V.K.R., V.N.B. & A.G.K. College of Engineering, Gudivada 521301, India
[2] Department of Mechanical Engineering,
Jawaharlal Nehru Technological University, Kakinada, India
[3] Department of Mechanical Engineering,
V. R. Siddhartha Engineering College, Vijayawada 520007, India

Corresponding author's e-mail address: gvnb.prabhakar@gmail.com

Abstract. In the present work, hybrid composites of Al5083-reinforced with CNT and fly ash were produced by friction stir processing (FSP) successfully. The microstrucutral observations indicated bands like structure in the stir zone. Higher hardness was noticed for the composites due to the grain refinement resulted from FSP and also due to the presence of reinforcements (CNT and fly ash). The variations were observed as higher within the measured hardness for the composites which can be claimed to the variations in the concentration of the reinforcements across the stir zone. Machining studies were done by carrying drilling experiments on the composites as well as on base material. Higher cutting forces were recorded for the composites. Additionally, decreased cutting forces were observed for both the base material and the composite with increase in the cutting speed. The overall observations suggest that Al5083-CNT-fly ash hybrid composites can be successfully developed by FSP which exhibit better mechanical properties and machining behaviour.

Keywords: FSP, MMCs, hybrid composites, machining, hardness.

1. Introduction

Metal matrix composites (MMCs) are a category of modern engineered materials in which matrix is a metal dispersed with suitable fibre materials in the form of continuous or discontinuous phase [1]. MMCs exhibit better properties than their constituting phases. Several properties are influenced by the type of dispersing phase, quantity, size and shape of the reinforcement. Mechanical properties, wear and corrosion properties, machining behaviour are the important bulk properties of MMCs influenced by the reinforcements [2]. If more than one dispersing phase is used in developing composites, such materials are called as hybrid composites. Each reinforcement phase serves a specific purpose and when combined with another dispersing phase, hybrid properties can be achieved.

H. Patle et al. [3] reinforced Al7075 alloy with B_4C and MoS_2 to develop hybrid composite by FSP (friction stir processing) and demonstrated improved mechanical properties and wear properties.

Particularly, presence of MoS_2 reduced the coefficient of friction due to the lubrication effect attained in the hybrid composite. Zayed et al. [4] developed hybrid composites of Al5083 dispersed with SiC and Al_2O_3 phases. Improved hardness up to 30% was measured for the hybrid composite and higher wear resistance was observed with the composite containing 50% SiC and 50% Al_2O_3. Similarly, Heiderpour et al. [5] also produced surface composites of Al5083-Al_2O_3-TiO_2 by FSP. Due to the dispersed phases improved ultimate tensile strength was observed for the hybrid composite. Furthermore, decreased coefficient of friction and better wear resistance were demonstrated for the hybrid composite. Machining composites arise critical issues unlike machining pure metals and alloys. Due to the presence of different phases with diversified properties, composites may exhibit abnormal behaviour during machining. This kind of machining behaviour depends on the type of composite and the fraction of the composite dispersed in the matrix. In our earlier work, due to the addition of graphene into Al5083 alloy, lower cutting forces were recorded which was claimed to the solid lubricant effect imparted into the composite due to the addition of graphene [6]. Understanding the machining behaviour of composite is crucial as most of the times, machining is a compulsory process to develop structures made of composites.

Among the available aluminium alloys, Al5083 contains magnesium as the main constituting alloying element. The applications of Al5083 can be found in manufacturing marine structures and components [7]. On the other hand, carbon nanotubes (CNTs) as dispersing phase have been widely used in developing MMCs due to their potential to promote better properties [8]. Fly ash is a waste product results from the coal burning in the thermal power plant which contains SiO_2 and Fe_2O_3 phases. Hence, in the present work, Al5083 alloy has been chosen as matrix material and combnation of fly/CNTs reinforcements were used to produce composites by FSP route. Microhrdness measurements were carried out for the composite and also subjected to machining by conducting drilling experiments to assess the machining behaviour of the composite.

2. Materials and methods

AL5083 aluminumalloy sheets of size 100 X 100 X 5 mm (Venku metals, Hyderabad, India) were used as base material. CNTs (single walled) were procured from nano Wings, Khammam, India. Fly ash was collected from Dr Narla Tata Rao Thermal Power station, Vijayawada, India. In order to develop hybrid composites, equal amounts of CNTs and fly ash (50% and 50%) were filled in the grooves machined on the surface of the workpieces. The grooves were produced on the surface of the workpiece of width 1 mm and depth 2 mm. The workpieces were then fixed on a specially fabricated attachment on the vertical milling machine work table. The milling machine is an automated type which can be given auto feed and speed. Then the FSP tool was fixed in the spindle of the milling machine. In the present study, the spindle of the milling machine does not have the vertical axis tilting facility and hence, the tool tilt angle was assumed to be 0 degrees. The geometrical dimensions of FSP tool are mentioned in our earlier work. Based on the preliminary experiments, processing was completed at 1400 rpm and 25 mm/min processing parameters. Before starting the process, the grooves were closed by using an FSP tool without having a pin to eliminate the escape of the filled reinforcing phases from the grooves. Then, FSP process was carried to develop the surface MMCs. Fig 1 shows the schematic illustration of composites development by FSP.

After processing, specimens were cut across the processed zone and also from the base materials to investigate the microstructure and micorahrdness. For microstructural observations, the samples were polished as per the standard metallographic procedure and chemically etched the polished surfaces with Keller's reagent. Then the optical microscope images (Leica, Germany) were obtained. Vickers indentation method (Omni Tech, India) was adopted to assess the hardness of the samples. Machining experiments were done by drilling holes in base alloy and the composite to study the machinability. The workpieces were fixed on the load cell (Kestler, Germany) which was fixed on the work table of the drilling machine. Drilling of holes was done by using a 6 mm dia thrust drill bit at two different rotational speeds and two feed rates. No lubricant was applied during the machining experiments. While drilling the composites, holes were produced in the middle of the stir zone. Measurement of cutting forces was started when the tool tip touched the workpiece surface and continued for a few seconds.

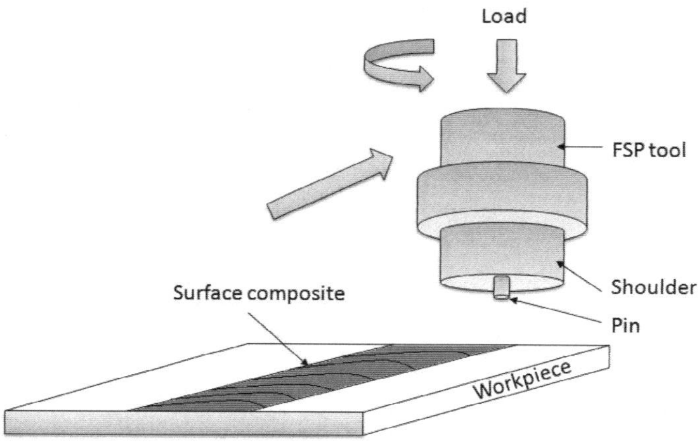

Fig 1 Schematic illustration of developing surface composites by FSP.

3. Results and discussion

From the macroscopic observations, defect free stir zone was observed after FSP. Fig 2 shows the microstructure of Al5083 alloy and produced composite. The base alloy has a typical microstructure that contains solid solution grains and intermetallics at the grain boundaries. The microstructure of composite has combination of gray and white regions which suggest the presence of reinforcements (CNTs and fly ash). During FSP, the material that is stirred within the nugget zone contains mixing of reinforcement and base material and results in a band like structure as shown in Fig 2 (b). This is similar to earlier reports where appearance of bands in the composites produced by FSP is a common finding. In the present work, CNTs and fly ash reinforcement has been incorporated within the matrix and the combination of dark and bright regions were appeared in the stir zone.

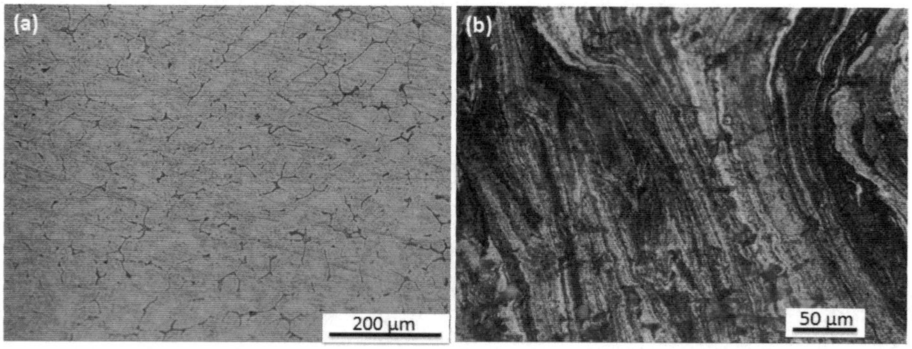

Fig 2 Microstructural observations: a) Al5083 and b) composite

The microhardness measurements obtained for the samples are presented in Fig 3. Form the average hardness values, increased hardness can be noticed for the composite and the increment is significant compared with Al5083. The distribution of the hardness within the composite was also observed as higher within the stir zone due to the presence of reinforcement and also due to the smaller grains which is usually found in FSPed materials.

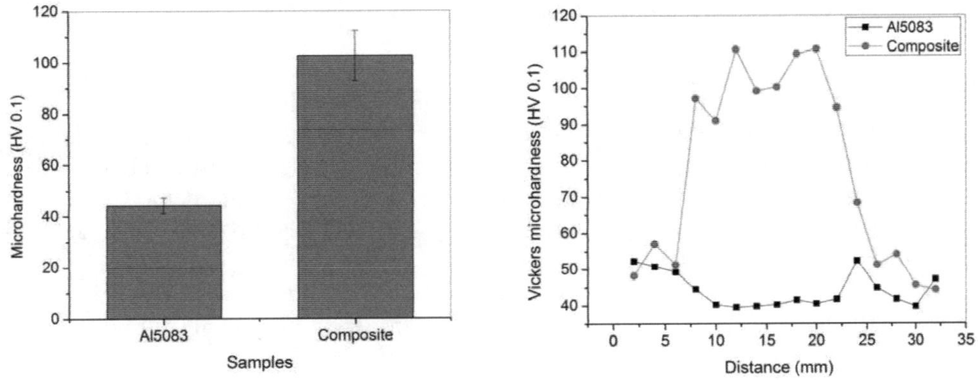

Fig 3 Microhardness data: a) average hardness values and b) hardness distribution of the samples.

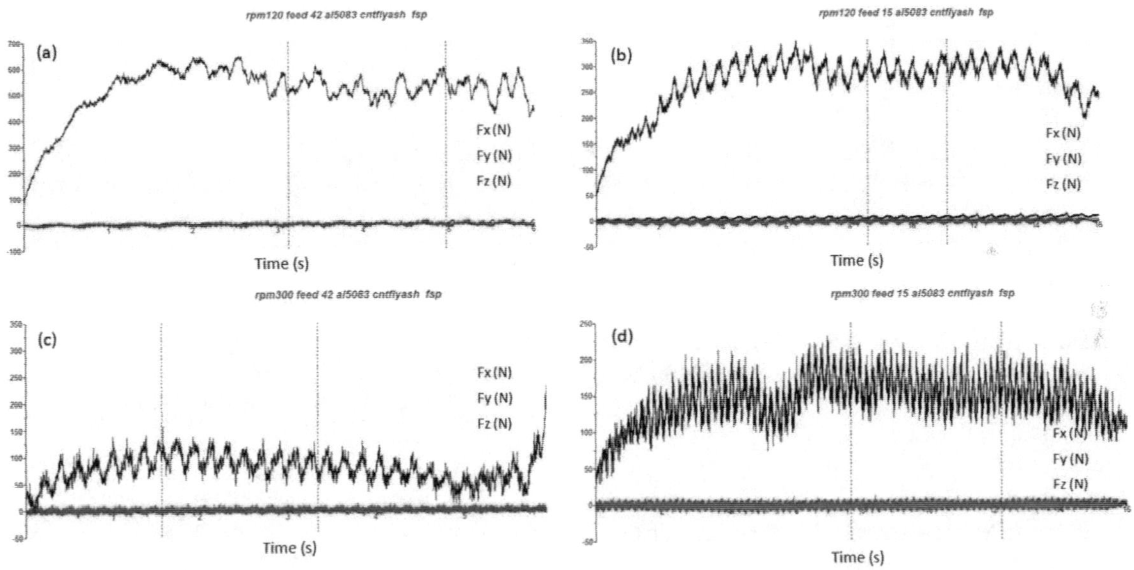

Fig 4 Cutting force profiles of the composites recorded during the drilling experiments: a) 120 rpm with 42 mm/min feed, b) 120 rpm with 15 mm/min feed, c) 300 rpm with 42 mm/min feed and d) 300 rpm with 15 mm/min feed.

The machining studies from the drilling experiments revealed that at the increased rotational speeds decreased cutting forces were observed. This can be understood by considering the generation of higher heat due to the increased RPM during drilling in the machining zone. Compared with the base alloy [9], the mean cutting forces were observed as slightly increased for the composites. This can be ascertained to the role of incorporated reinforcements into Al5083. The variations within the cutting forces were also observed as higher in the composites. During the machining of composites, presence of different phases leads to different level of resistance against the drill bit. Therefore, the cutting forces are observed with more variations. The preliminary studies from the Al5083 composite with CNTs and fly ash demonstrates higher hardness. Furthermore, higher cutting forces with more variations are important observations which help to assess the machining behaviour of the composite.

4. Conclusions

In the present work, hybrid composites of Al5083 dispersed with CNTs/fly ash were fabricated through FSP. Defect free stir zone was observed in the FSPed composite. Increased hardness due to the grain refinement and incorporated CNT and fly ash was measured for the composites compared with the base alloy. Machining studies by conducting drilling experiments demonstrated the role of added reinforcement to increase the cutting forces during drilling the composites. Variations within the measured cutting forces were also observed as higher in the composites due to the presence of different phases. The preliminary observations suggest that by incorporating CNTs and fly ash, Al5083 based structures can be fabricated with ease of machining with improved mechanical properties.

Acknowledgements

The funding to complete the present research work from AICTE RPS Project (File No. 8-188/RIFD/RPS (Policy-1)/2018-2019 is gratefully acknowledged.

References

[1] Nikhilesh C, Krishan C K 2003 *Metal Matrix Composites* (Springer, USA).
[2] Davis E A, Ward I M 1993 *An introduction to metal matrix composites* (Cambridge University Press, USA).
[3] Patle H, Mahendiran P, Ratna Sunil B, Ravikumar Dumpala, 2019 *Mater Res Express* **6** 086589.
[4] Zayed E M, El-Tayeb N S M, Ahmed M M Z, Rashad R M 2019 *Engineering Design Applications: Advanced Structured Materials* (Springer, USA) DOI: https://doi.org/10.1007/978-3-319-79005-3_2.
[5] Heidarpour A, Ahmadifard S, Kazemi S 2018 *Prot Met Phys Chem* **54**, 409-415.
[6] Prabhakar G V N B, Lingaraju Dumpala, Ravi Kumar N 2021 *Trans Ind Inst Met* **73**, 2955-2964.
[7] Davis Joseph R 1993 *ASM Specialty Handbook: Aluminum and Aluminum Alloys* (ASM International, USA).
[8] Coroş M, Pogăcean F, Măgeruşan L, Socaci C, Pruneanu S 2019 *Front Mater Sci* **13**, 23.
[9] Prabhakar G V N B, Lingaraju Dumpala, Ravi Kumar N 2021 *Mater Sci Eng Technol* (in press).

Hybrid composites of Al5083- graphene - fly ash by friction stir processing: machining studies

G.V.N.B. Prabhakar [1], *, Lingaraju Dumpala [2], N. Ravikumar [3]

[1] Department of Mechanical Engineering,
V.K.R., V.N.B. & A.G.K. College of Engineering, Gudivada 521301, India
[2] Department of Mechanical Engineering,
Jawaharlal Nehru Technological University, Kakinada, India
[3] Department of Mechanical Engineering,
V. R. Siddhartha Engineering College, Vijayawada 520007, India

Corresponding author's e-mail address: gvnb.prabhakar@gmail.com

Abstract. Hybrid composites are the modern engineered materials which exhibit presence of two or more than two reinforcements in a matrix. In the current work, Al5083 - Graphene and fly ash were dispersed in Al5083 alloy to develop hybrid composites by friction stir processing (FSP). Microstrucutral studies revealed band structures in the stir zone. Increased hardness was measured for the composites due to the grain refinement resulted from FSP and also due to the presence of reinforcements (Graphene and fly ash). The variations within the measured hardness values were observed as higher for the composites due to the variations in the concentration of the Graphene and fly ash in the stir zone. Lower cutting forces were recorded for the composites during drilling experiments conducted on the base materials and the composites. Here, the role of graphene as solid lubricant can be noticed as an important factor. The overall results demonstrate that Al5083-Graphene-fly ash hybrid composites can be successfully developed by FSP which exhibit better mechanical properties and machining behaviour.

Keywords: Machining, composites, FSP, hardness, drilling

1. Introduction

Friction stir processing (FSP) is a solid state processing route can be used to develop composites without melting the materials [1]. Recently, surface metal matrix composites (MMCs) are becoming a promising choice for various engineering applications where material properties are altered at the surface itself without altering the core [2]. Composite materials exhibit better properties compared with their constituting phases. Several properties such as mechanical, wear and corrosion resistance, machining behaviour are influenced by the type of dispersing phase, size, shape and fraction of the dispersing reinforcement [3]. Hybrid composites contain more than one dispersing phase and they exhibit excellent properties. Each of the reinforcements serves a special purpose in the hybrid composites and the combined effect of these reinforcements promotes performance of the composite particularly when subjected to extreme environment conditions during the application.

It was reported in the literature that Al7075 alloy with B_4C and MoS_2 reinforcements added by friction stir processing (FSP) has demonstrated improved mechanical properties and wear properties. Particularly, addition of MoS_2 reduced the coefficient of friction due to the lubrication effect attained in the hybrid composite [4]. Zayed et al. [5] also produced hybrid composites of Al5083 by dispersing SiC and Al_2O_3 reinforcements. Higher hardness (up to 30%) was observed for the hybrid composite and also higher wear resistance was noticed with the composite that contains 50% SiC and 50% Al_2O_3. Heiderpour et al. [6] also developed surface composites by dispersing Al_2O_3-TiO_2 into Al5083 matrix by FSP. Better mechanical properties were measured for the hybrid composite due to the dispersed phases. Additionally, coefficient of friction was measured as decreased and wear resistance was observed increased for the hybrid composite. Machining operations are important group of manufacturing processes which must be required to develop structures in various applications. Machining of composites is difficult due to the presence of combined phases compared with single phase materials. The machining characteristics of composite material depend on the type of reinforcement and the amount of dispersing phase. Due to the addition of graphene into Al5083 alloy, lower cutting forces were recorded for the composite which was claimed to the solid lubricant effect imparted into the composite by adding graphene [7].

Al5083 aluminum alloy that contains magnesium as the main constituting alloying element is widely used in marine applications compared with its relatively better corrosion resistance among all other aluminium alloy [8]. On the other side, Graphene; an isotope of carbon that has reported with high potential as advanced material with excellent properties. [9]. Fly ash is a waste product results from the coal burning in the thermal power plant which contains SiO_2 and Fe_2O_3 phases. In the current research work, hybrid composites made of Al5083 – grapheme – fly ash were prepared by FSP route. The effect of reinforcements on microhardness and machining behaviour of the composite were investigated.

2. Materials and methods

Al5083 aluminium alloy sheets (Venku metals, Hyderabad, India) of 100 X 100 X 5 mm size were used as base material. Graphene powder was procured from Nano Wings, Khammam, India. Fly ash was obtained from Dr. Narla Tataro Thermal Power Station, Vijayawada, India. Equal amount of grapheme and fly ash by wt.% was mixed and used as reinforcement to develop Al5083 based composite by FSP. The workpieces were placed on the worktable of a universal milling machine (Fig 1(a)). The workpieces were machined with a groove of 1 mm (width) and 2 mm (depth) and the groove was filled with the reinforcement mixture. Initially, the groove was subjected to surfacing by using an FSP tool that has no pin to avoid the escape of the reinforcement. The detailed procedure has been explained in our earlier work [7]. The surface treated workpieces were then subjected to FSP at 1400 rpm tool rotational speed with a tool travel speed of 25 mm/min. In the present study, the spindle of the milling machine does not have the vertical axis tilting facility and hence, the tool tilt angle was assumed to be 0 degrees. FSP process was carried to develop the surface MMCs. Fig 1 (b) shows the photograph of composite preparation by FSP.

Specimens were cut from the FSPEd samples across the processed zone and also from the base materials for the microstructural studies and micorahrdness measurements. Initially, the samples were polished by using coarse graded polishing paper and then followed by fine graded polishing papers. Then fine graded alumina paste was used to complete the polishing process. The polished specimens were then cleaned and chemically etched by using Keller's reagent. Microstructures were recorded by optical microscope (make: Leica, Germany) for both the samples. The hardness of the samples was assessed by Vickers indentation method (Omni Tech, India) by applying 100 g load. Drilling 6 mm holes was done for base alloy and the composite to study the machinability. A load cell (Kestler, Germany) was used to measured the cutting forces. Drilling experiments were done by using a 6 mm dia drill bit at two different rotational speeds and two feed rates without any lubricant. While drilling the composites, holes were produced in the middle of the stir zone. Measurement of cutting forces was started when the tool tip touched the workpiece surface and continued for a few seconds.

Fig 1 a) Photograph of vertical milling machine and b) developing surface composites by FSP.

3. Results and discussion

The FSPed regions in the composite were identified with defect free zone. The microstructures of the samples are shown in Fig 2. Al5083 alloy contains solid solution grains and intermetallic at the grain boundaries as shown in Fig 2 (a). Combination of black and white bands was observed in the composite. Due to the material flow and mixing within the stir zone of FSPed composite, the reinforcements are distributed in the nugget zone Fig 2 (b). This is similar to the earlier findings where bands of material flow can be seen [10]. In the present work, Graphene and fly ash reinforcement has been incorporated within the matrix and the combination of dark and bright regions were appeared in the stir zone. The distributed phases influence the bulk properties of the composite. FSP process also introduces grain refinement. The mechanism behind grain refinement has been explained elsewhere [1]. Due to the dynamic recrystallization duing FSP, smaller grains are produced.

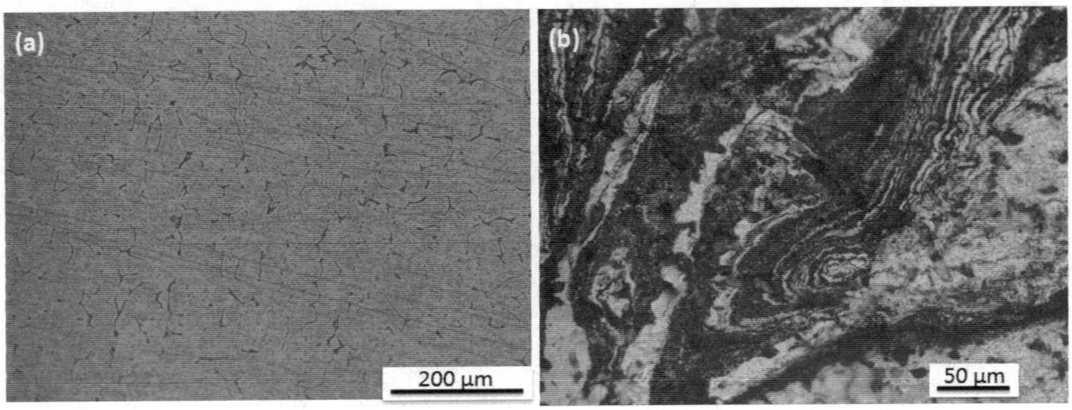

Fig 2 Optical microscope images: a) Al5083 and b) composite

From the microhardness measurements, higher hardness values were observed for the composite. Fig 3 shows the average hardness values and the hardness distribution. From the hardness data, composite was, found with higher hardness compared with the base alloy. Furthermore, the distribution of the hardness within the composite was also observed as higher within the composite. The presence of reinforcement and also due to the smaller grains, the hardness values were observed as higher in the composite. The increased hardness certainly influences the machining behaviour of the composite.

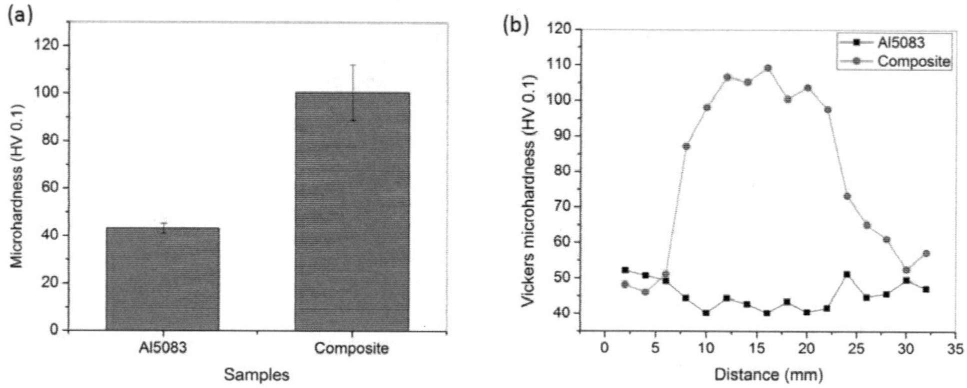

Fig 3 Microhardness measurements: a) average hardness values and b) hardness distribution of the samples.

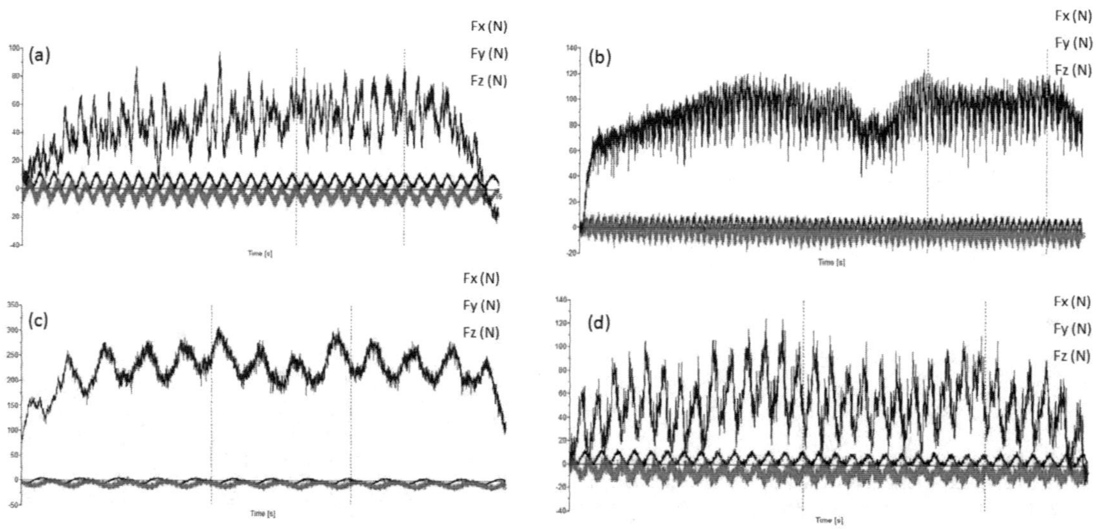

Fig 4 Cutting force profiles of the composites obtained from the drilling experiments: a) 120 rpm with 42 mm/min feed, b) 120 rpm with 15 mm/min feed, c) 300 rpm with 42 mm/min feed and d) 300 rpm with 15 mm/min feed.

Fig 4 shows the cutting force profiles of the compotes at different combinations of machiing parameters during the drilling operations. It can be seen that the increased rotational speeds has decreased the cutting forces. It is true that the heat generation is higher with the increased rpm. Increased heat within the machining zone reduces the cutting forces. The mean cutting forces observed in the composite were observed as slightly higher compared with the base alloy [9]. The incorporated Graphene and fly ash combination increased the material resistance against the cutting edges of the drill bit. Presence of different phases within the composite influenced the cutting forces and hence, the variations within the cutting forces were also observed as higher in the composites. The results from the machining studies of the composites confirm the influential role of added reinforcements on improving the mechanical properties and further machinability.

4. Conclusions

In the current research work, Al5083-graphene-fly ash composites were produced by FSP. No defects were observed in the stir zone of FSPed composite. Increased hardness was noticed for the composite due to the grain refinement and reinforced graphene and fly ash. Drilling experiments demonstrated the important role of added reinforcement to increase the cutting forces in the composite compared with the base alloy. Higher variations in the cutting forces were also noticed in the composites which is due to the added graphene and fly ash. The results suggest that by incorporating graphene and fly ash into Al5083, structures with higher hardness and improved machinability ca be produced.

Acknowledgements

The financial support obtained from AICTE - RPS Project (File No. 8-188/RIFD/RPS (Policy-1)/2018-2019) is gratefully acknowledged.

References

[1] Mishra R S, Ma Z Y, Charit I 2003, *Mater Sci Eng* A **341,** 307.
[2] Nikhilesh C, Krishan C K 2003 *Metal Matrix Composites* (Springer, USA).
[3] Davis E A, Ward I M 1993 *An introduction to metal matrix composites* (Cambridge University Press, USA).
[4] Patle H, Mahendiran P, Ratna Sunil B, Ravikumar Dumpala, 2019 *Mater Res Express* **6** 086589.
[5] Zayed E M, El-Tayeb N S M, Ahmed M M Z, Rashad R M 2019 *Engineering Design Applications: Advanced Structured Materials* (Springer, USA) DOI: https://doi.org/10.1007/978-3-319-79005-3_2.
[6] Heidarpour A, Ahmadifard S, Kazemi S 2018 *Prot Met Phys Chem* **54**, 409-415.
[7] Prabhakar G V N B, Lingaraju Dumpala, Ravi Kumar N 2021 *Trans Ind Inst Met* **73**, 2955-2964.
[8] Davis Joseph R 1993 *ASM Specialty Handbook: Aluminum and Aluminum Alloys* (ASM International, USA).
[9] Coroş M, Pogăcean F, Măgeruşan L, Socaci C, Pruneanu S 2019 *Front Mater Sci* **13**, 23.
[10] Prabhakar G V N B, Lingaraju Dumpala, Ravi Kumar N 2021 *Mater Sci Eng Technol* (in press).

Assessment of sludge formation in diesel storage tanks and eradication measures

Priyal Kumar [1, *], L K Durga Pradeep Raavi[1], Ashwin Anil Kumar [1], Ravikumar Dumpala[1]

[1]Department of Mechanical Engineering, Visvesvaraya National Institute of Technology, Nagpur440010, India

*Corresponding author: priyalkumars@gmail.com, +91 9834814389

Abstract. The sludge formation and its accumulation in fuel tanks cause several problems such as clogging of vent and suction pipes, reduction in fuel quality and tank capacity, and deterioration of tank material. Sludge is a complex composition of water, fuel, sand, waste oil, metals, bacteria, minerals, and other organic matter. It grows with time because of bacterial growth (that feeds on fuel and water), impurities, and the aerobic polymerization of hydrocarbons in the fuel. The main objective of the present work is to delay the sludge from forming in the diesel storage tanks. It aims at breaking down information collected from various sources on how sludge formation occurs with time and produce research results on the chemical analysis performed on a diesel sludge sample collected from a fuel depot. Two experimental cases have been set up to show how stirring affects sludge growth over a period of time. It is observed that the sludge growth is less in the sample that is agitated when compared with the sample without agitation. By taking this into account, a stirring design isproposed and developed. Also, an idea is provided for reducing the metal ions in the sludge, thus maintaining a high-quality fuel ready for dispensing at the fuel stations.

Keywords: Sludge; Diesel; Storage tanks; Chemical analysis; Growth prevention

1. Introduction

Petroleum sludge contains both organic components and heavy metals [1]. Diesel sludge (one of the subdivisionsof petroleum sludges) is not a single well-defined product, and its composition varies with location. Most representative hydrocarbons present in it arestraight-chain alkanes ranging from C10 to C28. Pristane and Phytane are the predominant branched alkanes identified [2]. Microorganisms homologous to Bacillus mycoides, Paenibacilluslautus, and Rhodobacter spp. are known to be present in diesel sludge [3]. To get an idea about the contaminants present, a typical diesel sludge sample was collected and tested in Maharashtra Enviro Power Limited (MEPL) in their NABL certified laboratory. Following report was generated, showing different parameters and their values (Table 1). Accumulation of large quantities of sludge affects the life of diesel filters, storage capacity, fuel quality, and even reduces tank life. This can also lead to leakages which result in groundwater and soil getting contaminated with petroleum [4].These days, diesel which is blended with biodiesel such as fatty acid methyl ester (FAME) shows significant operational problems due to sludge (microbial contamination) [5].The build-up of sludge in many storage tanks is due to the stagnant fuel, rusting of tank, scale formation, organic growth and presence of water content in it. Even the slightest amount of

IOP Conf. Series: Materials Science and Engineering 1185 (2021) 012006 doi:10.1088/1757-899X/1185/1/012006

water is enough to initiate the phase separation of diesel, leading to two layers of different densities at the the tank bottom. This water may enter the tank as rainwater, groundwater, or atmospheric moisture.Also, due to the temperature difference between night and day, water condenses and is built upon the inner surface of the tank [6].Water and other contaminants cannot be completely filtered out as carrying out filtration at a bulk level in depots is not economically feasible. A feasible solution to this problem is delaying the sludge formation by slowing down its growth.

The project aims at coming up with a technique to delay the sludge growth, resulting in a longer time interval between consecutive tank cleaning cycles. This reduction in sludge concentration also helps in avoiding problems in tank calibration and fuel storage, and prevents the sludge from chocking the vent and delivery pipes. Here a time-bound study has been done on commercially available diesel. Impurities were grown by adding water to the diesel samples. One sample was stirred regularly, while the other was left undisturbed. Based on the outcome of the experiments, a stirring design has been proposed for vertical cylindrical storage tanks in the fuel depots. Sludge also contains metallic ions, which because of their size, result in damaging the vehicle carburetors. An idea has been generated to enrich the customer experience at fuel stations.

Table 1. Chemical analysis of the diesel sludge carried out in NABL certified laboratory.

S.No.	Parameter	Value	CPCB:2009-2010
1	pH	7.33	4 - 12
2	Calorific Value(cal/gm)	1120.2	<2500 CAL/G
3	Loss on Drying (%)	76.90	NOT SPECIFIED
4	Loss on Ignition (%)	98.07	<20%
5	Ash content (%)	1.93	NOT SPECIFIED
6	Oil & Grease (%)	<1.0	<4
7	Cadmium as Cd (mg/L)	<0.1	<0.2
8	Chromium as Cr^{3+} (mg/L)	<0.1	NOT SPECIFIED
9	Copper as Cu (mg/L)	<0.1	<10
10	Iron as Fe (mg/L)	<0.1	NOT SPECIFIED
11	Nickel as Ni (mg/L)	<0.1	<3
12	Zinc as Zn (mg/L)	<0.1	<10

2. Methodology

Twoexperimental cases were set up to study the effects of stirring on sludge growth. Two steel containers were filled with 800ml of commercially available diesel and 15ml water each. One was kept undisturbed, and the other was stirred at 30rpm for 20 seconds once a day. The samples showed considerable black growth in the container with time. The composition of this black compound is not known and has been referred to as sludge here onwards. The growth of sludge has been recorded with time by capturing photographs of the samples at a regular interval of 20 days. Figure 1 shows the same for the stirred sample. Since the amount of sludge was quite small, quantitative chemical analysis was not feasible. To calculate the amount of sludge formed, the photos of the sample were tessellated into square grids of size 55x55, as shown in Figure 2. Container size was noted, and base diameter was kept as reference, formulation of the amount of volume that a square represented on the grid was carried out. All the calculations being relative to the base diameter, the influence of factors such as the

IOP Conf. Series: Materials Science and Engineering 1185 (2021) 012006 doi:10.1088/1757-899X/1185/1/012006

height of the camera above the sample and angle of tilt was eliminated. Likewise, calculations were done for a period of 100 days, and a graph was plotted between the volume of sludge in cm^3vs number of days,as shown in Figure 3.

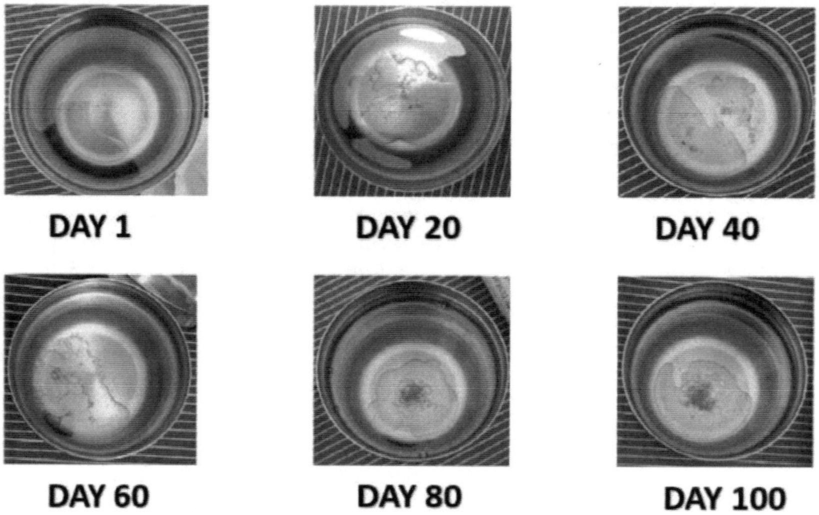

Figure 1. Photos of the stirred sample taken at a regular interval of 20 days.

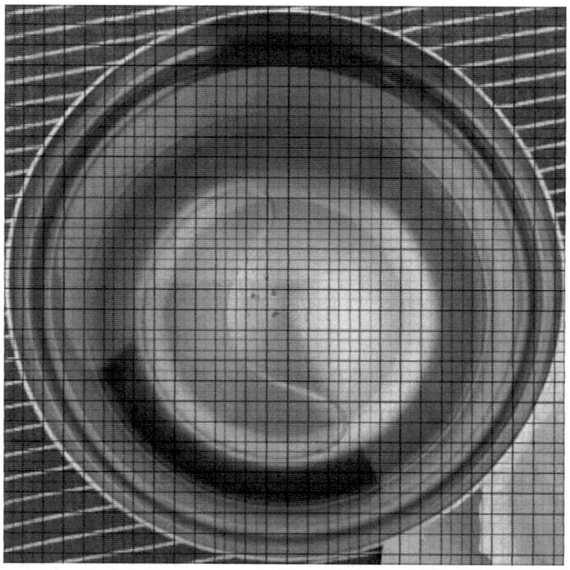

Figure 2. Tessellated photo of the sample (55x55) used for volumetric calculations.

Now in order to extrapolate results curve fitting was done with a 4th order polynomial (Quartic Fitting). Equation obtained is as follows-

$$y = -0.003630159 + 0.18888x - 0.000312631x^2 + 0.00000173518x^3 + 2.182292*10^{-8}x^4$$

Using this calculation can be done on how often the tank requires cleaning. After consulting tank cleaning service providers and fuel depots, it was noted that 1% sludge concentration in a storage tank is the maximum allowable level beyond which problems in functioning start to occur, safety is compromised, and sales at fuel stations is affected. Thus, before this point is reached, sludge needs to be removed. Finding this upper limit for our experiment, back substituting in the equation, 8ml sludge (1% concentration) will be formed in approximately 134 days.

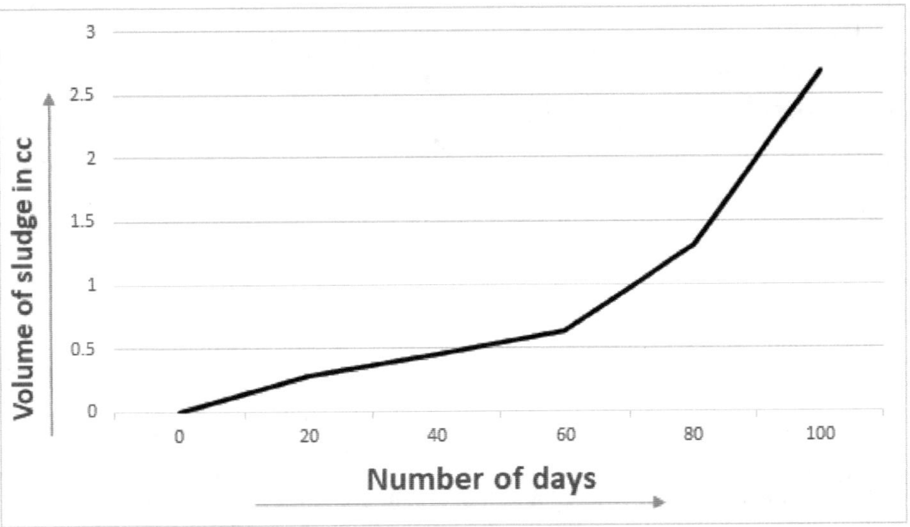

Figure 3. Graphical plot showing Volume of sludge grownvs Number of days.

Figure 4. Unstirredsample at the end of 60 days.

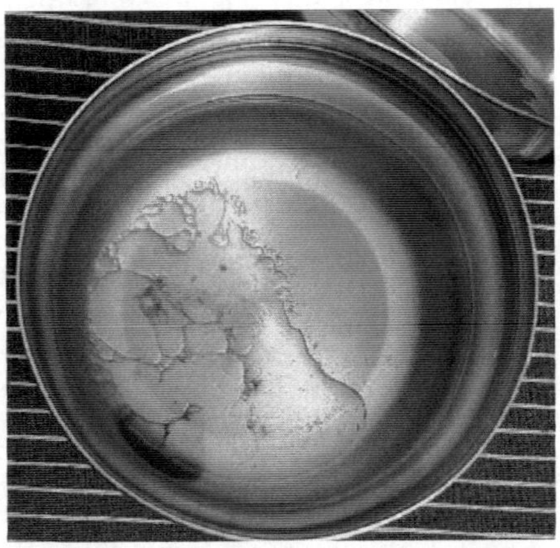

Figure 5. Stirredsample at the end of 60 days.

A comparison was carried out between the stirred and unstirred samples on the basis of the sludge growth, and it was found less in the sample which underwent stirring. Figure 4 and Figure 5 show the state of the unstirred and stirred samples at the end of 60 days, respectively. Figure 6 displays the sludge volume in cm^3 vs. time for both samples. Thus, stirring was considered to be a factor in slowing down the sludge formation. After further research on the same, agitation methods can be suggested to the depots and fuelstations.

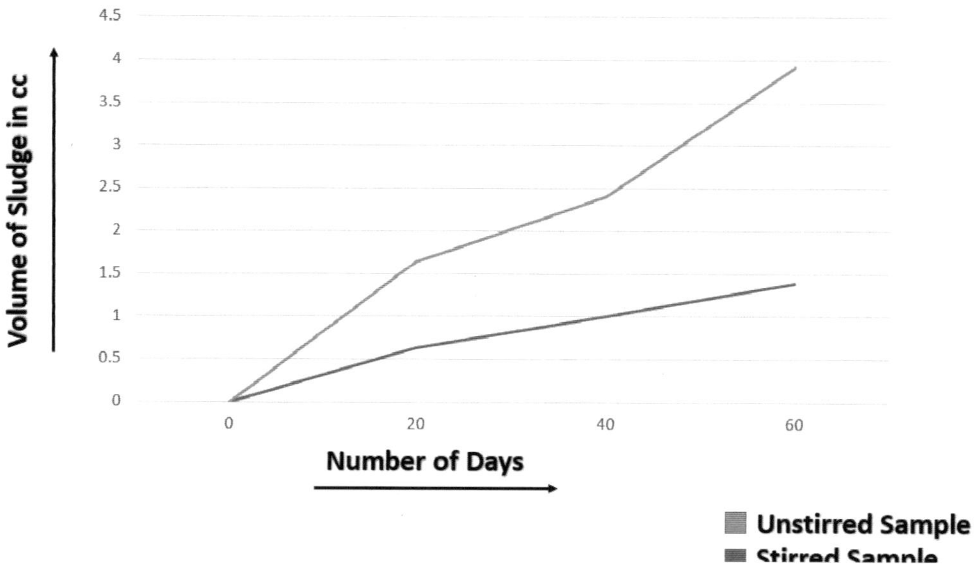

Figure 6. Sludge growth Comparison between stirred and unstirred samples.

3. Stirring approach for vertical cylindrical diesel storage tanks at depots

3.1 Design proposal for stirring

Considering stirring as a factor to delay sludge growth, a stirring design has been formulated for vertical cylindrical tanks in fuel depots. Twoways of fuel agitation can be employed,one in the horizontal plane and the other along the height of the tank.Circulation in the horizontal planemakes the sludge pile up at the centerof the tank and this could act as a fertile ground for further organic growth. Whereas circulation along the height disperses the contaminants throughout the tank. In the proposed design, circulation along the height has been incorporated. Figure 7 shows the proposed stirring pattern for a diesel storage tank of height 20 meters. Numbers 1,2,3,4,5, and 6 represent the pump numbers. The directionsof pumping are mentioned in Table 2.

Table 2. Type of circulation carried out by each pump.

Details of the pumps	Direction/Axis of pumping
Pump 1 and 6	Horizontal
Pump 2,3,4 and 5	Vertical

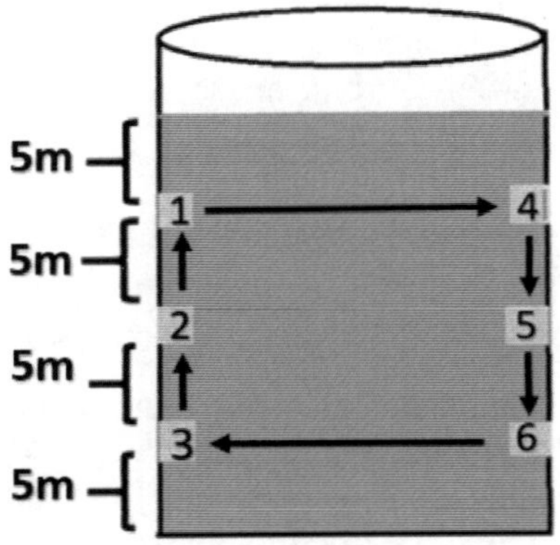

Figure 7. Proposed stirring pattern for the vertical cylindrical diesel storage tanks. Pumps 1, 2, and 3 are placed diametrically opposite to pumps 4,5, and 6.

To cover sufficient vertical distance and prevent undesirable turbulence, design is included with 6 pumps. Design is based onpump 3, which is most heavily loaded. Maximum pressure head required to be generated by pump 3, which is20 meters (15meters of static diesel head and additional 5 meters head required to pump diesel upwards).Therefore, the individual maximum pressure head required by each pump is 20 meters

Pressure to be handled by each pump = 850 x 9.81 x 20= 166.77 kPa

Specifications of a commercially available diesel pump were 220V, 750W. Pressure head developing capacity of the pump is 5m of diesel.Using model law for pumps,

$$\frac{H}{N^2 D^2} = \text{constant}$$

For the same rotational speed, to achieve a head of 20m, pumps have to be scaled by 2 times.Thus, mathematically it can be concluded that to achieve the proposed design of stirring the diesel, we will require pumps double the size of commercially available pumps.

3.2 Effect of stirring

The major drawback with stirring is dispersion of contaminants which would have otherwise settled at the bottom as sludge. These contaminants get carried to gas stations. This could adversely affect the life of filters that are used and may even affect customers vehicle life. So, it is better to reduce the concentration of some harmful contaminants in the depots or in the stations. Owing to their large size, heavy metals are the most harmful contaminants. Metallic concentration in the fuel may increase as a result of tank wear. As seen from table 1, diesel sludge contains metal ions like iron and nickel. To counter this issue, a suggestion has been provided using the concept of magnetic separation. Iron and nickel, though present in minute concentrations, greatly affects the life of an engine, destroys the carburetor, and costs heavyto the customer. The method of magnetic separation, which is usually used for iron ore processing, coolant sludge treatment, and separating magnets from clay, could be used to remove considerable amounts of iron, nickel, and mild steel from the contaminated fuel by allowing the fuel to pass through akidney loop filtration setup, where itwill be periodically circulated through a magnetic separator.The magnets attract the heavy ferrous metals, thus reducing the contamination in the fuel to some extent.

4. Conclusions

The rate of sludge growth is analyzed, and the results are extrapolated to find out a suitable tank cleaning time had the same setup been for a real tank. Experimental results have shown a delay in sludge formation when fuel is stirred. Stirring hence has been considered as a measure to slow down sludge formation.A stirring design has been proposed using 6 commercially available pumps. For the industrial implementation of the same, we need to scale up the pump size by a factor of 2. Chemical analysis of the sludge sample has been carried out to find its composition and based on that, and an idea is generated to minimize it.

References

[1] Olufemi A J and Augustine CA 2019 Petroleum sludge treatment and disposal: A review, *Environ. Eng. Res.***24(2)** 191-201

[2] Edelvio de B, Flavia L, Adriana L, Raquel S P, Alexandre S R and Nei P Jr. 2012 Microbial community dynamics in diesel waste biodegradation using sequencing batch bioreactor operation mode (SBR), *Afr. J. of Biotechnol.***11 (21)** 4899-4908

[3] Gordon V, William Sf, Helene f, John C D and Phillip F 2005 Characterization of adiesel sludge microbial consortia for bioremediation, *Scanning***27** 8–14

[4] Singh C, Lin J 2008 Isolation and characterization of diesel oil degrading indigenous microrganisms in Kwazulu-Natal, South Africa, *Afr. J. of Biotechnol.***7 (12)** 1927-1932

[5] Wolfgang S 2013 Microbial contamination in diesel fuel – are new problems arising from biodiesel blends? *Colloq. Fuels Conv. Future Energy Automob. Conf. 9*

[6] Khalid K, Grozescu I V, Tiong L K, Sim L T, and Mohd R 2003 Water detection in fuel tanks using the microwave reflection technique. *Meas. Sci. Technol.***14(11)** 1905–1911

Local & Average Heat Transfer and Friction Factor Characteristics of a Flow through a Helical Wire Coil for Turbulent Flow Conditions

Rahul Chaudhari, Prashant W. Deshmukh *, Vishal Bhalla, Subhash V. Lahane

Department of Mechanical Engineering, College of Engineering Pune, Wellesley Road, Shivajinagar-411005, Maharashtra, India

* Corresponding author e-mail: pwdeshmukh25@gmail.com

Abstract. Heat transfer augmentation is essential for numerous industrial applications. The flow through a helical coil is one of the most efficient and effective techniques of heat transfer enhancement due to the provision of a large surface area. The present study is focused on the thermo-hydraulic performance of turbulent flow through a helical coil, made up of copper with a curvature ratio of 0.0519 (*d/D*) and pitch equal to 50 mm using water as a test fluid. The objective of the present study is to reveal the influence of secondary flow, centrifugal force, and Reynolds number (*Re*) on local velocity, local heat transfer coefficient in a helical coil. The values of average heat transfer and friction factor at different *Re* for a flow through a helical coil are compared with corresponding values for a flow through a smooth tube. The numerical results are presented for the *Re* number range at inlet conditions of the helical coil from 5000 to 20000 in the increment of 2500 under constant heat flux conditions. Numerical results indicate that there is a circumferential variation of local velocity and local Nusselt number (*Nu*) values concerningthe inner and outer sides of the helical coil. The values of average heat transfer are observed to be 150 to 200% more than the smooth tube values with increased pressure drop.

Keywords. Helical coil, Turbulent flow, Thermo-hydraulic performance, Local heat transfer

1. Introduction

Heat transfer augmentation is essential for many industrial applications. There are different types of heat transfer augmentation techniques: active and passive. Inactive techniques external forces such as external vibration, surface vibration, the electric field are applied to the smooth tube to increase the heat transfer. In passive techniques specific geometries (such as helical coil, conical coil, twisted tape), tube inserts, fluid additives are used to increase the heat transfer. The helical path of the tube increases the residence time of fluid over the heated surface. The major disadvantage of a smooth straight tube is that it creates thermal expansion under heat transfer applications. Thethermal expansion associated with the helical coil is minimum due to the generation of secondary flow within the mainstream flow that causes enhancement in heat transfer. However, the secondary flow generateddue to the helical path of the fluid within the coil increases pressure drop and hence increasesthe pumping power in comparison with the straight tube [1].

IOP Conf. Series: Materials Science and Engineering 1185 (2021) 012007 doi:10.1088/1757-899X/1185/1/012007

Grindley et al. [2] and Eustice J. [3] reported the experimental data for pressure drop forthe flow of water through flexible helical coiled tubing of various radii of curvature. The authors reported that the pressure drop is increased due to increases in the flow length of the fluid. Dean W.R. [4,5] reported the theoretical study for the flow of the incompressible Newtonian fluids in the torus. The authors concluded that the pressure drop is observed to depend on the flow rates, the tube curvature. The authors presented the results of isothermal pressure drop in terms of dimensionless quantity Dean number.

Few researchers reported the study of flow through a helical coil for understanding the physical phenomenon of induction of helical fluid flow path. The fundamental studies explore the mechanism of helical flow inside helical as well as curved tubes. Mishra and Gupta [6] reported the study of flow characteristics for different geometrical parameters of the curved tubes. The various geometrical parameters for helical coils such as coil mean diameter, pitch, tube diameter, the thickness of tube that affect the thermo-hydraulic performance of the system. The authors also reported that changes occur within the fluid and flow characteristics significantly differ for configuration when pitch exceddsthe coil mean diameter, whereas when the pitch is smaller than the coil mean diameter the influence on the flow is observed to be negligible.

Seban and Mclanghlin [7] reported that the tube curvature, torsion, and secondary flow create difficulty in the measurement of thermo-hydraulic performance of the coil and flow. Bai et al. [8] reported that the curvature ratio of the helical affects the pressure drop, which further alters the intensity of the secondary flow generated within the mainstream flow. Many researchers [9-13] reported the effect of variation of *Re*on the pressure drop and average heat transfer values in helical coils. Ito [14] carried out a hydrodynamics study of fluid flow through coiled tubes for estimation of friction factor at different flow rates corresponding to turbulent flow regime. The authors concluded that the friction factor gradually reduces with an increase in *Re*of the flow. Hardik et al. [15] collected the experimental analysis of flow through the helically coiled tube, and the effects of several geometrical parameters like tube curvature, and flow parameters like *Re* on heat transfer with working fluid likewater. A brief assessment of work carried out on for a flow through a helical flow.

It is observed from the literature review that there is a complexity for the measurement of local *Nu* value at different locations of the helically coiled tube. Therefore, the results of average heat transfer and pressure drop are available in the literature. There is a need to measure the local wall temperature to estimate the circumferential variation of local heat transfer values over the surface of the helically coiled tube. The present study is focused on the thermo-hydraulic performance of turbulent flow through a helical coil, made of copper with a curvature ratio of 0.0519 (d/D) and pitch equal to 50 mm. The water is used as a working fluid.

2. Numerical Modelling Methodology

A three-dimensional model of the helical coil is modeled in the Solid Edge V19 packageshown in Figure 1. The geometric specifications of the tube such as the theouter diameter oftube d_o, thickness of the tube t, inner diameter of tube d_i,and the coil such as coil diameter D, pitch P, number of turns n, and length L are listed in Table 1. The material of the helical coil used is copper.

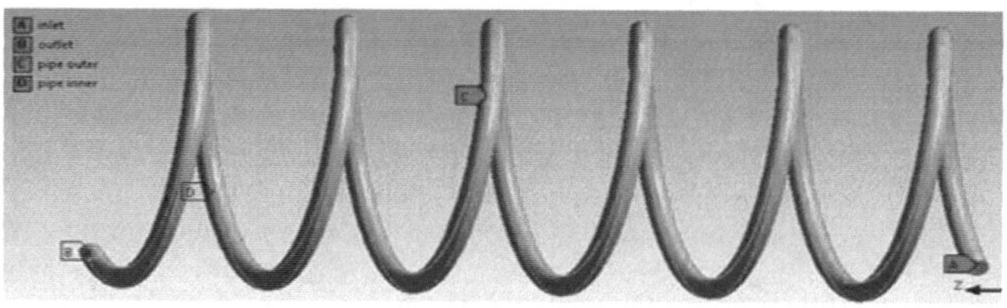

Figure 1. 3Dgeometry of the helical coil.

Table 1. Geometric parameters of the helical coil.

Tube parameters			Coil parameters			
Outer diameter, d_o (mm)	Inner diameter, d_i (mm)	Thickness, t (mm)	Pitch, P(mm)	No. of Turns, n	Diameter of Coil, D (mm)	Length of the coil, L (mm)
6.53	6.03	0.25	50	6	116.2	2238

2.1. Governing equations

In the present work, the investigation for friction coefficient and heat transfer for helical turbulent flow conditions for the flow of water with the *Re* range from 5000-20000 is undertaken in the step increment of 2500.The corresponding inlet velocity ranges from 0.834-3.336 m/sec. The selection of the *Re* number range for the present study is closely matched with most of the thermal applications. The study presented is for the steady, three-dimensional turbulent flow of Newtonian fluid i.e. water.

The rate equations of continuity, momentum, energy and standard k-ε model are:

Continuity equation

$$\frac{\partial(\rho u_i)}{\partial x_i}=0 \tag{1}$$

Momentum equation for turbulent flow

$$\frac{\partial(\rho u_i)}{\partial t}+\frac{\partial(\rho u_i u_j)}{\partial x_j}=-\frac{\partial(P)}{\partial x_i}+\frac{\partial}{\partial x_j}\left(\mu\left(\frac{\partial(u_i)}{\partial x_j}+\frac{\partial(u_j)}{\partial x_i}\right)-\frac{2}{3}\mu\frac{\partial(u_k)}{\partial x_k}\delta_{ij}\right) \tag{2}$$

Energy equation

$$\frac{\partial}{\partial x_j}\left(\rho u_j C_p T-k\frac{\partial T}{\partial x_j}\right)=u_j\frac{\partial P}{\partial x_j}+\left[\mu\left(\frac{\partial(u_i)}{\partial x_j}+\frac{\partial(u_j)}{\partial x_i}\right)-\frac{2}{3}\mu\frac{\partial(u_k)}{\partial x_k}\delta_{ij}\right] \tag{3}$$

Standard k-ε model (Launder and Spalding)

Turbulent kinetic energy rate equation (k)

$$\frac{\partial(\rho.k)}{\partial t}+div(\rho k U)=2\mu_t.S_{ij}.S_{ji}+div\left[\frac{\mu_t}{\sigma_k}grad\ k\right]-\rho\epsilon \tag{4}$$

Turbulent dissipation energy rate equation (ε)

$$\frac{\partial(\rho.\epsilon)}{\partial t}+div(\rho\epsilon U)=C_{1\epsilon}.\frac{\epsilon}{k}.2\mu_t.S_{ij}.S_{ji}+div\left[\frac{\mu_t}{\sigma_s}grad\ \epsilon\right]-C_{2\epsilon}.\frac{\epsilon^2}{k} \tag{5}$$

$$\begin{bmatrix}Rate\ of\\ change\ of\ k\ or\ \epsilon\end{bmatrix}+\begin{bmatrix}Transport\ of\\ of\ k\ or\ \epsilon\ by\ convection\end{bmatrix}=\begin{bmatrix}Rate\ of\ production\ of\\ k\ or\ \epsilon\end{bmatrix}+\begin{bmatrix}Transport\ of\\ of\ k\ or\ \epsilon\ by\ diffusion\end{bmatrix}-\begin{bmatrix}Rate\ of\ destruction\\ of\ k\ or\ \epsilon\end{bmatrix}$$

2.2. Meshing

The meshing decides solution accuracy in computational analysis. The goodness of the mesh not only depends on the size of the element but also depends on the parameters like aspect ratio, skewness, wrapping angle, orthogonal quality. The most affected parameter on the goodness of the mesh is the orthogonal quality.The suggested value of the orthogonal quality is nearly equal to one. The present study uses the mesh refinement grid with tetrahedrons elements as shown in Figure 2. The grid details are listed in Table2.

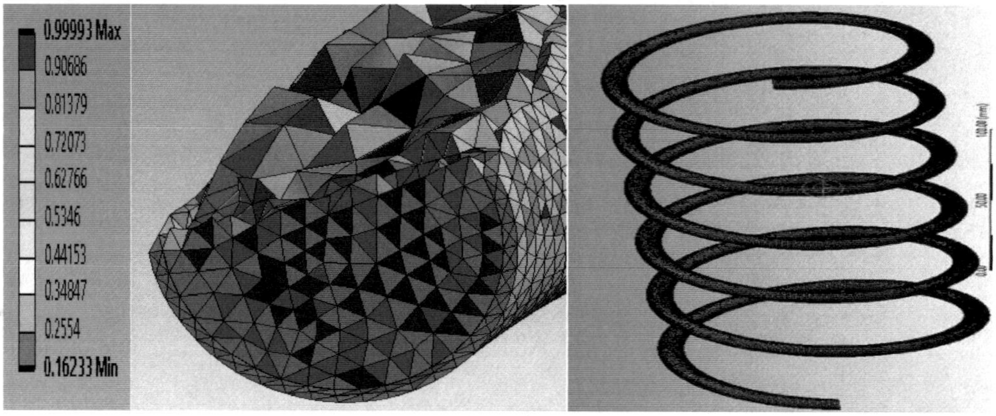

(a) (b)

Figure 2. Mesh of the model (a) orthogonal quality (b) whole grid

Table 2. Mesh details

Parameters	Statistics
Nodes	2430414
Elements	1701661
Element type	Tetrahedrons
Orthogonal quality	0.8

2.3. Boundary conditions

At the outer surface of the helical coil the constant heat flux condition is applied. In the present study the temperature drop between the outlet and inlet of the fluid is maintain nearly equal to 27°C. At the tube inlet the correspondence velocity is given as boundary condition. The operating conditions used in the simulations are mentioned in Table 3.

Table 3. Operating conditions

Sr. No.	Reynolds number, *Re*	Velocity, *V* (m/sec)	Heat Flux, *q”* (kW/m^2)
1	5000	0.834	58.5
2	7500	1.251	87.77
3	10000	1.668	117.03
4	12500	2.085	146.326
5	15000	2.502	175.59
6	17500	2.919	204.85
7	20000	3.336	234.12

2.4.Computational scheme

The pressure-based steady solver is used in the present study. All the governing equations listed above are solved by the boundary conditions using the finite volume method. The terms turbulent kinetic energy, turbulent dissipation rate momentum, aremodeled in the standard k-ε turbulent model using a second-order upwind scheme.

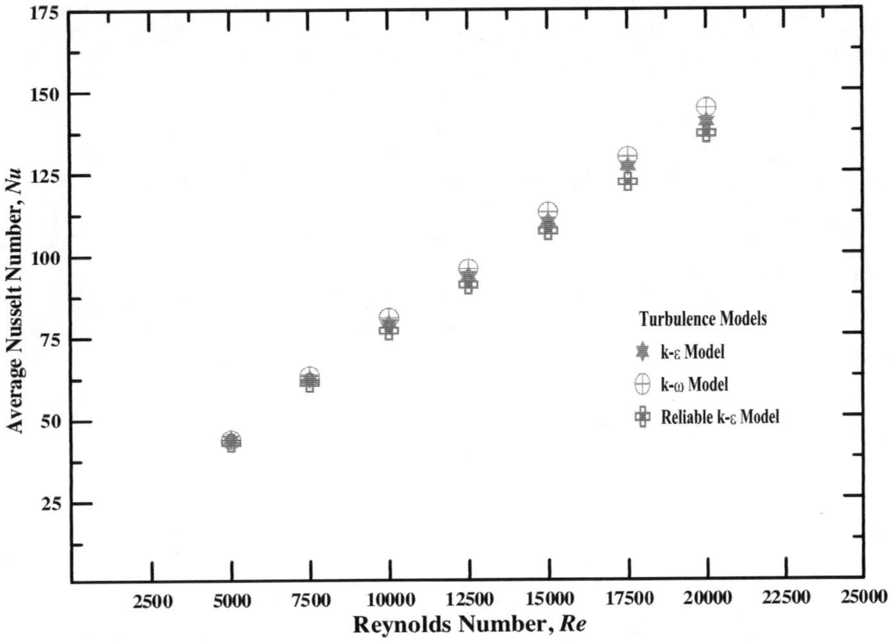

Figure 3. Variation of *Nu*with *Re* for various turbulence models

The turbulent intensity was selected as 5%. The temperature and heat flux terms are taken intoan account using an energy equation. The convergent criterion for all the residual variables wise *x* velocity, *y* velocity, *z* velocity, continuity is 10^{-3} and for energy,the equation is 10^{-6}. The pressure and velocity are coupled by using a COUPLED algorithm to avoid the limitation regarding the temperature that occurred in the case of SIMPLE, PISO algorithms.

To decide the suitable turbulence model which gives the correct result, the coil is simulated for three different turbulence models viz. realizable k-ε, standard k-ε, and standard k-ω. Figure 3 shows that the variation of *Nu*concerning*Re* for various turbulence models. It is concluded from Figure 3 that the standard k-ε model can be used as the variation is observed to be less than 5% in comparison with the other two models of turbulent flow.

2.5. Data reduction equations

The required thermo-hydraulic parameters are calculated from the obtained computational results using data reduction equations are listed below.

The friction coefficient for straight, f_s, and helical coiled flow, f_c

$$f_c = \Delta P \frac{d}{2\rho V^2 L} \tag{6}$$

Friction coefficient (Blasius) and *Nu*(Dittus-Boelter correlation) for turbulent flow

$$f_s = 0.079 \times Re^{-0.25} \qquad Nu = 0.023 \times Re^{0.8} \times Pr^{0.4} \tag{7}$$

Reynolds number,

$$Re = \frac{\rho \times v \times d_i}{\mu} \tag{8}$$

Heat transfer,

$$Q = \dot{m} C_p (T_{out} - T_{in}) \tag{9}$$

Heat flux,

$$q = \frac{Q}{\pi d L} \tag{10}$$

Heat transfer coefficient,

$$h = \frac{q}{T_s - T_m} \tag{11}$$

Nusselt number,

$$Nu = \frac{hd}{K_w} \tag{12}$$

3. Results and Discussion

The numerical results are presented in the form of thermohydraulic performance of the flow through a helical coil are as follows.

3.1. Variation of friction coefficient

Figure 4 depicts the variations of the friction coefficient concerning*Re*. The friction coefficient and average *Nu* values for the straight tube are obtained using equation(7). The average friction coefficient of the present study for the flow through the wire coil of geometrical parameters mentioned in table 1 is shown in Figure 4. The figure also includes the results of Ito[14] for the coil of the same configuration. It is observed that the friction coefficient decreases with increases in the *Re*. It is to be noted that the friction coefficient is nearly 13 times more than the straight tube which shows higher pumping power associated with the helical coil than the straight tube. The values of the friction coefficient of the present study are comparatively less than the values specified by Ito [14].

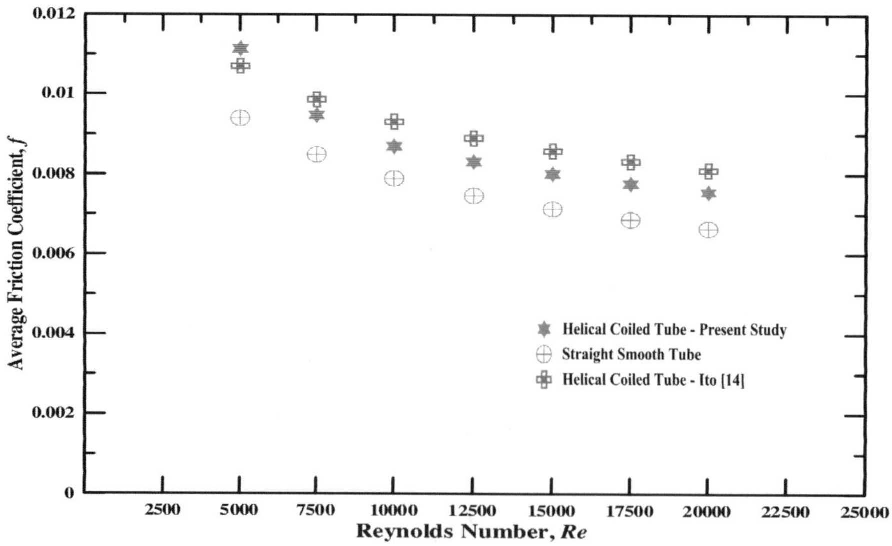

Figure 4. Deviation of friction coefficient with *Re*.

IOP Conf. Series: Materials Science and Engineering 1185 (2021) 012007 doi:10.1088/1757-899X/1185/1/012007

3.2. Visualization of secondary flow

The secondary flow is the key phenomenon to understand the mechanism of the helical flowpath of the fluid particles. It is shown in Figure 5.The streamlines are used for the visualization of induced secondary flow. The gap between the two halves of streamlines provides the evidence of the generation of the secondary flow.

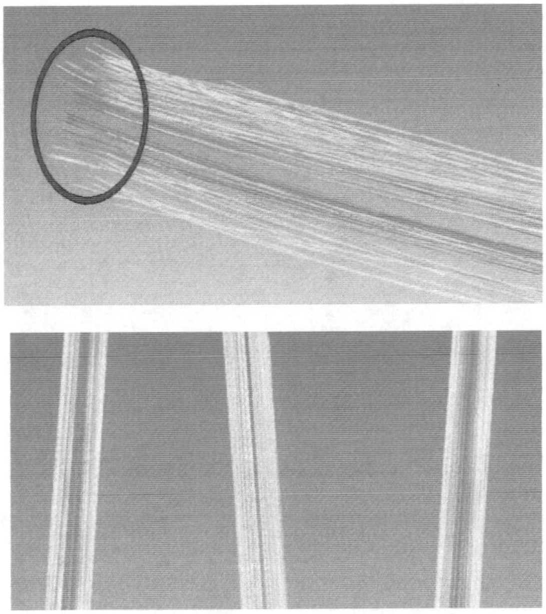

Figure 5. Secondary flow streamlines for *Re* = 17500

3.3. Variation of velocity

Figure 6 depicts the velocityvariation of the fluid particles in the circumferential direction due to induced centrifugal force as a result of the helical path of the fluid. It is observed clearly from figure 6 that the fluid particles at the outer side of the tube have more velocity in comparison with the fluidparticle at the inner side of the tube.

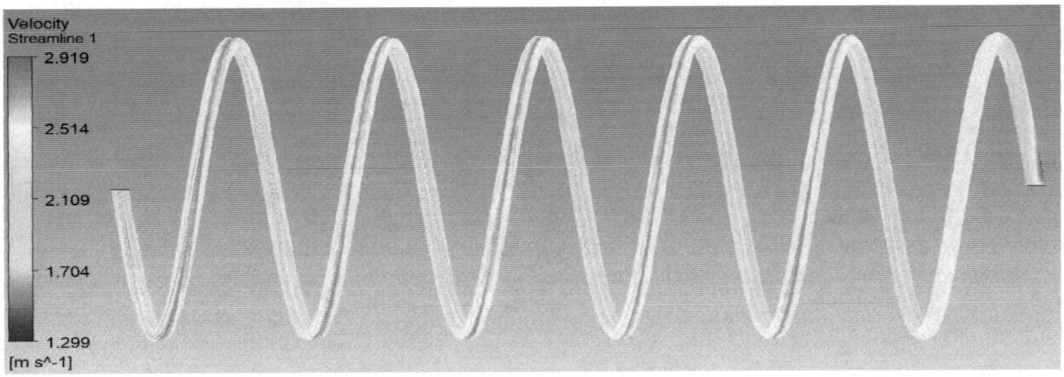

Figure 6. Circumferential variation of velocity for *Re*= 17500

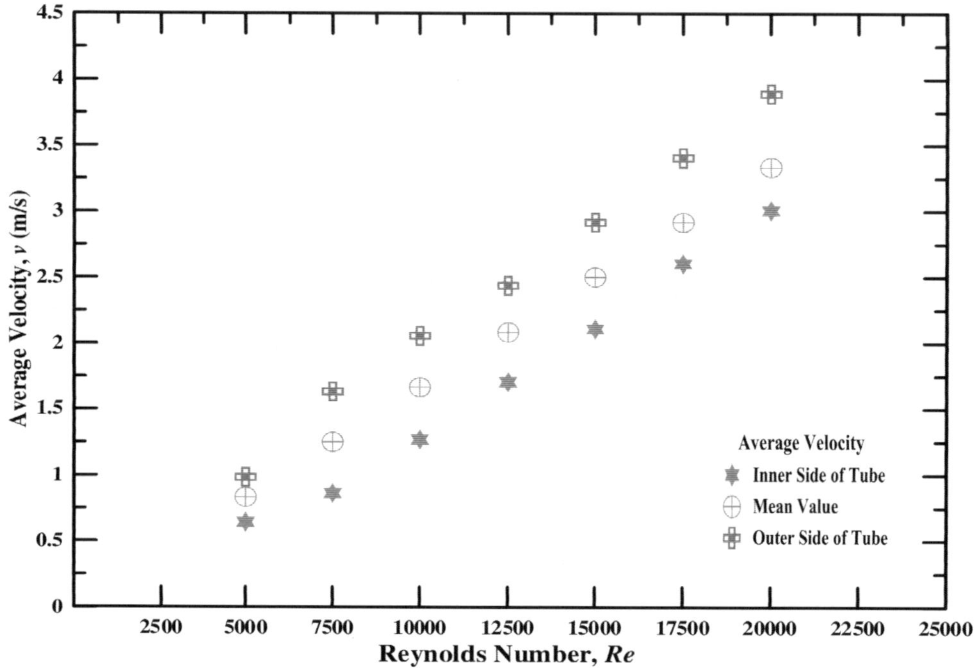

Figure 7. Velocity variation at different *Re.*

The variation of velocity is observed in a circumferential direction however, the velocity of fluid particles almost remains constant in the axial direction. Theaverage velocity at the outer side of the helical coilis observed to be around 20 % more than the mean velocity. The variation of velocity at the inner side, at the outer side, and the mean velocity concerning*Re*is shown in Figure 7.

3.4. Variation of local heat transfer

The variation of the local wall heat transfer coefficient is shown in Figure 8. It is observed that the increase in heat transfer at the outer side of the helical coil is more than that of the inner side. This is due to cooler mainstream fluid from the central region of the tube cross-section directed more vigorously towards the outer side of the helical coil due to the centrifugal force. It causes more transport of heat at the outer side than the inner side of the helical coil.

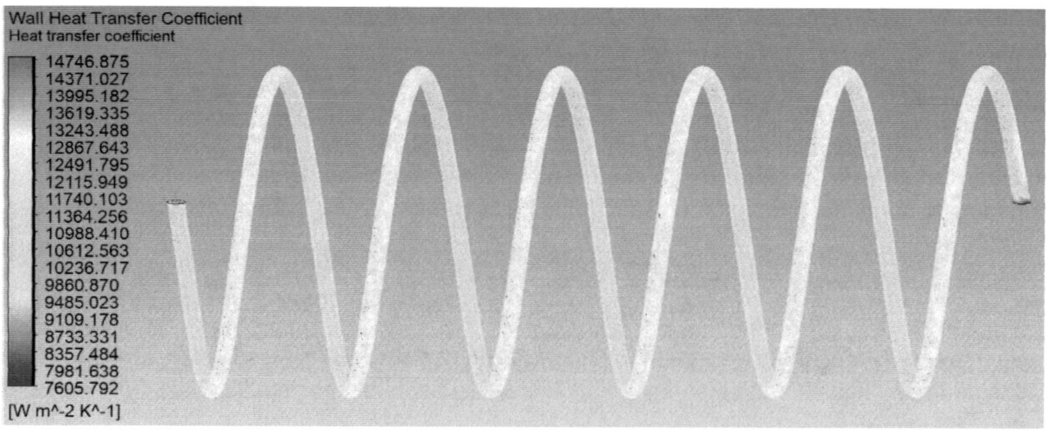

Figure 8. Contour plot of heat transfer coefficient for*Re* = 17500

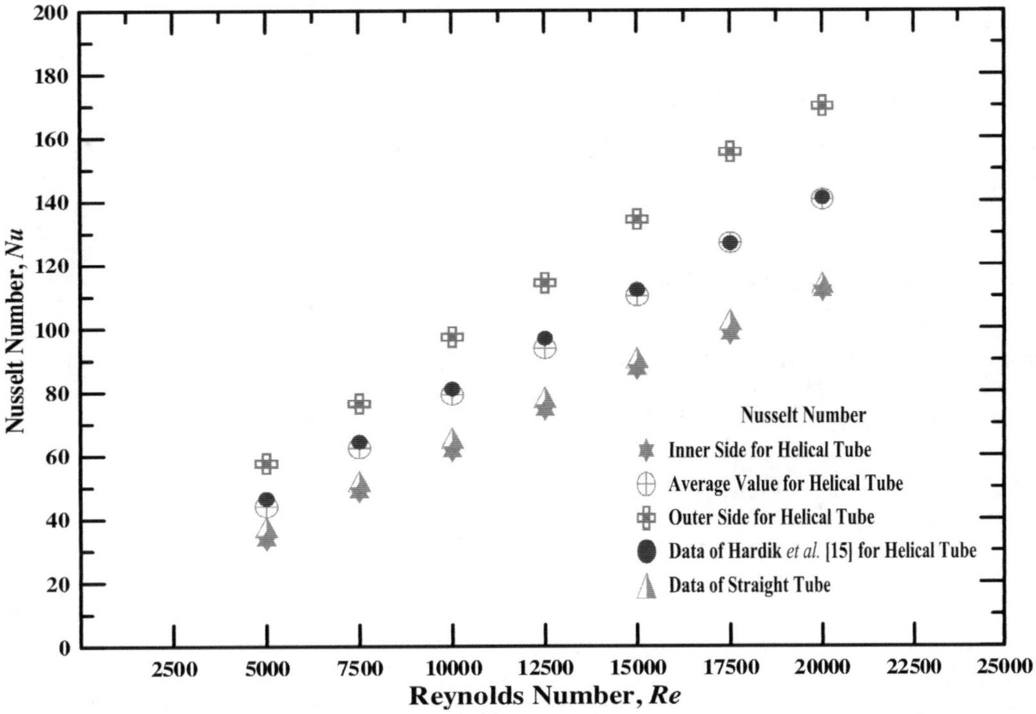

Figure 9. *Nu* variationwith *Re*

The variation of *Nu* at the outer, inner side, and their average is illustrated in Figure 9 at different *Re*. The average value reported by Hardik *et al.* [15] is also shown in Figure 10 to ensure the correctness of the results of the present study. Figure9 also contains the *Nu* values of smooth tube specified by Dittus-Boelter correlation at different *Re*. In all the cases it is observed that the heat transfer increases at higher flow rates. The figure shows the variation of the outer side, inner side, and average *Nu* values of the present study. The outer side *Nu*value for the flow through a wire coil is observed to be around 23% more than the average *Nu* value of the coil whereas the inner side *Nu* value is observed to be around 22% less than the average *Nu*value. It is to be noted that the enhancement in heat transfer is observed to be very poor at the inner side of the tube as the *Nu*is less than that of the smooth tube flow. The average heat transfer enhancement of 20% is observed for the geometry used in the present work.

4. Conclusion

The numerical work is presented for the study of the effect of *Re* and constant Prandtl number (*Pr*) equal to 5.0 at a coil curvature ratio (*d/D*) of 0.0519 on local *Nu,* and friction coefficient with water as working fluid. The friction coefficient is obtained from the pressure drop values while the *Nu*and local temperature distribution is obtained from temperature contours and wall heat transfer coefficient contours. The following inferences are extracted from the present study.

- Thefriction coefficient in the helical coil was more than the smooth tube for turbulent flow.
- There is a variation of the velocity of the fluid in the circumferential direction. The outer side fluid velocity in the helical coil is observed to be 20% more than the mean inlet velocity while the inner side velocity is less than the mean inlet velocity of the fluid.
- The average *Nu*value is observed to be 20% more than the corresponding smooth tube values.
- The outer and inner side local *Nu*valuesare observed to be 23% more and 22% less than the average *Nu* value.

Nomenclature

Symbol	Description
A	The surface area of the coil, (m^2)
V, U	Mean velocity of the fluid, (m/sec)
μ	Dynamic viscosity, (Pa.s)
ϵ	Turbulent dissipation energy rate, (m^2/s^3)

References

[1] Vashisth S., Kumar V., Nigam K.D.P.2008 *A* review on the potential applications of curved geometries in process industry, *Ind. Eng. Chem.* Res. **47**, 3291-3337.
[2] Grindley J.H., Gibson, A.H.1908 On the frictional resistance to the flow of air through a pipe. *Proc. R. Soc. London, Ser.* A **80** 114 –139
[3] Eustice J.1910 Flow of water in curved pipe. *Proc. R. Soc. London, Ser.* A **84**, 107–118.
[4] Dean W.R. 1927 Notes on the motion of fluid in a curved pipe. *Philos. Mag.* **4**, 208–233.
[5] Dean W.R. 1928 The streamline motion of fluid in a curved pipe. *Philos. Mag.* **5**, 673–695.
[6] P. Mishra, S.N. Gupta, Momentum transfer in curved pipes. 1. Newtonian fluids, 1979 *Indust. Eng. Chem. Process Des.* Dev. 18130–137.
[7] Seban R.A., Mclanghlin E. 1963 Heat transfer in tube coils with laminar and turbulent flow, *Int. J. Heat Mass Transfer* **6** 387–395.
[8] BaiB.,Guo L., Feng Z., Chen X.1999 Turbulent heat transfer in a horizontal helically coiled tube, *Heat Transfer Asian Res.* 28 395–403
[9] Kubair V., Kuloor N.R. 1966 Heat transfer to Newtonian fluids in coiled pipes in laminar flow, *Int. J. Heat Mass Transfer* **9** 63–75.
[10] Pawar S.S., Vivek K.S. 2013 Experimental studies on heat transfer to Newtonian and non-newtonian fluids in helical coils with laminar and turbulent flow, *Exp. Thermal Fluid Sci.* **44** 792–804.
[11] Pimenta T.A.,Campos J.B.L.M, 2013 Heat transfer coefficients from Newtonian and non-newtonian fluids flowing in laminar regime in a helical coil, *Int. J. Heat Mass Transfer* **58**676–690.
[12] Dravid A.N., Smith K.A., Merrill E.W., Brain P.L.T. 1971 Effect of secondary fluid on laminar flow heat transfer in helically coiled tubes, *Am. Inst. Chem. Eng.* J. **17** 1114–1122.
[13] Janssen L.A.M., Hoogenwdoorn C.J. 1978 Laminar convective heat transfer in helical coiled tubes,*Int. J. Heat Mass Transfer* **21**1197–1206.
[14] Ito H.1959 Friction factors for turbulent flow in curved pipes, *Trans. Am. Soc. Mech. Eng.* D **81** 123–134
[15] Hardik B.K., Baburajan P.K., Prabhu S.V. 2015 Local heat transfer coefficient in helical coils with single phase flow *Int. J. Heat Mass Transfer* **89**522-538.

Heat Transfer and Friction Factor Characteristics of Turbulent Flow through a Conical Heat Exchanger

Dulesh R. Pathode, Prashant W. Deshmukh, Vishal Bhalla, Subhash V. Lahane

Department of Mechanical Engineering, College of Engineering Pune, Wellesley Road, Shivajinagar-411005, Maharashtra, India

* Corresponding author e-mail: pwdeshmukh25@gmail.com

Abstract: In this present energy demand scenario, the use of a waste heat recovery device like regenerator, recuperator, reheaters in the industrial processes is gaining much more importance due to improvement in thermal efficiency, and consequently reduction in thermal pollution. In the power plant, the waste heat carried by hot flue gases through the chimney remains untapped which can be recovered through the use of a conical heat exchanger at the inner portion of the chimney. Also, few customized designs of the conical coil heat exchanger can have the potential to be used to recover the waste heat in other industrial processes like chemical reactors, electrical contractors, HVAC applications, condenser, boiler, and evaporator, etc. In the present study, different arrangements of the conical coil heat exchanger are analyzed using a numerical technique. A conical coil of different geometrical configurations viz. at three different cone angles 45°, 90°and 135°, and at three different pitches 20 mm, 25 mm, and 30 mm are considered. The results for heat transfer and friction factor are presented for the flow of fluid of different Reynolds numbers. The impact of geometrical parameters as well as flow parameters on pressure drop and heat transfer under constant wall heat flux conditions are presented. The average heat transfer and pressure drop values are observed to be 175 to 225% more than the smooth tube values with increased pressure drop. Also, the results indicate that there is a nonlinear increase in both Nusselt number and friction factor with the increase in cone angle and pitch of conical coil.

Keywords: heat recovery; heat transfer; conical heat exchanger; turbulent flow.

1. Introduction

Now a day's waste heat recovery in industrial processes like combustion is gaining much more importance. Waste heat recovery not only helps in improving thermal efficiency but also helps in lowering the thermal pollution caused because of industrial processes. In power plants regenerators, recuperators, reheaters serve the purpose of waste heat recovery. But the heat carried by flue gases through the chimney remains untapped. So, there is scope for heat recovery through the chimney by designing a conical shape of the heat exchanger.The same design of conical coil heat exchanger can also be used for waste heat recovery in other industrial processes like chemical reactors, electrical contractors, HVAC applications, condenser, boiler, and evaporator, etc.

2. Literature Survey

Most of the experimental and numerical studies has been conducted towards the development of simple helical coil heat exchangers, whereas conical coil heat exchanger has not gained so much attention. This may be because of its complex design and construction over the helical coil heat exchanger. Purandare et al. [1] carried out an experimental study on conical coil heat exchangersfor thermal analysis. The observations were reported for thermohydraulic characteristics at a different flow rate of cold and hot fluid. The various fluid flow and geometrical parameters like Nusselt number, friction factor,and effectiveness are estimated for the heat exchanger.Joshi et al. [2] Conjectured a primitive design of one conical heat exchanger with the purpose of exhaust gas heat recovery of twin-cylinder diesel engine. The experimentation brought out the relationship between temperature rise in water for various flow rates. A simple program developed for prediction of overall heat transfer coefficient is found to underpredict the values than actual leaving ample scope for characterizing heat transfer coefficient.Waghmare *et al.* [3] Conducted comparative numerical analysis of straight and conical coil heat exchangers. The results show enhanced heat transfer for the conical coil as compared to the straight coil heat exchanger.Prabhanjan et al. [4] reported an experimental study on fluid-to-fluid heat exchangers and compared the heat transfer rates betweena helically coiled heat exchanger and a straight tube heat exchanger. Experimental results showed that the heat transfer coefficient is a function of geometrical parameters and the temperature of the water bath containing the heat exchanger.

It is observed from the literature that additional secondary flow is developed within the mainstream flow. This additional flow causes the overall increase in turbulence level of the fluid which primarily responsible for the enhancement of heat transfer. The present study focuses on a simulation study of the conical heat exchanger of three different cone angles 45°, 90°and 135°, and at three different pitches 20 mm, 25 mm, and 30 mm. The results of the present study are presented for heat transfer and friction factors in the conical heat exchanger at different inlet flow conditions.

3. Numerical Model

3.1. Geometry

The geometry and different geometricalparameters of the conical heat exchanger used in the present study are shown in Figure 1 and Figure 2 shows 3D view of the conical coil

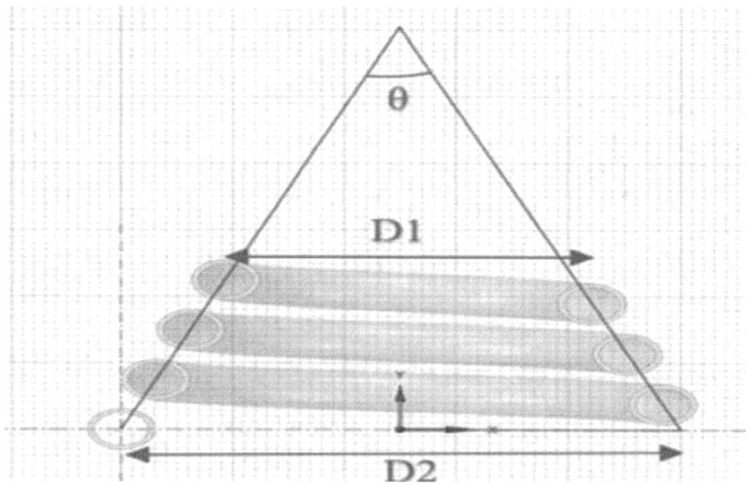

θ - Cone Angle D1 – Minimum Diameter D2 – Maximum Diameter
d – Tube Diameter Dm-Mean Diameter

Figure 1. A sectional view of conical coil

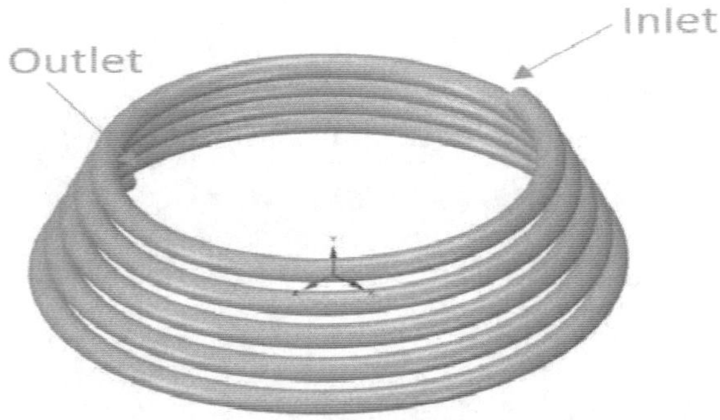

Figure 2. 3-Dimensional view of a conical coil (45° cone angle and 15 mm pitch)

Table 1. Geometricalelementsof a conical heat exchanger

Parameter	**Details**
Coil Mean Diameter	200 mm
Total Number of coils	9
Coil Pitch(mm)	15, 20, 25
Cone Angle (θ)	45°, 90°, 135°
Tube Size ID x OD (mm x mm)	10 x 12
Tube Length (mm)	3000

3.2. Assumption and Boundary Condition

The flow of the fluid through the conical coil is assumed to be three dimensional, incompressible, the steady-state flow of water. The Reynolds number at the inlet of the tube is varied from 4000 to 12000 in the steps of 2000. The boundary/zone conditions considered for analysis are listed below:

- Solver: Pressure Based Steady state
- Model: Viscous K- Epsilon Turbulence Model
- Fluid Domain: Water – Solid
- Solid Domain / Tube Material: Copper
- Tube / Coil inlet: Velocity Inlet type with Temperature Ti (=300K)
- Tube Coil outlet: Outflow Type
- Wall Boundary Condition: uniform wall heat flux at the exterior wall (such that tube outlet temperature remains To = 330K)
- All other walls: Stationary No Slip

3.3. Governing Equations

The rate equations for the cylindrical coordinate system are mass conservation, momentum equations, and energy equation.

3.3.1 Conservation of Mass:

$$\frac{\partial(\rho u_i)}{\partial x_i} = 0$$

3.3.2 Momentum Equation for turbulent flow:

$$\frac{\partial(\rho u_i)}{\partial t}+\frac{\partial(\rho u_i u_j)}{\partial x_j}=-\frac{\partial(P)}{\partial x_i}+\frac{\partial}{\partial x_j}\left(\mu\left(\frac{\partial(u_i)}{\partial x_j}+\frac{\partial(u_j)}{\partial x_i}\right)-\frac{2}{3}\mu\frac{\partial(u_k)}{\partial x_k}\delta_{ij}\right)$$

3.3.3 Conservation of Energy:

$$\frac{\partial}{\partial x_j}\left(\rho u_j C_p T-k\frac{\partial T}{\partial x_j}\right)=u_j\frac{\partial P}{\partial x_j}+\left[\mu\left(\frac{\partial(u_i)}{\partial x_j}+\frac{\partial(u_j)}{\partial x_i}\right)-\frac{2}{3}\mu\frac{\partial(u_k)}{\partial x_k}\delta_{ij}\right]$$

3.4. Model Validation and Grid Independency

Before the simulation for conical heat exchanger, a validity of simulation model is carried out by modeling the flow through a straight tube of length equal to 3000 mm. The flow through the straight tube is simulated at different values of Reynolds number. Figure 3 shows the comparison of the values of the present simulation study with the analytical results obtained using the Dittus-Boelter equation for the smooth tube. It is observed that the variation of the average Nusselt number at different Reynolds numbers is less than 5%. This ensures the correctness of the simulation methodology used in the present study.

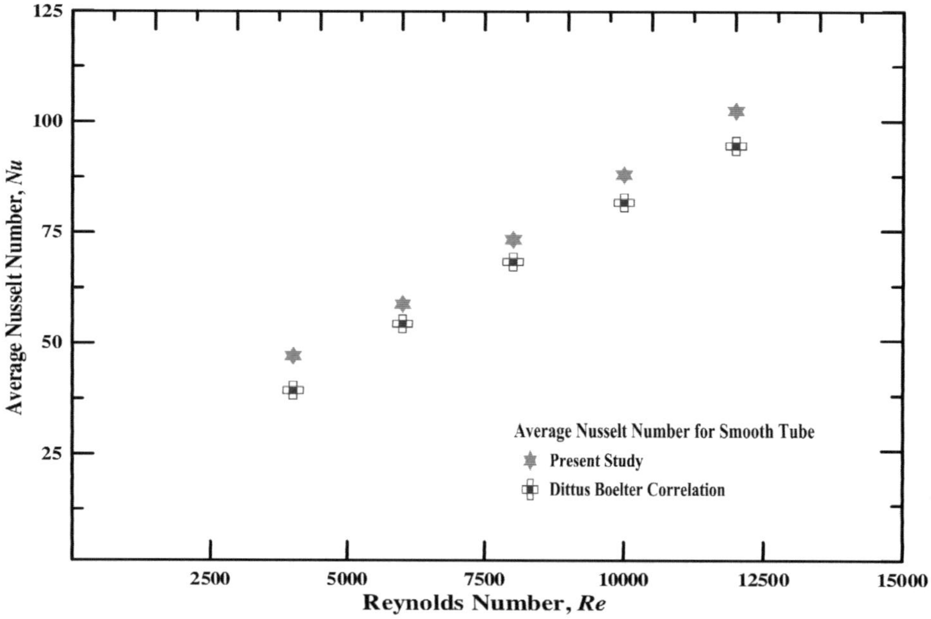

Figure 3. Validation of Nusselt number, *Nu* with Reynolds number, *Re*

It is reported in the literature that the K-epsilon model overpredicts the analytical results by around 7 to 8 % which is well within the acceptable range.The grid independence test for pressure drop is carried out. The results as shown in Figure 4 areplottedagainst the number of elements as other parameters (velocity and outlet temperature) remain almost constant for any number of elements.It is found that for the number of elements more than 75,95183 (for which the number of divisions in edge sizing is 60) the pressure drop almost remains constant and it is independent of the number of elements.

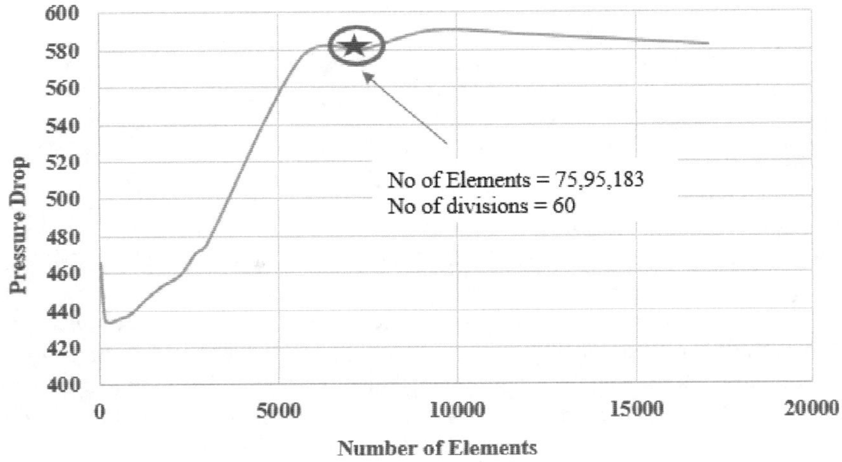

Figure 4. Grid independence check for pressure drop

3.5.Meshing

The meshing of the geometric model is done in ANSYS 2019 R3 academic version. Figure 5 shows the mesh model of 45 cone angle coil divisions. For meshing of geometry, edge sizing of the faces is done and this face mesh is swept all along the length of the tube to ensure the smooth flow of the element all over the mesh (Figure 6).

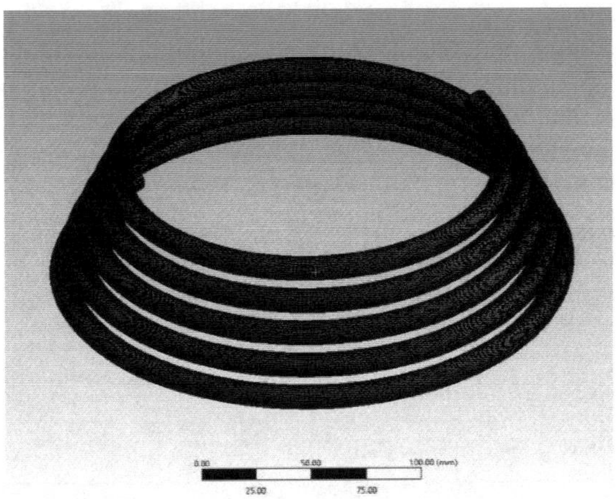

Figure5. Mesh model of 45 cone angle coil divisions

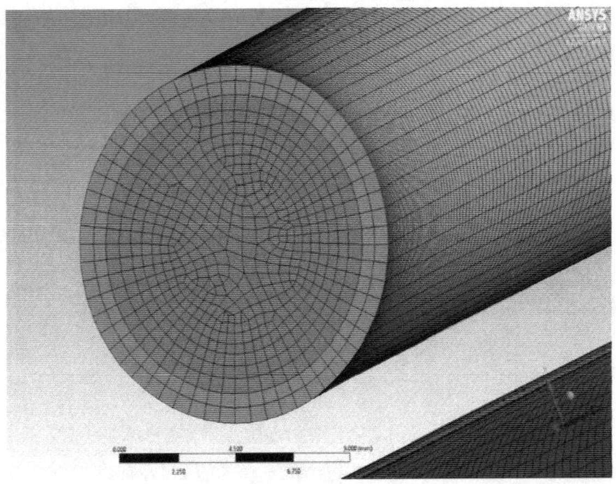

Figure 6. Face sizing with 60 number of

4. Data Reduction Method

The heat transfer and pressure drop parameters viz. Nusselt Number and friction factor and flow parameter like Reynolds Number, are evaluated using following well-established relations.

Tube Reynolds number,

$$Re = \frac{\rho V d}{\mu_w} \tag{1}$$

Average Nusselt number,

$$Nu = \frac{hd}{K_w} \tag{2}$$

IOP Conf. Series: Materials Science and Engineering 1185 (2021) 012008 doi:10.1088/1757-899X/1185/1/012008

Friction factor,

$$f_c = \Delta P \frac{d}{2\rho V^2 L} \tag{3}$$

5. Results and Discussion

5.1. Temperature and Pressure Contours:

The pressure and temperature contour of the conical coil at the outer wall are shown as follows. It is observed from Figure 7 that the pressure values are almost constant at a given cross-section. However, as it is a case of pressure-driven flow, it is observed that the value of absolute pressure reduces gradually from the inlet to the exit of the pipe. Therefore, it can be concluded that the flow through a conical pipe is associated with a large pressure drop. The contours of pressure are mentioned in Figure 8, shows that along the conical path of the fluid the surface temperature of the tube increases from inlet to the exit of the conical pipe. It is also observed that there exists a variation of the surface temperature of the tube in the circumferential direction.

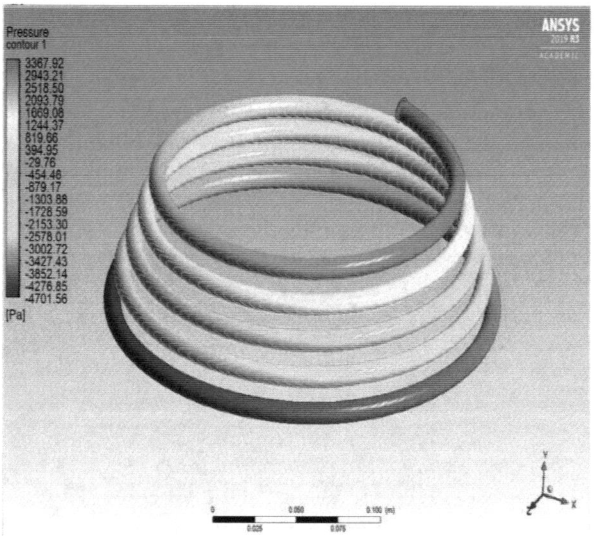

Figure 7. Pressure Contour

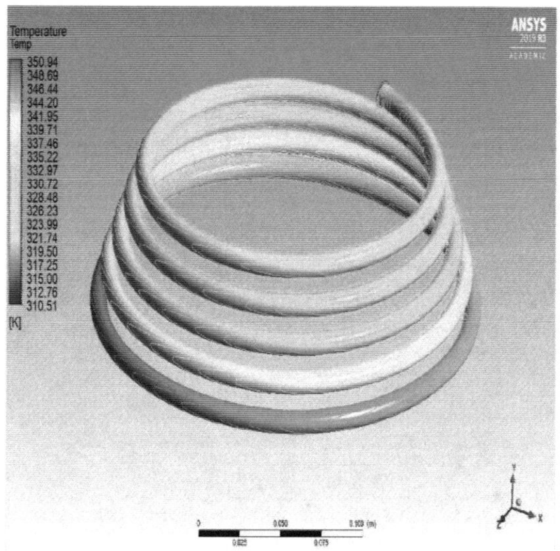

Figure 8. Temperature Contour

5.2. Variation of Nusselt number and friction factor

The change in average Nusselt number in terms of Reynolds number at the tube inlet is shown in Figure 9 for the conical coil having cone angle equal to 45° at three different pitches viz 15mm, 20 mm, and 25mm at the configuration of the coil. It is observed that the Nusselt number rises with the increase in Reynolds number. The values of the Nusselt number at a particular Reynolds number are observed to be very close for all three pitches. Hence, it can be concluded that the variation of the pitch does not affect heat transfer for conical coil configuration of cone angle equal to 45°. The generation of swirl flow for three different pitches is observed to be uniform maintaining the same heat transfer enhancement.

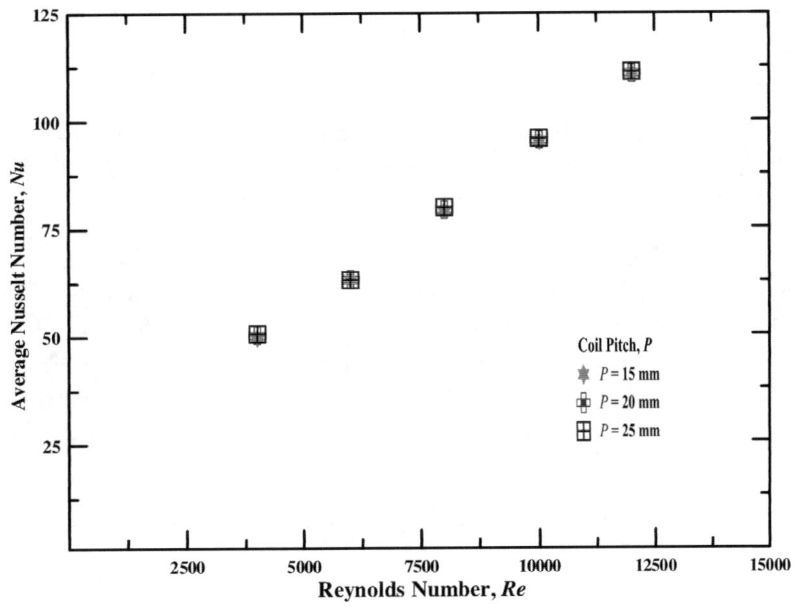

Figure 9. Deviation of Nusselt number, *Nu* with Reynolds number for 45° cone angle coil

Similarly, the change of average friction factor with tube inlet Reynolds number is shown in Figure 10 for the configuration of the coil of cone angle 45° at three different pitches viz 15mm, 20 mm, and 25mm. It is observed that the average friction factor gradually reduces with an increase in the value of the Reynolds number. The higher values of friction factor are observed at low values of Reynolds number while the variations in these values are observed to be less at higher Reynolds number. Another interesting feature can be interpreted from Figure 9 that the effect on friction factor for three different pitches is found negligible as these values are observed to be very close at a particular Reynolds number.

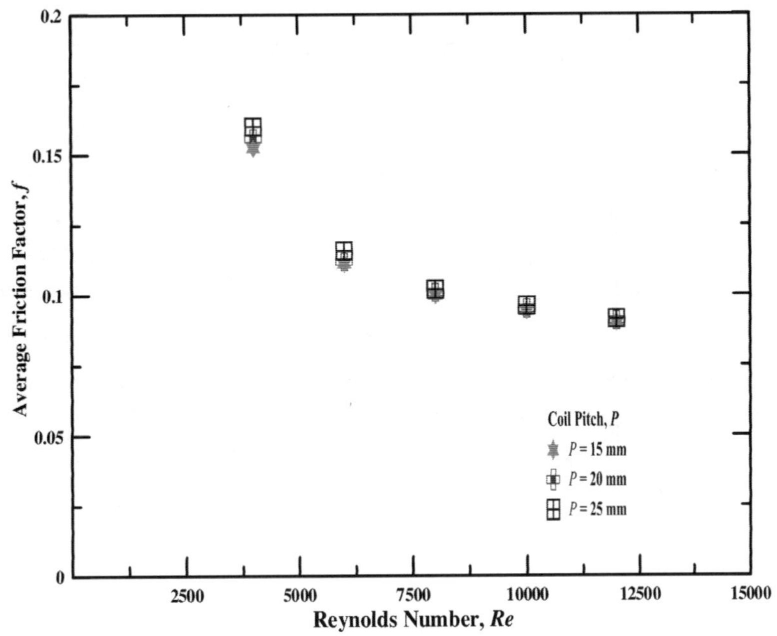

Figure 10. Deviation of friction factor with respect to Reynolds number for 45°coil cone angle

5.3. Effect of Cone angle

The change of average friction factor with the Reynolds number is shown in Figure 11 for the configuration of a coil having a pitch of 15 mm at three different cone angles 45°, 90°, and 135°. It is to be noted that the values of the average friction factor showslightly higher values at a higher cone angle. However, this is valid only for lower values of Reynolds number till 6000. At higher values of Reynolds number more than Re=6000, a negligible variation is observed with the cone angles. It can be concluded that the variation of cone angle does not change the turbulence level within the tube, at higher values of Reynolds number, resulting in the same values of average friction factor at three different cone angles. The higher values of the average friction factor are observed at low Reynolds numbers while these values are observed significantly low at higher flow rates.

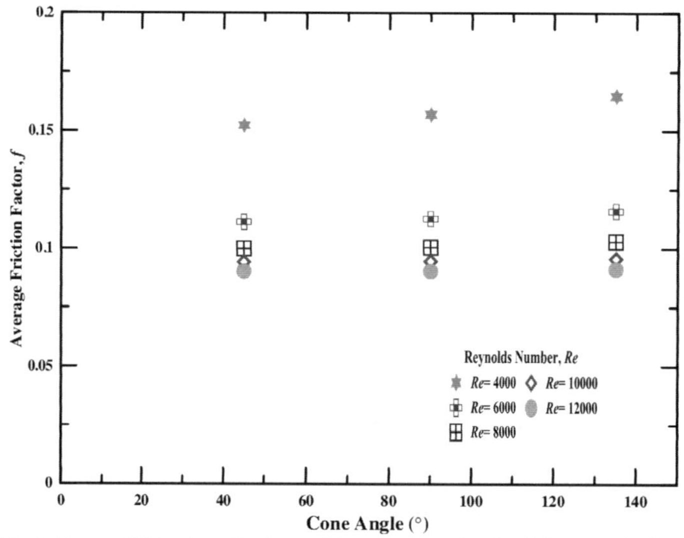

Figure 11. Variation of Friction factor with cone angle for 15 mm pitch coils

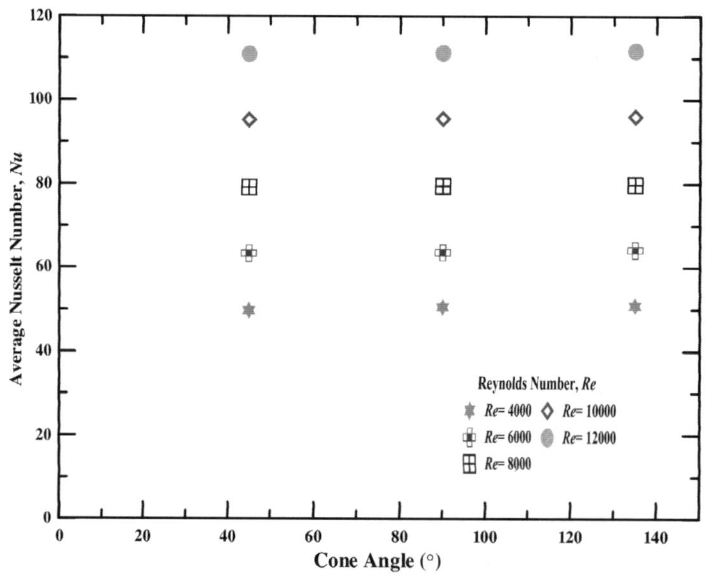

Figure 12. Variation of Nusselt number with cone angle for 15 mm pitch coil

The average Nusselt number with varying Reynolds numbers is shown in Figure 12 for a 15 mm configuration coil at three different cone angles 45°, 90°, and 135°. It is to be noted that the average Nusselt number shows negligible variation with the cone angles. It can be concluded that the variation of cone angle does not affect the flow characteristics significantly, therefore produces the same values of average heat transfer at three different cone angles. The higher values of heat transfer are observed at Reynolds number equal to 12000 while lower values are observed at Reynolds number of 4000.

5.4. Effect of Pitch

Figure 13 shows the variation of average friction factor for the configuration of the conical coil of cone angle 90°of three different pitches viz 15mm, 20 mm, and 25mm at same Reynolds number. It is noticed that the average friction factor almost remains constant with an increase in the value of coil pitch. This is observed at all the values of *Re* studied. It also can be confirmed from Figure 13 that higher values of friction factor are produced at low values of Reynolds number. These results bring the important conclusion that the effect of the pitch of a conical coil is minimum at a particular Reynolds number value.

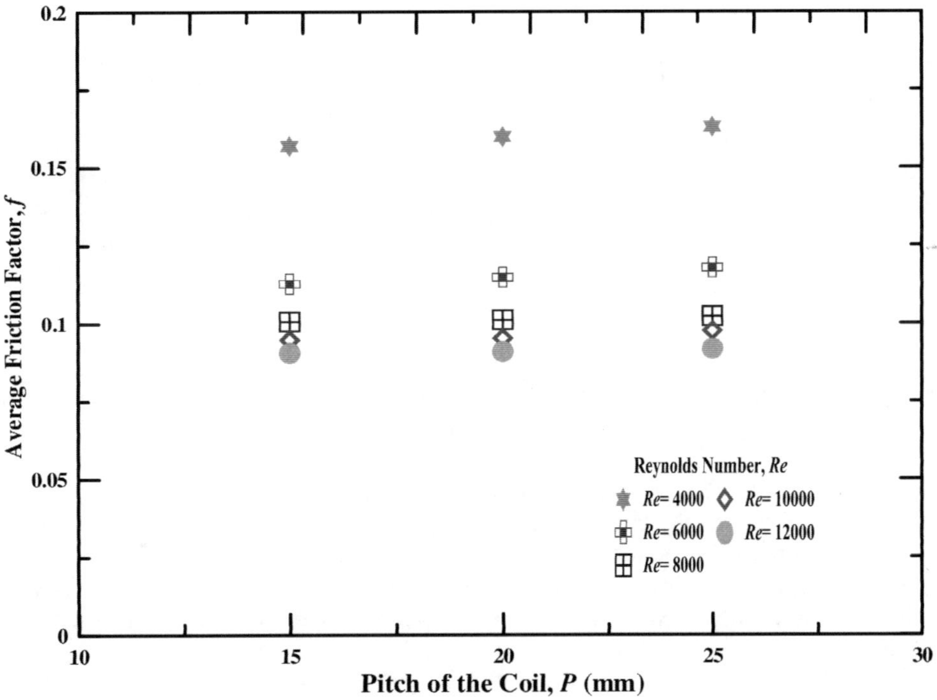

Figure 13. Variation of Friction factor with the pitch for 90° cone angles

Similarly, the variation of the average Nusselt number for the configuration of the conical coil of cone angle 90°of three different pitches at the same *Re* is shown in Figure 14. It is also observed that the configurations of different pitches provide uniform heat transfer enhancement at different coil pitches. This is observed for all flow conditions concerning different values of *Re* studied. It also can be confirmed from figure 13 that higher heat transfer is possible at higher values of Reynolds number and vice versa.

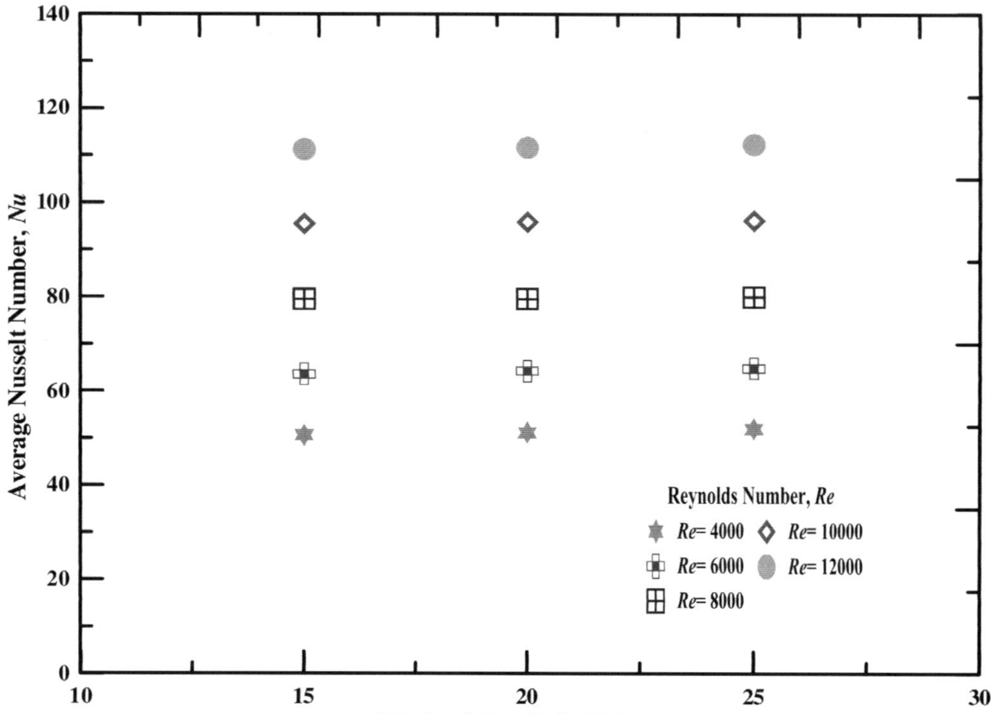

Figure 14. Variation of Nusselt number with the pitch for 90° cone angles

It can also be concluded from Figures 13, and 14 that the higher values of Reynolds number produce higher heat transfer at the expense of low values of friction factor. Thus, it is recommended that the flow rates through the conical coil can be maintained higher as lower flow rates produce less heat transfer at the expense of relatively high-pressure drops.

6. Conclusion

The various configurations of a conical heat exchanger are studied using a numerical technique. A conical coil heat exchanger of three different cone angles 45°, 90°,and 135°, and at three different pitches 20 mm, 25 mm, and 30 mm are presented for the study of heat transfer and friction factor characteristics at different flow rates. The influence of geometrical and flow parameters on pressure drop and heat transfer is presented for heat exchanger tube exhibited to constant wall heat flux condition. The average heat transfer values are observed to be 175 to 225% more than the corresponding smooth tube values with increased pressure drop. The numerical results reveal that a nonlinear increase in both Nusselt number and friction factor with the increase in cone angle and pitch of the conical coil.

Nomenclature

Symbol	Description
A	surface area of coil, (m^2)
C_p	Specific heat, (kJ/kg.K)
D	Mean Coil Diameter, (mm)
d_o	Outside tube diameter, (mm)

d_i	Inside tube diameter, (mm)
h	Heat transfer coefficient, (W/m^{2}K)
L	Length of wire coil, (mm)
K_w	Thermal conductivity of water, (W/m.K)
N	Number of turns
p	Pitch of the coil, (mm)
ΔP	pressure drop, (Pa)
Q	Heat transfer, (W)

References

[1] Purandare, P.S., Lele, M.M. and Gupta, R.K., 2015. Experimental investigation on heat transfer analysis of conical coil heat exchanger with 90 cone angle. *Heat and mass transfer*, *51*(3), pp.373-379.

[2] Joshi, S.M. and Anand, S.R., 2015, February. Design of conical helical coil heat exchanger for waste heat recovery system. In *2015 International Conference on Technologies for Sustainable Development (ICTSD)* (pp. 1-8). IEEE.

[3] Waghmare, R.S., Deshpande, H.N., Bhosale, S.Y., Magar, S.M. and Tagad, B.T., Comparative Numerical Analysis of Straight and Conical Coil Heat Exchanger. *Global Research and Development Journal for Engineering, ISSN*, pp.2455-5703.

[4] Naphon, P. and Wongwises, S., 2006. A review of flow and heat transfer characteristics in curved tubes. *Renewable and sustainable energy reviews*, *10*(5), pp.463-490.

[5] Prabhanjan, D.G., Raghavan, G.S.V. and Rennie, T.J., 2002. Comparison of heat transfer rates between a straight tube heat exchanger and a helically coiled heat exchanger. *International communications in heat and mass transfer*, *29*(2), pp.185-191.

[6] Lines, J.R., 1991. Helically coiled heat exchangers offer advantages. *Bulletin HHE, 30.*

[7] Acharya, N., Sen, M. and Chang, H.C., 2001. Analysis of heat transfer enhancement in coiled-tube heat exchangers. *International Journal of Heat and Mass Transfer*, *44*(17), pp.3189-3199.

[8] Salimpour, M.R., 2009. Heat transfer coefficients of shell and coiled tube heat exchangers. *Experimental thermal and fluid science*, *33*(2), pp.203-207.

[9] Cengel, Y.A. and Ghajar, A.J., 2007. Heat and mass transfer. *A practical approach.*

[10] Patil, R.K. and BW, S., 1982. Designing a helical-coil heat exchanger.

Design and Fabrication of Hybird Mini Scooter

G Srinu, G Sumithra, M Sampath, G Varun, V Sirisha, A Sai Chandan, M. Akhil

Department of Mechanical Engineering , Kakatiya Institute of Technology & Science, Warangal , Telangana – 506015 , India

* Corresponding author's e-mail address: gs.me@kitsw.ac.in, gst.me@kitsw.ac.in,

Abstract The present work intends to study and create a solution to be used as a means of transportation adapted to the metropolis environment. This solution is lies in projecting and building a prototype of an electrically assisted power vehicle, mainly directed to be used around the Last mile" concept. Now-a-days there are plenty of two-wheeler scooters causing many problems. Some of major problems are pollution, traffic, and reduction of non-renewable energy resources etc. The main purpose of the project to create a eco-friendly portable automobile which is easy to handle to emit 0% emission. Manual pedalling was used as main power source to reduce emission and solved fuel consumption problem. One of the major issues is Battery charging. Hence, in pedalling, energies converted from mechanical to electrical, which was used to charge the battery. Scooter was modelled in Fusion 360, simulation was performed in ANSYS and fabricated model could be folded easily. The model was unique, light-weight foldable electric scooter and eco-friendly to use.

Key words: Scooter, mini scooter, design, Ansys, and fabrication;

1. Introduction:

Electrical vehicles are claiming a place in several industries, especially in the fields of transportation. The application of electrical motors in bicycles and cars opens up new possibilities and a large number of advantages. Electric motor vehicles are a concept to take into account in the present and even more in the future, as they can open new possibilities or even replace the possibilities given nowadays by the common internal combustion engines. It presents a very small ecological footprint, specially comparing with cars, once that they are less or virtually non-pollutant. Another feature that increases this variation are the considerable different occupation rates.

Cars usually present occupation rates around 1 and 2 persons, representing 20 to 40 percent of its total capacity while bicycles employ its entire capacity rate, increasing efficiency and reducing the foot print. Kumar, et al. 993 [1] provided clearly information on Electric vehicle through their study. They powered the vehicle with Nickel metal battery and got a range of 100km between the recharges. When they replaced with polymer batteries, its range increased exponentially to 300 km, when they were implemented the power kits on existing bicycles, which had solved the many issues like cost, pollution. Thus help to the commutator of low range distances like medicine suppliers, education and premises in rural and urban areas. They used DC motor and motor controller for power controls and they were suggested and developed the controller configurations. Henry M. and Gannon, 2016 [2] study relates to multi- wheeled vehicle but not resisted to a bicycle. It consists of a standard conventional bicycle with multi-speed transmission, plus an electrical generating system and a solar charging arrangement. The arrangement was powered by combination of motor and pedaling, coupled in such a way that they provided power at any instance. Electrical System consists of DC hub motor, lead acid batteries and a hand lever operating a throttle means. Regenerative braking arrangement was studied to utilize the arrangement to generate energy that was charged the battery [3, 4].

One usual problem associated with the use of electric scooter is battery charging, it takes lot of time for full charging of the battery for that reason a model of hybrid concept is introduced in the present study. Battery is to be charged with help of pedalling to some extent and can also provide assistance to the rider through the journey, to help rapidly achieve speed or to let the rider move along the way. This concept can reveal to be also very beneficial to people with locomotion difficulties, as it can transform and upgrade a common scooter or similar vehicle to meet people's needs, helping them on transportation and increasing its mobility [5-9].

The major objective of this study is to consider the best alternatives to be used as a daily mean of transport to commute to effort. We should come up with a solution able to solve the problems inherent to the common urban means of transportation, as public transports, private cars or common bicycles; a new approach is adopted in the present study. Thus create a better alternative for this specific purpose, for such reason, the best alternatives is considered and is studied from the several hypotheses for an electrically assisted vehicle. This study also has the end of building a fully working prototype within the possibilities that are given; this is taking into consideration the time, capital available, access to building methods and materials among other constraints.

2. Materials and methods

2.1 Concept of Electric Bike (E-Bike):

An electric mini scooter is commonly known as E-Bike, which is as an integrated electric motor used for propulsion. Several types of e-bikes are existing in market, which have the small motor to assist riders pedal power.

2.2. *Chassis design*

The frame is the main component of a bike, it's the component that connects all the other parts and where these are fitted in. But it isn't just the component that connects all the parts, as it has extreme influence in the scooter performance, safety and nearly all aspects of the scooter.

2.3 Material selection

The most common material for the frame is used as the steel from the beginning of the any vehicle history, but there are several other ordinary alternative, such as aluminium alloys, titanium, carbon fibre and other composite materials. All the materials are having its own advantages, but the steel is most used one, due to high strength, relatively easy to work, cheap and reliable. Aluminium, in spite of its lower density compared to steel, is more difficult to work with and more expensive, Titanium, in spite of having a high strength to weight ratio and excellent corrosion resistance it's even more expensive and difficult to machine than both steel and aluminium. The carbon fibre, it has the advantage that can adapt to virtually any shape but it needs constant care and maintenance. According to the properties which are depicted in Table 1, chosen the material of the frame is AISI 1018 steel. Figure 1 shows the modelling of the chassis.

Table 1: Comparison table between AISI 1018 and AISI 4130 material

PROPERTIES	AISI 1018	AISI 4130
Density	7.8kg/	7.8kg/
Yield Stress	392.25MPa	799.803MPa
% Elongation atfracture	10.83%	10.83%
%Decrease in area	55.36%	30.57%
Breaking Strength	355.736N/	751.745N/
Elongation at break	24.130mm	28.730mm
Brinell Hardness	120	197

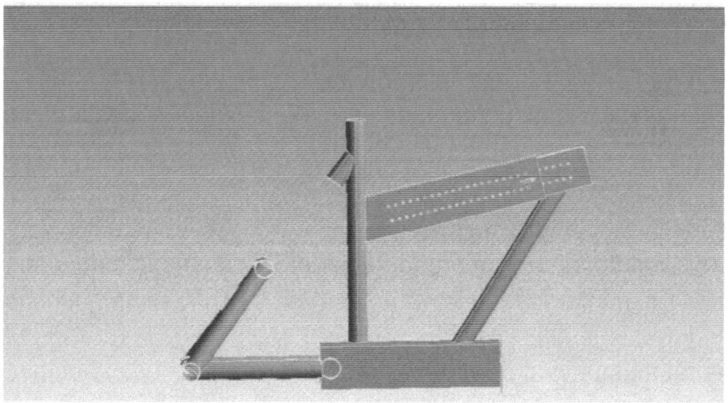

Figure 1 Chassis

2.4. Components used in mini scooter

Main components used in hybrid mini scooter are motor, Lithium ion battery, hand throttle, bicycle chain, chains pocket, pedal, handle bar, disk rotor, brake lever, brake caliper, headlight, saddle, suspension, telescopic fork, wheels.

3 Theoretical Approach

Figure 2 shows the approach adopted in the present work.

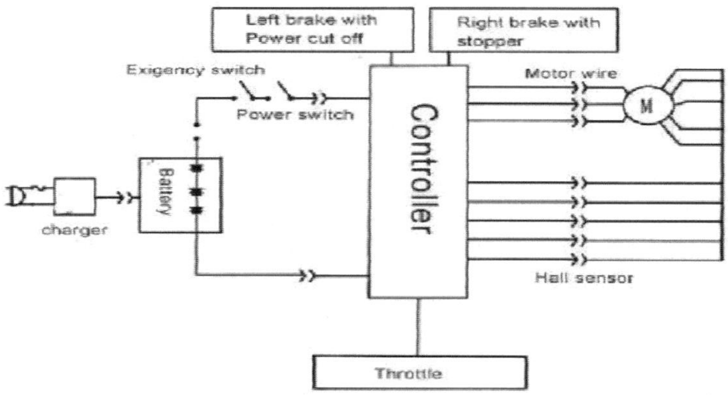

Figure 2. Electrical connections of motor

4. Results and discussion

4.1. Calculations

Power = 250 watt

Speed (N) = 360 rpm,

Voltage (V) = 24 V

Type of power transmission used: Chain Drive Reduction Ratio: - 1:2

$P = 2*\pi*N1*T1/60$

$T1 = 60*360/ (2*\pi*250)$

T1 = 6.63 Nm ; N1/N2 = T2/T1; T2 = 360*6.63/180 T2 = 13.2 Nm

T3 = 1.8 Nm

T4 = (360*1.8)/180

T4 = 3.6 Nm

N2 = 3000/2 = 1500 rpm

V = N2*circumference of rear wheel in meters

V = (180*3.14*0.508)/60

V max = 20 kmph

Here, P = Power at motor

N1 = rpm at motor shaft = 360 N2 rpm at wheel

T1 = continuous torque at motor shaft,

T2 = continuous torque at wheel shaft,

T3 = peak torque at motor shaft

T4 = peak torque at wheel shaft

V=velocity of the vehicle

4.2Analysis Considerations and Assumptions

The velocity of vehicle before hitting = maximum velocity of car

The time of collision = 0.4 seconds

Mass of the vehicle = 80 kg;

Final velocity = 0;

a = acceleration;

Where V = initial velocity = 20kmph

$$= 20\times\frac{5}{18} \text{ m/s} = 5.56 \text{ m/s}$$

From laws of kinematics $v = u + a \times t$

$0 = 5.56 + a\times0.4$

$A = 13.9 m/s^2$ (RETARDATION)

From Newton'slaw

Force = Mass × acceleration;

Force = 180×14

Force=1120 N

G-Force = Force/ (mass x acceleration due to gravity)

G-Force = 1120/(80 x .8)

G-Force = 1.42

4.3 Front impact test

Force = 4G,

Factor of Safety = Yield strength / working load
FOS= 440/354.5

FOS = 1.24

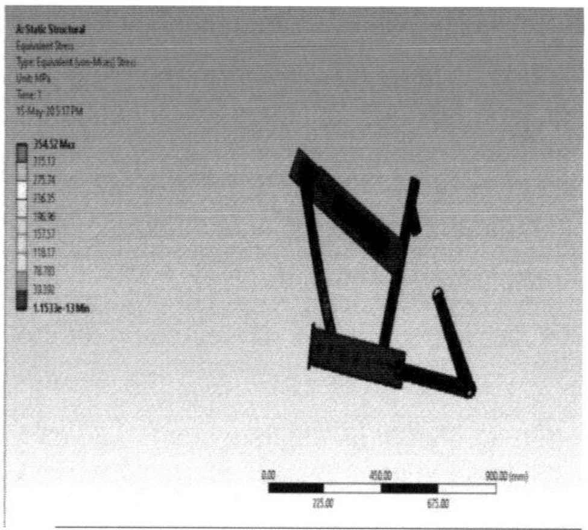

Figure 3. Front impact Equivalent stress

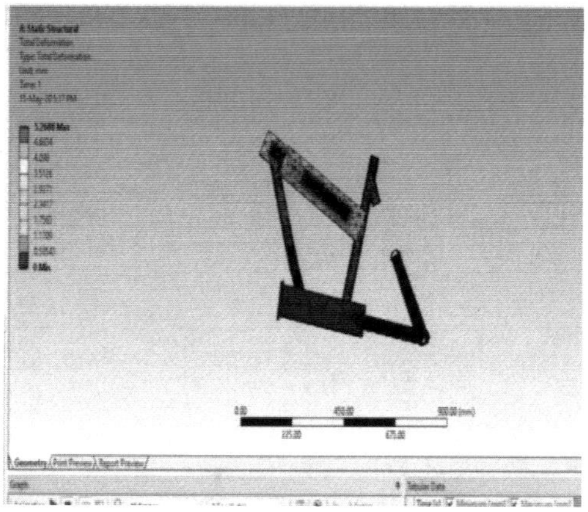

Figure 4: Front impact Deformation

From Figure 3 and Figure 4, it was observed that maximum Equivalent stress was obtained at 354.52MPa and deformation is 5.2688mm for front impact test.

4.4 Rear Impact Test

We considered the situation as in the previous case where opponent car is trying to overtake our kart and hit hours. Force = 3G FOS = 443/173 = 2.5

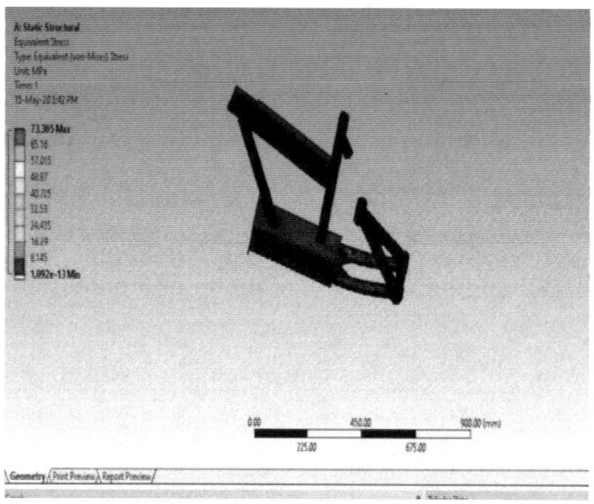

Figure 5. Rear impact Equivalent stress

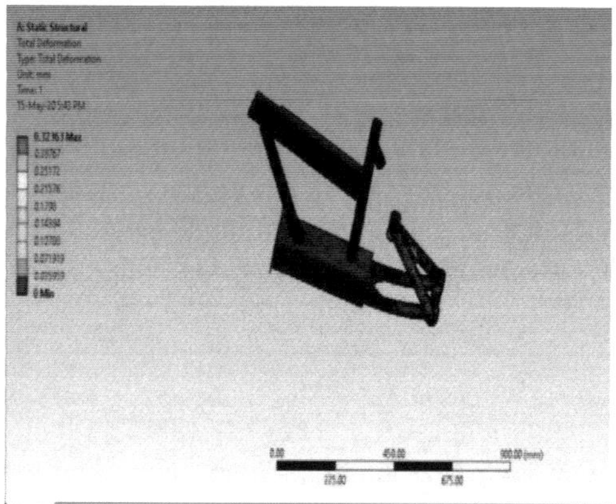

Figure 6. Rear impact Deformation

From Figure 5 and Figure 6, it was observed that maximum Equivalent stress was obtained at 73.305MPa and deformation is 0.32363mm for rear impact.

IOP Conf. Series: Materials Science and Engineering 1185 (2021) 012009 doi:10.1088/1757-899X/1185/1/012009

4.5 Side Impact Test

We considered 2G force

Force = 2G

FOS = 440/264.42 = 1.6

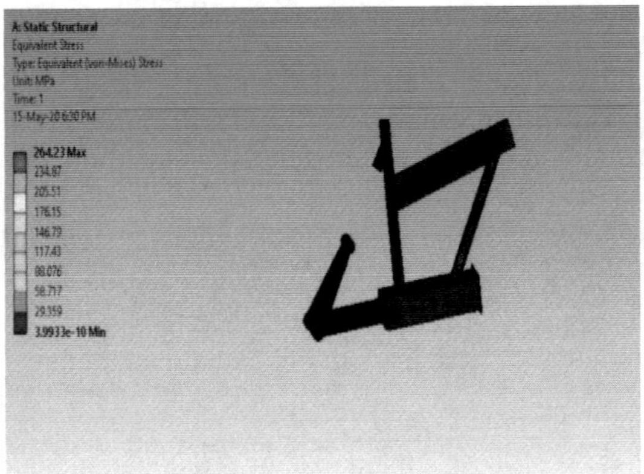

Figure7.Side impact Equivalent stress

Figure8. Side impact Deformation

From Figure7 and Figure 8, it was observed that maximum Equivalent stress was obtained at 264.23MPa and deformation is 4.6505mm for side impact.

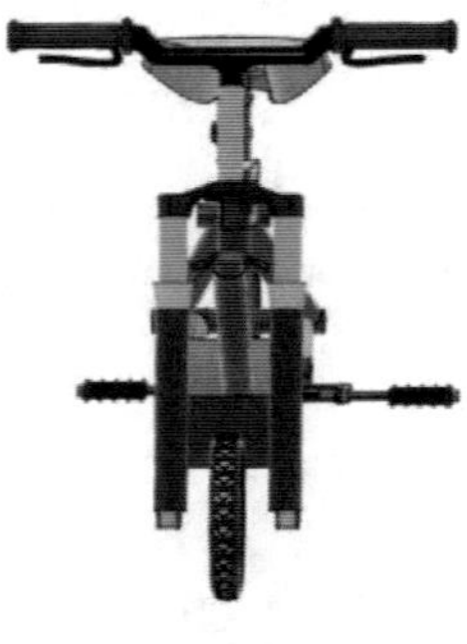

Figure 9. Front view

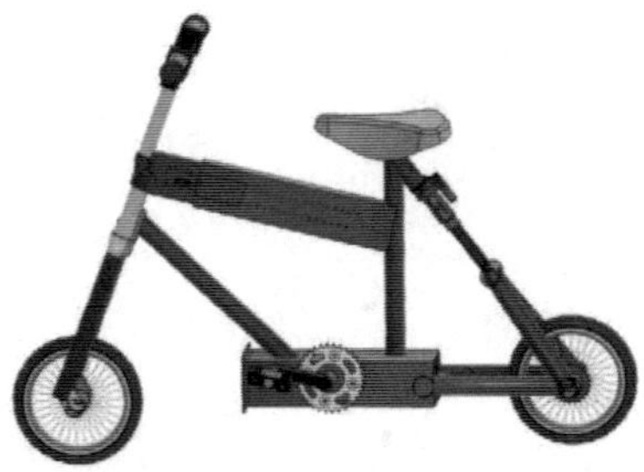

Figure 10. Side view

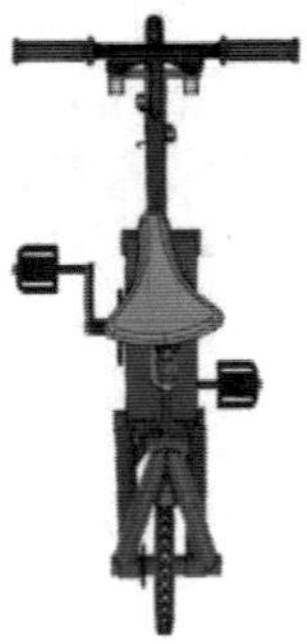

Figure 11. Top view

5. Conclusions

The study was thought to simplify and ease the transport in big city environments in general, yet, it was mainly aimed to be applied in the "Last mile" concept. It was made a study on the market of electrical bikes which revealed a large and exponential growth of sales of electrical bikes in the last years. In this study, the designing was done in Fusion 360 and parts were analyzed by applying some boundary conditions in ANSYS software. By applying static structural loads FOS was calculated at each front, rear and side of the bike and theoretical calculation was also performed. A future work and development of this project would be the improvement and reinforcement of the design as well as considering other alternatives, possible better suited, to hold the bike in the folding position.

References

[1] Kumar Binod, H. Oman, Power control for battery-electric bicycles, Aerospace and Electronics Conference, 1993. Proceedings of the IEEE 1993, National. IEEE,1993.

[2] Morchin, William C. Battery-powered electric bicycles, North con/94 Conference Record. IEEE, 1994.

[3] K. S. Yun, FOLDABLE BICYCLE. Seoul Patent US 2011/0193313 A1, 11 Aug2011.

[4] L. Ford Global Technologies, FOLDABLE ELECTRIC BICYCLE. MI/US Patent US 9,469,364, 18 Oct2016.

[5] Dumitrache et.al., E-bike electronic control unit, Design and Technology in Electronic Packaging (SIITME), IEEE 22nd International Symposium for. IEEE, 2016.

[6] J. Dill, G. Rose, 2012 Electric bikes and Transportation policy Insights from early adopters, *Transportation Research Record: Journal of the Transportation Research Board,* 2, 1.

[7] Henry Gannon, Electric and pedal driven bicycle with solar charging, patent no. 5316101, May 311994.

[8] F. L. Maestrelli, Bicycle frame optimization by means of an advanced gradient method algorithm, Technical report, Studio Maestrelli, 2008.

[9] P. D. Soden et.al. 1986 Loads, stresses and deflection in bicycle frames. Journal of strain analysis, **21(4)** 185-195.

Experimental and numerical analysis of orthogonal cutting of high strength aluminium alloy Al7075-T6

Ch Vasu[1,2,*], Atul Thaware[1], Gaurav Tiwari[1] and Ravikumar Dumpala[1]

[1]Department of Mechanical Engineering, Visvesvaraya National Institute of Technology, Nagpur, 440010, India

[2]Department of Mechanical Engineering, Rajiv Gandhi University of Knowledge Technologies, Nuzvid 521202, India

* Corresponding author e-mail: chvasu@rguktn.ac.in , Phone: +91-984-904-7429

Abstract. High strength Aluminum alloys are the key material class for the structural parts of aircraft, aerospace, military, and transportation industries because of their well-known performance, high strength, and light weight. In this present study, the Al7075-T6 tube was machined under orthogonal tool geometry conditions at different parametric combinations of cutting speeds and feed rates to investigate the chip morphology. Orthogonal cutting tests were performed with the high speed steel tool having a back rake angle of 8° and clearance angle of 7°, with varying cutting speeds and feed rates. The metallographic analysis was carried out on the chips obtained from experiments to find out the chip thickness and the type of serrations. Two dimensional finite element analysis was simulated by using ABAQUS/Explicit software.2D orthogonal plane strain model was created using the Johnson-Cook elastoviscoplastic material model and chip separation criteria were given by the Johnson-Cook damage model. Orthogonal cutting simulations were carried out with the same machining conditions. The orthogonal cutting operation was predicted with numerical simulation and the predicted chip thicknesses were close to experimental chip thickness.

Keywords: Al7075-T6; orthogonal cutting; finite element analysis; Johnson-Cook damage model; numerical simulation.

1. Introduction

Al alloys have been used for the structural bodies in construction for many years due to their good-established design systems, ability to easymanufacture, and consistent inspection methods. Aluminum metals are relatively light weight materials and cheaper compared to other structural materials[1]. Some of these Al alloysmay be heat treated to loadcomparatively high amount of stresses. Due to its exceptional properties such as excellent surface finish characteristics, excellent resistance to corrosion, easy workability, and good weldability, the Al alloysare extensively used in the aircraft and automotive industry[2]. There have been significant latest advances in Al alloys in all sectors. The aluminum 7000 series show better strength while compared to other series of Al materials and are particularlyused in the manufacture of stringers, wing skins, and frames in the aircraft industry. Al 7075 alloys can be age-hardened (artificially aged tempers T6 and T7) to attain high strength and these series of aluminum alloys were highly susceptible to stress corrosion cracking in aqueous and air mediums[3].

Although various researches have carried the work in the region of metal cutting and it is still sometimes revealed as one of the least known manufacturing processes due to itscomplications. The nature of chips in the machining process depends on the type of material, tool geometry,cutting parameters (speed, feed rate, and depth of cut), and type of lubrication (dry, wet, and minimum quantity lubrication). Chip morphology and its formation, usuallyknown as the effect of shear distortion in the machining zone.Different chip sizes (discontinuous or continuous) and shapes influencemanycharacteristics such as cutting forces, temperature, and chatter constancywhich affect surface finish, tool life, and efficiency of the metal cutting process. With the advent of computer technology and its widespread use, obtaining numerical solutions that are similar to approximation has become a more viable option. On the one side, this allows one to retain the problem's complexity while maintaining the preferred precision on the other. FEM is a standard numerical tool for analyzing the machining process that has evolved in recent years. Itdeliversclose displacement and velocity field outcomes based on the assumptions made while designing the model for the orthogonal machining process[4].

Daoud et al.[5] investigated the thermomechanical behavior of the material by the JC model in orthogonal cutting of two Al alloy Al2024-T3, Al6061-T6 using FEM. The chip morphology and the experimental flow stress were matched with the FEM simulation. Paulo et al.[6]analyzed the thermal and mechanical behavior in orthogonal cutting of Aluminum 7075 with PCD and cemented carbide tools, using alagrangianFEA based machining method and projected machining forces, heat distribution, plastic strain, vonmises stresses, andwas concluded that the polycrystalline diamond tool has got asuperior performanceduring the analysis. Filice et al. [7] investigated the orthogonal cutting experiments on an AISI 1045 steel tube workpiece with the axial feeding using the uncoated carbide (0° rake angleand4° relief angle) in a dry condition with the parameters as 100 m/min machining speed and a feed rate of 0.1 mm/rev, and the experimental results (temperature and cutting forces) were compared with the numerical results. Johnet al.[8]analyzed two FEM model formulations as lagrangian approach and eulerian approach during the orthogonal cutting of aluminum 2024-T361 and compared the stresses, strain fields, chip morphology, and cutting forces with all the cutting conditions and obtained a good agreement with lower cutting speeds for both dry conditions and wet conditions.Kara et al. [9] investigated the study ofmachining forces and temperature during orthogonal cutting of experimental and numerical analysis using Defrorm-2D on AISI 316L steel with coated and uncoated carbide inserts, and predicted the numerical machining temperatures by means of numerical forces using an artificial neural network model.Shaomin et al.[10]presented the mathematical model comprises of strain mechanism, cutting forces and kinematic response during machining ofaluminium alloy to examine the burr formation. Parida et al.[11] validated experimental and numerical results of machining forces, chip analysis (flow angle, morphology), surface quality, and cutting tool wear during turning of Ti alloy with the help of DEFORM software.Asit[12] studied the influence of parameters (speeds and feed rates) during machining of Ti alloy (Ti-6Al-4V) to validated both the experimental and numerical results on the characteristics like temperature, deformation, stresses, and cutting forces.Mabroukiet al.[13] investigated the numerical analysis and experimental procedure on chip damage and fracture energy during the orthogonal machining of aluminiumA2024-T351 and the authors used the numerical approach ABAQUS software to explain the fragmentation of the chip and the chip morphology. Kai et al. [14] studied the toolvariationson the size effect in the orthogonal cutting process and resulted that the strain effect was dominant at all machining conditions.

In this research work, a coupled thermo-mechanical finite element model was established in ABAQUS/Explicit, to assess the effect of different rotating speed of work piece (270, 443, and 650 rpm and at different feed rate 0.08, 0.12, 0.2, 0.28, and 0.32 mm per revolution) on chip morphology (chip thickness and serrations) of Al 7075-T6. As the machining was a short period process with substantial non-linear plastic strain,therefore explicit scheme was chosen in the presentinvestigation. A 2-D cutting experiment was carried out on the Al7075-T6 tube on the lathe machine to validate the numerical simulation data. The objective of this study was to find out the chip morphology in the Aluminum alloy (7075-T6) workpiece at different parametric conditions like machining speed and feed rate by means of experimental and numerical methods.

2. Experimental details

The orthogonal machining experiments were performed on the Aluminum alloy 7075-T6 tube with the dimensions of 34.11 mm outside diameter of the tube and 27.18 mm inner diameter of the tube. The chemical composition of the Al 7075-T6 alloy is shown in Table 1. HSS tool used in the cutting experiment was designed with 8°back rake angle and 7°clearance angle. Tube face turning was performed with the help of a conventional lathe machine and high speed steel tool. The Al alloy tube was locked on the machine tool (lathe) using the 4 jaw chuck and the HSS tool wastightened on the tool holder of the lathe machine. The orthogonal cutting experimental setup is shown in Figure 1.

Table 1. Chemical composition (wt.%) of Al 7075-T6 alloy.

Element	Si	Fe	Cu	Mn	Mg	Cr	Zn	Ti	Al
Al 7075-T6alloy(wt.%)	0.0526	0.1053	1.3518	0.0189	2.318	0.2243	5.3536	0.015	Balance

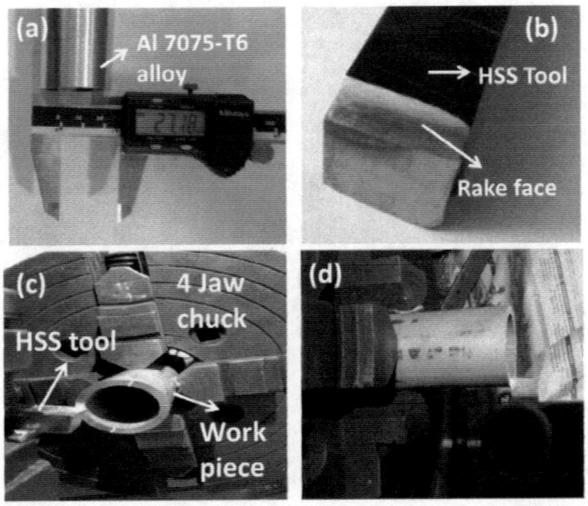

Figure 1. Photograph showing (a) Al 7075-T6 alloy, (b) HSS tool, (c) front view of theexperimental set up,and (d) top view of experimental set up of orthogonal cutting.

In the present work, speed in rpm, and feed rate in mm/rev are the parameters selected for the tube face turning operations. Machining experiments were performed at three different speeds (270, 443, and 650 rpm), and different feed rates (0.08, 0.12, 0.2, 0.28, and 0.32 mm/rev) as presented in Table 2. The tool feed rate was given in axial direction to the workpiece. Dry turning i.e. without lubrication orthogonal cutting was performed on the aluminum workpiece. Seven different cutting conditions were selected by varying cutting speeds and feed rates as shown in Table 2. After every machining condition, chips were collected from the machining zone for the study of chip morphology

Table 2. Cutting parameter of orthogonal cutting experiment.

Experiments	Cutting speed (rpm)	Feed rate (mm/rev)
1	270	0.08
2	270	0.12
3	270	0.28
4	443	0.08
5	443	0.12
6	443	0.32
7	650	0.2

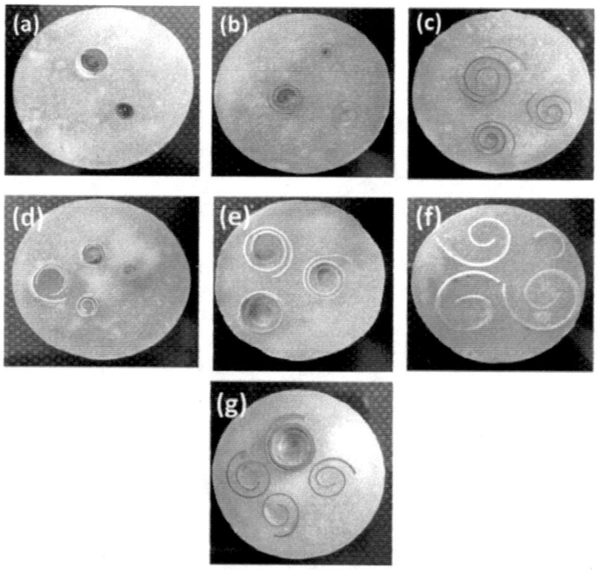

Figure 2. Chip mould images at different machining conditions (a) 273 rpm and 0.08 mm/rev feedrate, (b)273 rpm and 0.12 mm/rev, (c) 273 rpm and 0.28 mm/rev, (d) 443 rpm and 0.08 mm/rev, (e) 443 rpm and 0.12mm/rev, (f) 443 rpm and 0.32 mm/rev, and (g) 650 rpm and 0.2 mm/rev.

For the metallographic analysis of the chip, the mould preparation has been done by using the acrylic plus powder along with acrylic plus hardener solution. On the top of the plane glass, the chip samples were placed inside the circular shape of the empty mould and a small amount of grease is applied on the glass surface as well as inside the wall mould surface area in order to have easy removal of the mould after solidification. The acrylic powder was poured inside the mould above the sufficient height of the chip. After this process, the acrylic hardener solution was poured into the mould and in order to harden the setup was kept in the air for one hour. Finally, after the hardening process, the mould cases were removed and polished using grit papers of different sizes 300, 600, 1000, and 1200. The moulds are cloth polished and

IOP Conf. Series: Materials Science and Engineering 1185 (2021) 012010 doi:10.1088/1757-899X/1185/1/012010

examined in a microscope for further analysis to study the chip thickness and serrations. The chip moulds were as shown in figure 2.

3. Numerical simulation

In recent times, the finite element method (FEM) became the essential tool for the investigation of the machining process. In the recent advancement of FEM method a non-linear geometric boundaries, which includes free surfaces and chip-tool Interactions can be modeled easily. Therefore FEM based numerical analysis offers thorough qualitative and quantitative details into the chip formation process, which is very much essential for a detailedawareness of the effect of various process parameters. The computational models and simulation arecommonly employed out to decreasethe experimentation work.

Table 3. Al 7075T6 alloy properties and HSS tool materials.

Material Property	Values	Tool material properties	Values
Density (kg/m^3)	2800	Density (kg/m^3)	7800
Modulus of elasticity (GPa)	71.7	Conductivity(w/mk)	21
Poisson's Ratio	0.33	Young modulus(GPa)	207
Specific heat (J/kg°C)	960	UTS (MPa)	572
Thermal conductivity (W/m°C)	130		
T_{melt}(°c)	635		
$T_{transition}$(°c)	10		
Inelastic heat fraction	0.9		

In this present work,the FEM model was established in ABAQUS to model the chip formulation atthe orthogonal cutting of Al 7075-T6 using HSS tool to find the effects of various parametric combinations ofmachining speeds and feed rates on the chips.

Metal cutting is a highly complex method in which large geometric deviations occur. The Abaqus program was preferred since it treats temperatures and deformations as nodal parameters in combined temperature-displacement measurements, leading to higher mesh aspect ratio by appropriate meshing. The finite elementmodel was established in ABAQUS/Explicit software with Al7075-T6 as workpiece and high speed steel tool. Al7075-T6 and HSS tool dimensions, properties,and mesh were created in the modules. The is work piece is fixed in x and y direction and a prescribed displacement is given to tool in negative of x directions i.e. the length of cut. To specify speed of tool predefined field is used where speed is given in m/sec. Figure 3(a) represents the workpiece ina 2D cutting model with loading conditions.The meshed aluminium work part is shown in figure 3(b).Al 7075-T6 workpiece properties and high speed steel tool properties are shown in Table 3. After the completion of the model, ABAQUS/CAE creates an input/initial data for the analysis part, which produces the required output in post-processing. Lagrangian FEM formulation method was used in this research work [15].

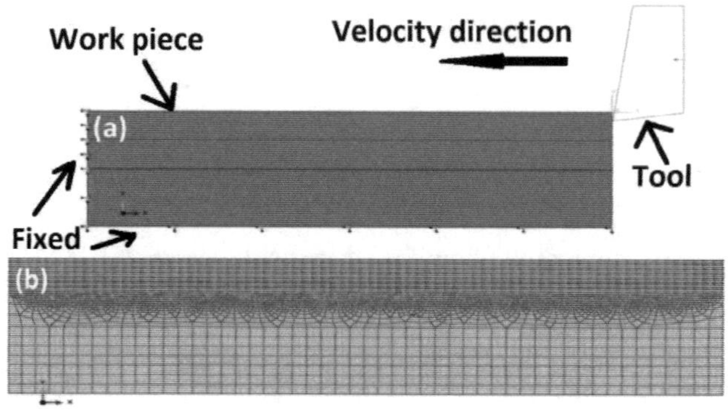

Figure 3. (a) 2D cutting model with loading conditions and (b) Meshed work piece

Johnson-Cook material model is an appropriate for high strain-rate variation over anenormous range, which is used for calculating large plastic deformation. This model defines the equivalent stress of the machinedsurface in viewof strain, strain-rate, and temperature distribution and the flow stress[5] is given by the equation (1).

$$\bar{\sigma} = \left(A + B\bar{\varepsilon}^n\right)\left[1 + C\ln\frac{\dot{\bar{\varepsilon}}}{\dot{\bar{\varepsilon}}_o}\right]\left[1 - \left(\frac{T - T_{room}}{T_{melt} - T_{room}}\right)^m\right] \quad (1)$$

Where, $\bar{\sigma}$ = equivalent stress, $\bar{\varepsilon}$ = equivalent plastic strain, n = work-hardening exponent, $\dot{\bar{\varepsilon}}$ = plastic strain rate, $\dot{\bar{\varepsilon}}_o$ = reference strain rate, A = yield stress, B = strain hardening constant, C = strain rate senstivity, m = thermal softening coefficient, T_{room}= ambient temperature, and T_{melt} = molten temperature.

Table 4 represents the values of various constants used in the existing simulations[16]. JC damage model is also used to define chip deformation founded measured by the value of equivalent plastic strain at element integration points. Damage and failure in the JC model are represented by a cumulative damage rule, which is given by equation (2).

$$D = \sum \frac{\Delta\bar{\varepsilon}}{\bar{\varepsilon}_f} \quad (2)$$

Where, $\Delta\bar{\varepsilon}$ represents that the increase in equivalent plastic strain

$\bar{\varepsilon}_f$ = equivalent strain corresponding to failure.

$\bar{\varepsilon}_f$ is given by the following equation (3)

$$\bar{\varepsilon}_f = \left[D_1 + D_2 \exp\left(D_3 \frac{P}{\bar{\sigma}}\right)\right]\left[1 + D_4 \ln\frac{\dot{\bar{\varepsilon}}}{\dot{\bar{\varepsilon}}_o}\right]\left[1 + D_5\left(\frac{T - T_{room}}{T_{melt} - T_{room}}\right)\right] \quad (3)$$

In the above equation, the 1[st]term denotes the pressure-dependent, the 2[nd]term denotesthe strain-rate influence and the 3[rd]term denotesthe heat effect. The $D_1$to D_5values of damage quantitieswere considered based on the literature as given in Table 5[17]. One of the most complex phenomena in the metal cutting process is the friction around the cutting tool surface and chip. In this work Coulomb's friction concept is used to define surface to surface contact. The penalty method was used to give friction of 0.5 [13].

Table 4. Johnson cook model for Al7075-T6.

A (MPa)	B (MPa)	n	C	M
546	678	0.71	0.024	1.56

Table 5. Johnson cook damage model of Al7075-T6.

D_1	D_2	D_3	D_4	D_5
-0.068	0.451	-0.952	0.036	0.697

4. Results and discussion

4.1Numerical analysis

Numerical analysis of the orthogonal cuttingprocess of Al7075-T6 at different cutting conditions wascarried out using ABAQUS and is shown in figure 4. From the results, it was observed that the highest von-mises stress occurs in the shear plane alongthe interface between chip and tool. When the stress value in the shear plane is more than the yield strength of the aluminum Al7075-T6, deformation of the chip occurs at the tool-chip interface. It was also observed that the von–mises stress values in the deformed chips increases with a rise in the feed rate. During machining of the tube at the cutting speed of 270 rpm, it was observed that the value of the maximum von-mises stresses increases when the feed rate rises from 0.08 to 0.28 mm/rev and it is greater than the ultimate tensile strength of the material.When the feed rate rises from 0.08 to 0.32 mm/rev with the 443 rpm cutting speed machining conditions, the value of the maximum von-mises stresses has been increased.

4.2 Metallographic analysis of chip

Experimental chip thickness was investigated using microscopic images of chip thickness at magnification factor 5X at all machining conditions. From the experimentation outcomes of chip morphology, it was observed that the serrated chips were generated. At a lowerfeedrate, these sawtoothappearance were low in the chip, but with the increment of feed rate, saw tooth appearance in the chip was also increased. This was due to the high machining forces and alternating large shear strain followed by low shear strain. The thickness of the chip was found and compared between numerical chipthickness with experimental chip thickness shown in above Figure 5 and Table 6 presents the chip thicknes comparison. It was seen that 2-dimensional plain strain orthogonal cutting operation was predicted with numerical simulation and the numerically predicted chip thickness was close to the experimental chip thickness. From the experimental analysis, the lowest chip thickness was found to be 0.1662 mm at 270 rpm cutting speed and 0.08 mm/rev feed rate. As the feed rate increases the chip thickness increases from 0.1662 mm to 0.5 mm at 270 rpm cutting speed and 0.203 mm to 0.625 mm at 443 rpm.

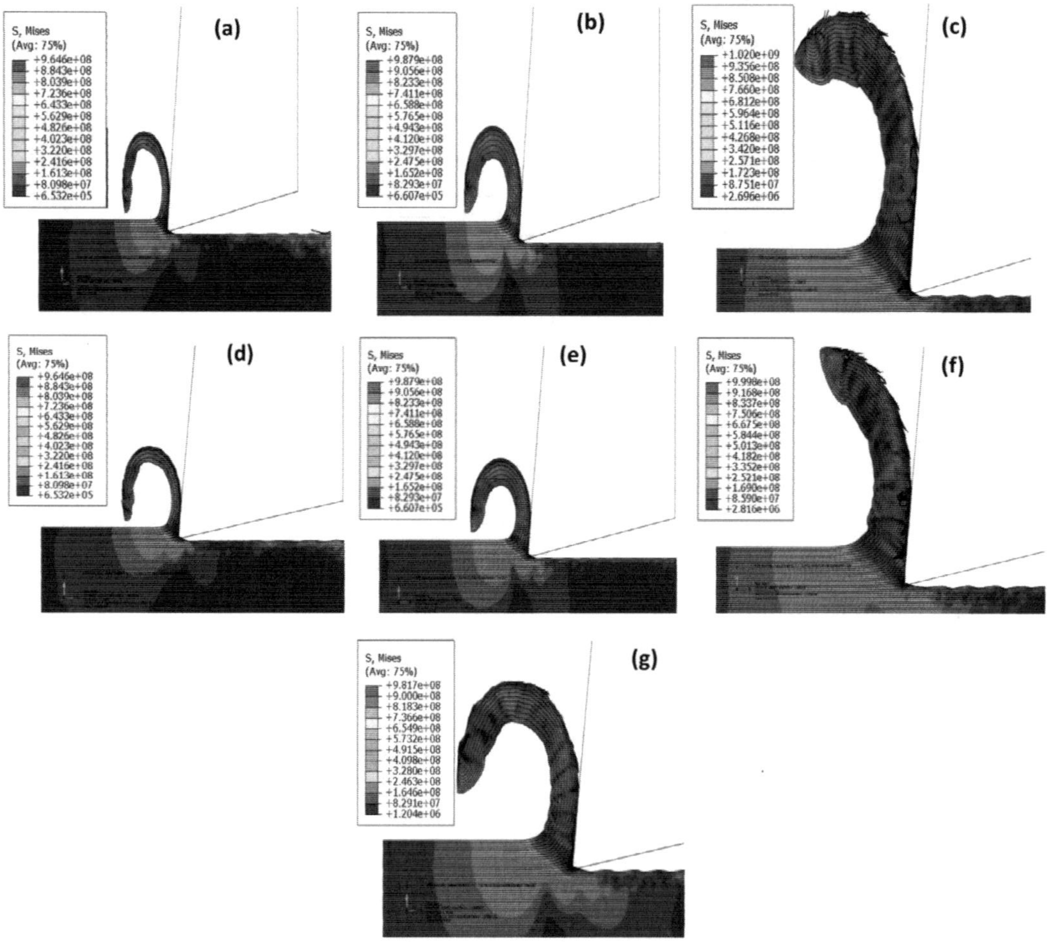

Figure 4. Photographs showing von-mises stresses at conditions (a) 273 rpm and 0.08 mm/rev feedrate, (b) 273 rpm and 0.12 mm/rev, (c) 273 rpm and 0.28 mm/rev, (d) 443 rpm and 0.08 mm/rev, (e) 443 rpm and 0.12mm/rev, (f) 443 rpm and 0.32 mm/rev, and (g) 650 rpm and 0.2 mm/rev.

The chip thickness was also observed to be increased with the rise of machining speed from 270 rpm to 650 rpm. The numerical analysis also resulted in a similar effect that the simulated chip thickness increases from 0.128 mm to 0.448 mm as the feed rate increases from 0.08 mm/rev to 0.28 mm/rev at 270 rpm cutting speed. And the simulated chip thickness was found to be increased from 0.1344 mm to 0.4992 mm with the increase of feed rate from 0.08 mm/rev to 0.32 mm/rev at 443 rpm cutting speed. From the numerical simulation analysis, the chip thickness grows from 0.128 mm to 0.2746 mm when the cutting speed raises from 270 rpm to 650 rpm. At 0.08 mm/rev feed rate, the chip thickness was found to increase from 0.1662 to 0.203 mm, as the speed rises from 270 to 473 rpm and a similar trend was observed with 0.12 mm/rev feed rate, as the speed rises from 270 to 473 rpm, the chip thickness changes from 0.301 to 0.3602 mm.

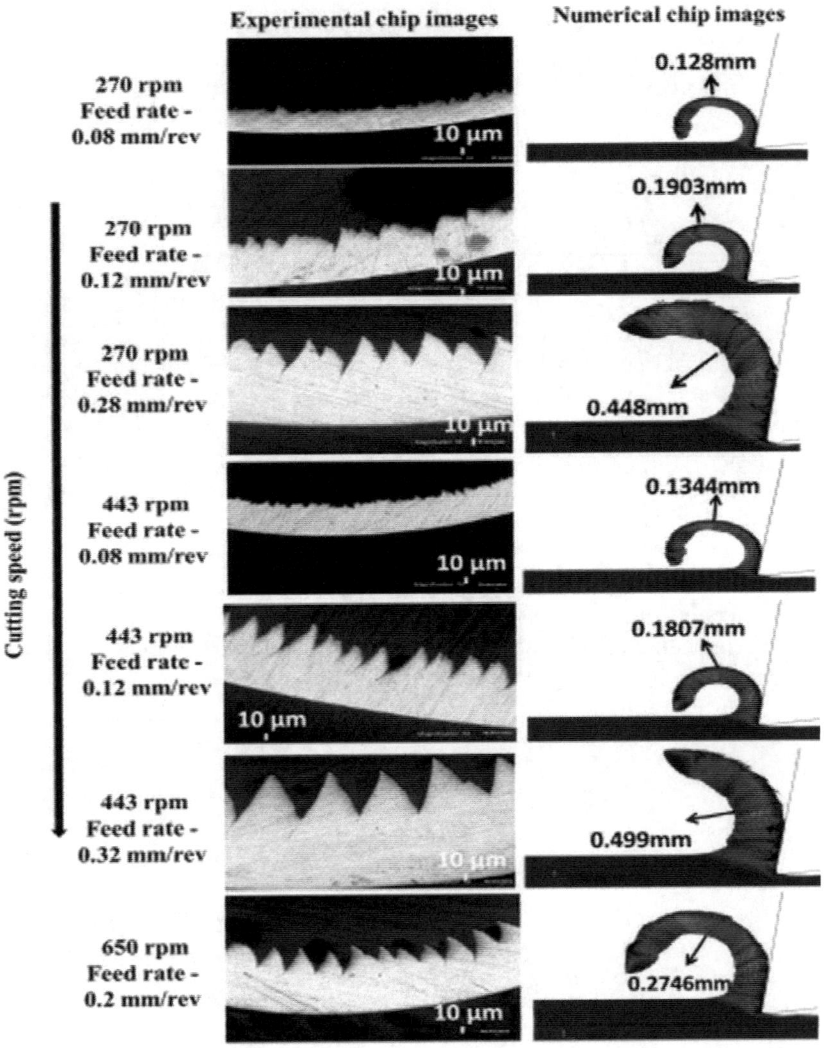

Figure 5. Comparision of experimental and numerical results of chip morphology at different machining conditions.

Table 6. Experimental and numerical chip thickness

SNo.	Cutting speed(rpm)	Feed rate (mm/rev)	Experimentalchip thickness (mm)	Simulated chip thickness(mm)	% error Approx
1	270	0.08	0.1662	0.128	22.98
2	270	0.12	0.301	0.1903	36.77
3	270	0.28	0.5	0.448	10.4

4	473	0.08	0.203	0.1344	33.79
5	473	0.12	0.3602	0.1807	49.83
6	473	0.32	0.625	0.4992	20.128
7	650	0.2	0.371	0.2746	25.98

5. Conclusions

In this study, the orthogonal machining of Al7075-T6 using high speed steel tool has been carried out at different parametric combinations using cutting speed and feed rate. FEM was used to modelthe predicted chip morphology using a Johnson-Cook material and Johnson-Cook damage model under a certain range of parameters in ABAQUS.

- FEM using ABAQUS provided satisfactory results compared with analytical results in terms of the main cutting speed and feed rate. Hence, the FE model and material model can be used to forecast the chip morphology analysis.
- The serrations on the free surface of the chips have been observed using the metallographic studies with all machining conditions. The saw-tooth appearance of the chip increases with the increase of feed rate. And also the serrations on the chip rises with the rise of cutting speed.
- From the experimental analysis and numerical analysis of chip morphology, it was found that the chip thickness rises with the increase of speed and feed rate.
- Finally, the chip morphology in the numerical analysis showed closeness to the experimental analysis.

References

[1] Miller W, Zhuang L, Bottema J, Wittebrood A J, De Smet P, Haszler A and Vieregge A 2000 *Mater. Sci. Eng., A*280 37-49
[2] Dursun T and Soutis C 2014 *Mater. Des.*56 862-71
[3] Li J-F, Peng Z-w, Li C-X, Jia Z-Q, Chen W-j and Zheng Z-Q 2008 *Trans. Nonferrous Met. Soc. China*18 755-62
[4] Bagci E 2011 *Int. J. Phys. Sci.*6 1267-82
[5] Daoud M, Jomaa W, Chatelain J and Bouzid A 2015 *Int. J. Adv. Manuf. Technol.*77 2019-33
[6] Davim J P, Reis P, Maranhao C, Jackson M, Cabral G and Gracio J 2010 *Int. J. Mater. Prod. Technol.*37 46-59
[7] Filice L, Micari F, Rizzuti S and Umbrello D 2007 *Int. J. Mach. Tools Manuf.*47 709-14
[8] Carroll III J T and Strenkowski J S 1988 *Int. J. Mech. Sci.*30 899-920
[9] Kara F, Aslantaş K and Cicek A 2016 *Appl. Soft Comput.*38 64-74
[10] Li S, Zhang D, Liu C and Tang H 2020 *J. Manuf. Processes*56 350-61
[11] Parida A, Rao P and Ghosh S 2020 *J. Braz. Soc. Mech. Sci. Eng.*42 1-14
[12] Parida A K 2018 *Int. J. Lightweight Mater. Manuf.*1 197-205
[13] Mabrouki T, Girardin F, Asad M and Rigal J-F 2008 *Int. J. Mach. Tools Manuf.*48 1187-97
[14] Liu K and Melkote S N 2007 *Int. J. Mech. Sci.*49 650-60
[15] Ceretti E, Fallböhmer P, Wu W and Altan T 1996 *J. Mater. Process. Technol.*59 169-80
[16] Ucun İ 2016 *J. Mech. Sci. Technol.*30 1843-50
[17] Brar N, Joshi V and Harris B 2009 *AIP Conf Proc.* (American Institute of Physics) pp 945-8

Friction stir processing of ZE41 Mg alloy: Optimizing the process parameters

B. Kiran Babu [1, 2, *], A. Jawahar Babu [3], G. Ranga Janardhana [4]

[1] Research Scholar, Department of Mechanical Engineering,
Jawaharlal Nehru Technological University kakinada, Kakinada 533003, India
[2] Department of Mechanical Engineering,
Usha Rama College of Engineering and Technology, Telaprolu 521109, India
[3] Department of Mechanical Engineering,
Gudlavalleru Engineering College, Gudlavalleru 521356, India
[4] Department of Mechanical Engineering,
Jawaharlal Nehru Technological University kakinada, Kakinada 533003, India

* Corresponding author's e-mail address: kiranbabu.battula58@gmail.com

Abstract. Friction stir processing is a novel processing route used to develop fine grained structures. Processing by FSP is influenced by tool rotational speed, tool travel speed, workpiece temperature, tool tilt angle, etc. Processing at optimum parameters is important to avoid defects and to develop grain refined materials by FSP. In the current research work, ZE41 magnesium alloy was processed by FSP to develop fine grained structure. FSP was done at different tool rotational and travel speeds. Decreased grain size and increased hardness was observed for all FSPed samples. However, variations were clearly observed within the measured grain size, hardness and nugget zone width of the FSPed samples. From Taguchi optimization studies using gray relation analysis; optimum set of parameters which yield smaller grain size and higher hardness were identified. From the results, 1400 rpm and 25 mm/min were observed as optimum. The design model presented in the current study helps to select optimized parameters in producing fine grained ZE41 for upcoming energy efficient applications. .

Keywords: Magnesium alloys, FSP, optimization, fine grains, optimization.

1. Introduction

Developing fine grained structures at the surface of metals by friction stir processing (FSP) is novel in the surface engineering [1]. FSP completely eliminates the complex problems elevated during materials processing which involve melting and solidification. However, there are several process parameters which influence the success of FSP which include tool tilt angle, rotational speed and tool travel speed, etc. [2]. Optimizing these parameters is crucial to achieve higher level of grain refinement without defects in the stir zone. There are several optimization techniques available to select the most appropriate process parameters which give better outcome. Gray relation analysis is such as simple optimization technique which helps to identify the optimized process parameters to achieve highest grain refinement [3].

Due to their excellent properties such as low weight and high strength, damping properties and good machinability; Magnesium (Mg) based materials gained popularity in the materials Engineering [4, 5]. Developing grain refined Mg alloys helps to produce structures made of Mg alloys with beter properties as the grain refined materials exhibit better properties. Producing fine grained Mg based alloys by liquid state processing methods is complex due to the nature of Mg. On the other hand, processing Mg within the solid state is difficult due to the low formability at the ambient conditions. In this context, FSP can be a viable solution to process Mg alloys to develop gain refined structures. Several other reports demonstrated using FSP to develop grain refined Mg alloys [6, 7]. Among the available Mg alloys, ZE41 alloy that has zinc (Zn), rare earths (RE) and zirconium (Zr) as the main alloying elements is used in automobile and aerospace applications [8, 9, 10]. Developing grain refined ZE41 Mg alloy helps to achieve improved performance in the structures produced by ZE41 Mg alloy. Hence, in the present research work, FSP of ZE41 Mg alloy was done at three different tool rotational and tool travel speeds and optimized process parameters were obtained to achieve smaller grains without defect in the stir zone.

2. Materials and methods

ZE41 Mg alloy billet was commercially procured (Exclusive Magnesium, India) and workpieces of size 100 X 50 X 5 mm^3 were machined form the billet. The ZE41 Mg alloy contains 3.9 % Zn, 0.56% Zr, 1.1 % RE, 0.003% Fe and balance being Mg (by wt.%). FSP was done on a vertical milling machine. The experiments were carried out by using a tool manufactured with H13 tool steel. 1120, 1400 and 1800 rpm rotational speeds and 16, 25 and 50 mm/min tool travel speeds were adopted. Since the spindle of the milling machine does not have tilt facility, tilt angle during FSP was considered as zero. FSP illustration and photograph of the process are shown in Fig 1. In the present research work, the process parameters were selected to form an L_9 full factorial orthogonal array (3^2). Table 1 lists the process parameters adopted to complete the experiments. Taguchi model using orthogonal array was adopted and from the gray relation analysis optimum set of parameters which yield smaller grain size and higher hardness were identified. In grey relation analysis; grey relation coefficients are calculated by measuring performance characteristics and normalizing them to a value between zero to one. The grey relation grades are obtained from the coefficients and the overall performance response is obtained. Details on carrying grey relation analysis can be referred elsewhere [3].

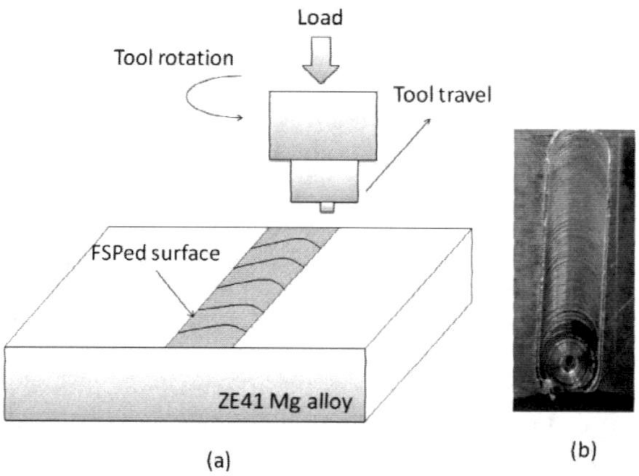

Fig 1a) Schematic diagram of FSP and b) photograph showing surface of FSPed ZE41 Mg alloy

For microstructural studies, the samples were metallographically polished by following a standard procedure which includes first polishing by using coarse grained sheets followed by fine grained sheets. Finishing of polishing is done by using alumina paste with the help of disc polishing machines.

The cleaned samples were then dried. Chemical etching (picric acid reagent) was done on the polished surfaces to reveal the microstructure. Microstructures were recorded by using optical microscope (Leica, Germany). Stir zone width was also measured and analyzed from the cross section of the samples. By conducting Vicker's indentation tests, hardness was measured by applying 100 g load was applied on the surface by a diamond indenter for 10 s dwell time. Measurements were done across the FSPed region of all the samples. From the grey relation analysis, optimum process parameters to obtain higher hardness, higher stir zone width and smaller grains size were obtained.

Table 1. FSP process parameters (tool rotational and travel speeds) selected in the current research work.

Factors	Rotational speed (rpm)	Travel speed (mm/min)
Level 1	1120	16
Level 2	1400	25
Level 3	1800	50

3. Results and discussion

Friction stir processing results in grain refinement in metals. During FSP, the heat that is generated in the stir zone helps to plastically deform the material and causes dynamic recystallizaiton (DRX) and produce fine grains [2]. The stirring action of the pin during FSP helps to deform the material and the heat that is generated during FSP dissolves the intermetallics into the solid solution grains. Therefore, several alloy which exhibit intermetallics at the grain boundary demonstrate decreased amount of intermetallics. In ZE41 Mg alloy, base microstructure reveals solid solution grains and intermetallics of Mg-Zn at the grain boundaries (Fig 2). The average grain size was measured as 107 ± 6.7 μm for base alloy Fig 2 (a). After FSP, a clear observation was made that confirms the grain refinement in all the samples. Typical microstructure of stir zone of FSPed alloy (1400 rpm & 25 mm/min) is shown in Fig 2 (b) which is evident that FSP leads to grain refinement.

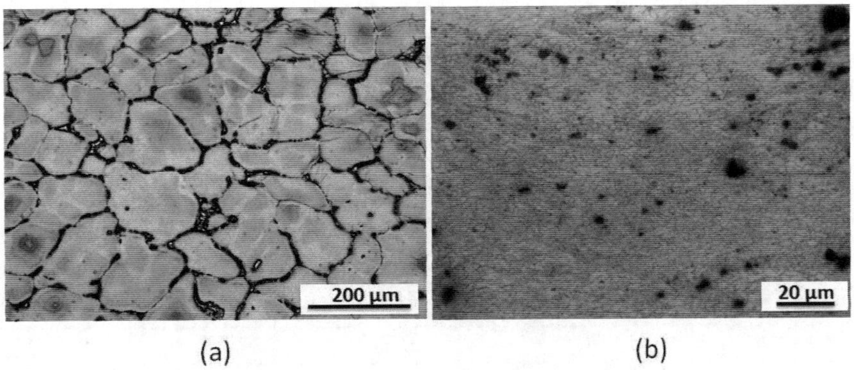

Fig 2 Microstructure of the samples: a) base alloy and b) FSPed alloy (1400 rpm & 25 mm/min).

The normalized values for grain size, width of the stir zone and hardness were obtained for the grey relation analysis as demonstrate elsewhere [11]. From the values, it is noted that for grain size, lower is better characteristic is considered. For hardness and stir zone width, higher is better characteristic is considered. Table 2 lists the average grain size, width of the stir zone width and average microhardness values of all the samples obtained at different FSP process parameters. Table 3 shows the corresponding grey relation normalized values, Table 4 lists grey relation coefficients and Table 5 shows the grey relation grades and their relative ranks.

Table 2: Average grain size, width of the stir zone and microhardness of the samples processed at different process parameters.

S. No.	Speed (rpm)	Feed Rate (mm/min)	Grain Size (μm)	Width (μm)	Micro-hardness (Hv)
1	1120	16	4.122	2357.599	96.26
2	1120	25	2.8	2140.067	90.4
3	1120	50	2.727	2643.196	89.3
4	1400	16	4.337	2519.256	112.5
5	1400	25	2.633	2316.477	128.2
6	1400	50	3.351	2175.144	121.7
7	1800	16	4.474	2054.16	97.3
8	1800	25	2.683	2272.132	95.8
9	1800	50	3.271	2006.216	102.1

Table 3 Grey relation normalized values

S.No.	Speed (rpm)	Feed Rate (mm/min)	Grain Size (μm)	Width (μm)	Micro-Hardness (Hv)
1	1120	16	0.1912	0.5516	0.1789
2	1120	25	0.9092	0.2101	0.0282
3	1120	50	0.9489	1	0
4	1400	16	0.0744	0.8054	0.5964
5	1400	25	1	0.4870	1
6	1400	50	0.6099	0.2652	0.8329
7	1800	16	0	0.0752	0.2056
8	1800	25	0.9728	0.4174	0.1670
9	1800	50	0.6534	0	0.3290

Table 4 Grey relation coefficients

S.No.	Speed (rpm)	Feed Rate (mm/min)	Grain Size (μm)	Width (μm)	Micro Hardness (Hv)
1	1120	16	0.3820	0.5272	0.3784
2	1120	25	0.8464	0.3876	0.3397
3	1120	50	0.9073	1	0.3333
4	1400	16	0.3507	0.7198	0.5533
5	1400	25	1	0.4936	1
6	1400	50	0.5617	0.4049	0.7495
7	1800	16	0.3333	0.3509	0.3862
8	1800	25	0.9484	0.4618	0.3751
9	1800	50	0.5906	0.3333	0.4270

Table 5 Grey relation grades (GRG) and their rankings

S. No.	Speed (rpm)	Feed Rate (mm/min)	GRG	Rank
1	1120	16	0.42924	8
2	1120	25	0.5246	6
3	1120	50	0.74689	2
4	1400	16	0.54131	5
5	1400	25	0.83121	1
6	1400	50	0.57208	4
7	1800	16	0.35686	9
8	1800	25	0.59516	3
9	1800	50	0.45032	7

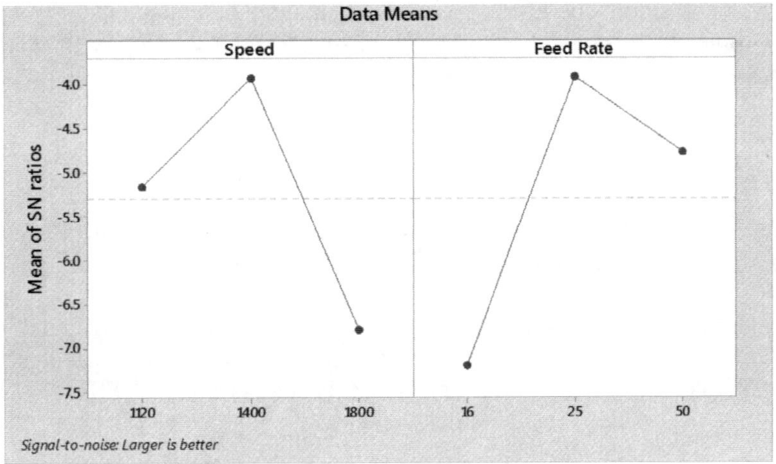

Fig 3 Main effects plot for SN ratios: Mean of SN ratios.

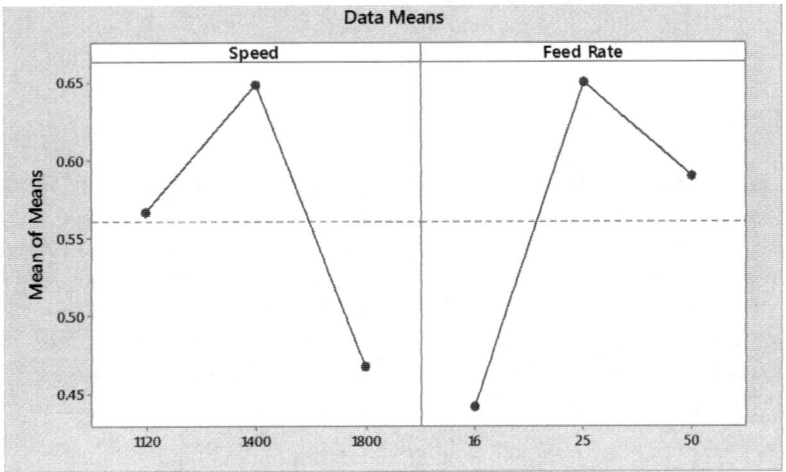

Fig 4 Main effects plot for means: Mean of means

Fig 3 and Fig 4 present the SN plots. From the GRG rankings, it can be understood that 1400 rpm & 25 mm/min parameters as the optimum set of process parameters to achieve smaller grain size. In terms of larger stir zone width, several set of process parameters were identified as prominent form the measured values as listed in Table 2. However, for higher hardness values, 1400 rpm & 25 mm/min were observed as optimum as also studied from GRG rankings. Other combination of process parameters (1120 rpm with 50 mm/min and 1800 rpm with 50 mm/min) were also observed as close to the optimized parameters (1400 rpm & 25 mm/min) to give relatively smaller grain size and higher hardness. This was also confirmed form the GRG rankings as listed in Table 5.

Achieving defect free stir zone and better grain refinement are the prime objectives behind optimizing the process parameters in FSP. Even though the FSP is a most convenient processin route to develop surface grain refined regions on metallic plates and sheets, it is very essential to optimize the process parameters for every new tool design and material systems. This essentially demands more optimization studies to know the best suited parameters to process. In addition to grain refinement, obtaining higher stir zone widths also helps to reduce the number of FSP passes carried out side by side to produce larger FSPed surface area. The grain refined region exhibits improved hardness compared with the base alloy. Therefore, the current analysis helps to understand the effect of tool rotational speeds, travel speeds on achieving grain refined ZE41 Mg alloy with better stir zone width and higher hardness.

4. Conclusions

In the present research work, FSP was done to produce grain refined surfaces on ZE41 alloy at different process parameters. Taguchi optimization analysis was used and the gray relation analysis was carried out to optimize the parameters. Defect free stir zones with smaller grain size and higher hardness with wider stir zone width was obtained at different process parameters. From the analysis, 1400 rpm & 25 mm/min parameters were found to be optimum. Hence, the present study helps to identify the best process parameters to develop fine grained ZE41 alloy by FSP.

References

[1] Mishra R S, Ma Z Y, Charit I 2003, *Mater Sci Eng* A **341,** 307.
[2] Ma Z Y, Feng A H, Chen D L, Shen J 2018, *Crit Rev Solid State Mater Sci,* **43(4),** 69.
[3] Yiyo Kuo, TahoYang, Guan-Wei Huang 2008, *Comput Ind Eng* **55(1),** 80.
[4] Fridrich H E, Mordike B L, 2006, *Magnesium Technology*, (Springer, Germany)
[5] Mordike B L, Ebert T, 2001, *Mater Sci Eng A* **302(1),** 37.
[6] Hemendra P, Ravikumar D, and Ratna Sunil B 2018, *Trans Indian Inst Met* **71** 951
[7] Saikrishna N, Pradeep Kumar Reddy G, Munirathinam B, and Ratna Sunil B 2016, *J Magnes Alloys* **4,** 68.
[8] Polmear I 2006, *Light Alloys: From Traditional Alloys to Nanocrystals* (Elsevier, USA).
[9] Praveen Kumar N, Naveen Ch, Sujatha D, Anandarao D, Naga Sireesha J, Anand Kumar S, Siva I, Venkateswarlu B, Jagannatham M**,** Ratna Sunil B 2018**,** *Matl Tech Adv Perf Matl* **33(9)** 603.
[10] Syam Prasad U, Kondaiah V V, Akhil K, Vijay Kumar V, Nagamani B, Jhansi K, Ravikumar Dumpala, Venkateswarlu B, Ratna Sunil B 2019, *Koroze A Ochrana Materiálu* **63(2)** 79.
[11] Girish B M, Siddesh H S, Satish B M 2019, *SN Appl Sci* **1**, 937.

Enhancing the wettability of pure titanium by shot peening for implant applications

Chelamalasetti Pavan Satyanarayana[1], **Lam Ratna Raju** [1,*], **B. Ratna Sunil** [2]

[1] Department of Mechanical Engineering Vignan's Foundation for Science, Technology & Research, Vadlamudi 522213, India.
[2] Department of Mechanical Engineering, Bapatla Engineering College, Bapatla 522101, India.

* Corresponding author e-mail: lamratnaraju@gmail.com, bratnasunil@gmail.com

Abstract. Metals are widely used for load bearing applications in manufacturing bone fixing implants in the biomedical engineering. Compared with polymers and ceramic materials, metals are suitable for load bearing applications in medical industry. Among the available metals, titanium based implants have become promising for load bearing implant applications. However, titanium suffers from bio-inert nature. The bioactivity of titanium surface can be improved by enhancing the wettability of the surface. Refining the microstructure at the surface of the implant increases the surface energy and improves the wettability. In the present work, pure titanium sheets were processed by implants can be or load bes applications. Horing implant for load beteels, alloys of cobalt and titanium are commonly used in the orthopedic peening by using steels balls of diameter 2 mm. The microstructure at the surface was observed as refined up to 100 μm. Microhardness of processed titanium was measured as increased up to 30% compared with base material. Lower water contact angles were observed for the processed titanium which suggests increased wettability after shot peening due to smaller grain size.

Keywords: titanium, bioactivity, bone implants, tissue bonding, wettability.

1. Introduction

Surface treatment of biomaterials is one of the important strategies to alter the surface properties of implants. Metallic implants suffer from the poor bioactivity when used in bone fixing applications [1]. Titanium based materials are one of the important metallic groups used to manufacture load bearing orthopedic implants. The surface of titanium can be treated by several ways. Providing bioactive coatings on the titanium surface, producing a self generated oxide layer and modifying the microstructure at the surface are the promising strategies to enhance the surface function of titanium [2, 3]. Microstructure of metals play very important role on altering the material properties [4, 5]. Reducing the grain size is one of the prime objectives in modifying the microstructure of metals. By adding inoculants or grain refiners within the liquid state of metals, smaller grains can be produced in metals and alloys during solidification [6]. On the other hand, there are several techniques can be applied to alter the microstructure within the solid state. Severe plastic deformation techniques are the best example for such solid state processing routes where no geometrical change is done for the material but the grain refinement is achieved [7, 8]. Higher amount of strain is introduced into the material to initiate more crystal defects and by doing so, sub grains are produced in the material. These grains with low angle grain boundaries further developed into new grains with continuous processing. Other paragraphs are indented (Body text Indented style).

Thirugnanam et al., [9] has demonstrated that the grain refinement achieved by groove pressing in pure titanium has shown higher wettability and bioactivity. Wettability is a desirable property for biomaterials which increases the bonding strength between the tissue and the implant to promote better healing of the tissue. Wettability can be increased by developing fine grained structure. Higher fraction of grain boundaries associated with smaller grains increase the surface energy of the implant and therefore, hydrophilicity is also enhanced which improves the wettability [10]. In the present work, pure titanium (medical grade) sheet was processed by shot peening to develop high wettable surface. The role of modified microstructure achieved at the surface on hardness and wettability was investigated.

2. Materials and methods

Commercially pure titanium sheets (2 mm thickness) of medical grade were procured from Midhani, Hyderabad, India. Then a sample size of 100 X 100 X 2 mm was prepared for shot peening experiment. Shot peening experiments were conducted by using tungsten carbide balls of size (2 mm diameter) for 30 min. Fig 1 shows the schematic illustration of shot peening process in which the balls introduce compressive stresses into the surface of the workpiece (Fig 1 (b)).

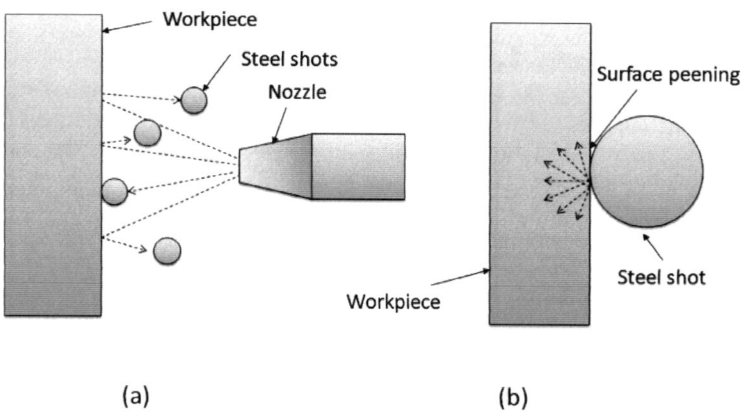

Fig 1 a) schematic illustration of spot peening: b) compressive stresses induced at the surface of the workpiece

The received material was examined by X-ray diffraction (XRD, D8 Bruker, USA) method between 2θ range from 20 to 80 ° with a step size of 0.1. The processed samples were subjected to microstuctural examination at the cross section. For microstrucutral studies, the samples were polished as per the standard metallographic procedure which includes polishing by emery papers of different grades followed by fine polishing by using diamond paste along with lubricant. After polishing, the samples were etched by using Kroll's reagent for 60 s. Then the microstructure was observed by using inverted optical microscope (Leica, Germany). Microstructures were obtained at the cross section of the processed titanium and also for the base material. Microhardness was measured for both the samples by Vickers indentation method by applying 100 g load. For every measurement, 1 mm distance was maintained while conducting the hardness measurements. For wettability studies, contact angle measurement studies were conducted by using distilled water (5 ml) as the solvent. The contact angles (θ) were measured and from the measured values, hydrophilicity of the titanium surfaces was assessed.

3. Results and discussion

Being a bioinert material, titanium (Ti) exhibits poor bonding kinetics with the host tissue. The level of bonding strength between the implant and the tissue impacts the healing rate of the fractured or deceased bone. In the present work, the level of tissue interactions at the interface of the implant and tissue are altered by microstructure modification at the surface. Fig 2 shwos the XRD pattern of commercial pure Ti (medical grade). All the peaks were identified and indexed by referring ICDD data. No peaks corresponding to impurities or other phases were identified. Fig 3 shows the microstrctures of base materials and processed Ti. It is evident from the observations that the average grain size of the base material as 47 ± 5.5 μm and shot peening resulted in grain refinement at the surface of the workpiece. The affected thickness at the surface of processed Ti was measured as ~ 100 μm. Next to the grain refined surface, several twins were appeared in the processed Ti. Twinning is a deformation mechanism usually observed in *hcp* metals such as Ti. In addition to slip, twinning helps to introduce plastic deformation in metals with lower number of slip systems. Compare with *fcc* and *bcc* crystal structures, *hcp* crystal structure has lower number of slip systems. Therefore, it is inevitable to form twins by which lattice deformation is achieved in *hcp* metals. Hence, it is a common finding in Ti when subjected to mechanical loads to observe number of twins as appeared in processed Ti (Fig 3 (b)).

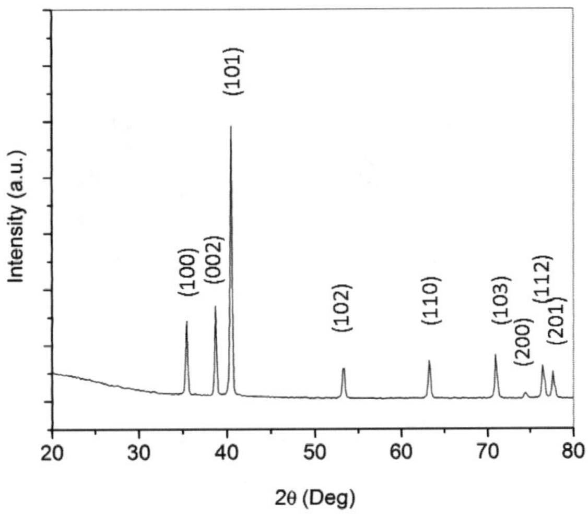

Fig 2 XRD pattern of pure Ti used in the present work.

From the microstructural studies (Fig 3), it can be noticed grain refinement up to an average value of 1.5 ± 0.7 μm after shot peening at the surface. The mechanism behind grain refinement in such severe plastic deformation techniques can be understood by considering the role of crystl imperfections in development of new grains in plastic deformation. Usually, in heat treatment, new grains are formed above a particular temperature which is known as recrystallization temperature. Whereas in plastic deformation of metals; several crystal imperfections are originated within the each grain due to high amount of mechanical loads. Dislocations are the important crystal defects usually developed in the absence of slip when the metals are subjected to mechanical loading. These dislocations are accumulated within the crystal and develop mismatch regions. Such regions with higher dislocation densities lead to develop cells within the gains which in turn evolve as sub-grains [11]. The sub-grains with low angle grain boundaries further develop as grains with different orientation and gives fine grains. This kind of recrystallization during plastic deformation is called as dynamic recrystallization which is the prime reason behind grains refinement in shot peening. These microstructural changes certainly introduce changes in the surface properties.

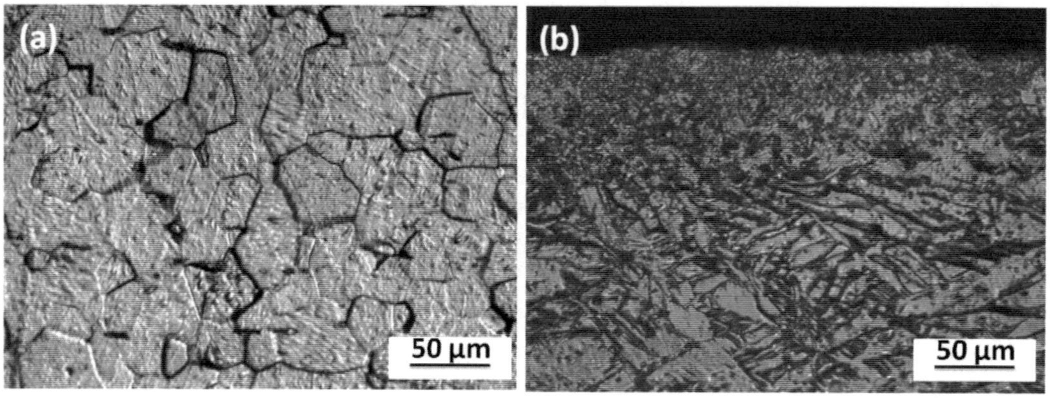

Fig 3 Optical microscope images: a) unprocessed Ti and b) processed Ti

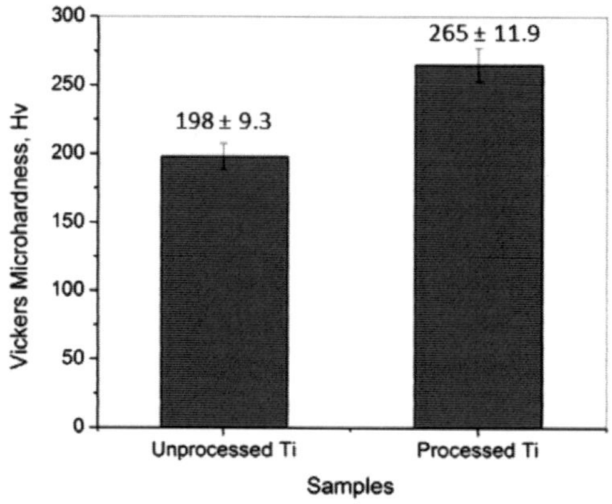

Fig 4 Microhardness of the samples.

Fig 4 shows the microhardness data of the samples. Compared with base materials, processed Ti has shown higher hardness. This can be explained by relating the gain refinement effect on improving the resistance to plastic deformation. Smaller grains increase the fraction grain boundary and introduces grain boundary strengthening effect which leads to increase the hardness of the processed Ti. Fig 5 shows the water droplets on the surface of the samples. Compared with base material (θ_1= 67.7 °), processed Ti exhibited lower contact angle (θ_2= 45.4 °). Lower contact angles indicate higher hydrophilicity of the surface. Usually, if the contact angle is more than 90 °, those surfaces are categorized as hydrophobic surfaces and with contact angle less than 90 ° are categorized by hydrophilic surfaces. Within the hydrophilic surfaces, lower values of contact angles strongly suggest higher hydrophilicity. Wettability of the surfaces is dependent of hydrophilicity. Higher wettability promotes better deposition of mineral phases which accelerate the cell response at the implant and tissue interface and leads to higher healing rtes. In the present work, lower contact angle was measured for processed Ti compared with base material. This indicates increased wettability for the processed Ti.

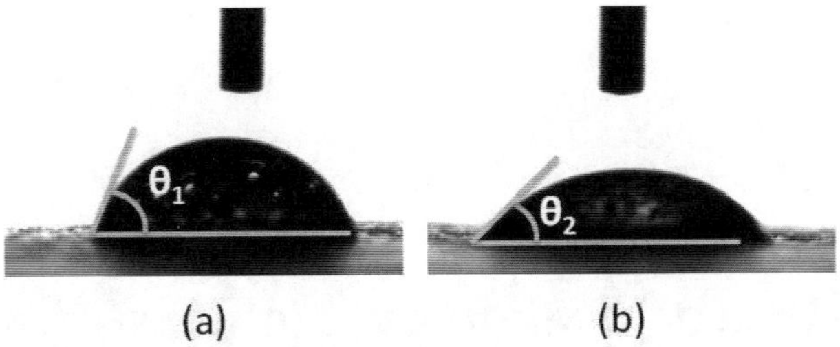

Fig 5 Contact angle measurements: a) water droplet on unprocessed Ti and b) processed Ti.

From the results it can be summarized that the grain refinement achieved by shot peening helps to increase surface mechanical properties and surface wettability in Ti. Medical implants used in orthopedic applciaitons where strong interactions are required to reduce the implant rejection can be surface treated to develop fine gain structure.

4. Conclusions

In the present work, commercial pure titanium was processed by shot peening to develop fine grained surface. A grain refinement from 47 ± 5.5 μm to 1.5 ± 0.7 μm was achieved after shot peening. Due to the grain refinement, increased hardness was measured for the processed Ti compared with base material. From the contact angle measurements, lower contact angle was measured for the processed Ti compared with base material due to the availability of more fraction of grain boundary. The results suggest that the fine grained surface can be successfully produced by shot peening which exhibits higher wettability that promotes higher healing rate for load bearing bone implant applications.

References

[1] Park J, Lakes R S 2007 *Biomaterials an Introduction* (Springer - New York, USA).
[2] Cochran D L, Schenk R K, Lussi A, Higginbottom F L, Buser 1998 *J Biomed Mater Research* **40** 1–11.
[3] Jemat A, Ghazali M J, Razali M, Otsuka Y 2015 *BioMed Research Intl*, 2015, 791725.
[4] George Dieter 1961 *Mechanical Metallurgy*, (McGraw Hill - USA).
[5] Meyers M A, Mishra A, Benson D J, 2006 *Prog Mater Sci*, **51(4)** 427-556,
[6] Sidney H Avner 1974 *Introduction to Physical Metallurgy* (McGraw Hill – USA).
[7] Segal V M 2002 *Mater Sci Eng A* **338** 331–344.
[8] Valiev R Z, Islamgaliev R K Alexandrov I V 2000 *Prog Mater Sci*, **45(2)** 103–189.
[9] Thirugnanam A, Sampath Kumar T S, Uday Chakkingal 2010 *Mat Sci Eng C* **30** 203–208.
[10] Ratna Sunil B, Thirugnanam A, Uday Chakkingal, Sampath Kumar T S 2016 *Mater Technol*, **31(13)** 743-755.
[11] Ratna Sunil B 2015 *Mater Manuf Proces* **30(10)** 1262-1271

Solid state surface deposition by friction surfacing: A review

N.M.K. Sarath Kumar [1, 2], G. Suresh [3], Ravikumar Dumpala [4], B. Ratna Sunil [5,*]

[1]Research Scholar, Department of Mechanical Engineering,
Vignan's Foundation for Science, Technology & Research,Vadlamudi 522213, India.
[2] Department of Mechanical Engineering,
Vignan's LARA Institute of Technology & Science, Vadlamudi 522213, India.
[3] Department of Mechanical Engineering,
Vignan's Foundation for Science, Technology & Research, Vadlamudi 522213, India
[4] Department of Mechanical Engineering,
Visvesvaraya National Institute of Technology, Nagpur 440010, India
[5] Department of Mechanical Engineering,
Bapatla Engineering College, Bapatla 522101, India

* Corresponding author's e-mail: bratnasunil@gmail.com

Abstract. Developing surface coatings and depositions for imparting specific properties at the surface of materials is an important strategy in surface engineering. Several processing routes are widely used in surface engineering which involve melting the coating materials and deposition on the substrate. On the other hand, surface coatings by solid state routes eliminate complex issues associated with liquid based processes. Friction surfacing is one of such potential solid state processes in which a consumable rod made of coating material is used to deposit a layer on the substrate. Several material systems have been used to produce surface coatings by friction surfacing. In this present review, the state of the art in developing several non-ferrous based coatings by friction surfacing is discussed. The role of process parameters and material properties influenced by the coatings are summarized and presented.

Keywords:friction surfacing, coatings, solid state, non-ferrous metals, surface engineering

1. Introduction

Surface engineering is a useful approach to improve the surface properties of a bulk structure without affecting the properties of the core of the structure. Developing several modern engineered materials addressed the existing issues in the materials engineering. However, inevitably, altering the surface properties or developing surfaces with special properties is a great demand for various applications where surfaces play very important role. Surface coatings has become an important area of research in recent years due to their significance associated with the surface of components, which needs a wide range of knowledge, including materials engineering, physics, and manufacturing sciences. Most of the components and structures fail or become disbanded due to fatigue, wear, and corrosion during functioning. Various surface modifications were developed to improve the surface qualities and extend the working life of components and structures, thereby decreasing risks associated with unprogrammed delays [1]. Surface engineering encompasses a wide range of techniques that modify the chemistry and properties of a thin layer of the substrate's surface, which include cladding process, laser treatment, cold spraying, liquid deposition methods, thermal spraying, anodizing, chemical and physical vapour deposition etc [2, 3]. Most of the liquid state processes results in several other complications such as cast microstructure, oxidation, development of unwanted intermetallics etc. On the other hand, several viable and novel alternatives to conventional techniques were developed in surface engineering such as friction stir processing (FSP) and friction surfacing (FS) [4].

Friction surfacing (FS) was first patented in 1941 and has now developed into a well-established machine tool-based technology that utilizes a single consumable tool that can produce a large variety of coating combinations over a substrate. Among surface engineering technologies, the FS process is becoming increasingly important that works on friction manufacturing principles. The expansion of those concepts resulted in one of the most substantial developments in the history of surface engineering [5]. A few combinations of dissimilar materials were processed in friction surfacing along with similar alloys with different combinations of process parameters and the results were promising [6- 8]. Wide range of materials such as steels, aluminium alloys, copper, inconel, sterlite etc have been used to develop surface coatings by FS as reported in the literature [9 - 12]. Friction surfacing of alloy steels on low carbon steel is the subject of majority of the reported research. Various commercial applications for friction-surfaced steel coatings are currently being considered. Additionally, the practicality of generating coatings by friction surfacing on various nonferrous substrates using a range of consumable rods is of significant interest due to the inadequate response of some nonferrous metals to greater loads and temperatures. In this present review, state of the art in developing several non-ferrous based coatings such as aluminium, magnesium, copper, and Inconel 800 by friction surfacing is discussed. These combinations hold promise for making bimetallic strips for a variety of industrial applications. The role of process parameters and material properties influenced by the coatings are summarized and presented

2. Friction surfacing

Friction surfacing (FS) is a solid-state deposition technique that utilizes the plastic deformation of a consumable rod. As illustrated in Fig.1, an axial load is applied to a revolving rod that is forced on the substrate. At the rod tip, frictional heat creates a viscoplastic boundary layer. Pressure and temperature conditions result in an interdiffusion process between the plasticized material and the base metal, forming a metallic connection [13]. As a result, the heat conducted into the base metal and the layer consolidates along with the contact, forming the connection between viscoplastic shearing and the revolving consumable rod and the deposited layer. The viscoplastic shearing contact slides away from the substrate surface as heat conduction increases the layer thickness. The viscoplastic material is deposited continuously onto the substrate surface by a transverse action. Notably, FS heats materials primarily by interfacial friction and plastic deformation, enabling processing at temperatures below the fusion. Due to the thermo-mechanical process, a continuous layer of fine-grained microstructures is deposited as the rod is consumed [14]. Additionally, the process is recognized by the formation of a revolving flash of material at the rod tip, which imparts a mushroom-shaped geometry to the rod as shown in Fig 1[15].

FS generates and processes a viscoplasticized solid-state region into a new layer on the surface. While this area remains solid, it exhibits a three-dimensional substantial flow pattern similar to a liquid, allowing for simple blending of various metals [16]. The concept of the 'third-body region' is usually used to represent this phenomenon. Mechanically, this third-body zone contains relatively lower flow stress and temperatures that are higher than the material recrystallization temperature but lower than its melting point [5]. Heat is generated by friction at the contacting surfaces and internally throughout the material flow in this solid-state process, which is purely governed by mechanical energy input

[17]. Since frictional heat dissipates to zero as the material approaches to fusion temperature, the maximum temperature reached inside the treated zone is physically restricted by the fusion temperature. As a result, all deformation is limited to the solid-state condition.

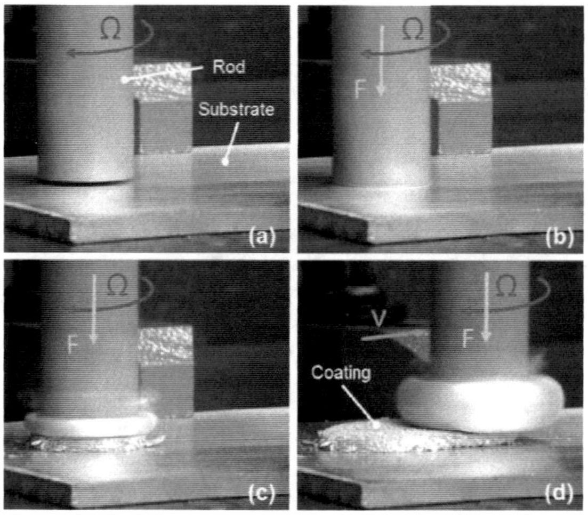

Fig. 1. Friction surfacing Process (a) Rotation start, (b) initial contact, (c) initial deformation stage and (d) deposition stage [Source: Gandra et al., [15])

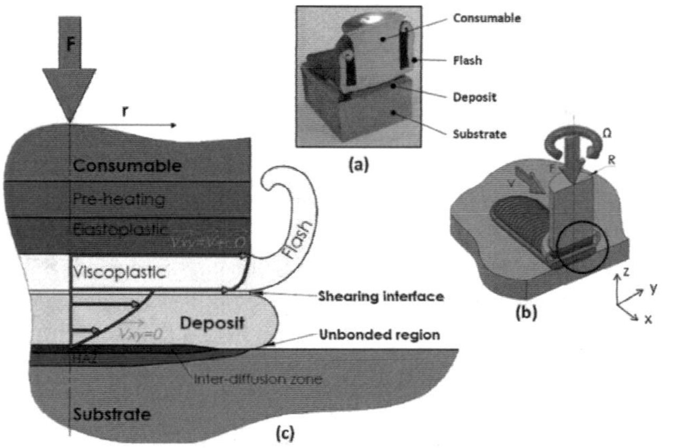

Fig. 2. Thermo-mechanics of friction surfacing. (a) Sectioned consumable, (b) process parameters and (c) thermo-mechanical transformations and speed profile (Source: P. Vilaça [5])

Depending on the metallurgical changes outlined, Fig 2 displays a possible model for the thermal and mechanical processes that are involved in FS [5]. The speed difference between the rotating viscoplastic material with the rod and the material that is successfully adhered to the substrate allows the deposit to detach from the consumable rod. During the operation, the primary source of heat is the viscous shearing friction between the deposition and the consumable [18]. Due to the lack of restraint at the lower end, the highly plasticized material flows outside the consumable rod diameter, producing in a rotating flash connected to the consumable rod end and poorly adhered coating edges. Additionally, unbounded patches correlate with greater rotation speeds at the consumable rod outer radius. At the deposit's relative speed, the substrate breaks the bonding contact and disrupts the ongoing diffusion bonding process. Nonetheless, flash and unbounded zones contribute significantly to the joining process by serving as temperature and pressure border conditions [4, 19]. The flash generated during the transitory starting phase of the FS process will expand and harden when the viscoplastic material transitions to the solid-state. This increases the necessary forging pressure into the viscoplastic layer, effectively closing the rod's viscoplastic zone. From then on, the flash will only raise because of excessive mechanical energy (through forging force and/or torque) imparted to the rod during the stationary phase of the operation. Thus, the flash formation rate along the process can be used to monitor the correct parameter selection.

3. Process parameters

Coatings are rated for thickness, appropriate bond width/extension, and bond strength for a particular consumable rod and substrate material, all of which are dependent on controlled process parameters [5]. The important process parameters, which influence the success of developing surface coatings by FS are given below.

i) Axial Force
ii) Rotational speed
iii) Travel speed
iv) Consumable rod diameter.

The bonding interface is subjected to shear by the relative speed of the deposit layer and the substrate and affects the bonding process during FS. As a result, excessive rotation and travel speeds can damage the joining by reducing the cross section's bonded width [14, 20]. It is reasonable to assume that increasing the rotational speed results in increased friction, which always has a beneficial effect on bonding efficiency. However, at lower rotational speeds, the slower relative speed of the coated material to the substrate can occasionally promote an effective diffusion process and enhance the contact area between the deposit and the rod [1, 4]. Process parameters have a complex and nonlinear impact on the end results, as they are highly dependent on the used material combination. However, the literature has established several general tendencies and effects [4, 20].

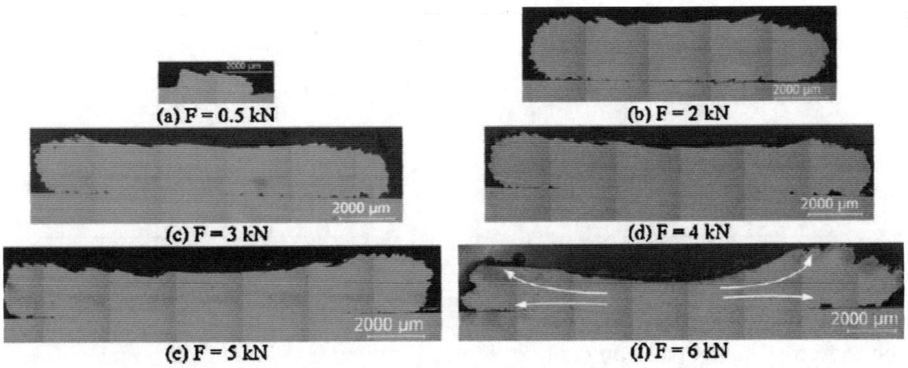

Fig. 3. Effect of depth of HAZ due to axial force [Source: Gandra et al., [4]]

Axial force promotes bond extension, resulting in broader and thin deposits [21, 22]. However, severe loads lead to an irregular deposition in the center of the material discharged from the consumable rod's tail zone. Inadequate forging forces result in suboptimal interface consolidation [4]. The increased axial force is also shown to result in deeper heat-affected zones in the substrate. The adherence of the deposit increases proportionately, resulting in a decrease in the region of the cold lap. On the other side, increasing the applied force decreases the deposit thickness, thus increasing their width [20, 21]. As illustrated in Figure 3, the increase in forging force decreases the deposit layer thickness.

Rotational speed has an effect on the adherence quality, coating width, and roughness as illustrated in Figure 4. Good bond quality is obtained with slow to moderate speeds, but flatter and regular deposits are produced at higher speeds with significant forging action. Coating width often reduces as rotation speed increases. Heat distribution between the substrate and consumable rod is controlled by both rotation speed and HAZ depth. At the same time, faster rotations were resulting in a deeper HAZ. This occurs because the contact area is reduced as the consumable rod rotates at high speed and the sub-strata advance speed increases [22, 23].

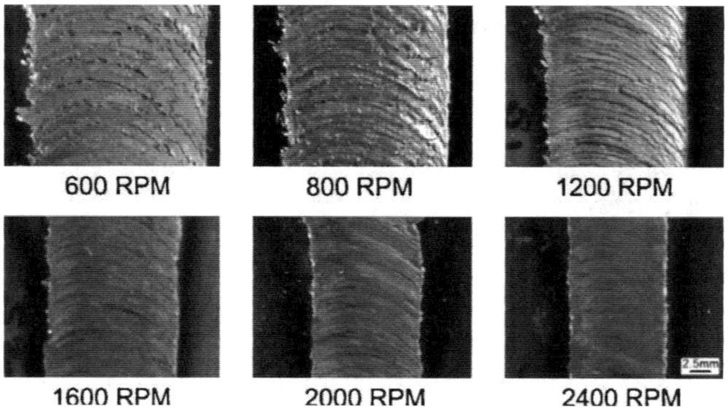

Fig. 4. Tool steel (H13) coatings on low carbon steel: Change of width and coating surface roughness due to rotating speed [Source: Rafi et al., [22]

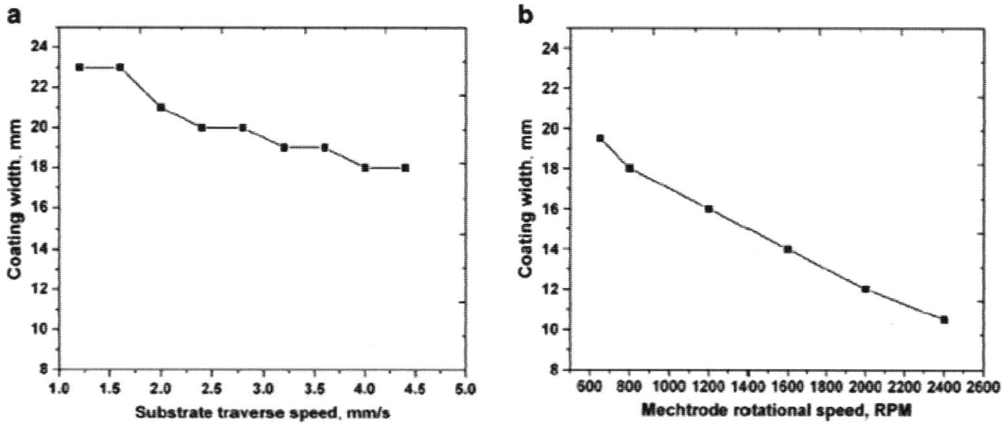

Fig. 5. Tool steel (H13) coatings on low carbon steel: a) The effect of travel speed on coating width and b) mechtrode rotational speed on coating width. [Source: Rafi et al., [22]

Travel speed has a significant effect on coating width and thickness, as it regulates material deposition rate. As a result, faster travel speeds produce thinner deposits. The heat exposure is short because of the increased travel speeds, resulting in less grain formation and fine microstructures [23]. The HAZ on the substrate diminishes with increasing travel speed. Increased travel speed has the potential to cause bonding at the edges of the deposition to degrade [21, 22]. In general, the required coating thickness is to maximize bonded width (with minimal deviations between the deposited and bonded widths) and decrease the under fill. Fig 5 demonstrates the decreased coating width by increasing the mechtrode travel speed and rotational speed while depositing tool steel (H13) coatings on low carbon steel [22]. Consumable rod diameter can be said to determine the deposit width; that is, larger the diameter of the rod, the larger the deposit. Consequently, when the diameter of this rod is larger, the deposit equipment potential will be greater. As a result, the equipment's potential is depending on consumable rod diameter [5].

4. Material systems

Several steels, aluminium, copper and their alloys are the important material systems used in friction surfacing in addition to some high entropy alloys. The coating phenomina, level of bonding between the surface coating and the substrate, coated layer characteristics are completely different for different combination of materials. For example, compared to steels, because of their fast softening, copper cannot sustain a high-temperature range. Furthermore, the parameter range is too limited narrowed for achieving better coatings of steel over a copper substrate [22]. Therefore, the balance between mechanical and thermal stability should be established. Prasad Rao et al., [18] conducted extensive study on the feasibility of different combinations of matrials as compared in Table 1.

Table 1 Feasibility of surface coatings in FS as demonstrated by Prasada Rao et al., [23]

Substrate	**Consumable rod**			
	Low carbon steel	**Copper**	**Aluminium Alloy(AA 6063)**	**Commercially Pure Titanium**
Commercially Pure Aluminium	Not possible due to drilling effect of steel consumable rod, but can be done by using startup plate.	Not Possible; Heat generation not sufficient to deform copper, hence copper rod deformed Al	Possible; Same Properties, Sufficient heat concentration at the contact surface	Not Possible Both the materials are softer
Commercially Pure Copper	Possible Higher thermal conductivity of copper cause heat concentration on steel side	Intermittent Coating; Limited heat concentration at the contact surfaces	Very thin Coating; Lack of heat concentration at the interface	Not Possible; Large difference in deformation Characteristics
Inconel 800	Possible; Comparable mechanical and thermal properties	Not Possible; Softens very fast due to its high ductility at elevated temperatures	Not Possible; Aluminium softened very fast due to high thermal conductivity	No continuous coating; It is highly instable at elevated temperature
Mg Alloy [ZM21]	Not Possible; Steel rod ploughed through Mg surface	Not Possible; Friction not sufficient to cause heating	Not Possible; Large difference in deformation behaviour	Not Possible Titanium rod skids over Mg surface
Ti-6Al-4V	Not Possible; Insufficient friction and oxide formation	Not Possible; Insufficient friction and oxide formation	Very thin coating; Forms more flash than coating due to faster softening	Intermittent Coating; Instability of Ti at elevated temperature prevents from continuous coating.

Friction surface coating o flow carbon steel over a copper substrate showed that the coated layer is uniform in thickness and free of gaps. At first, part of distortion is visible in a substrate, but this subsides as the coated layer progresses. The process of coating is separated into two stages: the immobile dwell period and the deposition phase, during which the substrate moves. During the dwell phase, friction between the consumable rod and the substrate generates heat. In the beginning, the frictional heat produced results in a localised softening of the copper substrate, as indicated by material displacement on either side. However, heat concentration on the interface increases. The steel rod preferentially heated due to dimensional difference and thermal conductivity between the substrate and the steel rod. This severely plasticizes the consumable rod's rubbing surface and the viscoplastic substance is transferred it into the copper substrate to produce a layer of coating. It is resulting in simply a difference in the coating width or thickness. The substrate travels at a steady speed during the deposition phase. Consequently, no direct contact at this stage between the steel rod and the copper substrate. Rather than that, the previously applied coating is in touch with the deformed steel rod. As a result, a regular coating is formed on the copper substrate with minimum distortion during the deposition phase. The coating-substrate interface is found to be fault-free and continuous. A physical interface of steel and copper suggests that the substrate underwent micro deformation during the deposition phase. Microhardness measurements across the coating-substrate interface reveal a rise in coating hardness relative to the substrate material. This occurs as a result of grain refinement of grains and is caused by dynamic recrystallization [24]

The high thermal strength of Inconel 800 makes it a good option as substrate and consumable for friction surfacing applications [25]. Due to the high thermal stability of both Inconel 800 and steel, it has been proven that coating is feasible. The interface micrograph of friction-surfaced layer of low-carbon steel on an Inconel substrate revealed a strong connection between the steel coating and the Inconel 800 substrate. There was no evidence of intermixing of the coated layer and the substrate. Due to its resilience to deformation, the Inconel 800 substrate exhibited no deformation in comparison with copper at high temperatures. When the thermomechanical properties of aluminium and magnesium were considered, the surfaces on both copper and Inconel 800 substrates could support friction surfacing. However, the surfaces on aluminium and magnesium substrates did not react well to the surfacing. All other consumable rods have left a cavity in the substrate in commercially pure aluminium except AA 6063. The rubbing surface of the consumable rod was undamaged because of low heat generation, indicating that the generated friction heat was minimum. AA6063 consumable rods were coated. The reason is that the materials have similar characteristics and a relatively high coefficient of friction between aluminium. Similarly, no coating was achieved on the magnesium substrate using any of the consumable rods [23].

A steel start-up plate with the same thickness as the aluminium substrate was secured close to the aluminium substrate, resulting in the plates forming a continuous surface. Consumable rod was initially in connection with the start-up plate and rotated until an adequate quantity of plasticized material was produced (dwell stage). After that, the table was permitted to move against the rotating consumable rod. When the steel rod reaches the aluminium substrate, the process becomes steady by developing a satisfactorily thick plasticized layer (deposition stage). When a thick layer of material begins to form, contact is established between the rod and the previously deposited coating. As a result, there is no direct contact between the rod and the substrate when the consumable rod reaches the aluminium substrate. The coating is created in this case by the simultaneous action of both forwarding motion and spreading action of the plasticized material due to relative movement between the consumable rod and the substrate and rotational movement of the consumable rod, respectively. Consolidation of the coating happens as the result of the continual action of axial force. A regular bead with no discontinuities was obtained.

FS can also be performed under water or any fluid to produce surface layers on immersed structures [26]. If the mechtrode is made of composite materials, FS can develop a composite surface layer on the substrates [27]. FS Fundamentals of FS have led to develop friction assisted additive manufacturing processes where the objective is to produce three-dimensional components by depositing number of overlapping layers [28, 29]. Several properties particularly at the surface were enhanced by producing surface layers through FS, which include mechanical, wear, corrosion etc [30 - 33]. It can be summarized that FS offers several advantages to develop solid-state coatings of similar and dissimilar materials with superior surface properties.

5. Concluding remarks

Among the available friction based processing routes, FS is a promising technique to develop solid state surface coating layers with enhanced mechanical, tribological and surface properties. FS has several advantages over conventional coating techniques. FS is environmentally friendly and reduces metallurgical flaws in the produced coating. FS eliminates or reduces the ill effects of intrinsic fusion-based procedures and limits HAZ. FS can be used to develop surface composite layers by using a composite rod as mechtrode. However, the technique has some drawbacks, including a high equipment cost, thin layers (0.5–6mm at most), and the removal of deposit borders that do not attach to the substrate (cold lap). On the other hand, developing coatings on the surface with complex geometries is difficult with FS. The reported literature also suggests the need of selecting feasible combination of materials to successfully develop surface coatings by FS. The process is new and still in its infant stage compared with several well-established surface-coating techniques. However, the solid-state nature and the superior properties exhibited by the surface coatings developed by FS make this process as a potential tool in near future for surface engineering applications.

References

1) M. L. K. Macedo, G. A. Pinheiro, J. F. Santos and T. R. Strohaecker, "Deposit by friction surfacing and its applications," Welding International, vol. 24, pp. 422-431, 6 2010.
2) M. S. Węglowski, "Friction stir processing – State of the art," Archives of Civil and Mechanical Engineering, vol. 18, pp. 114-129, 1 2018.
3) L. Quintino, "Overview of coating technologies," in Surface Modification by Solid State Processing, Elsevier, 2014, pp. 1-24
4) J. Gandra, R. M. Miranda and P. Vilaça, "Performance analysis of friction surfacing," Journal of Materials Processing Technology, vol. 212, pp. 1676-1686, 8 2012.
5) P. Vilaça, "Friction surfacing," in Surface Modification by Solid State Processing, Elsevier, 2014, pp. 25-72.
6) I. Voutchkov, B. Jaworski, V. I. Vitanov and G. M. Bedford, "An integrated approach to friction surfacing process optimisation," Surface and Coatings Technology, vol. 141, pp. 26-33, 6 2001.
7) V. I. Vitanov and I. I. Voutchkov, "Process parameters selection for friction surfacing applications using intelligent decision support," Journal of Materials Processing Technology, vol. 159, pp. 27-32, 1 2005.
8) U. Suhuddin, S. Mironov, H. Krohn, M. Beyer and J. F. D. Santos, "Microstructural Evolution During Friction Surfacing of Dissimilar Aluminum Alloys," Metallurgical and Materials Transactions A, vol. 43, pp. 5224-5231, 8 2012.
9) G. M. Reddy, K. S. Prasad, K. S. Rao and T. Mohandas, "Friction surfacing of titanium alloy with aluminium metal matrix composite," Surface Engineering, vol. 27, pp. 92-98, 3 2011.
10) K. P. Rao, R. Damodaram, H. K. Rafi, G. D. J. Ram, G. M. Reddy and R. Nagalakshmi, "Friction surfaced Stellite coatings," Materials Characterization, vol. 70, pp. 111-116, 8 2012.

11) R. Puli and G. D. J. Ram, "Microstructures and properties of friction surfaced coatings in AISI 440C martensitic stainless steel," Surface and Coatings Technology, vol. 207, pp. 310-318, 8 2012.
12) R. Puli and G. D. J. Ram, "Corrosion performance of AISI 316L friction surfaced coatings," Corrosion Science, vol. 62, pp. 95-103, 9 2012.
13) X. M. Liu, Z. D. Zou, Y. H. Zhang, S. Y. Qu and X. H. Wang, "Transferring mechanism of the coating rod in friction surfacing," Surface and Coatings Technology, vol. 202, pp. 1889-1894, 2 2008.
14) Z. Y. Ma, "Friction Stir Processing Technology: A Review," Metallurgical and Materials Transactions A, vol. 39, pp. 642-658, 2 2008.
15) J. Gandra, H. Krohn, R. M. Miranda, P. Vilaça, L. Quintino and J. F. Santos, "Friction surfacing—A review," Journal of Materials Processing Technology, vol. 214, pp. 1062-1093, 5 2014.
16) V. I. Vitanov, I. I. Voutchkov and G. M. Bedford, "Decision support system to optimise the Frictec (friction surfacing) process," Journal of Materials Processing Technology, vol. 107, pp. 236-242, 11 2000.
17) D. Govardhan, A. C. S. Kumar, K. G. K. Murti and G. M. Reddy, "Characterization of austenitic stainless steel friction surfaced deposit over low carbon steel," Materials & Design (1980-2015), vol. 36, pp. 206-214, 4 2012.
18) G. M. Bedford, V. I. Vitanov and I. I. Voutchkov, "On the thermo-mechanical events during friction surfacing of high speed steels," Surface and Coatings Technology, vol. 141, pp. 34-39, 6 2001.
19) T. Jujare, A. Kumar, S. V. Kailas and K. U. Bhat, "Friction Surfacing of Mild Steel by Copper: A Feasibility Study," Procedia Materials Science, vol. 5, pp. 1300-1307, 2014.
20) T. Shinoda, J. Q. Li, Y. Katoh and T. Yashiro, "Effect of process parameters during friction coating on properties of non-dilution coating layers," Surface Engineering, vol. 14, pp. 211-216, 1 1998.
21) H. Sakihama, H. Tokisue and K. Katoh, "Mechanical Properties of Friction Surfaced 5052 Aluminum Alloy," MATERIALS TRANSACTIONS, vol. 44, pp. 2688-2694, 2003.
22) H. K. Rafi, G. D. J. Ram, G. Phanikumar and K. P. Rao, "Friction surfaced tool steel (H13) coatings on low carbon steel: A study on the effects of process parameters on coating characteristics and integrity," Surface and Coatings Technology, vol. 205, pp. 232-242, 9 2010.
23) K. P. Rao, A. Sankar, H. K. Rafi, G. D. J. Ram and G. M. Reddy, "Friction surfacing on nonferrous substrates: a feasibility study," The International Journal of Advanced Manufacturing Technology, vol. 65, pp. 755-762, 5 2012.
24) H. K. Rafi, G. D. J. Ram, G. Phanikumar and K. P. Rao, "Microstructural evolution during friction surfacing of tool steel H13," Materials & Design, vol. 32, pp. 82-87, 1 2011.
25) M. Chandrasekaran, A. W. Batchelor and S. Jana, "Study of the interfacial phenomena during friction surfacing of mild steel with tool steel and inconel," Journal of Materials Science, vol. 33, pp. 2709-2717, 5 1998.
26) J. Q. Li and T. Shinoda, "Underwater friction surfacing," Surface Engineering, vol. 16, pp. 31-35, 2 2000.
27) D. Nakama, K. Katoh and H. Tokisue, "Fabrication of 6061 aluminum alloy/Al2O3 particle composites by friction surfacing," Journal of Japan Institute of Light Metals, vol. 58, pp. 299-304, 2008.
28) J. J. S. Dilip, S. Babu, S. V. Rajan, K. H. Rafi, G. D. J. Ram and B. E. Stucker, "Use of Friction Surfacing for Additive Manufacturing," Materials and Manufacturing Processes, vol. 28, pp. 189-194, 2 2013.
29) A. W. Batchelor, S. Jana, C. P. Koh and C. S. Tan, "The effect of metal type and multi-layering on friction surfacing," Journal of Materials Processing Technology, vol. 57, pp. 172-181, 2 1996.

30) J. Gandra, P. Vigarinho, D. Pereira, R. M. Miranda, A. Velhinho and P. Vilaça, "Wear characterization of functionally graded Al–SiC composite coatings produced by Friction Surfacing," Materials & Design (1980-2015), vol. 52, pp. 373-383, 12 2013.
31) J. Gandra, D. Pereira, R. M. Miranda, R. J. C. Silva and P. Vilaça, "Deposition of AA6082-T6 over AA2024-T3 by friction surfacing - Mechanical and wear characterization," Surface and Coatings Technology, vol. 223, pp. 32-40, 5 2013.
32) K. Fukakusa, "On the characteristics of the rotational contact plane - a fundamental study of friction surfacing," Welding International, vol. 10, pp. 524-529, 1 1996.
33) M. Chandrasekaran, A. W. Batchelor and S. Jana, "Friction surfacing of metal coatings on steel and aluinum substrate," Journal of Materials Processing Technology, vol. 72, pp. 446-452, 12 1997.

An Experimental Study on Self-compacting Concrete by Packing Density Method

A Nagaraju [1], **V.B. Reddy Suda** [2]
[1] Civil Engineering Department, CMR Technical Campus, Medchal 501401, Hyderabad, India.
[2] Civil Engineering Department, CMR Technical Campus, Medchal 501401, Hyderabad, India.

* Corresponding authors e-mail: [1]nagrajuavari@gmail.com, [2]vbsgen06@gmail.com

Abstract In the 21st century concrete had reached a big mile stone, invention of self compacting concrete (SCC), compaction of concrete was done by its own weight. SCC is a heterogeneous material; hence some problems are there to obtain the proportion of SCC. Some researchers have developed the design procedures to get the proportions of SCC, however very few studies found in establishment of relationship between water binder ratio (w/b) and compressive strength of concrete.Hence, more research has to be done on relationship between w/b ratio and compressive strength with different combination of materials with help of available mix design. From the all available design methods of SCC, Packing density method will give aggregate proportions appropriately and paste content also determined appropriately.In this study, experimental workwas carried on to develop a relationship between the compressive strength of concrete to w/b ratio using Packing density method. In First stage, packing density of aggregate was conducted for mixture of 20mm of aggregate with 10mm aggregate, River sand and Crushed sand individually. After that, packing density was determined for mixture 20mm aggregate, 10mm aggregate and crushed sand. Mix design was carried out based on packing density Method with w/b ratios 0.38, 0.40, 0.42. Fresh Properties of Self-compacting concrete was determined by Slum flow, L-box, V-funnel with different dosage of Super Plasticizer (SP) and compressive Strength was determined. Effect of w/b ratio and S.P dosage on compressive strength was studied through Graphs. Finally a relationship between the w/b ratios to Compressive strength of concrete was developed through linear Regression analysis in MS Excel sheet.

Keywords: Packing density; fresh properties; compressive strength, linear regression

1. Introduction

In concrete industry, Self-compacting concrete (SCC) is one of innovative invention. Self-compacting concrete means which flow without any external vibration and satisfying properties like flow ability, passing ability, filling ability and segregation to resistance. In recent times, the usage of SCC has been increasing rapidly and enhances the construction productivity. The fundamental principle to produce SCC is lower the coarse aggregate content, increase the powder content. However the problems due to heat of hydration will be reduced by incorporation of mineral admixture [1,2]. Apart from all these, SCC is a sustainable concrete [3].

The packing density is the ratio of volume of solids to the total bulk volume. This method is primarily fixing the optimum proportion of aggregates, and then paste quantity was determined to fill the voids. This method was introduced by the "Feet", in 1892[4-8]. The void content will plays a crucial role while determining strength of concrete. More voids will leads to less strength & poor quality of concrete. Therefore by good packing of aggregates will decrease the voids and increase the strength [9].

Even though many researchers are making SCC with different design methods, still investigations are going on to develop a correlationamong different parameters of concrete for different combination of materials was incorporated. In the 1990's Okamura and Ozawa have proposed a mix proportioning system for SCC by keeping the coarse aggregate and fine aggregate contents were constant and altering the w/p ratio and super plasticizer dosage. Later, JRMCA was proposed a "Standardized mix design method" of SCC to overcome the omkura's mix design. In 2001, Nan su proposed and a simple mix design for SCC [10, 11]. However, Particle Packing theory is also one of the most essential concepts, most of researchers focused on design the mix proportions of SCC. In packing density method, quantity of aggregate will increase and cement content will reduce without altering the strength, when compared to conventional method. M sriramachand et al. was suggested Compressible packing model was the best model to optimize the aggregate packing density for SCC[12].

Objectives of research:

- To determine the optimum quantity of Coarse aggregate and fine aggregate by packing density method
- To determine the dosage of super plasticizer for Self-compacting concrete with (water/binder) w/b ratios 0.38, 0.40 and 0.42
- To determine the compressive strength of SCC for w/b ratios 0.38, 0.40 and 0.42
- To determine the quality of concrete for w/b ratios 0.38, 0.40 and 0.42 by UPV
- To develop a relationship between the w/b ratios verses compressive strength of concrete

2. Materials Used

In this study, Ordinary Portland cement, river sand, crushed sand, 20mm down aggregate, 10mm down aggregate, super plasticizer Adva 960 was utilized and their physical properties are given in Table 1 and 2.

Table 1: Physical Properties of Cement

Property of cement	Test result
Normal Consistency	32
Fineness	6
Initial Setting time	45 min
Final setting time	600 min
Compressive Strength @ 28-days	53 MPa
Specific Gravity	3.12

Table 2: Properties of aggregates

Aggregate type	Fineness modulus	Specific Gravity
River Sand	2.67	2.45
Crushed Sand	2.54	2.60
20 mm down aggregate	7.5	2.8
10 mm down aggregate	2.66	5.6

- Super Plasticizer: A PCE based Super plasticizer was used for SCC and its Specific Gravity1.108.

- Ground granulated Blast Furnace Slag (GGBS): GGBS was taken from JSW Cements pvt.ltd and Specific gravity was 2.8
- No Viscosity Modifier admixtures (VMA) were used in the SCC preparation.

3. Methodology

In The present the research work is focussed on the influence of w/b ratio on properties of self-compacting concrete at different admixture dosage by using Packing Density Method. In the First Phase, Packing density of aggregate was determined by blending of 20mm coarse aggregate with 10mm coarse aggregate alone, with river sand alone and with crushed sand alone in different proportions. Based on results of above mentioned combinations, further, packing density was determined with the blend of 20mm coarse aggregate, 10mm coarse aggregate and crushed sand. Graphs were drawn between packing density verses percentage of fraction blended and optimum proportions of aggregate were determined. In Second Phase, Mix compositions were carried out by using packing density method. In binder content, 70% of OPC and 30% of GGBS were taken. Three w/b ratios (0.38, 0.40, and 0.42) are taken. All the mix compositions of SCC are given Table 3. The fresh properties of SCC were determined at 0.5%, 0.55%, 0.60% dosage of super plasticizer for 3 w/b ratios by Slump-flow test, V-funnel test, L-box test, J-ring test. The test results were verified with the European Federation for Specialist Construction Chemicals and Concrete Systems guidelines (EFNARC Guidelines).For each w/b ratio, at all S.P dosages, where SCC was achieved considered as the Optimum dosage of SCC for that w/b. Concrete cube specimens (150mmx150mmx150mm) were prepared and compressive strength was measured for 3-days, 7-days,-28days as per IS 14858:2000 in Compression Testing Machine. The Ultra sonic pulse velocity test was also performed to assess the Quality of concrete IS: 13311 (Part-II):1992. Further, Correlation study was done between w/b ratio and compressive strength of concrete by linear regression analysis. The w/b ratio values and compressive strength values were inserted in MS Excel sheets and graph was drawn. From regression analysis, one relationship between w/b ratio and compressive strength of concrete was achieved.

4. Results and Discussion

4.1 Packing Density test:

The packing density of aggregate mixture is defined as the solid volume in a unit total volume. Packing density was calculated based on Narasimha raj et.al [4].

$$\text{Packing density} = \frac{\text{Bulk density x weight of fraction}}{\text{Specific Gravity}}$$

Table 3 gives all the details of packing density of aggregates and Figure 1 shows a graphical representation of Packing Density of aggregates. Table 4 gives details of packing density of mixed aggregates (20mm aggregate sample, 10mm aggregate sample and crushed sand) and in Figure 2 graphical representation is shown.

Table 3: Packing density values of aggregates

% fraction	10mm mm aggregate (Kg/m^3)	river sand (Kg/m^3)	crushed sand (Kg/m^3)
0	523.7	514.6	364.9
10	562.0	634.8	384.5
20	569.7	675.7	430.8
30	576.8	675.5	435.6
40	584.0	690.4	452.3
50	597.3	698.2	470.0
60	582.7	681.1	462.9

70	586.7	684.6	420.9
80	579.2	654.2	411.6
90	573.5	650.7	399.8
100	534.8	608.7	351.3

From bulk density test of 20 mm and 10mm aggregate, maximum bulk density, minimum void content and optimum packing Density was obtained at 50:50 proportions. From bulk density test of 20mm and river sand,maximum bulk density, minimum void content and optimum packing Density was obtained at 50:50 proportions. From bulk density test 20mm aggregate and crushed sand,maximum bulk density, minimum void content and optimum packing Density was obtained at 50:50 proportions.

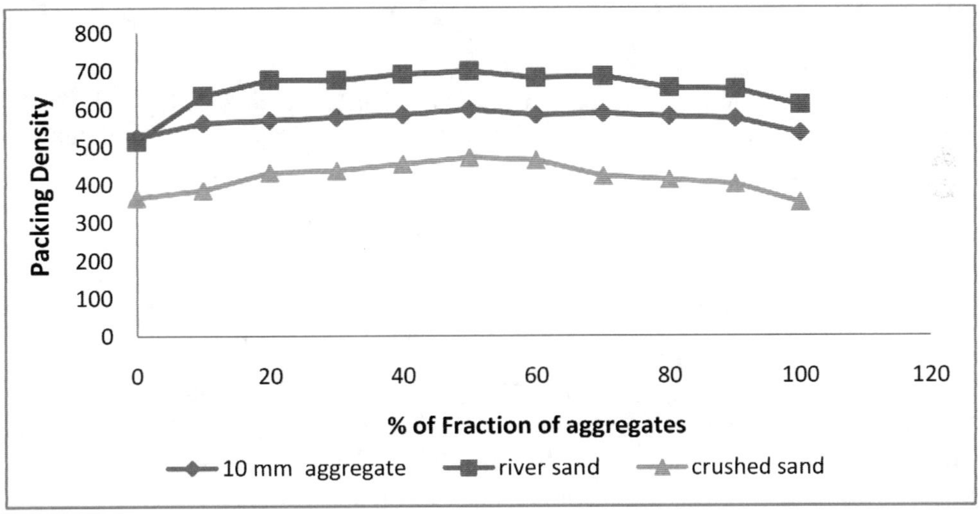

Figure 1: Packing Densities of all aggregates

Further, bulk density test was conducted for mixture of 20mm, 10 mm and with crushed sand. The 20mm and 10 mm aggregate mixture was taken as 50:50 and fraction of crushed sand was replaced from 0 to 100.

Table 4: Bulk density and packing density of mixed aggregates

S. No	proportion of 20 mm aggregate	proportion of 10 mm aggregate	proportion of crushed sand	Weight of both 20 mm + 10mm and Crushed Sand	Bulk Density	Packing Density	void content
1	45	45	10	35.52	1745.33	628.31	371.69
2	40	40	20	37.03	1846.00	669.60	330.40
3	35	35	30	37.58	1882.67	688.05	311.95
4	30	30	40	37.62	1885.33	694.18	305.82
5	25	25	50	37.47	1875.33	695.63	304.37
6	20	20	60	37.4	1870.67	699.02	300.98
7	15	15	70	36.47	1808.67	680.80	319.20
8	10	10	80	36.01	1778.00	674.12	325.88
9	5	5	90	35.6	1750.67	668.54	331.46

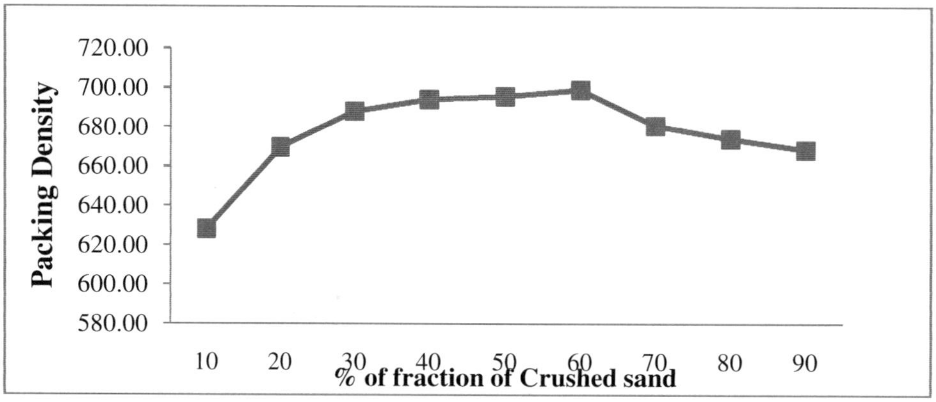

Figure 2: Packing density of mixed aggregates

4.2 Trail Mixes composition details

All trail mixes were designed by packing density method by considering w/b ratios 0.38, 0.40, 0.42at super plasticizer dosage (S.P) of 0.5%, 0.55%, and 0.6%. Details of all trail mixes composition is given in Table 5.

Table 5: Details of trail mixes composition

Mix No	Cement(Kg/m^3)	GGBS (Kg/m^3)	F.A (Kg/m^3)	C.A (20mm) (Kg/m^3)	C.A (10mm) (Kg/m^3)	Water (Kg/m^3)	SP dosage content
TM1	350	150	1030.74	343.58	343.58	190	0.5%
TM2	350	150	1030.74	343.58	343.58	190	0.55%
TM3	350	150	1030.74	343.58	343.58	190	0.6%
TM4	350	150	1030.74	343.58	343.58	200	0.5%
TM5	350	150	1030.74	343.58	343.58	200	0.6%
TM6	350	150	1030.74	343.58	343.58	200	0.6%
TM7	350	150	1030.74	343.58	343.58	210	0.5%
TM8	350	150	1030.74	343.58	343.58	210	0.55%
TM9	350	150	1030.74	343.58	343.58	210	0.6%

4.3 Fresh Properties of Self-compacting Concrete:

The fresh properties of SCC were conducted and verified as per European Federation for Specialist Construction Chemicals and Concrete Systems guidelines. In order to assess the Fresh properties of SCC, the Slump flow test, L-box test, J-ring test and V-funnel test was conducted. The test results data is shown in Table 6. TM1 –TM3 were prepared with w/b=0.38, TM4-TM6 were made with w/b=0.40 and TM7-TM9 were made with w/b=0.38.

Table 6: Fresh concrete properties test results of all mixes

Mix no	w/b ratio	SP dosage content	Slump flow (mm)	V funnel (sec)	L box	J ring (mm)
Range			650mm -800mm	8-12sec	0.8-1	0-10
TM1		0.5%	500	28	0.5	5
TM2	0.38	0.55%	550	13	0.4	3
TM3		0.6%	660	12	0.8	1
TM4	0.40	0.5%	600	46	2	1.5
TM5		0.55%	680	18	1	1

TM6		0.6%	700	13	0.8	1
TM7		0.5%	600	46	2	0.8
TM8	0.42	0.55%	650	26	2	1
TM9		0.6%	750	13	1	0.5

From the Fresh Properties test on Self-compacting concrete, for all three w/b ratios, the SCC properties are satisfied at S.P dosage 0.6%. However, at S.P dosage 0.55% was also given satisfactory results for w/b ratio=0.40 and w/b ratio=0.42. This is due to slighly higher water than w/b ratio = 0.38. The slump flow, V-funnel, L-box, J-ring values was increasing with w/b ratio. All mixes were given a satisfactory results in J-ring test. With 0.5% of S.P dosage, for all 3 w/b ratio mixes shown unsatisfactory resluts. With 0.55% of S.P dosage, only w/b ratio=0.38 mix shown unsatisfactory values in L-box and V-funnel test.As the water content increases, the SCC was formed at lesser S.P dosage. Is observed that, less than the 0.5% of S.P dosage flow characteristis was not form and more than 0.6% of S.P dosage leads to bleeding and segregation.Hence the optimum dosage of S.P to obtain SCC was taken at 0.6%.

4.4 Compressive Strength test

The Compressive Strength test results of 3-days, 7-days and 28-days was shown in Table 7.The strength variations of all 3 w/b ratios were shown seperatly. Figure 3, Figure 4 and Figure 5 will give the compressive strenngth details of w/b ratio =0.38, 0.40 and 0.42 for all S.P dosage.

Table 7: Compressive Strength of SCC

w/b ratio	S.P dosage	3 days (MPa)	7 days (MPa)	28 days (MPa)
0.38	0.5%	18.62	44.815	56.45
	0.55%	24.5	37.00	61.6
	0.6%	24.4	45.38	74.03
0.40	0.5%	21.50	51.46	65.16
	0.55%	22.9	52.91	69.40
	0.6%	23.1	49.45	70.00
0.42	0.5%	20.4	41.24	64.74
	0.55%	19.95	36.8	60.48
	0.6%	21.74	41.95	65.9

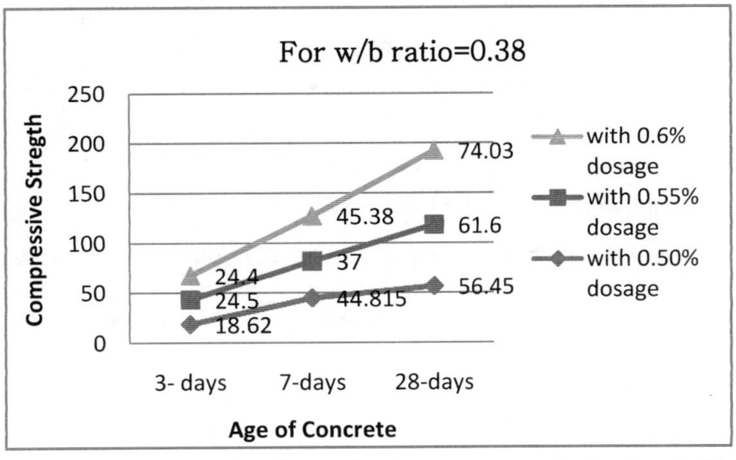

Figure 3: Compressive Strength of concrete graph of w/b = 0.42

From the Figure 3, the strength values of w/b=0.38 were increased as increasing S.P dosage. For 3-days strength was achieved in the range of 18 MPa to 24 MPa, for 7-days strength was achieved in the range of 32.2 M Pa to 45.38 M Pa and for 28 days strength was achieved in range of 56 MPa to 74 MPa. The increasing in percentage of strength of 28-days with 0.6% of S.P dosage was up to 23.7%.

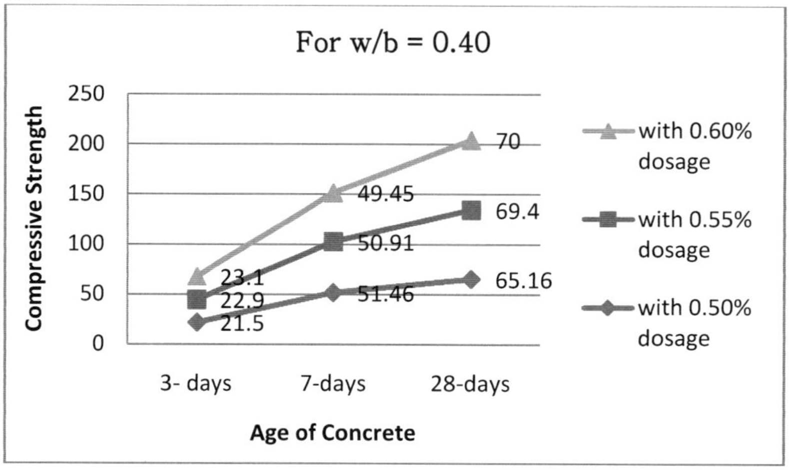

Figure 4: Compressive Strength of concrete graph of w/b = 0.40

From the Figure 4, the strength values of w/b=0.40 were increased marginally. For 3-days strength was achieved in the range of 22 M Pa to 23 MPa, for 7-days strength was achieved in the range of 49-50 MPa and for 28 days strength was achieved in range of 66 MPa to 70 MPa. The increasing in percentage of strength of 28-days with 0.6% of S.P dosage was up to 7.42%.

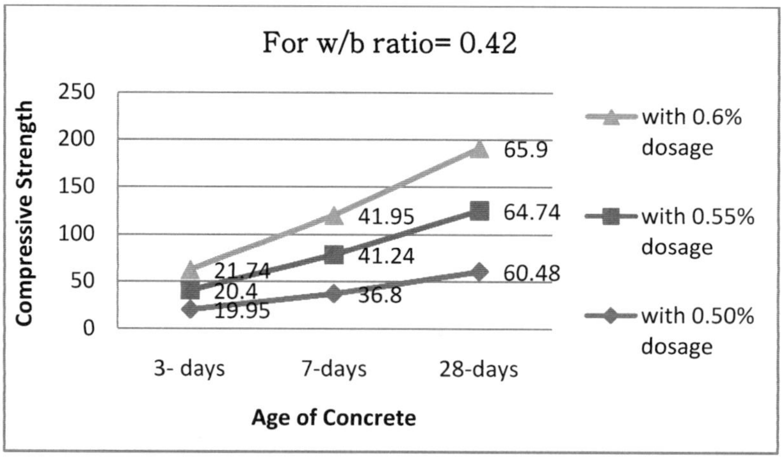

Figure 5: Compressive Strength of concrete graph of w/b = 0.42

From the Figure 5, the strength values of w/b=0.42 increased marginally. For 3-days strength was achieved in the range of 19 M Pa to 21 M Pa, for 7-days strength was achieved in the range of 36-42 MPa and for 28 days strength was achieved in range of 62 M pa to 66 M pa. The increasing in percentage of strength of 28-days with 0.6% of S.P dosage was up to 9%.

The increasing percentage of strength of 28-days with 0.6% of S.P dosage of w/b =0.38 is much higher than the w/b ratio=0.40, 0.42. For w/b ratio = 0.38, at lesser S.P dosages, doesn't exhibit the SCC characteristics, hence internal compaction was not done properly. Consequently, reduction in strength was high. For w/b ratio = 0.40 and 0.42, even at lesser S.P dosages, the flow characteristics were not met required value, still mix shown good characteristics up to some degree. Therefore, the increasing in percentage of strength of 28-days was less. For all w/b ratios, high strength concrete was achieved this is due to high binder (500 Kg/m^3) content.

4.5Ultra sonic Pulse Velocity Test:

Table 8: Ultra Sonic Pulse Velocity of SCC

Trail mix no	w/b ratio	3 days (km/s)	7 days (km/s)	28 days (km/s)
TM1		4.132	4.273	4.232
TM2	0.38	4.35	4.208	4.049
TM3		4.308	4.518	4.686
TM4		4.549	4.658	4.532
TM5	0.40	4.43	4.587	4.587
TM6		4.747	4.744	4.759
TM7		4.532	4.87	4.823
TM8	0.42	4.521	4.658	4.505
TM9		4.565	4.808	4.886

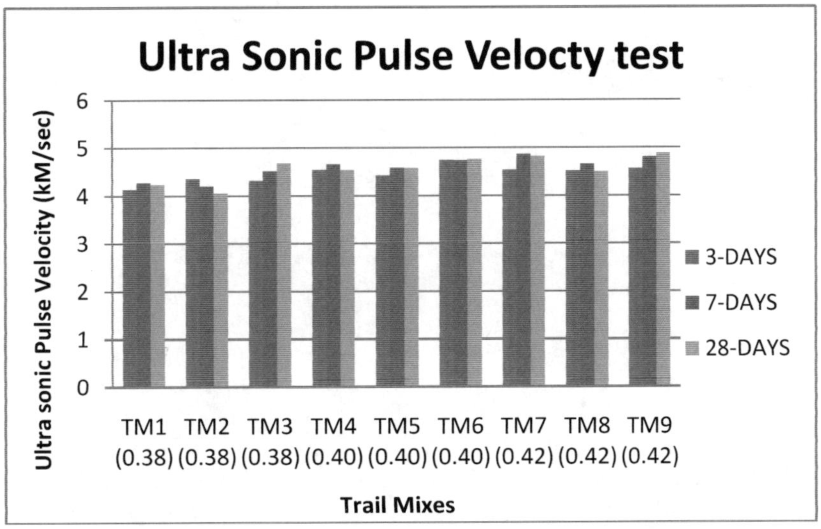

Figure 6: Ultra Sonic Pulse Velocity of Concrete test results

The Table 7 gives the details of Ultra sonic Pulse Velocity (UPV) test for all mixes of 3-days, 7-days, and 28-days. Figure 8 gives bar chat representation of UPV test results of all mixes.From the ultra sonic pulse velocity test results, the pulse velocity of mixes with w/b ratio = 0.38, for 3-days, 7-days, 28-days, it reaches from 4.1 km/s to 4.6 km/s. It is observed that, the samples which satisfy the SCC characteristics were shown good quality of concrete others shown less quality because of not forming Self-compact concrete. Only with S.P dosage 0.6% is given excellent quality of concrete. With w/b ratio = 0.40, all mixes shown above pulse velocity above 4.5 km/s. With w/b ratio = 0.42, all mixes shown above pulse velocity above 4.5 km/s. The quality of concrete samples of w/b ratio= 0.38 was shown less, because of less water content and internal compaction errors.

4.6 Correlation between w/b ratios versus compressive strength of Concrete:

In this correlation Study of w/b ratio versus compressive strength of concrete was developed for the mixes which weresatisfied the fresh properties of SCC, were taken into consideration. In Table 9, w/b ratio and corresponding strength values are shown. In Figure 7, a graph was drawn between w/b ratios and compressive strengths in MS-Excel sheets. By performing regression analysis a relationship was developed.

Table 9: w/b ratios and Compressive strength

w/b ratio	Compressive strength (MPa)
0.38	74
0.4	70
0.42	65

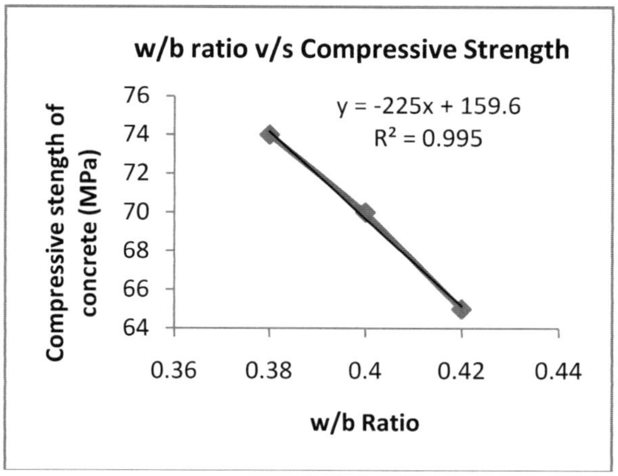

Figure 7: Relationship between w/b ratio v/s Compressive strength

From Figure 7, the relation betweenw/b ratio and compressive strength was linear and equation is $y = -225x + 159.6$. Muhammad Wihardi Tjaronge et.al were also developed an equation between w/c versus Compressive strength, $y = -200x + 135.6$ [14]. Both equations and graphsare having similar profile.

5. Conclusions

1. From the Packing Density method, the final proportion of aggregates arrived with mixture of 20mm aggregate, 10 mm aggregate and crushed sand is 20:20:60 receptively.
2. From experimental investigation, as the w/b ratio increases the flow characteristics was increased and among all w/b ratios, the optimum dosage of S.P was found at 0.6% of weight of cement.
3. From compression test results, it is observed that, mixes of three w/b ratios achieved high strength concrete and as the w/b ratio increases the compressive strength was decreased.

IOP Conf. Series: Materials Science and Engineering 1185 (2021) 012014 doi:10.1088/1757-899X/1185/1/012014

4. For all three w/b ratios, the excellent quality of concrete was shown at 0.6% of S.P dosage.
5. The equation y = -203.2x + 151.2 can be used to determine the compressive strength, For 500 kg/m^3 binder content with 30% of GGBS replacement and with 60:40 Fine aggregate to Coarse aggregate.
6. The research work can be useful in general construction of building where High strength SCC is required.

References

[1] Ma'aruf A, S I Abba and Nuruddeen M M 2017*Int. J. of InnovationTech. and Exploring Eng.***6(8)** 1-7
[2] Ajinkya Raut and Shruti P Shinde 2015*IOSR- J. Mechanical and Civil Eng.***06**27-35
[3] Sudarshan D kore and A K Vyas 2017 *J. Material& Eng.structures* 0**4**171-179
[4] Narasimha Raj, Suresh G Patil and B. Bhattacharjee 2014*IOSR- J. Mechanical and Civil Eng.***11(2)**34-46
[5] Santhosh K, Surendra P, Kotreppa Koppa, Ravikumar Patil, Sharanakumara, Shubham andPundlik Mane 2019*Int. Research J. Eng. & Tech.* **6(2)** 4448-54
[6] P N Nimodiya and H S Patel 2018*Int. J. of Civil Eng. and Tech.* **9(11)** 2126-31
[7] V Mallikirjuna Reddy, M. V. Seshagiri Rao 2014*IOSR- J. Mechanical and Civil Eng.***11(5)**15-21
[8] K Sudarshan and Vyas A K 2016Packing *Int. Conf. on Advances in Concrete Tech. Materials & Construction Practices Proc.*71-77
[9] Padmaja P kumar and P Rama mohan rao 2017*Int. J. of Civil Eng. and Tech.* **8(4)**1156–66
[10] Nan Su, Kung-Chung Hsub and His-Wen Chai 2001 *Cement& Concrete Research***31(12)** 1799-1807
[11] Paratibha aggarwal Rafat siddique Yogesh aggarwaland Surinder M Gupta2008*Leonardo Electronic J. of Practices and Tech.***7(12)** 15-24
[12] M sri rama chand, P Swamy naga ratna, K L Radhika, P Rathish Kumar and C Yedukondalu 2017*Indian Concrete J.***91(4)**20-32
[13] S S Patil, A S Sajane and R.B.Ladage 2017*Int. Research J. of Eng. and Tech.* **04(05)**2113-21
[14] Md Wihardi Tjaronge, A R Djamaluddin, Nocolaus R, Lemba and I Kalesaran 2006 *Media komunikasi teknik sipil***14 (01)**106-111

IOP Conf. Series: Materials Science and Engineering 1185 (2021) 012015 doi:10.1088/1757-899X/1185/1/012015

Influence of bentonite as partial replacement of cement in basalt fiber concrete

Patel Ashish[1] and Ch. Srinivasarao[2].

[1]PG student, Department of Civil Engineering, CMR Technical Campus,Hyderabad, India.
[2]Professor,Department of Civil Engineering, CMR Technical Campus,Hyderabad, India.

* Corresponding author e-mail: [1]ashishpatil5938@gmail.com, [2]chigurupatisrinivas@gmail.com

Abstract. Concrete is commonly accepted construction material in the construction field. Concrete is second in terms of per-capita consumption in the world that is about approximately 500 kilograms. This leads to increase in consumption of Ordinary Portland Cement(OPC). The environment is at risk in terms of damage caused by Carbon dioxide(CO_2) emission during OPC manufacturing process. Hence this is the hour to reduce consumption of OPC by partial replacement. The present study focuses on finding of optimum percentage substitution of OPCwith bentonite. As the concrete is weak in tension, basalt fiber is used in concrete to add the tensile properties to concrete. The investigation is carried out using cube, cylinders, and beams specimens with varying percentages of bentonite and basalt fiber by cement weight. The workability and mechanical properties like slump cone, compression, split tensile, flexure strengthis measuredfor different combinations of fiber and bentonitein conventional and trial mix concrete.Specimens are casted, cured, and tested for 3, 7 and 28 days as per Indian standards. The test results of conventional concrete and trial mix concrete are compared,and it is observed that 10% bentonite replacement and 1.5% basalt fiber replacement given optimum results.From the study benefit of bentonite as partial replacement material to cementis recommended.

Keywords: Bentonite, Basalt fiber, Workability, Mechanical properties.

1. Introduction

Concrete is the man-made construction material used worldwide. Now a day's urbanization is increasing rapidly attributable to movement of persons from rural areas to metropolitan areas for better standard of living. Hence this increases construction of new buildings which in turn increase consumption of concrete.The CO_2 emission increases due to many reasons one of the reasons is manufacturing of OPC. During manufacture of 1 metric ton OPC nearly around 0.9 ton of CO_2 is released in atmosphere. This results in reduction of OPC consumption by finding partial replacement materials.Someof the partial materials of cement namely fly ash, ground granulated blast furnace slag, metakaolin can be used. Organobentonite can be used as partial substitute material for cement [1]. There is a decrease in workability by the increase of bentonite content in concrete mix and recommended that partial use of ground granulated blast furnace slag and bentonite in cement [2]. It is observed that inclusion of basalt fiber in concrete improves split tensile strength, and compression strength of M20 concrete mix and recommended that optimum percentage use of basalt fiber for compression is 0.5% and for split tensile strength is 1.5% by cement weight [3]. Failure pattern of the specimen has more cracks than concrete without fibers than with the basalt fiber concrete and recommended that mechanical properties of concrete increases by inclusion of basalt fiber [4]. Found that basalt fiber addition gives higher compressive and tensile strength and recommended that 2% basalt fiber addition gives optimum results [5]. Basalt fiber is a potential material in construction sector to enhance the properties of concrete, and found it as a useful building material and can be recommended for use in construction [6]. Bentonite as a partial substitute material of OPC, and recommended 5% use of bentonite by weight of cement [7]. Addition of bentonite decrease in early age strength, use of 15% bentonite in cement replacement is recommended [8]. Sodium bentonite and fly ash can be used as partial substitute material for OPC, use of 25% fly ash and 15% bentonite by weight of OPC is recommended [9]. Bacteria enhance the strength of concrete, and recommended that use of bio concrete in general construction [10]. Bentonite and steel slag can be partially replaced in cement and fine aggregate [11]. Use of magnetized water for casting and curing of concrete there is 60% increases in strength [12]. In the present study author has chosen bentonite as partial replacement material to OPC which promotes use of natural product and reduces CO_2 emission.Concrete is brittle in nature, to increase the ductility property of concrete fibers are embedded in concrete to overcome this problem.Some of the fibers used in concrete are polypropylene, steel, asbestos, organic, carbon fibers are used. In the present study basalt fiber is used. Workability test is carried on fresh concrete and mechanical properties are carriedonhardened concrete specimens. Total of six trail mix were prepared and compared with conventional concrete.And it is recommended that 10% replacement of OPC with bentonite and 1.5% use of basalt fiber in concrete.Objectives of the present study are to analysis the fresh properties of concrete consisting bentonite and basalt fiber, to study the mechanical properties like compression strength, flexural strength, split tensile strength of concrete consisting bentonite and basalt fiber, finding optimum percentage replacement of bentonite and basalt fiber in concrete, to promote the usage of natural product as partial replacement of OPC and toreduceCO_2 emissionby use natural products.

2. Materials and Methodology

2.1 Ordinary Portland Cement(OPC):

Ordinary Portland Cement of 53 grade is used in this study as per IS: 269-2015. The physical properties of OPC are shown in Table 1.

Table 1. Physical properties of OPC.

Properties	Values	IS CODE
Consistency	33%	IS: 5513-1996
Initial setting time	48 minutes	IS: 4031 part-2-1998
Final setting time	252 minutes	IS: 4031part-5-1998
Fineness	1%	IS: 4031-part-1-1996
Specific gravity	3.13	IS: 2702-part-3

2.2 Fine aggregate(FA):

Manufactured sand conforming with IS: 383-2016 is used in this study. The properties of manufactured sand are showninTable 2.

IOP Conf. Series: Materials Science and Engineering 1185 (2021) 012015 doi:10.1088/1757-899X/1185/1/012015

Table 2. Physical properties of Fine aggregate.

Properties	Value	IS CODE
Specific gravity	2.34	IS: 2386 part-3-1963
Sieve analysis	Zone II	IS: 383-1970
Water absorption	6.4%	IS: 2386 part-3-1963

2.3 Coarse aggregate(CA):

20 mm size angular shape of coarse aggregate confirming IS:383-2016was used in this study.The physical properties of coarse aggregate are depicted in below Table 3.

Table 3. Physical properties of Coarse aggregate.

Properties	Value	IS CODE
Specific gravity	2.78	IS: 2386 part-3-1963
Water absorption	1.6%	-

2.4 Basalt Fiber(BSF):

In this present studychoppedbasalt fiber is used. The physical properties are shown in Table 4.

Table 4. Physical properties of Basalt fiber (Provided by Vaishnav Composites).

Properties	Value
Density of filament	2.7 kg/dm^3
Melting point	1350 ^{0} C
Diameter	13-20 μm
Length	12 mm
Moisture content	<0.3%
Loss on Ignition	<0.5%

2.5 Bentonite Powder(BN):

Bentonite is a partial replacement material used in this study. The physical properties of bentonite are shown in below Table 5.

Table 5. Physical properties of Bentonite powder (Provided byGantra Minerals and Mines).

Properties	Value
Moisture	12%
pH	8.5
Bulk density	1.2 tons/m^3
Sand content	3%
Colour	Brownish

2.6 Water:

Portable water available in the lab is used for casting and curing in this study.

2.7 Chemical Admixture(SP):

In the present study polycarboxylate ether based superplasticizer are used. The properties of superplasticizer are depicted in the Table6.

IOP Conf. Series: Materials Science and Engineering 1185 (2021) 012015 doi:10.1088/1757-899X/1185/1/012015

Table6. Specification of Superplasticizer (Provided by Fratellanza Chemicals).

Properties	Value
Specific gravity	1.08
pH	6
Colour	Light brown
Chloride content	Nil

Materials used in the study are shown in belowfigures.

Figure 1. Basalt fiber

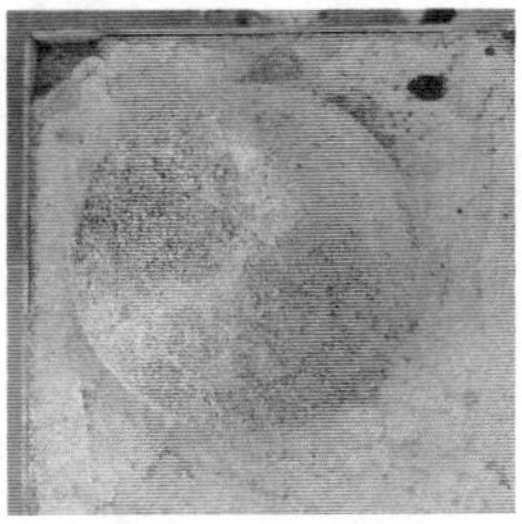

Figure 2. BentoniteFigure

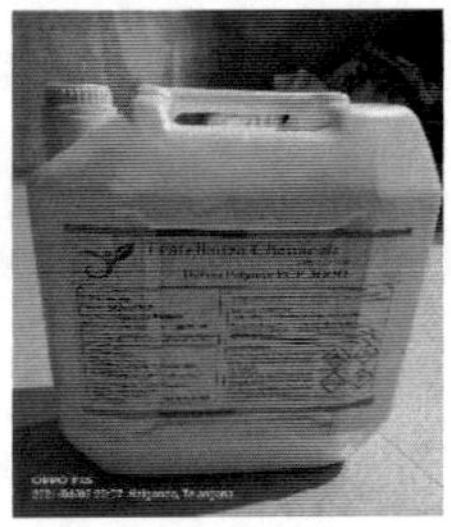

Figure 3. Superplasticizer

2.8 Design mix:

The design mix is carried as per IS: 10262-2019for M40 grade concrete. The materials quantity for 1 m^3 are shown inTable7.

Table7. Material quantity for conventional M40 Grade concrete(Kg/m^3).

Materials	OPC	Fine aggregate	Coarse aggregate	water	Superplasticizer
Weight	421	652	1197	169	2.95

In this study trial mix where prepared in which OPC is partially replaced by bentonite powder at 5%, 10% and 15%. The basalt fiber at 1% and 1.5% are considered. Table 8. explains the percentage replacement in detail.

Table8. Material quantity for trial mix M40 grade concrete (kg/m^3).

Mix Id	Mix Description	OPC	BN	FA	CA	Water	BSF	SP
A1	BN5%+BF1%	399.95	21.05	651.11	1196.84	168.35	4.21	2.95
A2	BN10%+BF1%	378.9	42.1	651.11	1196.84	168.35	4.21	2.95
A3	BN15%+BF1%	357.85	63.15	651.11	1196.84	168.35	4.21	2.95
A4	BN5%+BF1.5%	399.95	21.05	651.11	1196.84	168.35	6.32	2.95
A5	BN10%+BF1.5%	378.9	42.1	651.11	1196.84	168.35	6.32	2.95
A6	BN15%+BF1.5%	357.85	63.15	651.11	1196.84	168.35	6.32	2.95

Mechanical properties of concrete like compression, flexure, split tensile strength of both conventional and trial mix concrete are evaluated by casting cube specimens of size 150mm x 150mm x 150mm, beam specimen of size 500mm x 100mm x 100mm, and cylinder specimen of size 150mm in diameter and 300mm in height are casted and cured in portable water before testing for 3, 7 and for 28 days.

IOP Conf. Series: Materials Science and Engineering 1185 (2021) 012015 doi:10.1088/1757-899X/1185/1/012015

3. Test Results and Discussions

3.1 Workability Test:

Workability of fresh concrete is measured with the help of slump cone test as per IS: 1199-1959. The mix is designed for a targeted slump of 100 mm. It is observed that due to addition bentonite and basalt fiber to concrete workability of fresh concrete is decreasing which is then adjusted by addition of superplasticizer is depicted in below Table9.

Table9. Superplasticizer Dosage for 1 m^3 of Concrete.

Mix Id	Superplasticizer Dosage as per mix (ml)	Superplasticizer Dosage consumed (ml)	% Variation
A1	2731	2917	1.06
A2		3111	1.14
A3		4630	1.69
A4		5787	2.12
A5		6167	2.26
A6		6250	2.29

3.2 Compression Strength Test:

Concrete cube specimens were tested for compression strength as per IS: 516-2018. The concrete specimensare cured for 3,7 and 28 days before testing. Figure2.shows the compressive strength of concrete at 3, 7 and 28 days.

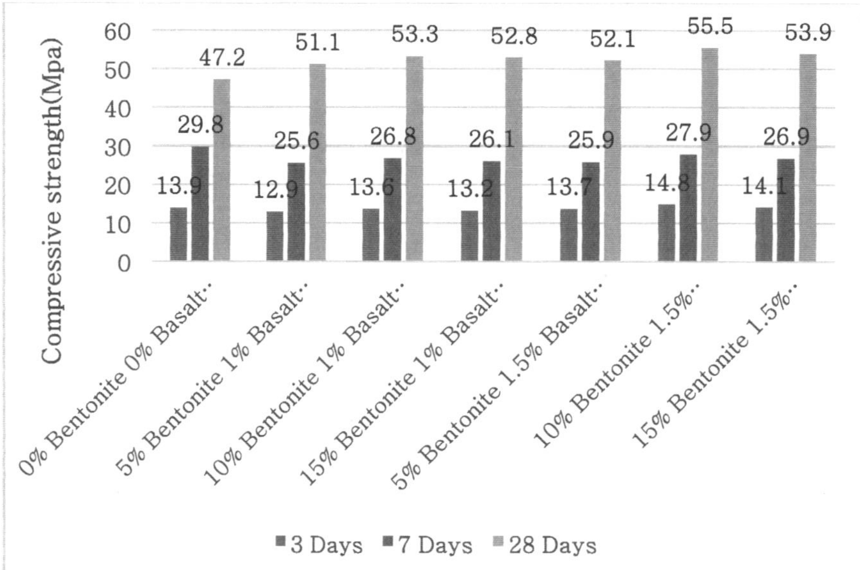

Figure 4. Compressive Strength Test Results.

It is observed from Figure2.that maximum compressive strength is achieved in mix A5 in which OPC is partially replaced by bentonite and basalt fiber at 10% and 1.5% respectively.Strength achieved in Mix A5 because of packing effect (size, shape) of bentonite particles. On the other hand,it is also observed that due to addition of bentonite there is reduction in early age strength while in comparison with conventional concrete.

3.3 Split tensile StrengthTest:

Concrete cylinder specimens were tested for split tensile strength as per IS: 516-2018. The concrete specimensare cured for 3,7 and 28 days before testing. Figure3. shows the split tensile strength of concrete at 3, 7 and 28 days.

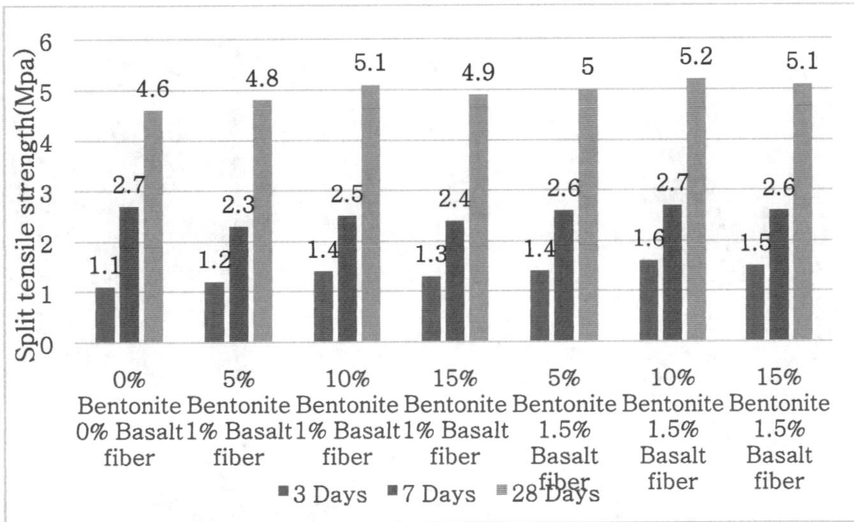

Figure 5. Split Tensile Strength Test Results.

The Figure3. shows the split tensile strength of concrete is optimum in mix A5. The optimum mix A5 shows there is an increase in split tensile strength when fiber is added to concrete while in comparison with conventional concrete. This is because of packing effect (size, shape) of bentonite particles and tensile properties of basalt fiber.

3.4 Flexural Strength Test:

Concrete beam specimens were tested for flexural strength test as per IS:516-2018. The concrete specimensare cured for 3,7 and 28 days before testing. Figure4. shows the split tensile strength of concrete at 3, 7 and 28 days.

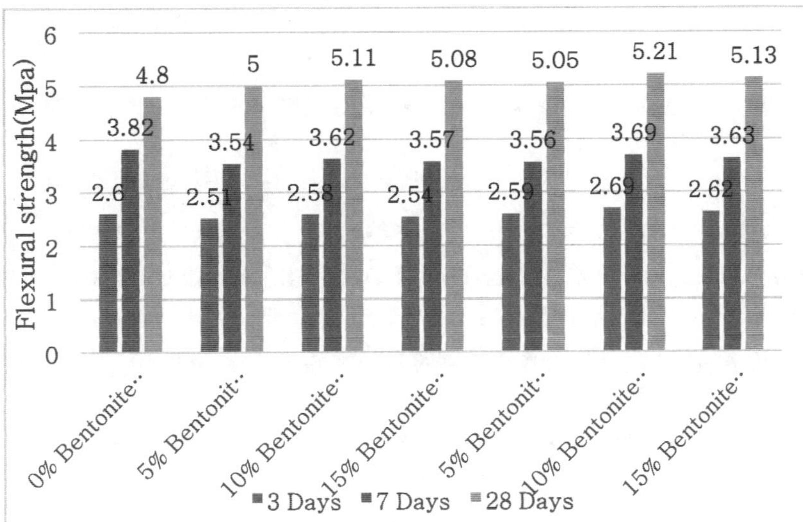

Figure6. Flexural Strength Test Results.

Figure4. shows that mix-A5 containing 10% bentonite and 1.5% basalt fiber showing maximum flexure strength due to addition of basalt fiberwhile in comparison with conventional concrete. This is because of packing effect (size, shape) of bentonite particles and tensile properties of basalt fiber.

Testing facilities of slump, compression, spilt tensile, flexure strength test setupis shown in below figures.

Figure 7. Workability

Figure 8. Compression

Figure 9. Split tensile

Figure 10. Flexure

5. Conclusions:

- Workability of fresh concrete decreases as the quantity of bentonite increases. The targeted slump of 100mm is obtained by adjusting the dosage of superplasticizer.
- Compressive, split tensile and flexural strength of concrete increases due to addition of basalt fiber in concrete.
- The optimum replacement of OPC with bentonite is10% by weight of cement.
- The optimum percentage usage of basalt fiber is 1.5% by weight of cement.
- It is recommended that 10% replacement of OPC with bentonite and 1.5% use of basalt fiber in concrete.
- The study revealed that use of natural product are beneficial in terms of engineering properties enhancement of concrete.
- Use of natural product reduces CO_2 emission.
- In all general constructions such School Buildings, Hospital, College, Shopping Complex, Residential Apartments, and Industrial Buildings.

Acknowledgment:

The experimental study is carried out in the laboratory of Advanced Concrete Technology, CMR Technical Campus, Hyderabad. Would like to thank to Vaishnav Composites and Ganatra Mines & minerals for supplying materials.

References

[1] Lima-Guerra, D.J., Mello, I., Resende, R. and Silva, R., 2014. Use of bentonite and organobentonite as alternatives of partial substitution of cement in concrete manufacturing. *International Journal of Concrete Structures and Materials*, *8*(1), pp.15-26.

[2] Sumitha, Y. and Abraham, R., Experimental Study on Bentonite Clay Powder with Silica Fume and GGBS as Partial Replacement of Cement in M40 Grade Concrete.

[3] Revade, A.B., 2017. Compressive and Split Tensile Strength of Chopped Basalt Fiber Concrete. *Int Res J Eng Technol*, *4*(6), pp.347-351.

[4] Raut, N. and Kawade, U., 2017. Conventional concrete by using basalt fiber. *Int Res J Eng Technol*, *4*, pp.696-699.

[5] Hawaldar, N. and Vishwanath, R.C., 2017. Effect of magnesium sulfate attack on basalt fiber reinforced concrete with partial to full replacement of natural sand by M-sand. *International Research Journal of Engineering and Technology (IRJET)*, *4*(07).

[6] Sim, J. and Park, C., 2005. Characteristics of basalt fiber as a strengthening material for concrete structures. *Composites Part B: Engineering*, *36*(6-7), pp.504-512.

[7] Sathyabama, M., Karthika, T. and Kavidha, R., AN EXPERIMENTAL STUDY ON CONCRETE WITH PARTIAL REPLACEMENT OF CEMENT BY BENTONITE POWDER.

[8] Reddy, G.V.K., Rao, V.R. and Reddy, M.A.K., 2017. Experimental investigation of strength parameters of cement and concrete by partial replacement of cement with Indian calcium bentonite. *Int J Civ Eng Technol*, *8*(1), pp.512-8.

[9] Rooban K R, Niyas Ali A, Shahul AhmeedDKS, Shankar Raj B and Syed A Z 2018 Eco-friendly concrete by partial replacement of cement by sodium bentonite and fly ash *International Research Journal of Engineering and Technology***5**pp 406-8

[10] Arasu N A, VP P, Ramasamy andHarvey D V 2020 A experimental analysis on bio concrete with bentonite as partial replacement of cement*International Research Journal of Engineering and Technology* **7** pp 689-95

[11] Mane D B andJadhav R M 2020 Evaluation of strength of concrete by partial replacement of C.A. by steel slag and cement by bentonite powder *International Research Journal of Engineering and Technology* **7**pp 2127-31

[12] Shamya S M and Shilpa V 2020Experimental study on the use of magnitized water basalt fiber reinforced concrete *International Research Journal of Engineering and Technology* **7** pp2476-82

Experimental Investigation on Ternary Concrete Containing of Granulated Blast furnace slag as a partial replacement to Fine aggregate.

V RohithKumar[1]and V B Reddy Suda[2]

[1]PG student,Department of Civil Engineering, CMR Technical Campus, Hyderabad, India
[2]Prof & Head Department of Civil Engineering, CMR Technical Campus, Hyderabad, India

*Corresponding author email: [1]rohithveldhi@gmail.com, [2]vbsgen06@gmail.com

Abstract. Concrete is a combination of cement, fine aggregate, coarse aggregate, water and admixtures. According to world coal association survey over 4.1 billion tons of Ordinary port land cement (OPC) was used across globally in 2020 and also use of OPC emits CO2 to the atmosphere.In order to overcome the problem, a search for alternative materials is the need of the hour and apart from cement, fine aggregate is also important additive to concrete. Due to the speedyconstruction rate deficiency of materials occurs reduction in natural aggregates causes problems like dredging of sand in large scale which creates environmental imbalance. The solution is utilisation of Fly ash, silica fume, Ground granulated blast furnace slag(GGBS) which are comes under Industrial by-products. The disposal of Industrial by-product should done properly, else it will cause land pollution. In this present study silica fume, fly ashes are replaced in OPC and fine aggregate with Granulated blast furnace slag(GBFS). The experimental work execute in order to determine mechanical properties such as compression, split tensile and flexural strength of concrete with age 3,7 and 28 days at various combinations of cement with fly ash varies 20%,30%,40% and constant 8%of silica fume and fine aggregate replaces with Granulated blast furnace slag of 30%,40% and 50% in M40 grade concrete with water-cement ratio 0.38. As per experimental results conclusions were drawn. As per the experimental results it is recommended that OPC is replaced with 8% of silica fume, 20% of fly ash and fine aggregate replaced with 40% of GBFS achieved optimum strength.

Key words: Concrete, Granulated Blast furnace slag, Industrial by-products, Mechanical properties, partial replacement.

1. Introduction:

Concrete is an established material for the purpose of construction. Utilization of large amount of OPC leads emission of $CO_2$and also extraction of high amount of natural aggregates leads to cause environmental imbalance. To overcome this type of problems need of substituent materials are required like Industrial-by-products. In this present study constant 8% of silica fume, fly ash varies 20%,30%,40% as replacement of OPC and 30%,40%,50% of GBFS in fine aggregate were used for the purpose of finding the mechanical properties of M40 grade concrete. Concrete takes place second the world in terms of using materials after the water. The high rate of manufacturing cement leads to emission of CO_2. According to some estimation by experts manufacturing of 1 ton ofOPC leads to 1 ton of CO_2 to atmosphere. To reduce CO_2 content in atmosphere need to go with the supplementary materials. Fly ash, micro silica,GGBS are some of cementitous materials which can use partial replacement of OPC [1].On the other hand, fine aggregate are important ingredient in concretelike river sand and Manufacturing sand.Utilization of large amount of natural aggregates, problems can be occurred such as soil erosion, effect of aquatic life, CO2 emission and the extracted industrial by-product should disposal properly, and else it may lead to environmental hazards. To overcome this problems the author tried an aim to investigate the optimum percentage of application ofindustrial by-product such asGranulated Blast furnace slag as fine aggregate[2]. Fly ash and GGBS are partially replaced in PSC and PPC along with M-sand replaces in fine aggregate it gives high strength in the range of 15-35% and 30-50% [3]. Steel slag is replaced 0-50% in fine aggregate and finally study concluded that 30% of steel slag gives optimum compressive strength [4]. OPC is replaced with micro silica and GGBSwith water-cement ratio 0.55. The study recommends that 10% micro silica and 30%of GGBS is required to get economical and durable [5].In M30 grade of concrete 25% of steel slag gives higher mechanical properties [6]. 10% of alccofine, 20% of bottom ash replaced in OPC and 30% of slag replaced in fine aggregate gives maximum strength in triple blended concrete [7].In M20 grade of concrete showed higher compressive strength when 40% of slag replaced in fine aggregate and 30% of slag replaced in coarse aggregate [8]. An experiment conducted on triple blended concrete getting higher strength when OPC replaces with 4% of fly ash and 16% of GGBS [9].

The objectives of present work is to dictatethefresh and hardened properties of concrete on ternary concrete consisting of silica fume and fly ash as a partial substitution of OPC, to determine the hardened concrete properties of ternary concrete consisting Granulated Blast furnace slag as a partial replacement of fine aggregate, to determine the optimum percentage utilization of Industrial-by-products in partial substitution of OPC and fine aggregate, to establish the application of industrial-by-products for the sake to reduce the CO_2 emission and conserving the natural resources.

2. Materials

2.1 Cement:

OPC 53 grade having specific gravity 3.13, consistency of 32% conforming to IS:269-2015 is used in this study.The brand of cement is DECCAN CEMENTS.

2.2 Fly ash:

Fly ash having specific gravity 2.12 used in this study according to IS:3812. This material data provided by ASTRRA CHEMICALS. The chemical properties of fly ash are mentioned below Table:1.

Table 1: Synthetic properties of Fly ash

Content	Percentage (%)
SiO2+Al2O3+Fe2O3	90.5
SiO2	58
CaO	3.6
SO3	1.8
Na2O	2

2.3 Silica fume:

The use of Silica fume in this study was conforming IS:15388-2003. Table 2 indicates that physical properties of material provided by GEOTECH SERVICES PVT.LTD. (soil&material testing laboratory).

Table 2: Material properties of silica fume

Properties	**Test results**	**Requirements as per IS:15388-2003**
Specific gravity	2.26	----
Fineness-45 micron sieve	0.94	Max 10
Compressive strength at 7 days(%)	98	Min 85

2.4 Fine aggregate:
Fine aggregate having specific gravity2.35, water absorption 6.38%, drawn ZONE II in sieve analysisconforming IS:383-2016.
2.5 Granulated blast furnace slag(GBFS):
Granulated blast furnace slag having specific gravity is 2.8, water absorption is 2.9% and sieve analysis is ZONE II. Chemical properties of materials provided byJSW CEMENTS LTD as per IS:12089. Table 3 indicates chemical compositions of GBFS.

Table 3: Chemical compositions of GBFS

Characteristics	**Requirement as per IS:12089**	**Test Results**
SiO2 (%)	----	33.04
Al2O3 (%)	----	20.66
Fe2O3 (%)	----	0.99
CaO (%)	----	36.20
MgO(%)	17 Max	7.74
Loss in ignition (%)	----	0.14

2.6Coarse aggregate:
Locally available crushed stones of size 20mm were used in this study as conforming to IS:383-2016.Physical properties of coarse aggregate were water absorption of 1.6%, specific gravity of 2.78 according to IS:2386-1963.
2.7 Super plasticizer:
KVRPCE-100 used as super plasticizer. Specifications of admixture are shown in Table.4 provided by GUJRAT POLYSOL CHEMICALS PVT.LTD.

Table 4: Specifications of super plasticizer

Appearance	Pale yellowish to brown
pH value	5.5-6.5
Solubility	Readily soluble in water
Solid	50%
Specific gravity	1.108
Chloride content	Below 0.02%

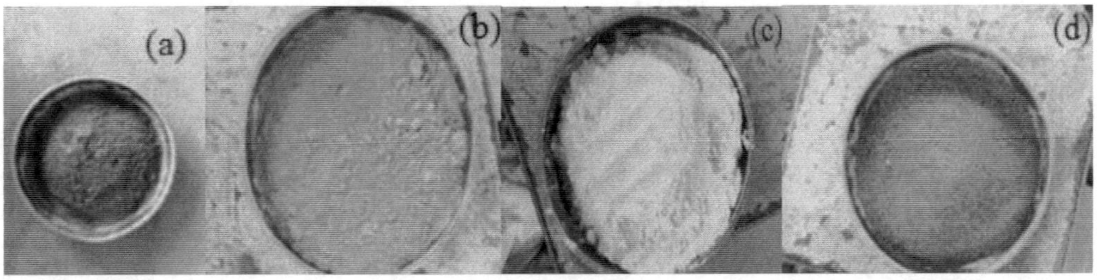

Figure1: Materials

The materials like Fly ash,OPC, Silica fume and GBFS are shown in Figure.1.

3. Results and Discussions:

3.1 Mix design:

The mix design for M40 grade tender as per IS:10262-2019. The quantity of concrete making materials furnished in Table5.

Table 5: Materials requirement of Conventional concrete to M40 (kg/m^3)

Material	OPC	Fine aggregate	Coarse aggregate	Water	Chemical admixture
Quantity	423	605	1270	160	3.6

In ternary concretesilica fume replaces constant of 8%,fly ash 20%,30%,40% in OPC and fine aggregate replaces with GBFS 30%,40%,50%. Future the mix id and mix descriptions are mentioned in Table 6.

Table 6: Mix ID and Mix description of Blended concrete

Mix ID	Mix descriptions
CC	Conventional concrete
MIX1	Silica fume 8%+Fly ash 20%+GBFS 30%
MIX2	Silica fume 8%+Fly ash 20%+GBFS 40%
MIX3	Silica fume 8%+Fly ash 20%+GBFS 50%
MIX4	Silica fume 8%+Fly ash 30%+GBFS 30%
MIX5	Silica fume 8%+Fly ash 30%+GBFS 40%
MIX6	Silica fume 8%+Fly ash 30%+GBFS 50%
MIX7	Silica fume 8%+Fly ash 40%+GBFS 30%
MIX8	Silica fume 8%+Fly ash 40%+GBFS 40%
MIX9	Silica fume 8%+Fly ash 40%+GBFS 50%

3.2 Workability:

The workability of concrete is measured by using slump cone apparatus. This test was carried out as per IS:1199-1959. The targeted slump for M40 design mix was 100mm. For the purpose maintain the targeted slump of 100mm the dosage of admixture (KVRPCE-100) is adjusted. Table 7 shows the Superplasticizer dosagesof mix proportions.

Table 7: Superplasticizer Dosage for 1m^3 of Concrete

Mix Id	Superlasticizer Dosage for Conventional mix (ml)	Superplasticizer Dosage for trail mix (ml)	Variation (%)
MIX1		2817	1.03
MIX2		2938	1.07
MIX3		3012	1.10
MIX4		3112	1.13
MIX5	2731	3541	1.29
MIX6		3687	1.35
MIX7		4215	1.54
MIX8		4630	1.69
MIX9		5022	1.83

From the above table, it is observed that percentage increase in GBFS, the chemical admixture dosages also increases it means the workability was decreasing because of water absorption of GBFS is less than fine aggregate.

3.3 Tests conducted:

Mechanical properties like compressive strength performed in this study as per IS:14858-2000. The mould size is 150mm length, 150mm width and 150mm height. Split tensile strength carried out as per IS:516-2018 mould size is 150mm diameter and 300mm length . Flexural strength of beam size is 100mm width, 100mm height, 500mm length carried out as per IS:516-2018. Testing set up of compression, split tensile and flexural strength are manifest in Fig.2.

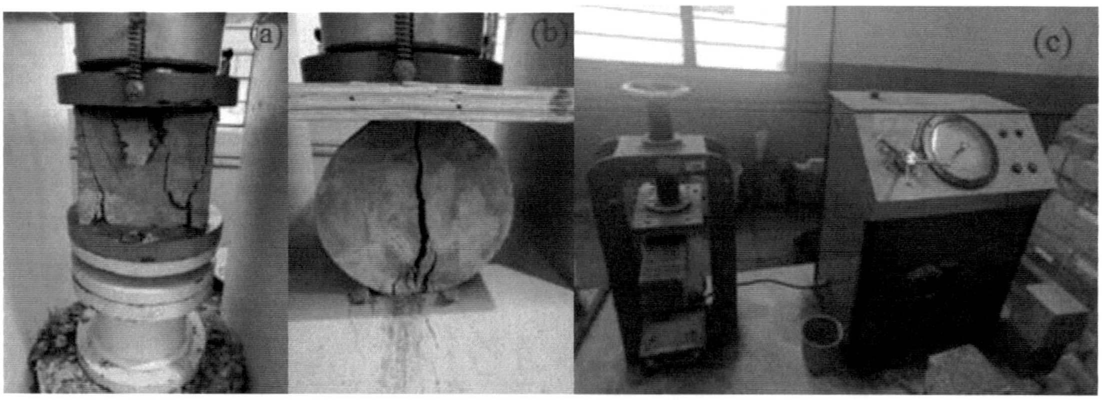

Figure 2: Experimental set up

3.3.1 Compressive strength:

The values of Compressive strength for 3,7 and 28 days are mentioned in Table:7. From this result observed the compressive strength of Mix-2 which is highest of all mixes with values of 41.54 N/mm^2, 42.33N/mm^2 and 58.76 N/mm^2 respectively and also improvement in early age strength due to adding of silica fume is observed, because the addition of fly ash lateral strength increases. After 28 days the specimen has been tested for compressive strength increased by 25% compared to conventional concrete.

Table 8: Compressive strength of Fresh Concrete and Blended concrete

Mix ID	Compressive Strength(MPa) Avg.		
	3 Days	7 Days	28 Days
CC	15.36	32.88	46.02
MIX1	23.88	39.40	52.84
MIX2	41.54	42.33	58.76
MIX3	38.89	41.23	53.57
MIX4	31.71	36.71	47.75
MIX5	33.12	40.53	51.64
MIX6	29.61	36.55	49.72
MIX7	28.49	38.98	46.81
MIX8	29.59	39.36	47.76
MIX9	26.56	35.57	44.03

3.3.2 Split tensile strength:

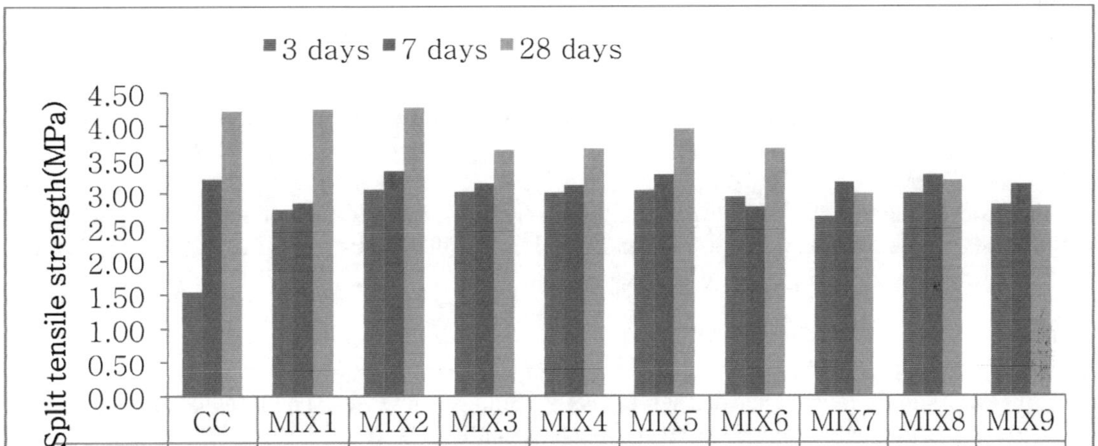

	CC	MIX1	MIX2	MIX3	MIX4	MIX5	MIX6	MIX7	MIX8	MIX9
3 days	1.54	2.76	3.06	3.03	3.01	3.04	2.95	2.66	3	2.83
7 days	3.21	2.86	3.33	3.15	3.12	3.28	2.8	3.16	3.27	3.13
28 days	4.22	4.24	4.27	3.64	3.66	3.95	3.66	3	3.19	2.81

Figure 3: Split tensile strength for 3, 7 and 28 days.

Figure 3 indicates the split tensile strength of triple blended concrete at 3,7 and 28 days. It is notice that from the above graphical representations the optimumstrength obtained at MIX2 contains 8% of silica fume, fly ash of 20% in OPC and 40% of GBFS in fine aggregate. Initial age strength increases due to the addition of silica fume, lateral strength is improved because addition of fly ash is observed.

3.3.3 Flexural strength:

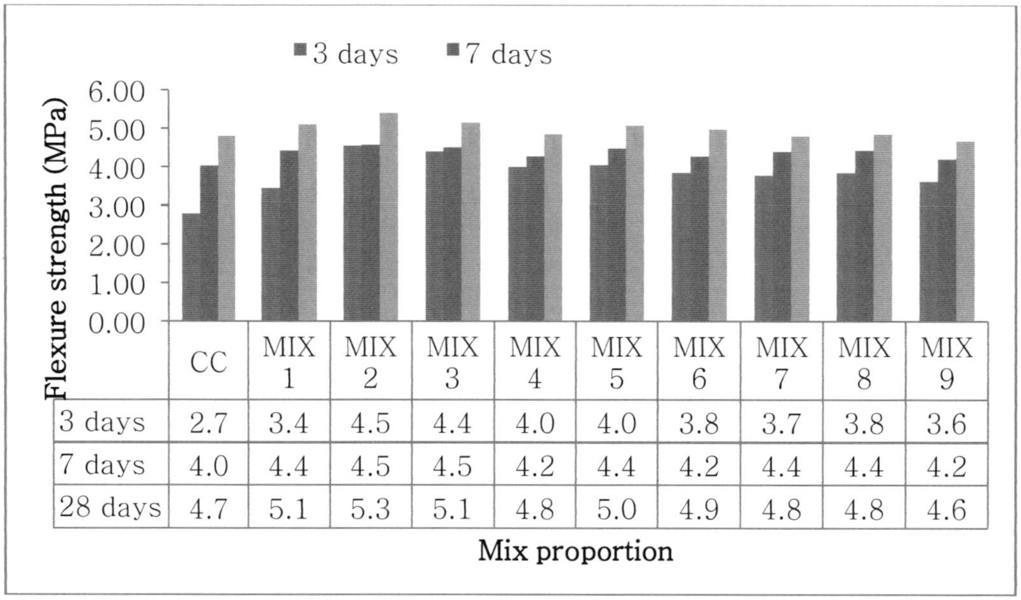

Figure 4: Flexural strength for 3,7 and 28 days

Figure 4indicates average flexural strength of M-40 grade concrete with 3,7 and 28 days. By the above figure it is observed that Mix-2 getting optimum strength compared to other mixes. The optimum containing 8% of silica fume, 20% of fly ash in OPC and 40% GBFS in fine aggregate. Due to the adding of silica fume early age strength increases.

Conclusions:

- The workability is decreased when GBFS percentage increases. The targeted slump of 100mm is carry off by adjusting the super plasticizer quantity.
- The addition of silica fume in concrete shows improvement in early age strength.

- Formed on the above test results the best mix proportion for M40 grade concrete is Mix-2 in whichthe OPC is substituted with20% of Fly ash, 8% of Silica fumeandfine aggregate is replaced with 40% of GBFS.
- The compressive strength of Mix-2 is 58.76 N/mm^2 which is highest of all mixes. The same tendency is noticed in split tensile and flexural strengths.
- The OPC can be replaced 20% fly ash, 8% silica fume. The fine aggregate is replaced with 40% of GBFS is recommended.
- By using Industrial by-products, the reduction in CO2 emission and saving in natural resources are established.
- The present study is applicable in all general constructions such as commercial buildings and bridges.

Acknowledgments
The experimental work was executed at the Advanced Concrete Materials lab, in CMR Technical Campus, Hyderabad. We would be glad to thank JSW CEMENTS LTD, GUJRAT POLYSOL CHEMICALS PVT.LTD and GEOTECH SERVICES PVT.LTD for providing the materials.

IOP Conf. Series: Materials Science and Engineering 1185 (2021) 012016 doi:10.1088/1757-899X/1185/1/012016

References

[1] Balakrishna Y, Naresh A, Lavanya Vand Vijaya Bhaskar Reddy S 2016*InternationalResearch Journal of Engineering and Technology* **3**pp 1500-07
[2] Abdulaziz I Al-Negheismish, Faisal H Al-Sugair and Rajesh Z Al-Zaid 1996*Journal of Environmental science of sustainable society* **1**pp 39-55
[3] Mithun R, Vasudev M V and Pallavi P 2018 *InternationalJournal of Research in Advent Technology*pp 97-103
[4] Rangaraj A and Ramkumar S2017 *InternationalJournal of Engineering Research & Technology***5**pp 1-5
[5] Balakrishna Y , Naresh A and Vijaya Bhaskar Reddy S 2016*InternationalResearch Journal of Engineering and Technology* **3**pp 1500-07
[6] Borole S T, Shinde R V, Mhaske R B, Pagare S S, Tribhuvam K S, Pawar N M, Tiwari V D and Sanehi A K*International Conferenceon Emerging Trends in Engineering and Management Research*pp 1582-89
[7] Dr Vijaya sekar reddy M, Ashalatha K and SasiK 2016*International Journal of Engineering Research and Technology***9**pp 111-20
[8] Subathra DeviVand Gnanavel B K 2014*ELSEVIER SciencDirect* Proc.**97**pp 95-104
[9] Pratap K V, Bhaskar M and Teja P S S R 2014*International Journal of Education and Applied Research***4**pp 54-58

Performance of high strength concrete consisting of alccofine and metakaolin as a partial replacement of cement and copper slag as fine aggregate

Itikela Durga Saikiran[1] and Ch. Srinivasa Rao[2]

[1]PG student, Civil Engineering Department, CMR Technical Campus, Hyderabad, India.
[2]Professor, Civil Engineering Department, CMR Technical Campus, Hyderabad, India.

* Corresponding author e-mail:
[1]durgasaikiran.itikela@gmail.com, [2]chigurupatisrinivas@gmail.com

Abstract. Concrete is used as building material from ancient times. The combination of Ordinary Portland Cement (OPC) with aggregates and water is called concrete. Due to rapid urbanization in India, a large quantity of OPC and fine aggregate are consumed in the construction industry, while manufacturing OPC large quantity of carbon dioxide is generated and exposed into the atmosphere. Continuous dredging of sand lead to improper infiltration of soil and soil erosion. To reduce the emission of carbon dioxide from the OPCmanufacturing industry and to minimize the fine aggregate usage there is a requirement for a new eco-friendly substitute material. In developing countries, there is a necessity for sophisticated buildings and High Strength Concrete (HSC). In this study, the HSC of grade M65 with a slump of 100mm and w/c of 0.28 is considered. The fine aggregate of the concrete is replaced by 40%, 60% and 80% with Copper Slag. OPC is partially substituted by Metakaolin with a constant 15% and Alccofine with 5% and 10% respectively. The HSC is evaluated for Workability and Mechanical properties like slump, Compression, Flexureand SplitTensile Strength for both conventional and trial mix concrete.The test results of trial mixes are compared with conventional concrete and conclusions were drawn. As per the experimental results, it is concluded that OPC replaced with 15% of Metakaolin, 10% of Alccofine and fine aggregate replaced with 60% of Copper Slag shows optimum results.

Keywords: High Strength Concrete, Alccofine, Metakaolin, Copper Slag, Workability, Mechanical Properties.

1. Introduction

Concrete is used as building material in the construction field. The components of concrete are aggregates, cement, and water. Cement and fine aggregate are the most used building materials. During the process of OPC manufacturing, a large number of raw materials like Lime and Clay are used, which are non-renewable natural resources and OPC Manufacturing factories release a massive volume of carbon dioxide at the stage of clinker formation, which is exposed to the atmosphere and causes global warming.Concrete also consumes a massive amount of fine aggregate which causes a reduction in non-renewable resources and ecological imbalance. Due to over dredging of sand problems like soil erosion, improper infiltration of soil and threats to aquatic life can occur. The products such as coarse aggregate, fine aggregate are non-renewable natural resources. The exhaustion of natural resources causes environmental hazards. For sustainable development, there is a requirement for the usage of eco-friendly materialsfor manufacturing concrete [1]. To eliminate the adverse effects on the environment, cement substitute materials like Metakaolin, Rice husk ash, GGBS,Silica fume and bentonite powder can be used [2]. Globally concrete industry uses 8-10 billion tons of fine aggregate per year [3]. Industrial by-product like Copper Slag can be used as a partial replacementfor fine aggregate[4].Replacement of cement with Alccofine shows increment in compression strength [5]. HSC consisting of Alccofine shows better Compression, Flexure and Split Tensile Strength than the normal concrete in which cement is partially substituted by silica fume [6]. Optimum strength of concrete is observed at 50% partial substitution of fine aggregate with Copper Slag [7]. After 28 days of curing, 10% replacement of Alccofine shows better compressive strength [8]. The concrete with dolomite replacement shows higher split tensile, compressive, flexure and bond strength than the recycled aggregate [9].The properties of SCC consisting of Alccofine and glass fiber shows optimum results than concrete with only glass fiber concrete. [10]. The strength of SCC increasing with an increase the Alccofine and decided that Alccofine is suitable as a partial replacement of cement in SCC [11]. In HSC as the percentage of Alccofine and foundry slag increased, also an increment in concrete compressive strength [12].Alccofine shows early strength gaining and high compressive strength in all ages compared to conventional concrete [13]. In the present study, HSC of grade M65 is considered. Cubes, beam and cylinder specimens were casted for both conventional and trial mix concrete. In the trial mix concrete,OPC is partly replaced with Alccofine and Metakaolin and fine aggregate is partially substituted by Copper Slag at various percentages as shown in Table11. Fresh and Hardened Concrete properties like slump, Compression, Split Tensileand Flexural Strength tests of both Conventional and Trial mix concrete are compared and results are evaluated. The objective is to study the Fresh and Hardened Concrete propertiesof HSC in which Ordinary Portland Cement(OPC) is partially substitutedby Alccofine and Metakaolin, to determine propertiesof HSC in fresh and hardened state which fine aggregate is partially substituted by Copper Slag, to determine the optimum replacement percentages of Alccofine, Metakaolin and Copper Slag in HSC and also to establish the percentage utilization of industrial by-products to reduce the Carbon dioxide(CO_2) emission and to conserve natural resources.

2. Materials and Methodology

2.1 Ordinary Portland Cement(OPC):

In the current study, OPC of 53 grade is used conforming IS:269-2015. The properties of OPC are depicted in Table1.

Table1. Physical properties of OPC.

Tests	Results	IS CODE
Initial setting time	46 mins	IS:4031-part 2-1998
Final setting time	250 mins	IS:4031-part 5- 1998
Consistency	32%	IS:5513-1976
Fineness	1%	IS:4031-part 1-1996
Specific gravity	3.13	IS:2702-part 3

2.2Alccofine1203(AF):

Alccofine is a microfine material with a very fine particle size distribution. The particle size distribution of Alccofine is very fine compared to any other microfine material. There are two types of Alccofine materials available they are Alccofien1203 which is used in High Strength and HPC andAlccofine 1101 is used for injection groutingand soil stabilization. In the present study, Alccofine 1203 is used as partial OPC replacement material. Alccofine1203 is provided by Counto Microfine Products Pvt.Ltd.Both chemical and physical properties of Alccofine1203 is given by the manufacturer are provided in Table2 and Table 3.

Table2. Chemical composition of Alccofine1203 (Provided by Counto Microfine Products Ltd).

Chemical composition	Mass by(%)	Requirement as per IS:16715-2018
MnO	1.24	5.5 Max
MgO	6.3	17 Max
SO_3	0.1	3 Max
Chloride content	0.020	0.1 Max
Glass content	>85	85 Min

Table 3. Physical properties of Alccofine1203 (Provided byCounto Microfine Products Ltd).

Specific gravity	Bulk density(kg/m^3)	Particle size distribution			
		D_{10}	D_{50}	D_{90}	D_{95}
2.9	600-700	1.4	4.2	9.2	10.9

2.3Metakaolin(MK):

Metakaolin is the most effective pozzolanic material used in concrete, it is the by-product formed in the process of extraction of china clay. Metakaolin is obtained during the calcination of kaolinite between 600^{0}C to 750^{0}C in a kiln. In the current study, Metakaolin is utilized as a partial substitute material for OPC.The properties of Metakaolin are listed in Table4.

Table 4. Metakaolinproperties (Provided by Shree Ram Minerals).

Properties	Results
Specific gravity	2.6
Ph	5-7
Bulk density	850 kg/m^3

2.4 Ground Granulated Blast Furnace slag(GGBS):

GGBS is a partial substitute material for OPC obtained from a blast furnace as a by-product of steel and iron making.In this study, GGBS is used as a partial substitute for OPC as per IS: 16714-2018. GGBS is utilized in the study is purchased from JSW Cement Ltd. The properties of GGBS are depicted in Table 5.

Table 5.GGBS properties(Provided by JSW Cement Ltd).

Properties	Results
Colour	White
Fineness in percentage	3
Water absorption in percentage	0.75
Specific gravity	2.77

2.5Fine Aggregate(FA):

In the present study, locally available manufactured sand is used as fine aggregate according to IS:383-2016. Various tests have been conducted on manufactured sand and the test results are presented in Table 6.

Table 6. Physical properties of Fine aggregate.

Tests	Results
Water absorption	6.4%
Sieve analysis	Zone II
Specific gravity	2.35

2.6Copper Slag(CS):

CS is a by-product obtained during copper extraction process, the process through which the Copper Slag is obtained is known as smelting. In the process of smelting, copper is heated at a temperature of 500^{0}C to 750^{0}C to separate impurities. These impurities separate from the copper and become Copper Slag when cooled. In the current study, fine aggregate is replaced partially by Copper slag.The properties of Copper Slagarerepresented in Table 7.

Table 7. Copper Slag properties(Provided by Ballast and India Pvt. Ltd).

Properties	Results
Specific gravity	3.5
Bulk density(g/cc)	2.1
Hardness	7 MOH
Water absorption	1%

2.7 Coarse aggregate(CA):

Coarse aggregate which is locally available of 20mm size confirming IS:383-2016 is utilized in the research. The coarse aggregate properties are mentioned in Table 8.

Table 8. Coarse aggregate properties.

Tests	Results	IS CODE
Water absorption	1.1%	-
Specific gravity	2.78	IS:2386-PART 3- 1963

2.8 Water:

For casting and curing of specimens, potable water is used. Water available at the laboratory which is free from lubricants, acids and other contamination is used as per the guidelines of IS:456-2000.

2.9 Chemical admixture(SP):

To achieve the desired slump chemical admixture is added to the concrete. In this study polycarboxylate etherbased superplasticizer is used.Table9 shows the properties of the superplasticizer.

IOP Conf. Series: Materials Science and Engineering 1185 (2021) 012017 doi:10.1088/1757-899X/1185/1/012017

Table9. Properties of superplasticizer(Provided by Fratellanza Chemicals).

Properties	Results
Specific gravity	1.08
pH	6
Colour	Light brown
Chloride content	Nill

Materials utilized in the current study like Alccofine, Copper Slag, Metakaolin and superplasticizer are shown underneath.

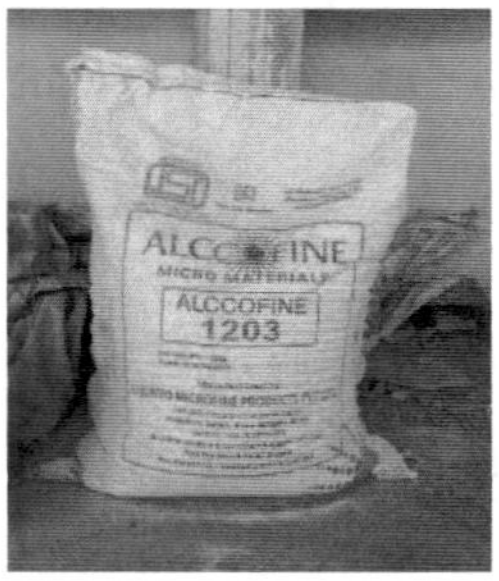

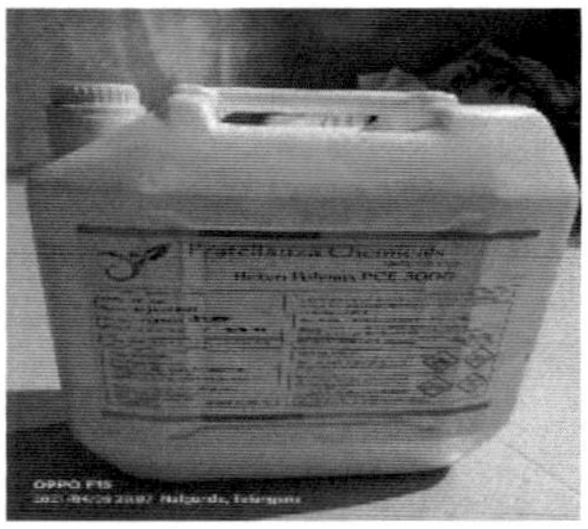

Figure1. Alccofine. **Figure2.** Copper Slag. **Figure3.** Metakaolin. **Figure4.**Super Plasticizer.

2.10 Mix design:

In thiscurrent study mix design of M65 grade is done as per the guidelines of IS:10262-2019.The design of conventional concrete mix (CC)is depicted in Table10.

Table10. Material quantity required for M65 gradeconcrete(kg/m^3).

Materials	Quantity(kg/m^3)
OPC	446
GGBS	78.6
Coarse aggregate	1201
Fine aggregate	596
Water	147
Superplasticizer	2.63

And a replacement mixis considered as a trial mix in which fine aggregate of the concrete is replaced by 40%, 60% and 80% with Copper Slag and Ordinary Portland Cement is partially substituted by Metakaolin with a constant of 15% and Alccofine with 5% and 10% by weight of OPCrespectively.Table11 explains the percentage of replacements in trial mix concrete.

Table 11. Trial mix and material quantities for M65 grade concrete (kg/m^3).

Mix Id	Mix Description	OPC	MK	AF	FA	CS	CA	WATER	SP
M1	MK15%+AF5%+CS40%	419	78.6	26.2	357	238	1201	147.87	2.63
M2	MK15%+AF5%+CS60%	419	78.6	26.2	238	357	1201	147.87	2.63
M3	MK15%+AF5%+CS80%	419	78.6	26.2	119	476	1201	147.87	2.63
M4	MK15%+AF10%+CS40%	393	78.6	52.4	357	238	1201	147.87	2.63
M5	MK15%+AF10%+CS60%	393	78.6	52.4	238	357	1201	147.87	2.63

M6	MK15%+AF10%+CS80%	393	78.6	52.4	119	476	1201	147.87	2.63

cube specimen of dimension 150mm x 150mm x150mm, cylinder specimen of diameter and height are 150mm and 300mm respectively and beam specimen of dimension 500mm x 100mm x 100mm were casted to perform a Compression, Split Tensile and Flexure Strength Test for both conventional and trial mix concrete.The specimens are cured in potable water before testing for 3,7 and 28 days.

3. Tests results and Discussions:

3.1 Workability Test:

The Fresh concrete is targeted for a slump of 100mm and is measured with the help of a slump cone confirming IS:1199-1959. As the Copper Slag replacement increases the slump of concrete decreases so the workability is adjusted by adding superplasticizer dosage. The adjusted value of the superplasticizer is depicted in Table 12

Table 12. Dosage details of superplasticizer for 1m^3 of concrete to achieve 100mm slump.

Mix ID	Superplasticizer (ml)		Variation (%)
	As per design mix	Consumed	
M1	2435	3581	1.47
M2		3961	1.62
M3		4219	1.73
M4		3576	1.46
M5		3952	1.62
M6		4211	1.72

From Table12 it is observed that due to an escalation in the percentage of Copper Slag there is a diminution in the slump of concrete, thus a 100mm targeted slump of fresh concrete has been obtained by adjusting the superplasticizer dosage.

3.2 Compressive Strength Test:

The test was conducted as per IS:516-2018. The compressive strength results are shown in Figure 5.

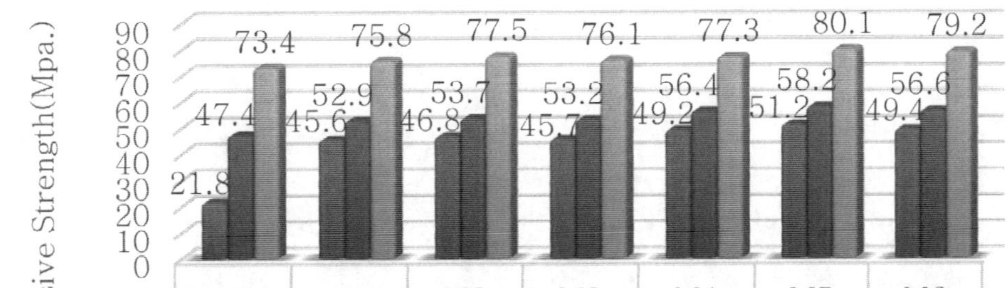

	CC	M1	M2	M3	M4	M5	M6
3-DAYS	21.8	45.6	46.8	45.7	49.2	51.2	49.4
7-DAYS	47.4	52.9	53.7	53.2	56.4	58.2	56.6
28-DAYS	73.4	75.8	77.5	76.1	77.3	80.1	79.2

Figure 5. Compressive Strength Test Results.

From Figure 5,it is observedall the trial mixes whose compressive strength is tested after 3-days of curing show very high early age strength when compared with conventional concrete. After 7-Days of curing mix M5 shows 22% more strength than normal concrete.After 28-Days of curingboth mix proportions, M5 and M6 show highcompressive strength of 80.1Mpa and 79.2 Mpa than that of the normal concrete, But mix-M5 consisting of 15% Metakaolin and 10%Alccofinc as partial replacements of OPCand 60% Copper Slagas the partial substitution of fine aggregate shows the strength of about 80.1Mpa which is 9.12% more than the normal concrete.Due to Alccofine, which is a microfine material that helps in pore refinement and gives strength to concrete and Metakaolin also increases the compression strength of concrete.Thusthe compression strength of concreteis increased because of the existenceof Alccofine and Metakaolin. The early strength of concrete is due to the existence of Alccofine.

3.3 Split Tensile Strength Test:

The test was performed as per IS:516-2018. The test results are depicted in Figure 6.From Figure 6, after 3-Days of curing, the specimen of mix M5 shows the optimum split tensile strength of 4.65Mpa,we can observe that trial mixes show early strength than conventional concrete.

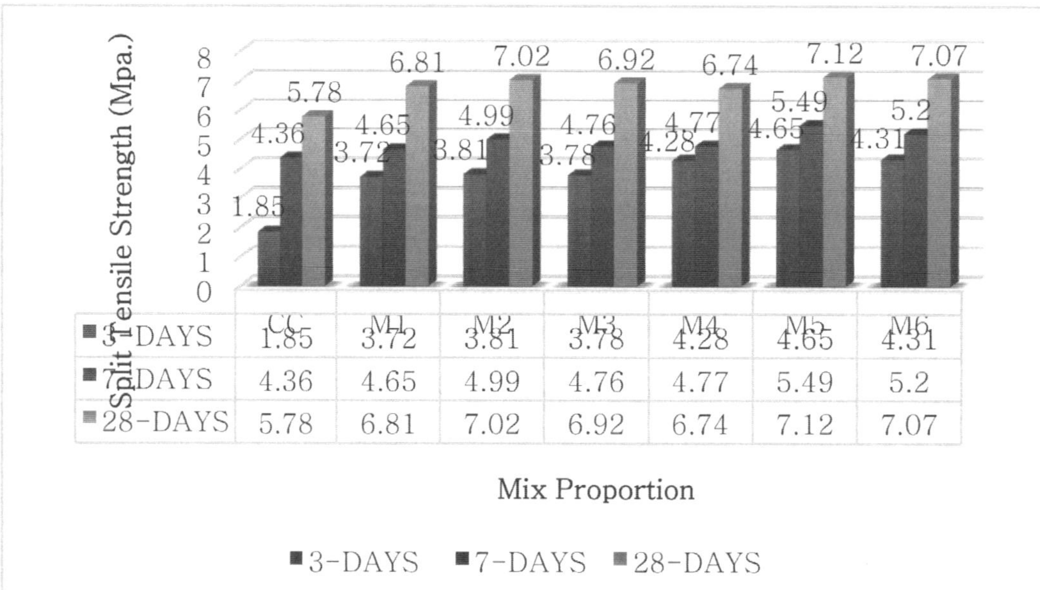

Figure 6. Split Tensile Strength Test Results.

The 7-Days strength results of conventional and trial mix concrete are very similar but mix M5 shows a slightly better strength of 5.49Mpa which is higher than conventional concrete. after 28 days of curing the mix proportions, M5 and M6 show similar split tensile strength of 7.12Mpa and 7.07Mpa the concrete of specimen of mix-M5 consisting of 10% Alccofine and 15% Metakaolin as partial replacement of OPC and 60% of Copper Slag as fine aggregate shows optimum strength of 7.12Mpa which is 28% greater than the conventional concrete. The early strength of concrete is increased due to the existence of Alccofine. Metakaolin, Alccofine and Copper slag as partial substitutes of OPC and fine aggregate are the reason for increased split tensile strength.

3.4: Flexural Strength Test

The flexural strength test was performed as per IS: 516-2018. The test results of flexural strength are shown in Figure 7.The flexure strength of trial mix concrete after 3-Days of curing is very high compared to conventional concrete. After 28-Days of curing concrete specimen of mix-M5 containing 10% of Alccofine and 15% of Metakaolin as a partial replacement in OPC, and 60% of Copper Slag as partial fine aggregate replacement showsmaximum flexural strength of 6.29Mpa which is almost 3% more than conventional concrete. The packing effect of Alccofine is the reason for higher and early flexure strength.

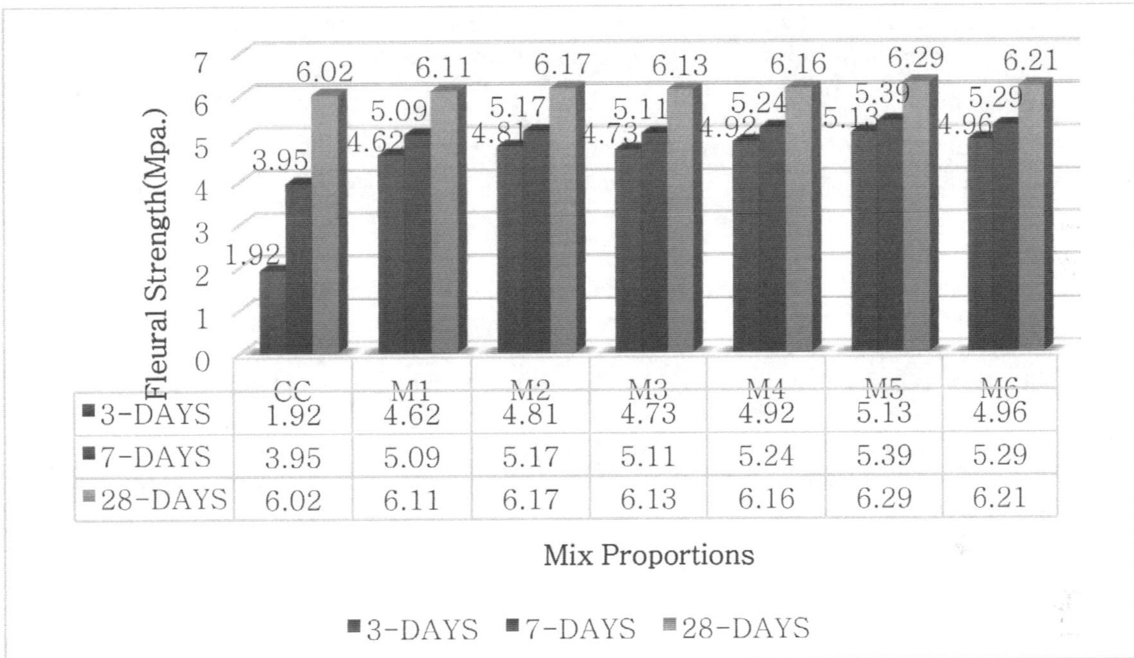

Figure 7. Flexural Strength test Results.

Testing facilities of Workability and strength of concrete like Compressive Strength, Split Tensile andFlexural Strength as shown beneath

Figure 8. Workability. **Figure 9.** Compression. **Figure 10.** Split Tensile. **Figure 11.** Flexure.

4. Conclusion:

The Mechanical properties and workability of conventional and trial mix concrete are compared and the following conclusions are drawn:

- For the fresh concrete, workability decreases with an increase in the percentage ofCopper Slag. The target slump of 100mm is achieved by adjusting the superplasticizer dosage.
- Due to addition of alccofine, it is observed an increase in early age strength in all trial mixes when compared with conventional concrete.
- HSC with 15% Metakaolin,10% Alccofine and 60% Copper Slag show superior results in Compression, Split Tensile and FlexuralStrength as compared to conventional concrete.
- As per the experimental investigation 15% of Metakaolin, 10% of Alccofine and 60% of Copper Slag are recommended in HSC.
- By using Industrial by-products such as Metakaolin,Alccofine and Copper Slag in the concrete leads to saving natural resources and reduction in CO_2 emission.
- The present study is applicable in all general and HighStrength Constructions such as Commercial Buildings,Hospitals, Shopping Complexes,Skyscrapers and Bridges.

Acknowledgment:

The work is carried out in the laboratory of Advanced Concrete Technology, CMR Technical Campus, Hyderabad. Would like to thank Counto Microfine Products Pvt. Ltd for providing the material.

References

[1] Goyal P, Singla R K and Kaur P 2019 Experimental study: alccofine as strength enhancer for geopolymer concrete*International Journal of Advance Research, Ideas and Innovations In Technology***5**pp227-31
[2] Subbamma V and Reddy D K C 2016 Experimental study on compressive strength of plain cement concrete with partial replacement of cement by flyash & metakaolin *International Journal of Civil Engineering and Technology* **7** pp82-9
[3] Tu T Y, Chen Y Y and Hwang C L 2006 Properties of HPC with recycled aggregates *Cement and Concrete Research* **36**pp943-50
[4] Al-Jabri K S, Hisada M, Al-Oraimi S K and Al-Saidy A H 2009 Copper slag as sand replacement for high performance concrete *Cement and Concrete Composites***31** pp483-8
[5] Koshti A A, Darshan U, Tushar B, Kunal D, Sourabh M and Shubham A 2020 Study of effect of alccofine on strength characteristics of concrete *International Journal for Research in Applied Science and Engineering Technology***8** pp337-42
[6] Gaba R, Sharma D and Sharma S 2020Effect of alccofine and silica fume on strength properties of concrete*International Research Journal of Engineering and Technology* **7** pp109-15
[7] Gupta R S, Kheshwala C B, Mom S R, Varaliya M A and Amin K 2020 Partial replacement of copper slag as fine aggregate*International Journal of Innovative Research in Technology***6** pp83-6
[8] Mohyuddin N V, Mehta N S and Patel H V 2020Effect of alccofine on mechanical property of concrete*International Research Journal of Engineering and Technology***7** pp1305-10
[9] Emad M, Soliman N M and Bashandy A A 2019Recycled aggregate high-strength concrete *International Journal of Civil Engineering and Technology* **10** pp128-46
[10] Kareem B M and Jamal S M 2019Effect of alccofine on glass fiber reinforced self compacting concrete*International Research Journal of Engineering and Technology***6** pp5691-3
[11] Issac L and Tom C A 2018A review on self-compacting concrete incoperating alccofine*International Research Journal of Engineering and Technology* **5** pp756-60
[12] Gupta S, Kumar A, Sharma S and Sharma D 2017Effect of alccofine and foundry slag on compressive strength of high strength concrete *International Journal of Engineering Research* **6** pp406-9
[13] Gautam M and Sood H 2017 Effect of alccofine on strength characteristics of concrete of different grades-A review *International Research Journal of Engineering and Technology* **4** pp2854-7.

Performance of Binary Steel Fibre Concrete Consisting of Copper Slag as Partial Replacement to Fine aggregate

D. Srinath[1], V. B. Reddy Suda[2]

[1]PG Student, CMR Technical Campus, Department of Civil Engineering, Hyderabad.
[2]Professor, CMR Technical Campus, Department of Civil Engineering, Hyderabad.

* Corresponding author: [1]srinathhoney6@gmail.com, [2]vbsgen06@gmail.com

Abstract. Concrete is the combination of H_2o, cement, fine aggregate and coarse aggregate. Concrete consumptionis extremelyhigh and its demand is increasing rapidly. OPC production leads to emission of harm full gas. 1 Ton of OPC production leads about 1 Ton of Co_2 emission. In the present situation there is deficiency in natural river sand due to ban of dredging of river sand in most of the states. To overcome these problemsit is essential to search for substitute materials.GGBS, Metakaolin, Copper slag,Silica fume, Steel slag etc are few Industrial by-products are identified as alternative materials. Here M40 concrete grade mix is used, in which 30% and 50% of GGBS is partially replaced with OPC and copper slag of 40% and 60% is utilized as fine aggregate. Concrete is brittle material, to make concrete as ductile material, steel fibres are used.The hardened concrete properties and fresh concrete properties such as compressive split tensile and flexure strengthdeterminedrespectively as well asit is compared through regular concrete.As per the experimental investigation results 50% of GGBS is used as partial substitute to OPC, copper slag about 60%is used to partial substitute to fine aggregate along with steel fibre of 1 % is recommended.

Key words: Mechanical Properties Ordinary Portland Cement(OPC), Fresh concrete properties, Ground Granulated Blast Furnace Slag (GGBS) and Copper Slag.

1. Introduction

Due to its high strength and stability, concrete is most popular construction material.Protecting the environment has become a serious problem in our society under the present conditions.During the manufacturing of 1 ton of OPC leads to emission of 1 ton of Co_2 in to the atmosphere. To reduce the Co_2 emission in the atmosphere, industrial by-products were used in construction industry as an alternative to conventional materials[1].

Due to its inherent cementitious properties, GGBS as a partial substitute for OPC is utilized. GGBS is Industrial by-product derived from the production of pig iron.

At present, there is arestrictionindredging of river sand which leads to deficiency to get the river sand. By dredging the river sand the environmental imbalance such as infiltration, soil erosion problem will occur.To conquer the above problem alternative materials are utilized.Slag, an industrial by-product, is obtained from the manufacturing process when copper is produced.In this study, the author considered GGBS as partial substitute to OPC and copper slag as partial substitute to aggregate (fine) along with 1% of steel fibre. Environmental pollution can be reduced by using Copper Slag as partial substitute to aggregate (fine) without compromising the properties of aggregate (fine)[2],40% slag is more economical[3], 40% GGBS gives durability, life span of structure and optimum strength[4], 10% GGBS and 30% Copper Slag gives the optimum value while testing mechanical properties[5], Compressive strength is increased by 20% substitute of OPC when compared with conventional mortar[6].To revise the standard grade properties like fresh and hardened with steel fibre concrete in which GGBS is partially substituted for OPC.The objectives of present study is to cram fresh propertiesand hardened properties of standard concrete grade of steel fibre in which fine aggregate is partially substituted by Copper Slag.To decide the optimum percentage of Industrial by-products can be utilized.To decide the optimum utilization percentage of industrial by-products as partial substitute for OPC and aggregate (fine).To save the natural resources and reduction in Co_2emission.

2.Methodology and Materials

2.1 Materials

Manufactured sand, cement and coarse aggregate, water, GGBS, Copper Slag and chemical admixture are utilized in the present study. All these materials are locally available. The properties of the material are determined as per BIS.

2.1.1 Cement

According to IS: 269-2015, in the present research 53 grade of cementis used. The physical properties, OPC 53 grade is depicted in Table 1.The representive sample of cement is publicized in Figure 1.

Table 1 Cementphysical properties

Test Performed	Result	IS Codes
Consistency (%)	32	IS: 5513-1976
Setting time initial (min)	45	IS: 4031 Part 5-1988
Setting time final (min)	250	IS: 4031 Part 5-1988
Fineness (%)	1	IS: 4031 Part 1-1996
Specific Gravity	3.13	IS: 2720 Part 3

2.1.2 GGBS

In this investigation GGBS is brought from the JSW Cement Pvt Ltd confirming to IS: 16714-2018. The physical property of GGBS is provided by JSW Cement Pvt Ltd isfurnished in Table 2.The representive sample of steel GGBS is shown in Figure 1.

Table 2 GGBSphysical properties

Properties	Physical Component
Water Absorption	0.75
Colour	White
Specific Gravity	2.77
Fineness	3%

2.1.3Fine aggregate

a) Manufactured sand

The manufactured/ robo sand confirming to IS 383-2016 is utilized in the present study. Robo sand physical properties are depicted in Table 3.The representive sample of steel fibre is shown in Figure 1.

Table 3. Robo sandphysical properties

Test	Result	IS Codes
Sieve Analysis	Zone II	IS383-2016
Water Absorption	6.38%	IS 2386 Part 3-1963
Specific Gravity	2.35	IS 2386 Part 3-1963

b) Copper Slag

Throughout the process of extraction of copper, an industrial by-product named copper slag is obtained. In this study fine aggregate is replaced with slag as substitute material. Copper slag physical properties are provided by manufacturer (Blastine India Pvt Ltd) and its properties are furnished in Table 4.The representive sample of slag is publicized in Figure 1.

Table 4. Copper slag physical properties

Properties	Physical Component
Grain shape	Angular, Multifaceted
Colour	Black & glassy
Hardness	7 MOH
Bulk Density at 25°c	3.25 x 10^{3}
Specific Gravity at 25°c	3.5
Conductivity at 25°c	4%
Moisture Content	<0.001%

2.1.4 Coarse aggregate

20mm size aggregate is locally available confirming to IS 383-2016is used.Coarse aggregate physical properties are depicted in Table 5.The representive sample of coarse aggregate is revealed in Figure 1.

Table 5 Coarse aggregate physical properties

Test	Result	IS Codes
Specific Gravity	2.78	IS 2386 Part 3-1963
Water Absorption (%)	1.6	-

2.1.5 Steel Fibre

Hooked end steel fibres confirming to ASTM A-820M 06 are used in this investigation. Manufacturer named Vaishnav Composites provided steel fibre properties which are depicted in Table 6 and the representive sample of fibre is exposed in Figure 1.

Table 6. Properties of Steel Fibre

Properties	Result
strength of wire (tensile) mpa	>1100
Diameter (D) mm	0.55
Strain at failure (%)	<4
Camber of Fibre	Max. 5% of L
Length (L) mm	30
Number of Fibres Per kg	16924
Aspect Ratio	54.55

2.1.6 Super Plasticizer

PCE based super plasticizer (KVRPCE-100) has been utilized in the study.Super plasticizer physical propertiesbe depicted in Table 7. The properties of super plasticizer are obtained from Gujarat Polysol Chemicals Pvt. Ltd. The representive samples of all the materials bemade known in Figure 1.

Table 7 Physical Properties of Super Plasticizer

Properties	Physical component
Appearance	Pale yellowish to brown
pH value	5.5-6.5
Solubility	Readily soluble in water
Solid (%)	50
Specific Gravity	1.108
Chloride Content	< 0.02%

Figure 1. Cement, GGBS, Aggregates, Copper Slag, Steel Fibre and Super Plasticizer

2.1.7 Design Mix

Design mix for M40 concrete grade is agreed for everyIS10262-2019. Material required 1m^3 conventional concrete (CC) are furnished in Table 8 and also the materials required for binary concrete with different replacement aredepicted in Table 9.

Table 8. Materials quantityrequired for CC (kg/m^3)

Material	Quantity
Cement	415
Fine aggregate	613
Coarse aggregate	1172
Water	174.3
Super plasticizer	83

Table 9. Materials quantityrequired for Binary concrete (kg/m^3)

Mix ID	Mix Description OPC%+GGBS%+CS%+SF %	Cement	GGBS	FA	CS	CA
M1	30+ 40+ 1	290.5	124.5	367.8	245.2	1172
M2	30%+ 40+ 1.5	290.5	124.5	367.8	245.2	1172
M3	30+ 60+ 1	290.5	124.5	245.2	367.8	1172
M4	30+ 60+1.5	290.5	124.5	245.2	367.8	1172
M5	50+40+1	207.5	207.5	367.8	245.2	1172
M6	50+40+1.5	207.5	207.5	367.8	245.2	1172
M7	50+60+1	207.5	207.5	245.2	367.8	1172
M8	50+60+1.5	207.5	207.5	245.2	367.8	1172

3. Results and Discussion

3.1 Workability

According to IS 1199-1959slump cone test is carried out. The targeted slump is of 100mm. The workability grades are furnished in Table 10. Workability is decreased when the slag percentage is increased. By adjusting the dosage of chemical admixture targeted slump is achieved.

Table 10. Workability test results

Mix Designation	Super Plasticizer Dosage (ml)	
	Actual	Adjusted
M1	238.7	328.7
M2	238.7	335.5
M3	217.6	307.6
M4	217.6	315.0
M5	155.4	245.4

M6	155.4	250.5
M7	145.6	235.6
M 8	145.6	240.5

3.2 Test on hardened concrete

The concrete hardenedproperties, for example compressive, flexure and split tensilestrength were performed.

3.2.1Compressive strength

As per IS 14858-2000, cube specimens are tested at 3 days, 7 days and 28 days of curing for compressive strength. The compressive strength of binary concrete is depicted in Table 11. The optimum packing effect of concrete is observed in Mix 7when compared with erstwhile mixes. The graphical image of the optimum mix is exposed in Figure 2.

Table 11. CC and Blended concrete Compressive results

Mix Designation	GGBS (%)	Copper Slag(%)	Steel Fibres (%)	Compressive (Mpa.)		
				3	**7**	**28**
CC	0	0	0	16.4	31.5	47.3
M 1	30	40	1.0	29.8	54.5	70.4
M 2	30	40	1.5	25.9	35.6	57.6
M 3	30	60	1.0	31.0	38.5	51.5
M 4	30	60	1.5	18.3	30.3	44.3
M 5	50	40	1.0	27.2	37.3	50.9
M 6	50	40	1.5	26.9	31.7	49.8
M 7	50	60	1.0	45.1	55.6	72.2
M 8	50	60	1.5	40.3	46.5	60.3

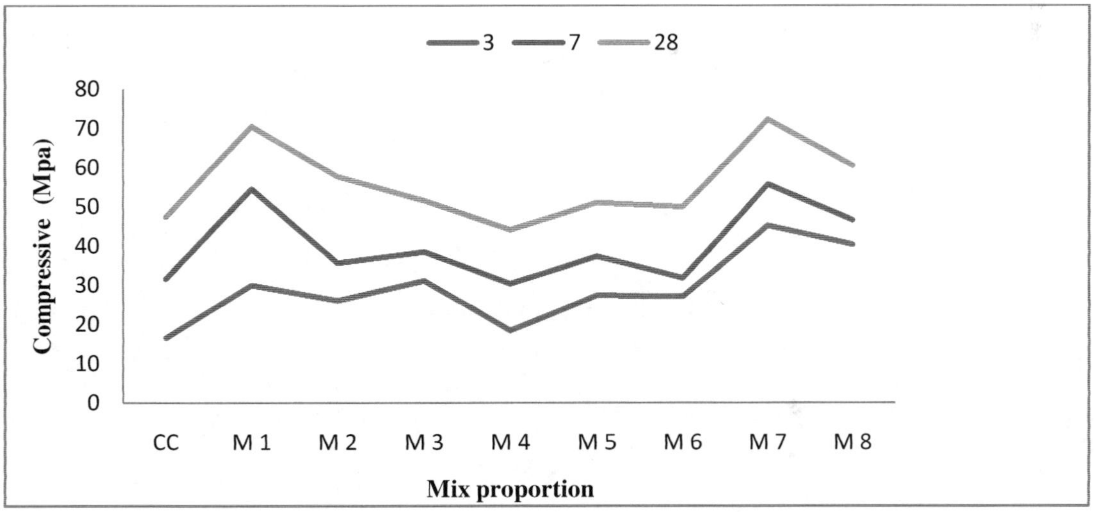

Figure2. Concrete compressive strength for 3 days,7 days and 28 days

From the given Figure.2, 3 days compressive strength result is increased by 36.3%,7 days compressive strength result is increased by 56.66% and 28 days compressive strength result is increased by 65.42% when compared with conventional concrete respectively.

3.2.2 Split tensile

According to IS 516-2016 cylinder specimens are tested for 3 days, 7 days and 28 days. Binary concrete split tensile strengthis depicted in Table 12.Optimum pore refinement of concrete is observed in Mix 7 when compared with former mixes. The graphical representation of the optimum mix up is shown in Figure 3.

Table 12.CC and Blended concrete Split tensile results

Mix Designation	GGBS %	Copper Slag %	Steel Fibres	Split Tensile(Mpa.)		
				3	7	28
CC	0	0	0	1.34	3.30	4.40
M 1	30	40	1	2.73	4.55	6.57
M 2	30	40	1.5	2.21	3.27	5.35
M 3	30	60	1	2.61	3.52	4.56
M 4	30	60	1.5	1.86	2.88	3.82
M 5	50	40	1	3.00	3.25	4.63
M 6	50	40	1.5	2.72	3.0	4.09
M 7	50	60	1	3.85	5.24	6.98
M 8	50	60	1.5	3.52	3.96	4.96

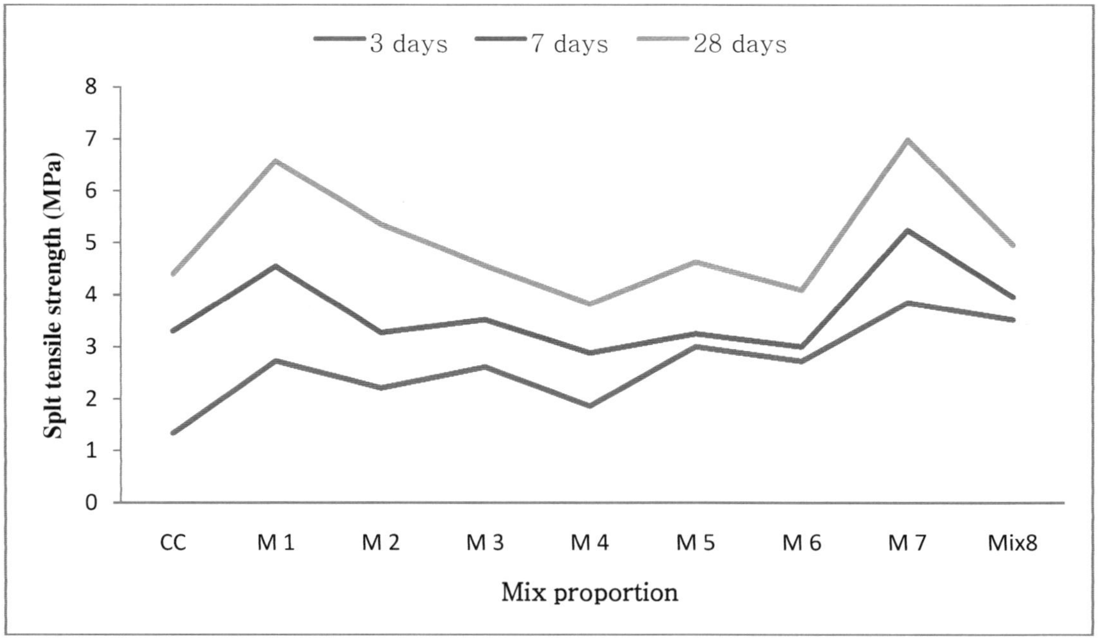

Figure3. Concrete split tensile for 3, 7 and 28

3.2.3 Flexure strength

As per IS 516-2016, prism moulds are tested for 3 days, 7 days and 28 days. The flexure strength of binary concrete is furnished in Table 13. Optimum packing effect of concrete is observed in Mix 7 when compared by other mixes. The graphical representation of optimum mix is shown in Figure 4.

Table 13. Flexure strength results of CC and Binary concrete

Mix Designation	GGBS (%)	Copper Slag(%)	Steel Fibres (%)	Flexure (Mpa.)		
				3	7	28
CC	0	0	0.0	4.1	4.2	4.5
M 1	30	40	1.0	4.0	5.2	5.9
M 2	30	40	1.5	3.6	4.2	5.4
M 3	30	60	1.0	3.9	4.4	4.9
M 4	30	60	1.5	3.0	3.9	4.7
M 5	50	40	1.0	3.8	4.3	5.1
M 6	50	40	1.5	3.6	3.9	4.9
M 7	50	60	1.0	4.9	5.3	6.1
M 8	50	60	1.5	4.5	4.9	5.5

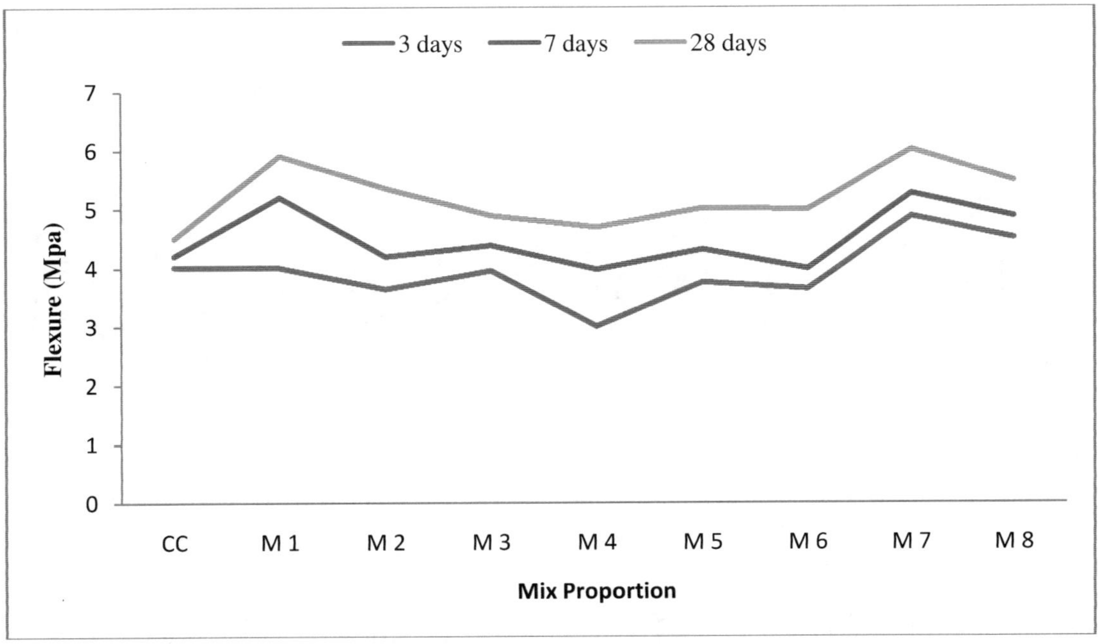

Figure 4. Concrete flexure strength for 3, 7 and 28 days

4. Conclusion

1. Based on above tests results, the strength compressive of Mix 7 is 72.15 Mpa. This is highest of all mixes.
2. From the above test results, the split tensile of Mix 7 is 6.98 Mpa. This is highest of entire mixes.
3. From the above test results, the flexure of Mix 7 is 6.01 Mpa. This is highest of the entire mixes.
4. As the slag percentage increases the workability decreases. The targeted 100mm slump is gained by adjustingquantity of super plasticizer.

5. From the above experimental results, it is recommended that OPC is partially replaced with 50% ofGGBS, fine aggregate is partially substitute with 60% of slag along with 1% steel fibre in concrete.
6. By using the GGBS and copper slag, reduction in Co_2 emission and conserving natural resources is established.
7. This study is applicable in general construction, runways, bridges etc.

References

[1] Vidhya K and Kanaka Mahalakshmi K*2019 Int.ResearchJ.Eng. and Tech.***06** 2128-31.
[2] Mahesh Bhimarao B, Akash Shrikant S and Dr Santosh Patil K 2020*Int.Research.J.Eng and Tech.***07**4201-04.
[3] Govindhan S and Papitha P2019*Int.Research.J.Eng. and Tech.* **06** 257-65.
[4] Venu Gopal C Suresh A and Gokul Nath V 2017*Int.Research.J.Engg and Tech.***03**313-22.
[5] Pavan Kumar M and Mahesh Y 2015 *J.Material.CivilEng.* **12**51-56.
[6] Freeda Christy C and Tensing D 2010*Indian J. Eng. and MaterialsScience.***17**140-44.

Effect of Welding Parameters on Mechanical Properties of Friction Stir Welding of Dissimilar Metals- A Review

M Sindhuja[1], S Neelakrishnan[2]and Benjamin Shiloh Davidson[3]

[1] Research Scholar, Department of Automobile engineering, PSG College of Technology, Coimbatore, Tamilnadu, India
[2]Professor, PSG College of Technology, Coimbatore, Tamilnadu, India
[3]Assistant Professor, PSG College of Technology, Coimbatore, Tamilnadu, India

* Correspondance e-mail: munnangisindhuja77@gmail.com

Abstract. This review article summarizes the extensive research works carried out on welding parameters (tool rotational speed, traverse speed, pin profile, axial force and tilt angle) effect on mechanical properties of FSW of dissimilar alloys. Extensive studies were being carried out by joining different materials using friction stir welding for different automobile and aerospace applications. FSW process takes place in the solid stage below the melting point of the metals to be joined and numerous materials including, aluminum, bronze, copper, titanium, steel, magnesium, and plastic can be welded. FS welded joints gives more strength and toughness when compared to fusion welded joints. FS welding technique is being used in many research institutes and production companies in the world.

Keywords: Friction stir welding, tool rotational speed, traverse speed, tilt angle, axial force, mechanical properties.

1. Introduction

The Welding Institute (TWI), England, U.K. developed FSW process which is used for joining lightweight materials [1]. FSW is a novel process to weld numerous metals and alloys in solid state condition providing good mechanical properties when compared to the welds formed by fusion processes. The properties of the FS welded samples depend upon the parameters like tool rotational speed, traverse speed and geometric conditions of tool. [2, 3]. A conventional NC operated friction stir welding machine (FSW-3T-NC) used for FSW process is shown in Figure 1.

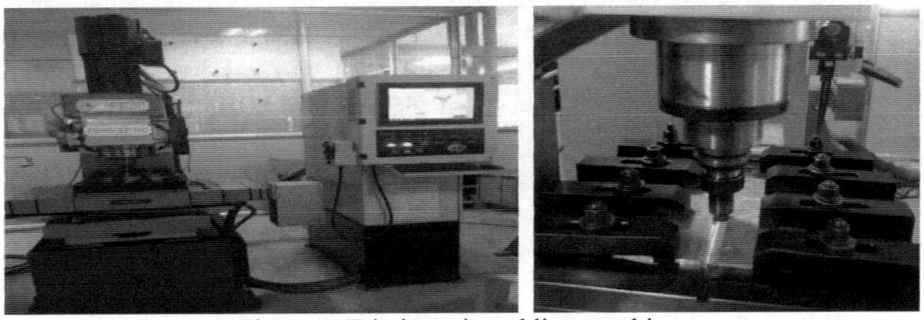

Figure 1: Friction stir welding machine.

Process parameters are to be examined critically since they are key factors for optimizing the process [4] . Material mixing, weld integrity and mechanical behaviour are influenced by the welding process parameters [5]. Plasticization and the material flow in FSW is controlled by the mechanical properties of base material underneath the action of rotating tool [6]. The molten material movement is determined by the probe design which also amends weld strength and characteristics. Tool design plays a key role in gaining weld strength and in tool geometry optimization [7]. AA1xxx metal matrix composite and the AA6xxx alloy were welded using FSW and the effect of FS weld parameters on the strength and properties of the welded joints were investigated [8]. Higher hardness is obtained than that of base materials in the investigation of FS welding of AA6xxx and AA5xxx alloys. The material flow which decides the strength of the dissimilar joints is directed by the pin profile of the tool [9]. FSW of heat treatable AA7xxx and AA6xxx alloys was done and had been found that material mixing and mechanical behaviour are influenced by the process parameters [5]. 2xxx and 6xxx alloys were joined using FSW in similar and dissimilar combinations. Results showed that fracture took place in the partially deformed zone of alloy 6xxx side where the sudden drop in hardness was observed. This occurred due to localized plasticity on the partially deformed zone of 6xxx, which took place due to joining of two dissimilar material with large strength mismatch [10]. AA 1xxx plates were joined using single and double pass friction stir welding. Highest hardness is seen in Double pass bottom flat plates due to fine grain refinement which is improved by thickness ration of plate and rotational speed [11]. Aluminium alloys AA5xxx and AA7xxx were fabricated by FSW and it was observed that parameters like plunge depth of the tool, rotational speed, weld traverse speed and D/d ratio influence the ultimate tensile strength [12]. Highest strength of the welds was obtained at 6000 rpm, 45 mm/min, 11kN and 1.5^0 using two tungsten-based tools in FS butt welding of AISI 316 L stainless steel plates [13]. Double-sided FSW was done with aluminium alloy plate under various rotational and welding speeds. High quality joints are formed with a steady flow of plastic material at a constant welding speed and lower rotational rate [14]. Increase in rotation and travel speeds in a constant proportion leads to higher tensile strength and ductility of the welded joints. The weld material exhibits higher strength than ductile base materials at the Mg/Al interface due to the presence of intermetallic compounds [15]. The joint strength will be low if the material flow is not good and liquation is high which occurs due to more heat generation through increased rotational and traverse speeds [16]. Rolled AA7075 aluminium plates of 12mm thickness were friction stir welded and it is found that higher fracture toughness properties are observed for the welded joints treated with solutionizing followed by artificial aging cycle [17]. Decrease in fracture toughness is observed in friction stir welded TA5 titanium alloy due to the crack propagation related to microstructure and texture of the joint [18]. Increase in fracture toughness is observed with decrease in input of heat and speed of the spindle in FS welded API X80 [19]. Table 1 exhibits the result of strength of FS welded joints effected by different welding parameters.

2. Applications of Dissimilar friction stir welding

Current applications involve use of different types of material based on its functional performance. Industries like marine, aerospace, and transportation which requires high strength lightweight structures can be benefited from dissimilar FSW. Design of aerospace and ship structures can be flexibly designed by FSW of dissimilar high strength and corrosion resistant alloys [20][4]. Many constructions need for joining of dissimilar materials and clad material. FSW has good possible for joining dissimilar aluminium alloys [21]. Aerospace industry uses FS welded high strength Al alloys as large tanks for satellite launch vehicles. Light alloy wheels and fuel tanks of high volume production in automotive industry uses FSW [22]. Modern Trent aero engines that drive Airbus A380 and the Boeing 787 uses friction stir welding [23]. The body of airplane is 20% of the aluminium alloy and AA2024, AA6061 and AA7075 are three dissimilar aluminium alloys which are commonly used in aerospace applications [24].

3. Effect of welding parameters on mechanical properties of friction stir welded joints

3.1. *Effect of tool rotational speed*

Rotational speed of the tool is a key weld process variable because it tends to influence the translational velocity [25]. It is observed that mixed flow region in FS welded AA5xxx-H111 and AA6xxx-T6 joints is affected by rotational speed and traverse speed. Uniform hardness disparity across the joint through metal plastic flow and mixing of base material alloys gives higher strength to the weld joint [26]. Tensile strength increases at rotational speed of 800 to 100rpm and with constant weld traverse speed of 120mm/min in FS welded AA5052-O alloy plates and the strength of the weld decreases as the rotational speed is increased from 1000 to 3000rpm [27]. Higher rotational speeds with a constant weld traverse speed generates heat input rapidly. Cavities are formed due to the turbulence flow caused by fluidity of the metal because of excess heat input. [28]. Higher rotational and traverse speeds will leads to defected weld joints because of lesser metal flow and higher liquation [29]. Higher rotational speeds with lower weld traverse speeds will lead to excessive heat generation which caused the weld metal to melt [16]. The rotation of the tool in counter clockwise direction was not able to produce a defect free and complete joint [30]. Higher rotational speed up to a certain value will improves impact energy of the welded joints and further increase will cause the strength to decrease as the heat generation causes grain refinement at weld zone [31]. The tensile strength (TS) of the FS welded joints depends upon the tool rotational speed [32]. 5083-H116 aluminium alloy and HSLA- 65 steel sheets were butt welded and observed that the lesser rotational speed with high weld traverse speed lead to poor material flow around the rotating pin and contributed to lack of fill void formation [33]. Different pin profiles and rotational speeds were used to FS weld the IS:65032 Al alloys. Higher tensile properties are obtained for the joints fabricated at 1300 rpm than the joints produced with other rotational speeds, without the concernment of weld traverse speed and pin profile [34]. Rotational speed with other parameters like pin profile, weld traverse speed and tilt angle are considered as insignificant in case of FS welded AA5454 and AA7075 alloys [12]. Tensile strength which exhibits variation through weld traverse speeds and rotational speeds seems to be related to the energy of the welded joints found in the investigation of EN AW-2024 T351 aluminium alloy welded joints [35]. Increasing the rotational speed leads to improvement of tensile and impact energies. Higher rotational speed and low feed rate leads to rapid solidification of the grains because of rapid generation of heat in FSW of AZ31B Mg alloys [36]. FSW on the dissimilar AA5058 alloy and polycarbonate (PC) sheets had been investigated and found that tool's rotational speed mainly controls the joint strength and the fracture mode [37]. Insufficient heat and inadequate material flow and mixing is caused due to departing of tool quickly which is occurred through low heat generation caused by Low tool rotation and higher feed rate [38]. Rotation of the tool generates the frictional heat due to stirring effect and allows the base materials to mix and plasticized. Tool spindle speed effects the material deformation rates of tool shoulder and pin [39]. When the rotational speed of the tool was increased, the generated heat was higher and excess material flush was noticed along with deposition of some base material to the FSW tool shoulder and cavities formation can be seen in the weld zone of the joints [40][41].

3.2. *Effect of tool traverse speed*

The traverse speed shows more impact on efficient strength of f FS welded AA8011 alloy components. Substantial increase in traverse speed leads to good quality welded joint with excellent joint properties [25].Generation of heat, joining of materials and straining amount in FSW are influenced by the tool traverse speed [42]. Plastic flow region development is influenced by the welding speed and mixed flow region is not appeared in the welded joints fabricated at highest and lowest weld traverse speeds [43]. Welded joints with good mechanical and microstructural properties can be produced with low traverse speed high heat treating effect [44]. Better joining of the metals with less material straining condition is obtained with weld traverse speeds at low rate and generation of heat due to the friction and distortion was high and vice versa [42]. The welding parameters, in particular the welding speed, have a great influence in the type of material flow that is produced [45]. Weak joints are formed due to generation of less heat at low rotational speed to weld traverse speed ratio, and increases as it results in less material flow [46]. With increasing the tool traverse speed, cross sectional defects like steel large fragments are caused which leads to lower tensile strength. Further increase of traverse speed leads to tunnel and voids kind of defects which gives less tensile strength [30]. More feed rate in FSW causes poor material flow resulting in a deep groove at the joint of lower strength [40]. Tensile strength is improved to a maximum value with high weld traverse speed and will be decreased with further increase in weld speed as the frictional heat generation is increased and decreased respectively. Voids in the joints are presented because of insufficiency in heat and less material flow in weld areas with short exposure period with higher weld traverse speeds [47]. Intermetallic compounds layer thickness will be reduced through shortening of high temperature period which occurs through high weld traverse speed [48]. AA6061 and AA7075 aluminium alloy plates are joined successfully in butt position using friction stir welding technique with two stir pins which have different geometry (cylindrical and triangular). Tensile strength linearly increased with increasing traverse speed [49]. Decrease in downward force (Fz) and increase in longitudinal force (Fx) is attained with increase in tool rotational speed which does not affect transverse force (Fy) [27]. Strong joining of the materials can be obtained due to the metal flow from rear side to front side of the tool pin as the pin eliminates the metal surface caused by the tool rotation. [42].

3.3. *Effect of tool pin profile*

Tools used in FSW with different pin profiles are shown in Figure 2. Weld joint strength and characteristics are varied by tool pin design which controls the movement of the liquified material. Tool design in FSW is significant as it helps in improving the strength of the welded joints. FSW tool consists of body, shoulder and pin (probe) and tool design also optimizes the geometry of the tool [45]. FSW tools with different pin profilesare used to weld AA5xxx-H111 and AA6xxx-T6 alloys and highest tensile strength of 273MPa is obtained at 950rpm with tool which has straight square pin tool due to mixing of alloys and high plastic flow which leads to homogeneous disparity of hardness along the joint [43]. Al–Mg–Cr–Mn (AA 5052-O) alloy was FS welded and was found to be defect free having good tensile strength produced by using a straight cylinder profiled tool. Stirring and mixing of material and increasing its temperature around the rotating tool pin is affected by the tool rotational speed [27]. J. Guo mentioned that the microstructural evolution and the material flow during welding were complex and mainly dependent on the tool geometry and the process parameters [8][50][2] . Sadeesh P et al. found that maximum tensile strength and defect free welds were obtained for the joint formed with cylindrical threaded and squared pin out of the profiles used (cylindrical threaded pin, straight cylindrical pin, taper pin, straight squared pin and straight stepped pin) [51]. Higher temperatures were produced with conical probe with three grooves and uniform material mixing in the weld zone is occurred with square frustrum tool pin in welded joints of 6xxx and 7xxx alloys. Generation of heat in FSW is influenced and affected by the tool pin profile and geometry which produces good material flow [52].

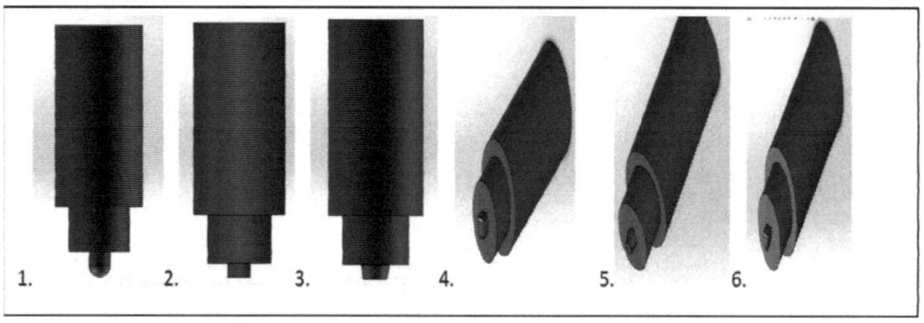

Figure 2: Tools with different pin profiles

3.4. *Effect of axial force*

Axial force plays a key part in FSW of dissimilar alloys as mentioned by Ramamoorthi et al. As the tool pin rotates and moves the plasticised metal towards retreating side, the tool shoulder pressure affects the tool pin penetration leading to good material flow. So, as the axial flow increases, the strength of weld increases and decreases with further increase in the force [53]. Gerbson de et al. mentioned that the production of flash can occur with high axial force, as this improves the heat input and allows the formation of a larger amount of plasticized material, and because of the instability of the application of axial force [54]. Application of axial force leads to plasticization and flow of the material with appropriate temperature which in turn produces good and defect free welded joints as mentioned by R. Srinivasan et al [55]. Axial force applied and tool pin movement and direction influences the flow of material which experiences the plastic deformation [56].

3.5. *Effect of tilt angle*:

The tool tilt angle is significant in dipping the tool pin with appropriate depth into the materials to be joined improving the devolution of the plastically deformed material by producing the sufficient heat required [57]. Tool should be tilted to enable a gradual increase of forging pressure [56]. The tool tilt angle also influences the surface appearance of the weld [58]. Shallow plunge depths produced due to large tilt angles reduce the contact area between tool shoulder and base materials so that the plasticized material cannot be transferred effectively from front to rear side of the tool with less heat generation leading to decrease in the weld strength and defective welds [59][60].

3.6. *Effect of welding parameters on microstructure*

Different microstructural zones are formed in FSW because of its asymmetric nature [56]. The different zones are heat affected zone (HAZ), thermo-mechanically effected zone and nugget zone or weld zone which is shown in Figure 3.

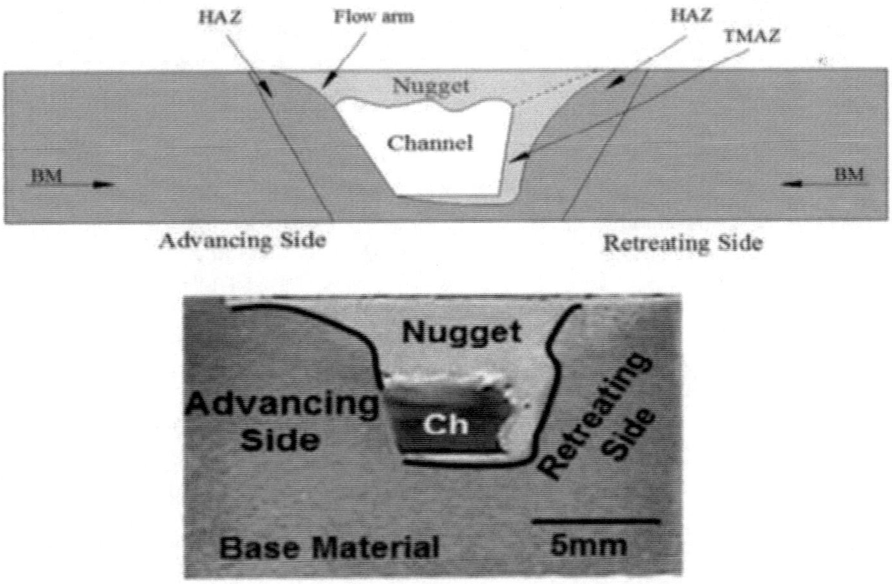

Figure 3: Zones in friction stir welding.

The stirring action of the FSW has produced a fine recrystallized structure at the stir zone. Homogeneous distributions of the fine grains are found at the stir zone. The intermetallic compounds which are appeared in the weld zone significantly influence the weld joint strength [61]. The equiaxed grain structure in the weld nugget zone is depicted in Figure 4.

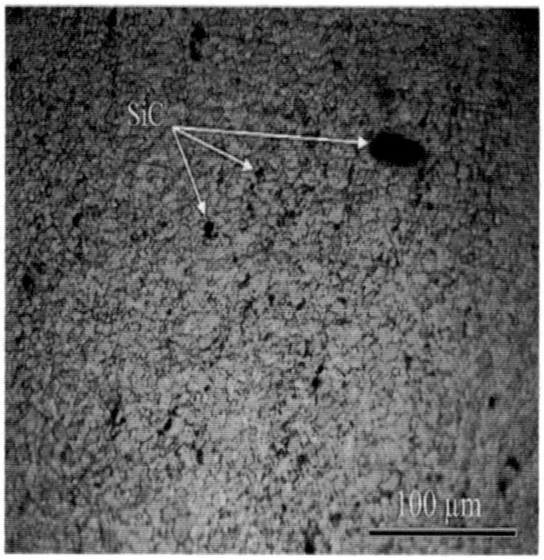

Figure 4: weld zone microstructure.

Sevvel et al have found that, the solidification of the weld zone was occurred rapidly due to higher rotational speed and low feed rate during welding of AZ31B-Mg plates. This rapid solidification leads to strengthening of weldments without defects [36]. Umasankar Das et al. welded AA6101 and AA6351 plates and stated that too much rise in the rotational speed declines the impact energy because of the high produced heat and grain refinement at the weld zone [31].

Table 1.welding parameters effect on strength of FSW joints

S.No	Authors and year	Material used	Inference	References
1	Peng chen et al. (2021	2195-T8 Al-Li alloy	Grain refinement strengthening is decreased with increase intool rotational speed.	[62]
2	Yadong Zhao et al. (2021)	1060 aluminum/AZ31B magnesium	An increase in the overlap ratio from 50 % to 75 % caused the maximum tensile shear force of the joint at 70mm/min tool traverse speed and 1500rpm tool rotational speed.	[63]
3	M. Sucharitha et al. (2021)	AZ31 magnesium alloy	Tool rotational speeds of 1120rpm and 900rpm produces welded joints with higher tensile strength and hardness.	[64]
4	Devender kumar et al. (2020)	AA6061/SiC Aluminum Matrix Composite	Tilt angle of the significantly influences the tensile strength of the weld joint besides tool rotational and travers speeds	[65]
5	S. Dharmalingam et al. (2020)	AA7475-AA8011	Defect free welded joints are attained at higher rotational and traverse speeds.	[66]
6	D. Wu et al. (2020)	Al-Cu aluminium alloys	Sound weld joint is obtained by adjustable-gap bobbin tool pin and tool traverse speed of 120-240mm/min.	[67]

7	H. I. Dawood et al. (2015)	AA6061	Tool pin profile with triangular pin produces the best metallurgical and mechanical weld properties among the other pin profiles used in FSW.	[68]
8	M. Felix Xavier Muthu et al. (2015)	AA1100H14 alloy and pure copper	The material flow and heat input in FSW of materials with different yield strengths is influenced by tool pin profile.	[69]
9	H. M. Rao et al. (2014)	AM60B Mg and AA6022-T4	Lower lap-shear failure loads were occurred with decrease in the width of the weld nugget zone resulted through higher tool rotational speeds.	[70]
10	S. Shashi Kumar et al. (2019)	AISI316L austenitic stainless steel	The tool tilt angle significantly effects the weld stir zone volume, increase in temperature and development of shear band in the weld zone.	[32]
11	Ling Long et al. (2017)	AA6061-T6 aluminium alloy	Wormhole defect on the advancing side can be avoided by the tool tilt angle which improves the material flow.	[71]
12	R. Srinivasan et al. (2018)	Squeeze cast A413 aluminium alloy	Good microstructural and mechanical properties were obtained in the welded joints produced with an axial force of 5kN.	[55]

4. Conclusion:

In this review, mainly the influences of welding conditions on mechanical properties have been investigated.Major conclusions drawn includes as follows.

1. The tool rotational speed has a significant effect on tensile strength of the FS welded joints.
2. Excess material liquation and less material flow gives low strength welded joints.
3. Formation of plastic flow region is affected by traverse speed and welded joints produced at lowest and highest tool traverse speeds does not show the mixed flow region.
4. Strength of the weld, movement of molten metal were altered and determined by the tool pin design.
5. The tool pin plunge depth into the base material during welding is achieved through applied axial force which influences the temperature and quantity of plasticised metal during FS welding.
6. Defective welds are produced due to large tilt angles which reduces the generation of heat and movement of metal flow.
7. The microstructural zones in friction stir welding are weld zone, heat affected zone and thermo-mechanically effected zone.
8. The strength of the FS weld is also affected by the microstructure which is affected and altered by rotational speed.

Acknowledgements
The Management and department of Automobile Engineering, PSG College of Technology is acknowledged for providing the necessary facilities for this research.

References

[1] Thomas, W. and E. Nicholas, Friction stir welding for the transportation industries.*Materials & design,* 1997. **18**(4-6): p. 269-273.
[2] Aval, H.J., S. Serajzadeh, and A. Kokabi, Evolution of microstructures and mechanical properties in similar and dissimilar friction stir welding of AA5086 and AA6061.*Materials Science and Engineering:* A, 2011. **528**(28): p. 8071-8083.
[3] Mahoney, M.W. and R.S. Mishra, Friction stir welding and processing. 2007: *ASM International.*
[4] Gite, R.A., P.K. Loharkar, and R.J.M.T.P. Shimpi, Friction stir welding parameters and application:A review. 2019. **19**: p. 361-365.
[5] Giraud, L., et al., Investigation into the dissimilar friction stir welding of AA7020-T651 and AA6060-T6. *Journal of Materials Processing Technology,* 2016. **235**: p. 220-230.
[6] Rajakumar, S. and V. Balasubramanian, Establishing relationships between mechanical properties of aluminium alloys and optimised friction stir welding process parameters.*Materials & Design*, 2012. **40**: p. 17-35.
[7] Das, J., et al., Review on process parameters effect on fatigue crack growth rate in friction stir welding. 2019. **18**: p. 3061-3070.
[8] Guo, J., P. Gougeon, and X.-G. Chen, Microstructure evolution and mechanical properties of dissimilar friction stir welded joints between AA1100-B4C MMC and AA6063 alloy.*Materials Science and Engineering:* A, 2012. **553**: p. 149-156.
[9] Ilangovan, M., S.R. Boopathy, and V. Balasubramanian, Microstructure and tensile properties of friction stir welded dissimilar AA6061–AA5086 aluminium alloy joints.*Transactions of Nonferrous Metals Society of China,* 2015. **25**(4): p. 1080-1090.
[10] Bharathi, S.S., et al., Mechanical Properties and Microstructural Characteristics of Friction Welded Dissimilar Joints of Aluminium Alloys. *Transactions of the Indian Institute of Metals*, 2018. **71**(1): p. 91-97.
[11] Sahu, P.K. and S. Pal, Mechanical properties of dissimilar thickness aluminium alloy weld by single/double pass FSW.*Journal of Materials Processing Technology*, 2017. **243**: p. 442-455.
[12] Elnabi, M.M.A., et al., Influence of friction stir welding parameters on metallurgical and mechanical properties of dissimilar AA5454–AA7075 aluminum alloys.*Journal of Materials Research and Technology*, 2019. **8**(2): p. 1684-1693.
[13] Kumar, S.S., N. Murugan, and K. Ramachandran, Influence of tool material on mechanical and microstructural properties of friction stir welded 316L austenitic stainless steel butt joints. *International Journal of Refractory Metals and Hard Materials*, 2016. **58**: p. 196-205.
[14] Xu, W., et al., Mechanical behavior of 7085-T7452 aluminum alloy thick plate joint produced by double-sided friction stir welding:Effect of welding parameters and strain rates.*Journal of Manufacturing Processes*, 2018. **35**: p. 261-270.
[15] Mohammadi, J., et al., Friction stir welding joint of dissimilar materials between AZ31B magnesium and 6061 aluminum alloys: Microstructure studies and mechanical characterizations. *Materials Characterization*, 2015. **101**: p. 189-207.
[16] Pourahmad, P. and M. Abbasi, Materials flow and phase transformation in friction stir welding of Al 6013/Mg.*Transactions of Nonferrous Metals Society of China*, 2013. **23**(5): p. 1253-1261.
[17] Sivaraj, P., et al., Fracture toughness properties and characteristics of Friction stir welded high strength aluminium alloy by post weld heat treatment (PWHT). 2021. **43**: p. 2150-2155.
[18] Gao, F., et al., Fracture toughness of friction stir welded TA5 titanium alloy joint. 2020. **776**: p. 138962.
[19] Tribe, A. and T.W.J.E.F.M. Nelson, Study on the fracture toughness of friction stir welded API X80. 2015. **150**: p. 58-69.

[20] Singh, U.K. and A.K. Dubey, Welding of dissimilar Mg alloys using indigenously developed friction stir welding set-up.*Materials Today: Proceedings,* 2020.
[21] Karlsson, L., E.-L. Berqvist, and H.J.W.i.t.W. Larsson, Application of friction stir welding to dissimilar welding. 2002. **46**(1): p. 10-14.
[22] Kallee, S., Industrial applications of friction stir welding, in Friction stir welding. 2010, *Elsevier.* p. 118-163.
[23] Mishra, A.J.A.a.S., Friction stir welding of dissimilar metal: a review. 2018.
[24] Boşneag, A., et al. Friction Stir Welding of three dissimilar aluminium alloy used in aeronautics industry, *IOP Conference Series: Materials Science and Engineering*. 2017. IOP Publishing.
[25] Davidson, B.S. and S. Neelakrishnan, Influence of Friction Stir Welding Parameters on Tensile Properties of AA8011 Aluminium Alloy Plate. J*ournal of Computational and Theoretical Nanoscience*, 2018. **15**(1): p. 93-98.
[26] Palanivel, R., et al., Effect of tool rotational speed and pin profile on microstructure and tensile strength of dissimilar friction stir welded AA5083-H111 and AA6351-T6 aluminum alloys. *Materials & Design*, 2012. **40**: p. 7-16.
[27] Moshwan, R., et al., Effect of tool rotational speed on force generation, microstructure and mechanical properties of friction stir welded Al–Mg–Cr–Mn (AA 5052-O) alloy.*Materials & Design* (1980-2015), 2015. **66**: p. 118-128.
[28] Gaafer, A., T. Mahmoud, and E. Mansour, Microstructural and mechanical characteristics of AA7020-O Al plates joined by friction stir welding.*Materials Science and Engineering: A*, 2010. **527**(27-28): p. 7424-7429.
[29] Azizieh, M., et al., Mechanical properties and microstructural evaluation of AA1100 to AZ31 dissimilar friction stir welds.*Materials Chemistry and Physics*, 2016. **170**: p. 251-260.
[30] Yazdipour, A. and A. Heidarzadeh, Effect of friction stir welding on microstructure and mechanical properties of dissimilar Al 5083-H321 and 316L stainless steel alloy joints.*Journal of Alloys and Compounds,* 2016. **680**: p. 595-603.
[31] Das, U. and V. Toppo, Effect of Tool Rotational Speed on Temperature and Impact Strength of Friction Stir Welded Joint of Two Dissimilar Aluminum Alloys. *Materials Today: Proceedings,* 2018. **5**(2): p. 6170-6175.
[32] Kumar, S.S., N. Murugan, and K. Ramachandran, Microstructure and mechanical properties of friction stir welded AISI 316L austenitic stainless steel joints.*Journal of Materials Processing Technology*, 2018. **254**: p. 79-90.
[33] Wang, T., et al., Friction stir butt welding of strain-hardened aluminum alloy with high strength steel.*Materials Science and Engineering:* A, 2018. **737**: p. 85-89.
[34] Rao, M.S., B.R. Kumar, and M.M. Hussain, Experimental study on the effect of welding parameters and tool pin profiles on the IS: 65032 aluminum alloy FSW joints.*Materials Today: Proceedings*, 2017. **4**(2): p. 1394-1404.
[35] Milčić, M., et al., Experimental investigation of fatigue properties of FSW in AA2024-T351.*Procedia Structural Integrity*, 2018. **13**: p. 1977-1984.
[36] Sevvel, P. and V. Jaiganesh, Characterization of mechanical properties and microstructural analysis of friction stir welded AZ31B Mg alloy thorough optimized process parameters.*Procedia Engineering,* 2014. **97**: p. 741-751.
[37] Derazkola, H.A. and M. Elyasi, The influence of process parameters in friction stir welding of Al-Mg alloy and polycarbonate.*Journal of Manufacturing Processes,* 2018. **35**: p. 88-98.
[38] Raturi, M., A. Garg, and A. Bhattacharya, Joint strength and failure studies of dissimilar AA6061-AA7075 friction stir welds: Effects of tool pin, process parameters and preheating. *Engineering Failure Analysis,* 2019. **96**: p. 570-588.
[39] Kumar, S.S., N. Murugan, and K. Ramachandran, Identifying the optimal FSW process parameters for maximizing the tensile strength of friction stir welded AISI 316L butt joints.*Measurement*, 2019. **137**: p. 257-271.

[40] Prasad, B.L., et al., Joining of AZ91 Mg alloy and Al6063 alloy sheets by friction stir welding. *Journal of magnesium and alloys*, 2018. **6**(1): p. 71-76.
[41] Jassim, A.K., D.C. Ali, and A. Barak, Effect of Tool Rotational Direction and Welding Speed on the Quality of Friction Stir Welded Al-Mg Alloy 5052-O.
[42] Ramachandran, K., N. Murugan, and S.S. Kumar, Influence of tool traverse speed on the characteristics of dissimilar friction stir welded aluminium alloy, AA5052 and HSLA steel joints. *Archives of civil and mechanical Engineering,* 2015. **15**(4): p. 822-830.
[43] Palanivel, R., et al., Mechanical and metallurgical properties of dissimilar friction stir welded AA5083-H111 and AA6351-T6 aluminum alloys.*Transactions of Nonferrous Metals Society of China,* 2014. **24**(1): p. 58-65.
[44] Gungor, B., et al., Mechanical, fatigue and microstructural properties of friction stir welded 5083-H111 and 6082-T651 aluminum alloys.*Materials & Design* (1980-2015), 2014. **56**: p. 84-90.
[45] Infante, V., et al., Study of the fatigue behaviour of dissimilar aluminium joints produced by friction stir welding.*International Journal of Fatigue*, 2016. **82**: p. 310-316.
[46] Firouzdor, V. and S. Kou, Al-to-Mg friction stir welding: effect of positions of Al and Mg with respect to the welding tool.*Weld. J,* 2009. **88**(11): p. 213-224.
[47] Palanivel, R. and P.K. Mathews, The tensile behaviour of friction-stir-welded dissimilar aluminium alloys.*Materiali in Tehnologije*, 2011. **45**(6): p. 623-626.
[48] Liu, X., S. Lan, and J. Ni, Analysis of process parameters effects on friction stir welding of dissimilar aluminum alloy to advanced high strength steel. *Materials & Design*, 2014. **59**: p. 50-62.
[49] Sarsilmaz, F., N. Ozdemir, and I. Kirik, Evaluation of microstructure and fatigue properties of dissimilar AA7075/AA6061 joints produced by friction stir welding.*Kov. Mater*, 2012. **50**: p. 259-268.
[50] Ahmadi, H., et al., Influence of pin profile on quality of friction stir lap welds in carbon fiber reinforced polypropylene composite. *International Journal of Mechanics and Applications*, 2012. **2**(3): p. 24-28.
[51] Sadeesh, P., et al., Studies on friction stir welding of AA 2024 and AA 6061 dissimilar metals.*Procedia Engineering*, 2014. **75**: p. 145-149.
[52] Aval, H.J., Influences of pin profile on the mechanical and microstructural behaviors in dissimilar friction stir welded AA6082–AA7075 butt joint.*Materials & Design,* 2015. **67**: p. 413-421.
[53] Ramamoorthi, R., et al., An investigation of the impact of axial force on friction stir-welded AA5086/AA6063 on microstructure and mechanical properties butt joints.*Materials Today:* Proceedings, 2020.
[54] de Queiroz Caetano, G., et al., Influence of rotation speed and axial force on the friction stir welding of AISI 410S ferritic stainless steel. 2018. **262**: p. 430-436.
[55] Srinivasan, R., A. Ramesh, and A.J.M.T.P. Athithanambi, Effect of axial force on microstructure and mechanical properties of friction stir welded squeeze cast A413 aluminium alloy. 2018. **5**(5): p. 13486-13494.
[56] Venu, B., et al., A review on Friction Stir Welding of various metals and its variables. 2019. **18**: p. 298-302.
[57] Anand, R. and V.J.M.T.P. Sridhar, Studies on process parameters and tool geometry selecting aspects of friction stir welding–A review. 2020. **27**: p. 576-583.
[58] Rajendran, C., et al., Effect of tool tilt angle on strength and microstructural characteristics of friction stir welded lap joints of AA2014-T6 aluminum alloy. 2019. **29**(9): p. 1824-1835.
[59] Kumar, S.S., N. Murugan, and K.J.M. Ramachandran, Effect of tool tilt angle on weld joint properties of friction stir welded AISI 316L stainless steel sheets. 2020. **150**: p. 107083.
[60] Zhai, M., C. Wu, and H.J.J.o.M.P. Su, Influence of tool tilt angle on heat transfer and material flow in friction stir welding. 2020. **59**: p. 98-112.
[61] Moradi, M.M., H.J. Aval, and R.J.J.o.M.P. Jamaati, Effect of pre and post welding heat treatment in SiC-fortified dissimilar AA6061-AA2024 FSW butt joint. 2017. **30**: p. 97-105.

[62] Chen, P., et al., Effect of rotation speed on microstructure evolution and mechanical properties of nugget zone in 2195-T8 Al-Li alloy friction stir welding joints. 2021: p. 111079.
[63] Zhao, Y., et al., Effect of welding parameters on the microstructures and mechanical properties of double-pass aluminum/magnesium dissimilar metal friction stir lap welding joint. 2021. **26**: p. 102132.
[64] Sucharitha, M., B.R. Sankar, and P.J.M.T.P. Umamaheswarrao, Experimental investigations on the effect of tool rotational speed on mechanical properties and microstructure of friction stir welded AZ31 Mg alloy. 2021.
[65] Kumar, D., et al., A parametric study of friction stir welded AA6061/SiC AMC and its effect on microstructure and mechanical properties. 2020.
[66] Dharmalingam, S. and K.J.M.T.P. Lenin, Effect of friction stir welding parameters on microstructure and mechanical properties of the dissimilar (AA7475-AA8011) joints. 2021. **39**: p. 105-109.
[67] Wu, D., et al., Effect of an improved pin design on weld formability and mechanical properties of adjustable-gap bobbin-tool friction stir welded Al-Cu aluminum alloy joints. 2020. **58**: p. 1182-1188.
[68] Dawood, H., et al., Effect of small tool pin profiles on microstructures and mechanical properties of 6061 aluminum alloy by friction stir welding. 2015. **25**(9): p. 2856-2865.
[69] Muthu, M.F.X. and V.J.T.o.N.M.S.o.C. Jayabalan, Effect of pin profile and process parameters on microstructure and mechanical properties of friction stir welded Al–Cu joints. 2016. **26**(4): p. 984-993.
[70] Rao, H., et al., Effect of process parameters on mechanical properties of friction stir spot welded magnesium to aluminum alloys. 2015. **66**: p. 235-245.
[71] Long, L., et al., Finite-element analysis of the tool tilt angle effect on the formation of friction stir welds. 2017. **30**: p. 562-569.

A review on Progressive failure analysis of composites

Ch. Srilakshmi [1], G. Sambasivarao[2], J. Suresh Kumar[3]

[1] Research Scholar, Dept. of Mechanical Engineering, JNTUH, Hyderabad, India.
[2] Professor & Principal, Sir C. R. Reddy college of Engineering, Eluru,India
[3] Professor, Department of Mechanical Engineering, JNTUH, Hyderabad, India

Corresponding author e-mail: srilakshmi.settipalli9@gmail.com

Abstract. This article presents a review of the progressive failure theories for analysing composites by using computational analysis methods like finite element method. Failure severity can be represented by the degree to which the elastic properties of a ply are damaged. Fibre failure is usually assumed to cause the failure of the matrix. So these fibre ply-to-ply failure mechanisms will be studied and analysed for better usage of the models for respective analysis. Sudden degradation models study the ply loading history with increase in loading, elastic properties will be reduced by a discrete factor resulting to successive failure mechanisms. Such models and failure mechanisms were studied and a detailed review was presented with possible simulation models in progressive failure of composites.

Key words: Progressive failure, composites, damage tolerance, RVE model, Multi Scale Model.

1. Introduction

Carbon fibrereinforced plastics (CFRPs) are being developed and optimized for better performance and usage over the past 50 years. Composite structures have been evaluated to ensure the integrity of a structure they were incorporated in. Fiber failure is usually assumed to cause the failure of the matrix. So these fibre ply-to-ply failure mechanisms will be studied and analysed for better usage of the models for respective analysis. Sudden degradation models study the ply loading history with increase in loading, elastic properties will be reduced by a discrete factor resulting to successive failure mechanisms.

2. Evolution of Failure models

Composite structures have been evaluated to ensure the integrity of a structure they were incorporated in. Fibre failure is usually assumed to cause the failure of the matrix. So these fibre ply-to ply failure mechanisms will be studied and analysed for better usage of the models for respective analysis.With the help of several models, testing methodologies and structural analysis theories, mechanical behaviour and characteristics of a composite material with minimal cost were developed in course of time [1]. In such a view point, if we go through the sequence of development in the failure theories Tsai-Wu tensor criterion would be an earlier one [2]. They are usually the concepts which describes the phenomena in conjunction with the models under certain assumptions to predict the strength of materials like composites. They are usually considered as semi-empirical models. These models have certain assumptions with composite as a material.

One of such assumption is that in a single-ply the composite exists with homogeneous properties. At macro level the failure of a material can be defined on the basis of stress-strain data which gives an engineering emphasis due to its wide usage from Hook's law.

But when we go into deeper levels to master a material for desired outcome, there is a need to study deeper than just what it looks to be at macro scales. So certain theories evolved to state micro and meso scale of attainment with the mechanism of how a material works and fails at last in particular state of scaled position, how it behaves in different scales of constituent level. For explaining the failure of a composite along with its mechanical behaviour at a constituent level many studies have been carried over and focused on failure mechanisms. In such effort, there came an evolution of non-empirical theories like Hashin criterion [3], Puck's criterion which explains the action planes and strength [4]. A criteria that can predict matrix and fiber failure with better accuracy, without the curve-fitting parameters have been emerged, and named as LaTCO3 criterion [5]. It is capable to suggest a crack initiation in the matrix failure due to fiber compression by a combination of five interactive conditions [6]. All these failure theories mainly used for macro-level analysis of stresses and material behaviour by considering the failure components [8]. When characterization becomes difficult, there will be a necessity of considering one or more additional constitutive or phenomenological interactive parameters. Similarly in failure behaviour study and its characterization such interactive parameters were required. Researchers like Chamis et.al [9], described failure modes based on modified distortion energy (MDE) of stressed laminates which gave rise to more scope in micro-mechanics based models of failure. In this model the macro-effective strength can also be determined by using homogenized constituents of strength in a laminate.

With these studies, a general consecutive statement can be given as, the micro-level events leads to the consequence of macro-level failure mechanisms. To predict the failure aimed at the constituents of the composite material became more intuitive so that fully determined micro-mechanical failure methods can be established. Disadvantage of such model is inefficient in modelling and numerical computation difficulties arises [10, 11].

Another way of characterization is to consider the representative volume elements (RVEs) to determine the properties and so the behaviour of the material in study along with its constituents (fibre and matrix). Using these RVEs, Bouaoune [12] and Beicha [13] studies the effects of periodic and random distributions of fibers in composites. Global effective elastic properties can be captured by using periodic RVEs. With these models there is a defect where the effective properties of composites with high fiber fractions are affected by micro-structures.

So to make a probable situation for better analysis and estimating a proper failure of the composite the model should possess both the advantage of the efficiency of macro models and the accuracy of micro models [14, 15]. For such a case, new models have been evolved taking advantage of combining the macro models with the RVE models forming to generic study of multi-scale approaches.

3. Need of Multi-Scale strategy in failure modelling

There are certain studies carried out by world-wide failure exercise (WWFE), their results concluded that there should be further development in the failure models and testing methodologies as the existing knowledge on failure models lacks certain features like predicting tolerable limit. Predicting an experimental response is also a task to be fulfilled. As the difference between a predicted value and the experimental response were not unusual in practice, the order of magnitude definitely doesn't match.

So researchers are continuously working to fill this gap. LaRCO3 model might be the first of its trials with predicted failure envelops nicely correlated with the experimental results, in the case of matrix or fibre failure under compression [5].

Studies of Benbouras [7] states that, Tsai-Wu criterion overestimates the succession and first macroscopic failure whereas Hasin, LaRCO3, Tsai-Hill and maximum strain theories were well established along the lines. Their studies also provided a strong correlation between the different failure criteria tested. They proposed an algorithm in which the deflection causing the damage allows the calculation of the force applied using the nonlinear approach of Venetis [16].

For studying the physical sense of micro-strength properties of a composite Zhidong Guan [8], recorded an experimental data of sensitivity analysis with description as shown in the table 1. It describes about the micro strength in modified – micromechanical failure (MMF3) criterion and their influence over macro strength with sensitivity analysis considerations. This reference study can make an adjunct point for the further development of new models towards multi-scale approach.

Table 1. The matrix-dominant micro strength for the sensitivity analysis [8].

Number	Tm/(MPa)	Cm/(MPa)	Sm/(MPa)	Description
1	138.7	353.3	145.1	Reference value
2	208.1	353.3	145.1	Tm is increased by 50%
3	92.5	353.3	145.1	Tm is decreased by 50%
4	138.7	530.0	145.1	Cm is increased by 50%
5	138.7	235.5	145.1	Cm is decreased by 50%
6	138.7	353.3	217.7	Sm is increased by 50%
7	138.7	353.3	96.7	Sm is decreased by 50%

The influence of matrix-dominant micro strength of the composite over the macro level off-axis strength is in accordance with their physical definitions. Here we can notice that the micro tensile properties are affected with relatively to the off-axis angles. Also the compressive strength of the matrix gets involved with the strength affecting the off-axis strength at the small off-axis angle. Authors also noted that, the micro compressive and shear strength of the matrix in a composite does actually have a little influential over the macro level relative strengths (compressive and tensile) at the off-axis angles. The influence is more over the shear strength at macro-level.

4. Multi-Scale Failure models

Gosse et al. [17] proposed a new theory based upon micro-strains and named it as the strain invariant failure theory (SIFT) which was based on the micro-strains. Upon later development to this theory Tay et.al. [18] Combined the element-failure method (EFM) with the SIFT. Tay's method studied the damage propagation of the composites. Based upon SIFT micro-level failures were studied on open-hole compression strength of laminates by Cai et.al. [19]. Stacking sequence plays a crucial role in SIFT for finding the critical values that need to be determined by testing the laminates. Here the laminates were arranged in a specially designed stacking sequences that can determine the micro-properties.

In further developments, notched and un-notched laminates based on a multi-axial mixed model continuum damage model (MMCDM) which is considered as a 3-D extension of Hasin criterion was proposed by Stier et.al. [20]. This method based upon the average strains within the RVE determines damage initiation in a laminate. Another set of researchers collaborated to draw a set of conditions for predicting failure of composites by utilizing average stresses within the matrix and fibre constituents raising to give Multi-Continuum Theory (MCT). Mayes et.al. [21, 22] proposed MCT, where micro-level property changes were considered.

For countering the demerit of these models where localized peak stresses in a laminate can be difficult to distinguish the damage initiation and degradation; Ha et.al., [23] proposed a new theory, micromechanics of failure criterion to predict the ultimate strength of a laminate [24]. Efforts of Muthusamy [25], Huang et. Al [26] made a new path towards multi-scale approaches with MMF criterion. Here several researchers studied about failure and life of composites utilizing the benefactor of MMF multi-scale models [28, 29].

The drawback of this model is that it has two significant demerits: 1) under any longitudinal tension there exists a premature failure of a matrix. 2) Shear strength will be predicted much lower than it used to be.

In this setting, the present study focussed the development and advantages of multiscale models. The thermal residual stress adding upto multilevel strategy inclusion to MMF3 criterion where the damage initiation was defined and degradation of the properties of constituent elements. The fundamental stand with the modification of the micromechanics-based criterion focuses directly on defining the matrix damage.

The off-axis loading using multiscale model were analysed and studied for knowing the matrix-dominant mechanical behaviours [8]. The RVE model really played a major role in the development of composite failure analysis. If square and hexagon are considered for the representative of different fibre distributions, then a quantitative relationship between the micro and macro level stresses were determined.

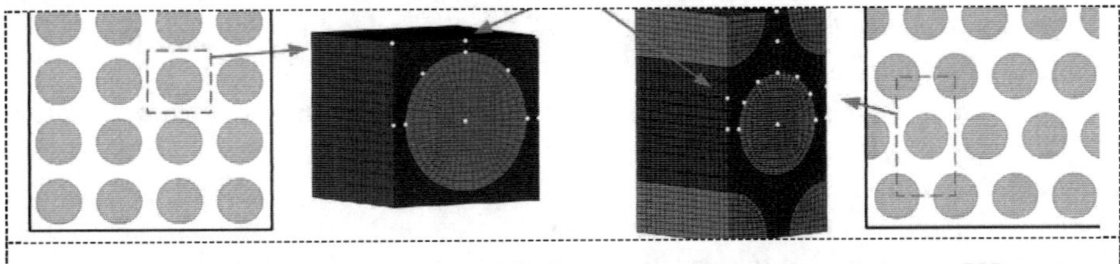

Figure 1. A sample RVE model with Square and Hexagon elements [8].

There is a continuous development in the field, further studies can be contributed towards different RVE models evaluated upon modified multi-scale damage method applied to progression of damage in the composites with long and short fibres is recommended.

5. Conclusion

A comprehensive review on the development of multi-scale strategy has been presented. Macro-scale models and micro-scale models were discussed towards how and why multi-scale models have been evolved. The RVE model really played a major role in the development of composite failure analysis. If square and hexagon are considered for the representation of different fiber distributions, then a quantitative relationship between the micro and macro level stresses were determined. In the multi scale strategy first stress were calculated and later the micro strength was reverse calculated from the macro strength of laminates. When the properties of damaged elements are observed to be degraded at fiber or matrix constituent level then, the more full-fledged multiscale models which were established are applied for studying the failure analysis of the composite. In the literature the studies with multi-scale strategies are effective with the constituent failure analysis of a laminate.

Also with the thermal residual studies, the effect cannot be negligible before damage initiation. This is because matrix failure under shear loading will lag which attains an overestimated strength not feasible to fit the fitment criteria. But as the load increases thermal residual stress value tends to descend.

References

[1] Orifici, A.C.; Herszberg, I.; Thomson, R.S.,2008, Review of methodologies for composite material modelling incorporating failure, *Journal of Composite Structures*,**86**, 194–210.

[2] Tsai, S.W.; Wu, E.M., 1971, A general theory of strength for anisotropic materials. *J. Compos. Mater.*, **5**, 58–80.

[3] Hashin, Z., 1980, Failure criteria for unidirectional fiber composites. *J. Appl. Mech.*, **4**, 329–

334.
[4] Puck, A.; Schürmann, H.,1998, Failure analysis of FRP laminates by means of physically based phenomenological models. *Compos. Sci. Technol.*, **58**, 1045–1067.
[5] Carlos G. Dávila et.al.,2005, Failure Criteria for FRP Laminates, *Journal of Composite Materials*, 39(4).
[6] C. Davila, N. Jaunky, and S. Goswami, 2003, "Failure Criteria for FRP Laminates in Plane Stress," in 44th AIAA/ASME/ASCE/AHS/ASC Structures, *Structural Dynamics, and Materials Conference*,Seattle,WA, USA,.
[7] Y. Benbouras, et. al., 2019, A review and application of the failure criteria on a graphite epoxy under three point bending test, *MATEC Web of Conferences***286**, 02007.
[8] Zhidong Guan et.al.,2018, Multiscale Analysis of CFRP Laminates with MMF3 Criterion under Different Off-Axis Loading Conditions, *Materials*, **11**, 2255.
[9] Davila, C.; Jaunky, N.; Goswami, S., 2003, Failure criteria for FRP Laminates in plane stress. In Proceedings of the 44th AIAA/ASME/ASCE/AHS/ASC Structures, *Structural Dynamics, and Materials Conference*, Norfolk, VA, USA.
[10] Chamis, C.C.; Abdi, F.; Garg, M.; Minnetyan, L.; Baid, H.; Huang, D.; Housner, J.; Talagani, F., 2013, Micromechanics-based progressive failure analysis prediction for WWFE-III composite coupon test cases. *J. Compos. Mater.*, **47**, 2695–2712.
[11] Han, G.; Guan, Z.; Li, X.; Ji, R.; Du, S., 2017, The failure mechanism of carbon fiber-reinforced composites under longitudinal compression considering the interface. *Sci. Eng. Compos. Mater.*, **24**, 429–437.
[12] Meng, Q.; Wang, Z., 2016, Micromechanical modeling of impact damage mechanisms in unidirectional composite Laminates. *Appl. Compos. Mater.*, **23**, 1099–1116.
[13] Bouaoune, L.; Brunet, Y.; El Moumen, A.; Kanit, T.; Mazouz, H.,2016, Random versus periodic microstructures for elasticity of fibers reinforced composites. *Compos. Part B Eng.*, **103**, 68–73.
[14] Beicha, D.; Kanit, T.; Brunet, Y.; Imad, A.; El Moumen, A.; Khelfaoui, Y., 2016, Mechanics of materials effective transverse elastic properties of unidirectional fiber reinforced composites. *Mech. Mater.*, **102**, 47–53.
[15] Wang, Y.; Huang, Z., 2018, Analytical micromechanics models for elastoplastic behavior of long fibrous composites: A critical review and comparative study. *Materials*, **11**, 1919.
[16] Y. Benbouras, A. Maziri, E. Mallil, and J. Echaabi, A, 2017, Nonlinear Analytical Model For Symmetric Laminated,*International Journal of Technology*, **8(6)**,437–447, 2017.
[17] Gosse, J.; Christensen, S., 16–19 April 2001, Strain invariant failure criteria for polymers in composite materials. In Proceedings of the 42th AIAA/ASME/ASCE/AHS/ASC Structures, *Structural Dynamics, and Materials Conference and Exhibit*, Seattle,WA, USA,.
[18] Tay, T.E.; Tan, S.H.N.; Tan, V.B.C.; Gosse, J.H., 2005, Damage progression by the element-failure method (EFM) and strain invariant failure theory (SIFT). *Compos. Sci. Technol.* **65**, 935–944.
[19] Cai, H.; Miyano, Y.; Nakada, M., 2009, Long-term open-hole compression strength of CFRP laminates based on strain invariant failure theory. *J. Thermoplast. Compos. Mater.*, **22**, 63–81.
[20] Stier, B.; Naghipour, P.; Pineda, E.J.; Hansen, L.; Arnold, S.M.; Bednarcyk, B.A.; Waas, A.M. 5–9 January 2015, Multiscale static analysis of notched and unnotched laminates using the generalized method of cells. In Proceedings of the 56th AIAA/ASCE/AHS/ASC Structures, *Structural Dynamics, and Materials Conference*, Kissimmee, FL, USA.
[21] Mayes, J.S.; Hansen, A.C., 2004, A comparison of multicontinuum theory based failure simulation with experimental results. *Compos. Sci. Technol.,***64**, 517–527.
[22] Mayes, J.S.; Hansen, A.C. 2004, Composite laminate failure analysis using multi continuum theory. *Compos. Sci. Technol.,* **64**, 379–394.
[23] Ha, S.K.; Jin, K.K.; Huang, Y., 2008, Micro-mechanics of failure (MMF) for continuous fiber reinforced composites. *J. Compos. Mater.***42,** 1873–1895.

[24] Ha, S.K.; Huang, Y.; Han, H.H.; Jin, K.K.,2010, Micromechanics of failure for ultimate strength predictions of composite laminates. *J. Compos. Mater.*,**44**, 2347–2361.

[25] Muthusamy, P.; Srinivasan, S.M.,2014, Micromechanical modeling of onset and progression of damage in composite structures. *Int. J. Comput. Methods Eng. Sci. Mech.*, **15**, 184–191.

[26] Huang, Y.; Jin, C.; Ha, S.K.,2012, Strength prediction of triaxially loaded composites using a progressive damage model based on micromechanics of failure. *J. Compos. Mater.*, **47**, 777–792.

[27] Huang, Y.; Xu, L.; Kyu Ha, S., 2012, Prediction of three-dimensional composite laminate response using micromechanics of failure. *J. Compos. Mater.*, **46**, 2431–2442.

[28] Sun, X.S.; Tan, V.B.C.; Tay, T.E., 2011, Micromechanics-based progressive failure analysis of fibre-reinforced composites with non-iterative element-failure method. *Comput. Struct.*, **89**, 1103–1116.

[29] Sihn, S. 2008, Strength prediction of composites using micromechanics-based failure. In Strength & Life of Composites; Tsai, S.W., Ed.; *JEC-Composites*, Stanford University: Stanford, CA, USA, 629–651.

[30] Jin, K.K.; Huang, Y.; Lee, Y.H.; Ha, S.K.,2008, Distribution of micro stresses and interfacial tractions in unidirectional composites. *J. Compos. Mater.*, **42**, 1825–1849.

[31] Li, X.; Guan, Z.; Li, Z.; Liu, L., 2014, A new stress-based multi-scale failure criterion of composites and its validation in open hole tension tests. *Chin. J. Aeronaut.*, **27**, 1430–1441.

[32] Polymer Matrix Composite 32., 2000, American Society for Testing Materials. Standard Test Method for Tensile Properties of Materials; ASTM D 3039/D 3039M-00; *ASTM International*,West Conshohocken, PA, USA,.

A Review on Damage Modelling and Analysis of Composites

Ch. Srilakshmi [1], G. Sambasiva Rao[2], J. Suresh Kumar[3]

[1] Research Scholar, Dept. of Mechanical Engineering, JNTUH, Hyderabad, India
[2] Professor & Principal, Sir C. R. Reddy college of Engineering, Eluru,India
[3] Professor, Department of Mechanical Engineering, JNTUH, Hyderabad, India

Corresponding author e-mail: srilakshmi.settipalli9@gmail.com

Abstract. This article presents an overview of advances and some results of the authors research interest in the modeling and analysis of progressive damage of fiber-reinforced composites and joining operations viz., lap joint. Similarly different models were studied and presented in comprehensive to the progressive damage of composites in meso scale level. A complex multi-scale and interactive nature of the damage at mesoscale level were studied and presented with possible sophisticated modelling and analytical techniques considering simplified assumptions to bring model simplification from unduly complexities. Such models with test data were studied for damage recognition patterns.

Key words: Progressive failure, composites, damage study of mechanical joints.

1. Introduction

Composite is a combination of fibre and matrix as constituents forming a new material with better strength and mechanical properties than that of its constituents individually. So in course of time, there had been a continuous development for better composites for light weight, corrosive resistant, better performance and so on. As the demand increases for better quality and sustainability of materials, the look into their properties, characteristics, behavioural attributes, and failure mechanisms increased. In such an attempt, progressive failure and damage modelling methods came into existence. Failure analysis of constituent is one part and failure analysis of a component/ assembly of components is another part to consider in view of failure models.

In the present study, a mesoscale to microscale progressive damage modelling procedures that evolved with respect to failure in component/ assembly of components level will be discussed. Several failure criteria like Hashin criterion, LaRCO3 criterion, Tsai-Wu criterion, MMF3 criterion etc., are in existing models.

These failure criteria can be grouped into two main groups:

- Failure criteria neglecting interactions between different stress components.
- Failure criteria considering interactions between different stress components.

In our present study, both the cases will be touched to view how the progressive damage will be occurring in composite plates with lap joint or adhesively bonded joint, fastened joint, notched and un-notched composites, cracking of a laminate, pin-loaded analysis of composites, hole and openings of composites.

ASTM D5961 [1] and D7332 [2] are widely used for mechanical joint tests of a composite laminate with different types of joints defined as per the standards. For a single lap joint, the tension load that occurs in the joint is transferred through the fastener in pull direction as well major load is carried over through shear load transfer. Certain parameters like joint area and dimensions, type of a joint, loads on the joint, and type of fastener used need to be considered to study the behaviour and analyse the structure as per design requirements. For features like hole, openings in middle are kind of complications while analysing a composite structure.

The stress concentration factor (SCF) is considered widely for analysing joint strength and its failure modes in macro level [3]. The distributed load condition on a pin loaded hole was investigated by Chang [4]. As well many other researchers have studied this face of composite laminate joints by looking at its characteristic curves [5,6]. For deriving these characteristic curve, we need to define some parameters for which additional tests like compression test, open hole test are required to get the mechanical properties of a laminate.

2. Modeling and Analysis Methods

There are micro, meso and macro levels of modelling exists in the literature for which this paper will try to define few parameters required in modelling and analysis with definitive models. By mainly focussing on three kinds of damage mechanisms, making complex things simple for future studies and investigations. Fibre breaking, debond between fibre – matrix, matrix cracks are considered to be major damage modes which is the central of discussion. Fibre breaking is due to failure at σ_{11}, represents crack in a fibre. In direction along the fibre, if failure with σ_{12} or σ_{13} exists due to the growth of debond failure in between fibre and matrix which is assumed to be a strong interface. Debonding is modelled by considering the maximum length of the crack is equal to the cell length in the direction of thickness/ say depth. Matrix cracks are also sometime called as radial cracks which grows along the fibre direction and modelled as a radial crack in fibre direction. These damage modes have been pictorially represented as shown in Figure 1.

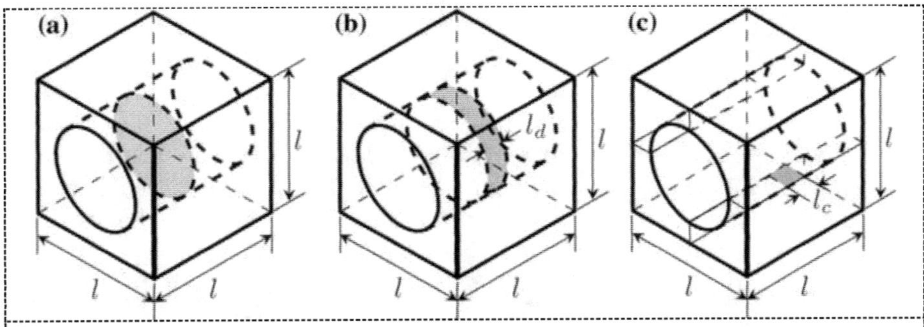

Figure 1. a Fiber breakage. b Fiber-matrix debond. c Matrix cracks [8].

3. Progressive Damage Modelling

Based on the earlier SCF research, a later more refined progressive damage failure analysis (PDFA) was correlated and developed using detailed three dimensional FEM and analysis results. PDFA was carried over single and double lap joints of composite laminates [7,8].

GUL APALAK et.al. [9], investigated about the progressive damage modelling of the composite lap-joint where initiation and propagation of damage zones studied for adhesively bonded single-lap joint composite plates under tension at the mesoscale level. In their paper, composite plates were adhesively bonded with a single lap joint kept in tension and studied its mesoscalebehaviour based on Hashin failure criterion. It was observed that the lower plate interface initiates the damage to the adhesive layer where the first ply along the adhesive layer propagates at free edge, so distributed and spreads along the surface through the adhesive neighbor plies of the lower plate of single lap joint.

IOP Conf. Series: Materials Science and Engineering 1185 (2021) 012021 doi:10.1088/1757-899X/1185/1/012021

When first ply loads decreaseitt increases the ply fiber angle, resulting in delamination due to fiber – matrix failure. It was noted that critical region exists at the right free edge of the lower plate–adhesive interface. In a meso-scale level their work presented an accurate prediction with failure modes and damage zones where the propagation initiated in the composite plates as adhesively bonded with a single lap joint under tension.
To predict the damage propagation of composite fastened joints Mattewset. al. [10], made an attempt to study the case using three dimensional approach of finite element modelling (FEM). Strength of mechanically fastened joints at bearing, tension and shear – out modes were studied. Damage was estimated by using internal state variables in conjunction with global failure criterion for numerical predictions of damage propagation, joint stiffness, and ultimate strength of the joints. They made an attempt to utilize explicit modelling concepts within FEM. This attempt made them to predict joint fracture along with failure loads at maximum hole deformation and occurrence of matrix cracking.

In bearing failure modes, damage propagation tend to occur below the washer, but at the first load drop-off strength of bearing specimen was over-predicted. They suggested the load that occurs here is a failure load after reducing the joint stiffness. It was said to be incapable to model delamination. By using interface elements between the layers, adhesive interface the extended result could have given a delamination response for bearing strength. By including the interface elements a number of assessments can be possible viz., to study the effect of stacking sequence, predicting the clamping pressure at the joint.

Harris et.al. [11], studied the growth of matrix cracks and fiber fracture with notched and un-notched specimens. They believed that penetration notches help to predict the residual strength of the laminate with an appropriate progressive damage methodology. Their discussion is based on the non-linear damage dependent constitutive model where center notched tension panel was designed for testing their model. They named this panel as trans-laminate and the fracture investigation over trans-laminate fracture with a capability of residual strength prediction. They observed a damage growth resistance in their study which define ply bridging. Ply bridging mean a fracture fibre of one ply will be over intact with the fibres of adjacent ply. They employed Allen-Harris constitutive model for their FE model to analyse the notched end stresses exists in the center notched tension panel to predict the matrix cracking and fiber fracture.

The damage progression in the composites has been studied and the progression of the damage can be simulated by using computer finite element formulations. One of such formulation have been developed by NASA into a computer code in the name of Computational Mechanics Testbed (COMET). The scheme of the progressive failure analysis has been shown in Figure 2, which is presented by Timothy et.al [11]. In this flow chart, the first block depicts the inputs required for the model analysis. In every finite element code there exists a mathematical set of arguments for calculating, processing and assembling the elemental stiffness and global stiffness matrices. This computational work at background represented in block 2 and 3. In block 4 damage dependency has been implemented through damage resultant force (DRF) into the finite element analysis code. Using damage growth incremental (DGI) parameters, the damage growth can be predicted and thus laminate failure can be estimated and simulated through the integrated finite elemental code with in the computational mechanics testbed.

There are researchers who later studied the cracks and defined cracks as to be assumed for which it initiates from a nucleus created due to divertive loading by localized fiber debonding and matrix cracking [12]. By modelling a crack as defined above the transverse cracking was studied with an elastic fibrous composite ply. Its low crack density was explored. The interaction of cracks with different plies properties was studied.

Tay et.al. [13] investigated PFA of composites using micromechanics models infused with a novel element failure method (EFM), with a pin loaded in a notched composite laminate analysis. They validated their results with various methods like Tsai-Wu criterion to MMF criterion. Their results were on par excellent experimental results. The more hybrid EFM-cohesive element approach can be utilized for individually fiber damage analysis of micro scale to component level macro scale. They have illustrated the potential of their model with an analysis of double-notched composite laminate.

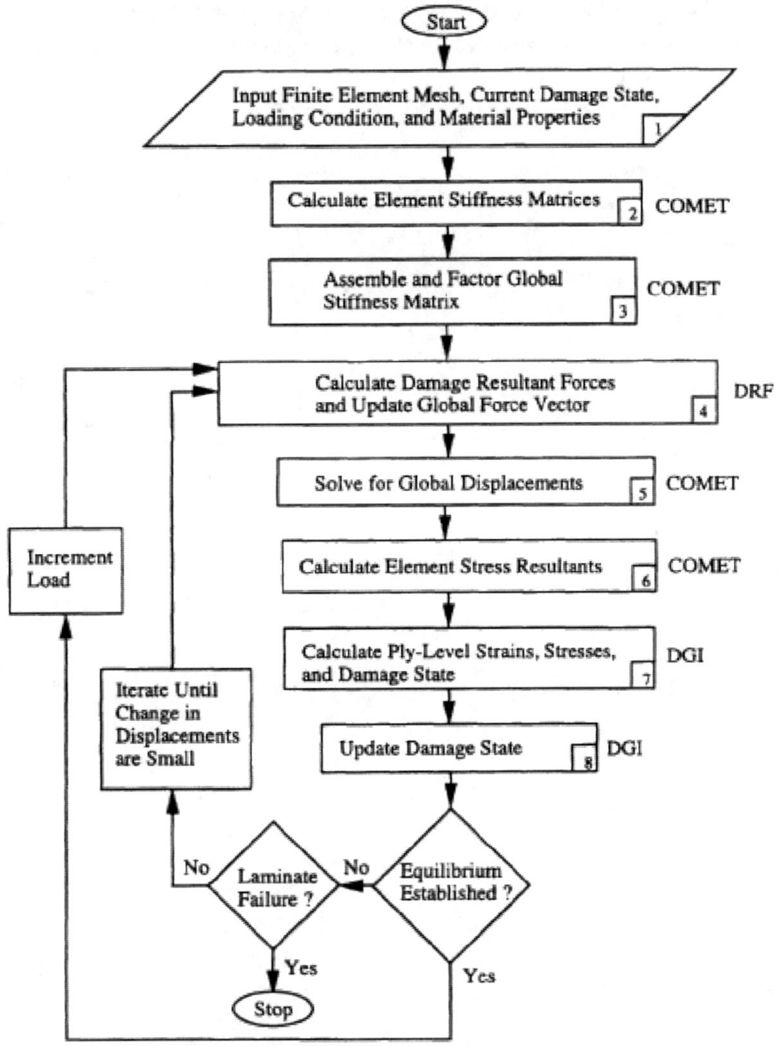

Figure 2. A flow chart for Progressive Failure Analysis [11].

Tan et.al. [14], investigated the features of micro damage with matrix cracking and fibre breakage under shear crippling or micro buckling. Their results were well correlated with X-ray radiographic experimental results. They extended their work form tensile loading to the compressive loading to study micro damage and characterizing the properties at micro and meso level. This model was not studied for delamination with openings and holes on a composite plate, which can be considered as a demerit of such accomplished and simplified model.

Seng C. Tan [15], investigated PFA model on composite laminates containing openings. He stated that, load increment have a little effect on ultimate strength. He had done an extensive parametric study with experimental investigation to provide a more scientific approach. In this work stress concentrations under in-plane loading were studied for PFA of composite laminates. They used ply-by-ply and element by element approach for analysing the matrix cracking and fiber breakage. They outlined that, damage of lamina signifies the stiffness degradation. So damaged lamina properties can be found by the respective constitutive relations. To reduce the complexity, they used stiffness degradation factors, introduced stable finite elements for meshing, defined load step increment relation with the ultimate strength. There is a continuous research in the field, the further studies would be on sudden degradation and gradual degradation methods and investigating on a practical problem with evaluations relative to sun [25] and lee [24] models. Table 1 summarizes the matrix failure models with respect to 3D finite element method.

Table 1. Degradation models for matrix failure in 3D FEM

Model	E_{11}	E_{11}	E_{11}	G_{23}	G_{13}	G_{12}	υ_{12}	υ_{23}	υ_{13}	υ_{21}	υ_{31}	υ_{32}
Nagesh [26]	-	0.1	0.1	0.1	-	-	0	-	0	na	na	na
Tsai&Wu [18]	-	0.45	na	na	na	0.35	0.3	na	na	na	na	na
McCarthy [27]	-	0.1	0.1	0.1	-	-	-	0.1	-	na	na	na
Camanho [28]	-	0.2	-	0.2	-	0.2	-	na	-	na	-	-
Camanho	-	0.4	na	0.4	na	0.4	na	na	na	na	na	na
Apalak [29]	-	0	0	0	0	-	0	-	-	na	na	na
Tsepes-I [30]	-	0	-	-	-	-	0	-	-	na	na	na
Tserpes-II [31]	-	0.2	-	0.2	-	0.2	-	-	-	na	na	na
Tserpes-III [32]	-	0.4	-	0.4	-	0.4	-	-	-	na	na	na

4. Conclusions

Comprehensive procedures for testing and analysing a composite laminate mechanical joints using progressive damage failure analysis was discussed and reviewed critical papers under the scope and thread of PFA of laminated composites with mechanical joints. Different analysis and methods proposed by the researchers were discussed. Micro, meso, and macro scale of looking the failure and failure modes have been noted for further study and future investigations on failure of mechanical joints and their prediction through progressive damage modelling and failure analysis.

References

[1] ASTM D 5961-08. Standard test method for bearing response of polymer matrix composite laminates. *Annual Book of ASTM Standard*, American Society for Testing and Materials.

[2] ASTM D 7332-07. Standard test method for measuring the fastener pull-through resistance of a fiber-reinforced polymer matrix composite. *Annual Book of ASTM Standard*, American Society for Testing and Materials.

[3] Hart-Smith LJ., 1977, Bolted joints in graphite-epoxy laminates, *NASA CR-144899*. Long Beach (CA): Douglas Aircraft Company;

[4] Chang F-K, Scott RA, Springer GS.,1982, Strength of mechanically fastened composite joints. *J Compos Mater.*, **16**:470–494.

[5] Chang F-K, Scott RA, Springer GS., 1984, Failure of composite laminates containing pin loaded holes – method of solution. *J Compos Mater.*, **18**, 255–278.

[6] Aktas A, Lmrek H, Cunedioglu Y., 2009, Experimental and numerical failure analysis of pinned joints in composite materials. *Compos Struct.*, **89,** 459–466.

[7] Miracle DB, Donaldson SL., 2001, ASM handbook. *Composites*, **21,** 692–697.

[8] Ireman T., 1998, Three-dimensional stress analysis of bolted single-lap composite joints. *Compos Struct.*,**43**,195–216.
[9] Z. GUL APALAK et.al.,2006, Progressive Damage Modeling of an Adhesively Bonded Unidirectional Composite Single-lap Joint in Tension at the Mesoscale Level, *Journal of Thermoplastic Composite Materials*, **19**, 671-702.
[10] P. P. Camanho and F. L. Matthews,1999, A Progressive Damage Model for Mechanically Fastened Joints in Composite Laminates, *Journal of Comps. Mater.*, **33**,2248-2280
[11] Timothy W. Coats and Charles E. Harris, 1999, A Progressive Damage Methodology for Residual Strength Predictions of Notched Composite Panels, *Journal of Comps. Mater.*, **33**, 2193-2224.
[12] George J. Dvorak and N. Laws,1987, Analysis of Progressive Matrix Cracking In Composite Laminates II. First Ply Failure, *Journal of Comps. Mater.*, **21**,309-329
[13] T. E. TAY et.al,2008, Progressive Failure Analysis of Composites, *Journal of Comps. Mater.*, **42**, 1921 – 1966
[14] Seng C. Tan and Jose Perez,1993, Progressive Failure of Laminated Composites with a Hole under Compressive Loading, *Journal of Reinforced Plastics and Composites*, **12**, 1043 – 1057
[15] Seng C. Tan, 1991, A Progressive Failure Model for Composite Laminates Containing Openings, *Journal of Comps. Mater.*, **25**, 556 – 577.
[16] Sung Kyu Ha et. al.,2008, Micro-Mechanics of Failure (MMF) for Continuous Fiber Reinforced Composites, *Journal of Comps. Mater.*, **42(18),**1873-1895.
[17] Jeong Sik Kwon et.al.,2018, A mechanically fastened composite laminate joint and progressive failure analysis, *Journal of Advanced Composite Materials*, **27(4),** 439–456.
[18] Tsai, S.W.; Wu, E.M., 1971, A general theory of strength for anisotropic materials. *J. Compos. Mater.* **5**, 58–80.
[19] Hashin, Z.,1980, Failure criteria for unidirectional fiber composites. *J. Appl. Mech.*, **4**, 329–334.
[20] Puck, A.; Schürmann, H., 1998, Failure analysis of FRP laminates by means of physically based phenomenological models. *Compos. Sci. Technol.*, **58**, 1045–1067.
[21] Chamis, C.C.; Abdi, F.; Garg, M.; Minnetyan, L.; Baid, H.; Huang, D.; Housner, J.; Talagani, F., 2013, Micromechanics-based progressive failure analysis prediction for WWFE-III composite coupon test cases. *J. Compos. Mater.*, **47**, 2695–2712.
[22] Tay, T.E.; Tan, S.H.N.; Tan, V.B.C.; Gosse, J.H., 2005, Damage progression by the element-failure method (EFM) and strain invariant failure theory (SIFT). *Compos. Sci. Technol.*, **65**, 935–944.
[23] Cai, H.; Miyano, Y.; Nakada, M., 2009, Long-term open-hole compression strength of CFRP laminates based on strain invariant failure theory. *J. Thermoplast. Compos. Mater.*, **22**, 63–81.
[24] Lee, J. D., 1982, Three Dimensional Finite Element Analysis of Damage Accumulation in Composite Laminate,*Comput. Struct.*, **15(3),** 335–350.
[25] Hwang, W. C., and Sun, C. T., 1989, Failure Analysis of Laminated Composites by Using Iterative Three-Dimensional Finite Element Method,*Comput. Struct.*, **33(1),** 41–47.
[26] Nagesh, 2003, Finite-Element Analysis of Pressure Vessels With Progressive Degradation,*Def. Sci. J.*, **53(1),**75–86.
[27] McCarthy, C. T., McCarthy, M. A., and Lawlor, V. P., 2005, ProgressiveDamage Analysis of Multi-Bolt Composite Joints With Variable Bolt-HoleClearances, *Composites, Part B*, **36(4),** 290–305.
[28] Camanho, P. P., and Matthews, F. L., 1999, A Progressive Damage Model forMechanically Fastened Joints in Composite Laminates,*J. Compos. Mater.*,**33(24),**2248–2280.
[29] Gul Apalak, Z., Kemal Apalak, M., and Serdar Genc, M., 2007, Progressive Damage Modeling of an Adhesively Bonded Composite Single Lap Joint Under Flexural Loads at the Mesoscale Level,*J. Reinf. Plast. Compos.*, **26(9),**903–953.
[30] Kermanidis, Th., Labeas, G., Tserpes, K. I., and Pantelakis, Sp., 2000, Finite Element Modeling of Damage Accumulation in Bolted Composite Joints Under Incremental Tensile Loading,

Proceedings of the Third ECCOMAS Congress, Barcelona, Spain, Sept. 11–14.

[31] Tserpes, K. I., Papanikos, P., and Kermanidis, Th., 2001, A Three- Dimensional Progressive Damage Model for Bolted Joints in Composite Laminates Subjected to Tensile Loading, *Fatigue Fract. Eng. Mater. Struct.*, 24 (10),663–675.

[32] Tserpes, K. I., Labeas, G., Papanikos, P., and Kermanidis, Th., 2002, Strength Prediction of Bolted Joints in Graphite/Epoxy Composite Laminates,*Composites, Part B*, 33(7),521–529.

[33] David C. Lo, Timothy W. Coats, Charles E. Harris, and David H. Allen,1996, Progressive Damage Analysis of Laminated Composite (PDALC), A Computational Model Implemented in the NASA COMET Finite Element Code, *NASA Technical Memorandum* 4724.

Parameter Optimization of EDM Characteristics on Ti-6Al-4V Using Different Electrodes

P.Sumanth [a], M. Vaishnav Reddy [a], Imran Shaik [a], Jagadeeswara Rao Maddu [b]

[a] CVR College of Engineering, Hyderabad, Telengana, India

[b] University College of Engineering, Osmania University, Hyderabad, Telengana, India

* Corresponding author e-mail: jaganmuddu@osmania.ac.in

Abstract. A throughout research has been conducted and enumerated on the process parameters of machining of Ti-6Al-4V alloy in Die sinker EDM . EDM is a well-known UCM measures which is suitable for machining hard and not ease to machine materials of difficult shapes. The method selected for optimization is Taguchi for this paper to examine the characteristics of Ti-6Al-4V alloy and the selected parameter for study is material removal rate. The electrode material selected as brass bronze and copper. The Flushing pressure, current, pulse time-on, pulse time-off and electrodes parameters percentage of contribution and significance effect are analysed with ANOVA. With the observational results, it is evident that current shows significant effect on the response parameter MRR.

Key words: -ANOVA, EDM, MRR, S/N, Ti-6Al-4V, Taguchi,

1. Introduction.

EDM is a one of the most consistent and economical machining processes for titanium and its alloys for past decades. A wide research has been reported in the stream of EDM for machining titanium and its alloys, and several commercially available standard technologies [1]. Thought titanium has advantageous properties like high strength, stiffness at greater temperatures but they are inferred as complex to machine materials by considering the following reasons.1. poor conductivity, 2. Require large shear forces, 3. High springy when compared to steel, 4. Speed limit in machining of titanium and its alloys ranges around 350 FPT. Hence EDM plays a major role in overcoming difficulties in machining titanium and its alloys. EDM is an electro thermal process suitable in machining conductive materials for manufacturing different automobile and biomedical applications [2]. A large number of electrical sparks was generated between cathode and anode in Electrical discharge machining at a span of seconds. The material removal from work piece takes place with high intensity heat produced by electrical sparks, therefore evident to machine tough materials [4]. As there is no contact between anode and cathode the effects such as mechanical vibrations and stresses are low. In spite of many advantages, instability occurs in the process due to low surface finish, MRR, HAZ and energyconsumption [7]. EDM is depicted in Fig. 1, which narrates mechanisms cut any electrical conductive material irrespective of its hardness.In industrial applications, the Taguchi method has been used to originate the experimental layout to evaluate the outcome of each process parameter on the output characteristic and to depict optimal selection for particular machining parameter [9]. It is crucial that we need to select the proper process parameter to achieve optimum machining performance. The selection of optimal combinations of machining parameters is often challenging as various number of parameters exist and in studying their effects & contributions many researchers showed their interest. Anil Kumar et al (2010), EDM is a modern method of machining used to create geometrically shapes of rigid materials that are very difficult to machine by traditional machining

IOP Conf. Series: Materials Science and Engineering 1185 (2021) 012022 doi:10.1088/1757-899X/1185/1/012022

process are processed. This technique of non-contact machining is continuously proving in manufacturing tools and dies in micro level. In recent years, research has been conducted on emphasising the improved efficiency of machining combined with surface treatments for better surface finish [8]. R. Venkata Rao et al, The primary Emphasis is held on the aspects of Optimisation of different parameters of the modern processes of machining and techniques for Optimisation were involved. The evaluation time and the research consideration considered is from 2006 to 2012. Similar modern processes like electric discharge process considered in this work and hybrid versions of these processes. Considering many processes at a time and thus the work of this analysis could be available at one location, the ready knowledge can be very useful in determining their course for the subsequent researchers from studies [11]. Ankit Jain et al, were carried out the experiments by best choosing l27 orthogonal array drilled Titanium alloy using electric discharge drilling. for experimentation the input process parameters and output parameters selected discharge current, pulse on time, pulse off time, dielectric pressure and hole circularity, hole taper respectively. The data from the experiments used for the establishment of multi regression for output parameters. The selected parameters are optimised using genetic algorithm MATLAB tool. The multi-objective outcome of the genetic algorithm optimisation shows progress in the output parameters [10]. V. S. Nimbalkar, In order to consider the effect of three significant EDM parameters, Taguchi analysis is used namelyPulse on time and discharge current, and difference voltage. In this survey, using Analysis of Variance (ANOVA) and the effect & contribution of process parameters on the responses is studied and contrasted with the mathematical models for validationTi-6Al-4V (Titanium alloy) and Ti-6Al-4V (Titanium alloy) have been studied. Strong Copper and hollow copper electrodes were used as the instrument. Using Taguchi study, the output response is designed using the experimental values obtained [12]. From the observations of previous studies not much focus spent on the effect of parameters and optimization of parameters. Here our study focus on finding optimum parameters, using ANOVA the level of significance of output parameters such as MRR and SR was studied and best combination set of parameters selected using Taguchi L18 design. In order to examine the effects of process parameters Design software Minitab 16 was employed.Todetermine the significant factors as well as to develop of the EDM response MRR the analysis of variance (ANOVA) were performed.

2. Methods and materials.

The work in this investigation was done on EDM machine CREATER Make. The work starts with selection of material, input and output parameters. Electrode as brass bronze and copper was selected shown in fig 2 and Ti-6Al-4V alloy chosen as the job material. The EDM parameters chose were flushing pressure, current, pulse time-on, pulse time-off and electrodes. Table 1indicatesthe design of experiments with parameters and their levels. Table 2 describes the matrix design on full factorial 18 investigations. The output parameter chosen is MRR. Experiments were directed depending on the design of experiments approach. ANOVA technique is used to find out significance level of parameters and their effect on MRR. For each test run, rectangular samples of 20x20x8 mm measurements were utilized for a set of indicated input parameter combinations shown in fig 2. The MRR was measured utilizing the formula.

MRR = (Weight before Machining – Weight after Machining) \ Time of machining

Table 1. Selected process parameters and respective levels.

S. No	Parameters	Level- 1	Level- 2	Level- 3
1	Flushing pressure	0.35	0.7	-
2	Current	10	20	30
3	Pulse on Time	200	400	600
4	Pulse off Time	40	80	120
5	Electrodes	Copper	Bronze	Brass

Electrical Discharge Machining

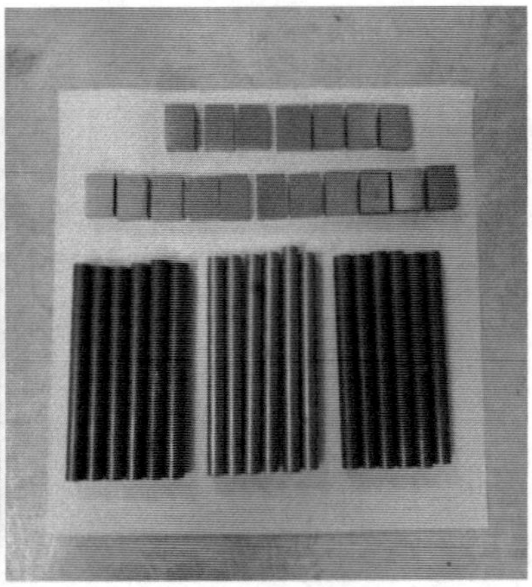
Work piece and Electrodes

Figure 1 Experimental setup

To analyze the outcomes of test runs and each parameter significance to be assessed with ANOVA for the outcome of MRR. Minitab 16 was utilized to perform the ANOVA test for knowing the importance of parameters. Taguchi shows S/N ratio which became the objective of DOE. these objectives are categorized as larger, smaller and nominal the better. The DOE and the outcome obtained from the experiments were tabulated in table 2. It is observed from the matrix that the set of input parameters combinations Flushing pressure (0.7 Kg/cm2), current (10 amp), pulse time-on (400), pulse time-off (40) and electrode (brass), experiment 13 results maximum MRR.

Table 2. Set of 18 experiments and resulting MRR and S/N Ratio.

S.No	Flushing pressure	Current	Pulseontime	Pulse off Time	Electrodes	M.R.R	S/N Ratio
1	1	1	1	1	1	1.16479	1.32492
2	1	2	2	1	2	0.96614	-0.2992
3	1	3	3	1	3	0.19639	-14.138
4	1	1	1	2	2	3.2754	10.3053
5	1	2	2	2	3	0.23928	-12.422
6	1	3	3	2	1	0.79233	-2.0219
7	1	1	2	3	1	0.62077	-4.1414
8	1	2	3	3	2	1.50113	3.52836
9	1	3	1	3	3	1.81242	5.16515
10	2	1	3	1	3	1.41761	3.03112
11	2	2	1	1	1	0.82299	-1.6921
12	2	3	2	1	2	0.40406	-7.871
13	2	1	2	2	3	3.65237	11.2515
14	2	2	3	2	1	2.47404	7.86814
15	2	3	1	2	2	0.21494	-13.354
16	2	1	3	3	2	0.7088	-2.9895
17	2	2	1	3	3	0.85779	-1.3324
18	2	3	2	3	1	2.06321	6.29085

IOP Conf. Series: Materials Science and Engineering 1185 (2021) 012022 doi:10.1088/1757-899X/1185/1/012022

3. Results and discussions

Material Removal Rate Analysis.Table 2 exihibits the MRR experimental results with the orthogon al array and its S/N ratio values. ANOVA's Primary aim is to know the influence of selectedparameter s. MRR results were calculated by ANOVA and percentage of contribution of each parameter were reported in table 3. From the table it is evident that current has high influence with percentage 57.98%, other parameters pulse off time, electrodes, flushing pressure and pulse-on-time with 22.44, 11.56, 3.56, 1.74% respectively.

Table 3: Analysis of Variance for SN ratios

Source	DF	Seq SS	Adj SS	Adj MS	F	P	%
Flushing pressure	1	10.74	10.736	10.736	0.12	0.740	3.56
Current	2	166.65	166.650	83.325	0.92	0.438	57.98
Pulse on time	2	5.02	5.022	2.511	0.03	0.973	1.74
Pulse off time	2	64.50	64.502	32.251	0.36	0.711	22.44
Electrodes	2	33.25	33.251	16.625	0.18	0.836	11.56
Residual Error	8	7.26	7.26	3.633			2.53
Total	17	287.42					

Table 4: Response table for Signal to Noise ratio.

Level	A	B	C	D	E
1	-1.41094	3.13032	0.06954	-3.27400	1.27140
2	0.13367	-0.72486	-1.19854	0.27124	-1.77994
3	-4.32136	-0.78691	1.08684	-1.40738	
Delta	1.54461	7.45168	1.26809	4.36084	3.05134
Rank	4	1	5	2	3

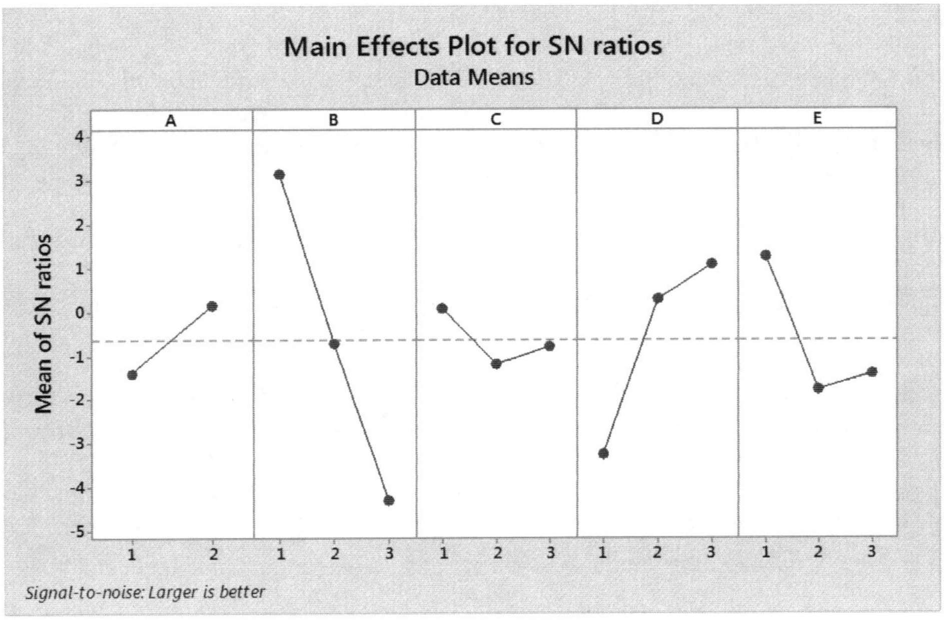

Figure 2: S/N Response graph for MRR

4. Conclusions:

EDM of Ti-6Al-4V alloy specimens has an impact of machining process factors on MRR studied analytically. The tests were conducted according to the factorial design of Taguchi L18 to test the impact of factors including flushing pressure, current, pulse time-on, pulse time-off and electrodes are machined. ANOVA has been performed in order to assess the significance of parameters. The following conclusions that emerged based on the test results were reported from the findings. The key significance parameters are current, and pulse-off time with percentage of contributions 57.98 and 22.44% respectively. Flushing pressure (0.7 Kg/cm2), current (10 amp), pulse time-on (400), pulse time-off (40) and electrode (brass), are the optimal conditions for MRR.

References

[1] Gill, S. S., & Singh, J. (2010). Effect of deep cryogenic treatment on machinability of titanium alloy (Ti-6246) in electric discharge drilling. *Materials and Manufacturing Processes*, 25(6), 378–385.

[2] Garg, M. P., Jain, A., &Bhushan, G. (2014). Multi-objective Optimization of Process Parameters in Wire Electric Discharge Machining of Ti-6-2-4-2 Alloy. *Arabian Journal for Science and Engineering*, 39(2), 1465–1476.

[3] Kumar, A., Kumar, V., & Kumar, J. (2014). Microstructure analysis and material transformation of pure titanium and tool wear surface after wire electric discharge machining process. *Machining Science and Technology*, 18(1), 47–77.

[4] Prakash, C., Kansal, H. K., Pabla, B. S., &Puri, S. (2017). Experimental investigations in powder mixed electric discharge machining of Ti–35Nb–7Ta–5Zrβ-titanium alloy. *Materials and Manufacturing Processes*, 32(3), 274–285.

[5] Prasad Arikatla, S., Tamil Mannan, K., &Krishnaiah, A. (2017). Parametric Optimization in Wire Electrical Discharge Machining of Titanium Alloy Using Response Surface Methodology. *Materials Today: Proceedings*, 4(2), 1434–1441.

[6] Oliver Nesa Raj, S., &Prabhu, S. (2017). Modeling and analysis of titanium alloy in wire-cut EDM using grey relation coupled with principle component analysis. *Australian Journal of Mechanical Engineering*, 15(3), 198–209.

[7] Goyal, R., Sngh, S., & Kumar, H. (2018). Performance evaluation of cryogenically assisted electric discharge machining (CEDM) process. *Materials and Manufacturing Processes*, 33(4), 433–443.

[8] Krishna, P. V., Reddy, V. V., & Kumar, B. S. (2017). Experimental Investigation on EDM of Titanium Alloy using Taguchi Method. 5(05), 134–139.

[9] Ram Prasad, A. V. S., Ramji, K., &Kolli, M. (2019). An experimental investigation on machining parameters of titanium alloy using WEDM. *Materials Today: Proceedings*, 18, A12–A16.

[10] Pramanik, A., &Basak, A. K. (2019). Effect of wire electric discharge machining (EDM) parameters onfatigue life of Ti-6Al-4V alloy. *International Journal of Fatigue*, 128(February), 105186.

[11] Rao, R. V., &Kalyankar, V. D. (2014). Optimization of modern machining processes using advanced optimization techniques: A review. *International Journal of Advanced Manufacturing Technology*, 73(5–8), 1159–1188.

[12] Nimbalkar, V. S., &Shete, P. M. T. (2017). Experimental investigation of machining parameters using solid and hollow electrode for EDM of Ti-6Al-4V. *International Research Journal of Engineering and Technology* (IRJET), 04(05).

Investigation on mechanical properties of biodegradable polymers

Veeresh Kumar G B[1]*, Krishna Kanth V[1,2], Pramod R[3]

[1]Mechanical Engineering Department, National Institute of Technology-Andhra Pradesh, Tadepalligudem, West Godavari district, Andhra Pradesh State, India.
[2]Mechanical Engineering Department, Raghu Institute of Technology, Visakhapatnam, Andhra Pradesh, India.
[3]Mechanical Engineering Department, Amrita School of Engineering, Bengaluru, Amrita Vishwa Vidyapeetham, India.

*Corresponding author'se-mail: veereshkumargb@nitandhra.ac.in

Abstract. Bioplastics are plastics made-out of biomass such as sugarcane, corn etc. These compounds are getting more attention to conserve fossil fuels, reduce plastic waste, and reduce CO_2 emissions.These substances are being increasingly spotlighted as means to save fossil fuels, plastic wastes and reducing CO_2 emission. Biodegradability of these bioplastics have been extensively publicized in the society and demand for packaging is quickly increasing among the food industry and retailers' at large scale. The plastics which are currently available in market are very dangerous as it is non biodegradable. Hence, it is the order of the day that the biodegradable plastics should be produced at large scale and used to replace non-biodegradable plastics. The present articleemphasis all these points regarding their applications, types, production, challenges, fermentation, sustainability, process development and use of cheap substrates for bioplastics production.The latter section of the project endeavours to educate a novel method in the production of biopolymers using Gelatin. Dissimilarities in parameters like plasticizer, polymers, hydrolysis and curing times were extensively tested and the optimum combination was obtained.

Keywords: Bioplastics, Gelatin, Biopolymers, Biodegradability, Mechanical Characterization.

1. Introduction

The standardised method for manufacturing starch-based bioplastics, according to the Royal Society of Chemistry, entails hydrolysis of the starch using an acid. [1] Abdolreza et al. (2010) identified the physiological, rheological, and thermal properties of acid hydrolyzed starch in their paper, demonstrating that the amylose content rises at first, but then decreases due to persistent hydrolysis.[2] KarntaratWuttisela et al. (2008) made similar points, emphasising that the amylose content was responsible for the starch's plastic growth. Plasticizers were used to improve the bioplastic samples' mouldability and flexibility. [3] J. Zhou (2008) debated the effects of common types of plasticizers and their effects on a variety of bioplastic film properties such as water vapor permeability, elongation, and tensile strength. [4] Y. Chen et al. (2019) addressed bioplastics applications, especially

IOP Conf. Series: Materials Science and Engineering 1185 (2021) 012023 doi:10.1088/1757-899X/1185/1/012023

in the packaging industry, where bio-based polymers were used as a factor in food packaging and various techniques for enhancing barrier properties, mechanical properties, and permeability values of multi-layered bio-based plastics were discussed.K. Kanimozhi (2014) [5] explored the research and characterisation of bio-plastic formation and identified the study of bioplastic formation and their composition using FTIR spectroscopy in different phases.

Two different types of technologies were developed for the fabrication of bioplastics from starch and agrowaste. The first technology is the direct transformation of inedible agrowaste into bioplastics and the second technology is elastomerization of agrowaste into robust bioelastomers. The first technologyutilizes an organic acid to convert the cellulosic agrowaste into amorphous plastic[6]. Later, the acid may be recycled in a closed system of production. The second technology utilizes micronized agrowaste powder with starch dispersed in Silicone (Si) based polymer precursor to produce bioelastomers containing less than 50% by weight vegetable-based ingredient. Typical agrowaste products that may be used were parsley and cacao pod husks, spinach stems, rice hulls, orange peels, starch, and oat hulls. Bioplastic is a type of plastic manufactured primarily from sustainable biomass sources such as fermented sucrose (which is altered to form polylactide), gelatin, vegetable oil, corn starch, pea starch, or cellulose.The plant material needed to make bioplastic can be grown continuously. In contrast, petrochemical-based polymers will eventually run out (i.e. our current rate of use exceeds the time taken for petrochemicals to form). [7] Additionally, many Bioplastic products are biodegradable which means they can be easily broken-down into the CO_2 and water by micro-organisms. Some may be put into an industrial composting method and will break-down by a amount of 90% less than six months. If prepared with a corn-starch biopolymer, the molecules gradually absorb water and will swell up, triggering them to break apart into tiny fragments that bacteria can digest easily.

The actual state of bioplastics and their prospects, since large-scale manufacturing in industries is prohibitively expensive, they have not been commonly used. Bioplastics manufacturing was dominated in the twentieth century by countries such as Western Europe, North America, Japan, and others, and it is expected that by 2013, Brazil will be among the world's leading bioplastics producers.By 2013, Japan's bioplastics production may be many times higher than 178000 metric tonnes, and China has planned to produce 100,000 metric tonnes of bioplastics at the same time [8]. In Southeast Asian countries, the bioplastics industry is also in its early stages.According to research conducted by the BCC, the bioplastics industry valuation reached 541 million pounds in 2007 and is expected to hit 1.2 billion pounds by 2012. Several varieties of biodegradable plastics, such as resins, polylactic acid, polyesters, and others, accounted for nearly 90% of bioplastics production in 2008.Nondegradable plastics may be replaced by environmentally friendly biodegradable plastics. The use of bioplastics would reduce CO_2 emissions associated with nondegradableplastics.One of the main concerns is that bioplastics could wreak havoc on current recycling systems. The cost of producing bioplastics is currently too high, and this is one of the major issues facing bioplastics growth.The main aim of this work is to develop biodegradable plastics and test for its physical, mechanical and biodegradability properties.

2. Methodology

The preparation methodology are being discussed in (6-10) and in this work studies the gelatin based bio degradable polymer. Partially hydrolyzed collagen derived from the skin, bones, and connective tissues of animals such as domesticated beef, chicken, pigs, and fish produces gelatin, which is a combination of peptides and proteins. [11] The natural molecular bonds between individual collagen strands are broken down during hydrolysis, allowing them to rearrange more easily.Its chemical structure resembles that of its parent collagen in several ways. Cattle bones and pig skin are often used to make photographic and medicinal gelatin. The polypeptide chain of gelatin contains proline, hydroxyproline, and glycine. The amino acid glycine is responsible for the chain's tight packaging. [12] The presence of proline causes the conformation to be limited.This is crucial for gelatin's gelation properties. Gelatin dissolves easily in hot water and sets to a gel when cooled. However, it does not

dissolve well when applied directly to cold water. [13] Many polar solvents are also soluble in gelatin.Shampoos, skin masks, and other cosmetics contain gelatin; fruit gelatin and puddings (such as Jell-O) contain gelatin; candies, marshmallows, cookies, ice cream, and yoghurts contain gelatin; photographic film contains gelatin; and vitamins contain gelatin as a coloring and as tablets contain gelatin.
The process starts with heating 125ml water to 80^0C, at this stage gelatin is add of 24g and need to stir the solution at 1200rpm using magnetic stirrer, until it becomes a clear solution. The solution has to be maintained 80^0C and further a 5g of glycerine is added. Using the magnetic stirrer mix the solution at a rate of 1000rpm until it starts foaming and gradually decreases the speed of the sitter to avoid further foaming of the solution. This solution with minimum foam has to be poured in the mould and allows it to cool at room temperature. The stages of procedure and cured bioplastic can be observed in the below Figure 1.

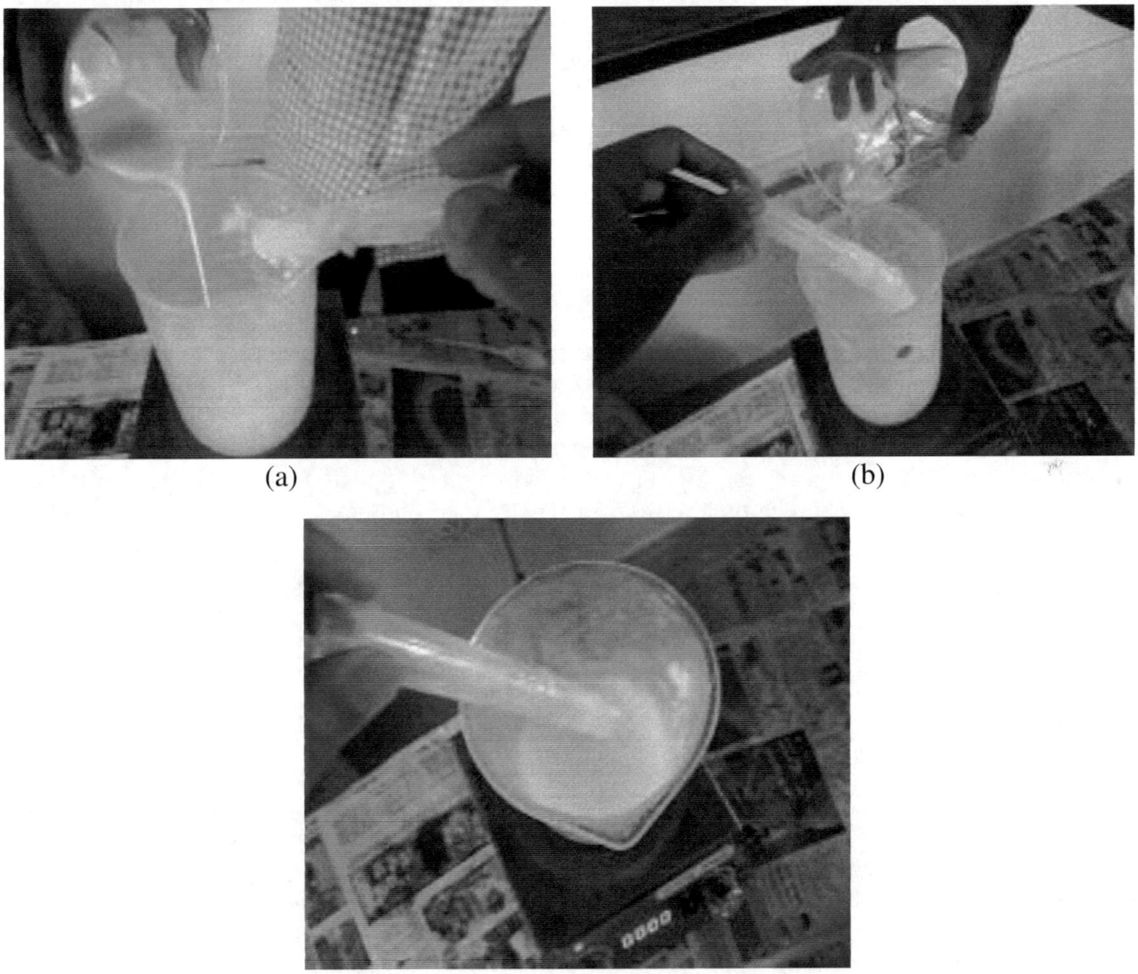

(a) (b)

(c)

Figures 1 (a,b,c): shows the preparation of bio degradable polymer

IOP Conf. Series: Materials Science and Engineering 1185 (2021) 012023 doi:10.1088/1757-899X/1185/1/012023

3. Experimentation

Experimental investigations are carried out in order to estimate the behavior of gelatin based biodegradable polymer. The American Society of the International Association for Testing and Materials (ASTM) is an international standards body that was established in 1898 as the American Society of the International Association for Testing and Materials. Other standards organisations such as the BSI (1901), IEC (1906), DIN (1917), ANSI (1918), AFNOR (1926), and ISO (1927) were established before it (1947). ASTM standard of various test are tabulated in the table 1.

Table 1: ASTM Test Specifications

Test	ASTMStandards	Specimen Sizes
Water Absorption	D570-98	40mm X 40mm
TensileStrength	D882-02	100mm X 25mm
Shore Hardness Test	D2240-00	20mm X 20mm

Biodegradable tests

As environmental concerns have grown in importance on a national and global scale, further attempts have been made to save our capital. Sustainable products are becoming increasingly important in industry. As a result, it is important that students understand the significance of these content. Biodegradable polymers are among such material.For an undergraduate materials laboratory course, a laboratory technique for testing polymer biodegradability was created. Students may obtain a thorough understanding of plastics by testing their biodegradability.Biodegradable testing as part of a programmepromotes constructive learning through hands-on testing and allows students to partake in lifelong learning and keep up with new content.This paper outlines a study that looked at the biodegradability of biodegradable plastics by tracking carbon dioxide levels and physical property degradation. A low-cost approach for determining biodegradability is also defined. The surface of the cured polymer is studied using SEM analysis.

4. Results and Discussions

4.1 Water absorption

The water absorption was determined using ASTM D570-98. The initial weight of dried sheets was by weighing them before placing them in a bath of purified water at room temperature. After 24 determined hours, the sheets were removed, the water on their surface was cleaned off, and the sheets were weighed again. The following equation was used to calculate the average of five replicates:

$$\textit{Water absorption percentage (\%)} = (W_i - W_f) / W_f \times 100\%$$

Where Wi is the original weight (g) before water immersion and W_f is the final weight (g) after water immersion.The average water absorption rate observed is 21.23%. This value is comparatively a better one in comparison with cassava and corn starch bio degradable polymers. The test results are showed in the Figure 2.

IOP Conf. Series: Materials Science and Engineering 1185 (2021) 012023 doi:10.1088/1757-899X/1185/1/012023

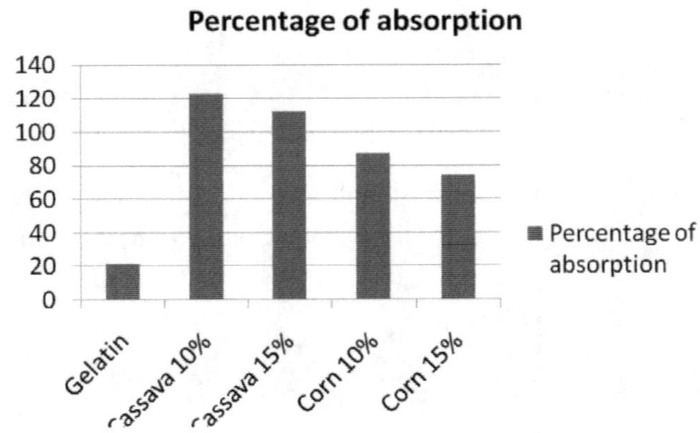

Figure 2: shows the comparison of water absorption test of Gelatin, Cassava and Corn Starch based polymer. Ref (14-17)

4.2 Tensile Test

Tensile testing is fundamental test in which material sample is subjected to controlled tension until breakdown of the sample. In this test the specimen of given standards is subjected to axial load until it gets fractured or get broken. Theelongation of the specimen for different loads is noted down and plotted in a graph elongation vs. load. For engineering applications of different materials these results are used. In material specifications tensile strength is one of the important specifications and is often compared with that of the other newly developed materials. Tensile properties are used to see the behaviour of the material under uniaxial loading. Tensile test is performed on Universal Testing Machine. The belowFfigure 3 represents a graph of tensile strength and also comparisons are made with others.

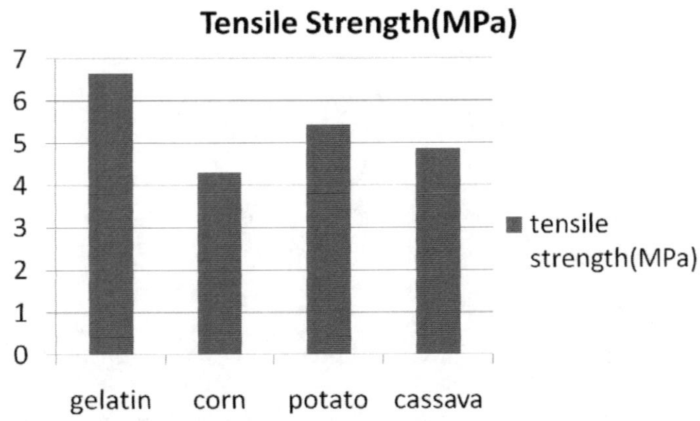

Figure 3: Shows the compares the tensile strength of Gelatin, Corn, Potato, Cassava based polymers.Ref (14-17)

Gelatin based biodegradable polymer shows potential values of tensile strength when comparing with corn, potato and cassava. So gelatin based polymers can be adopted and can be replaced the above three polymers. The below Figure 4 graph shows the comparison of young's modulus in MPa.

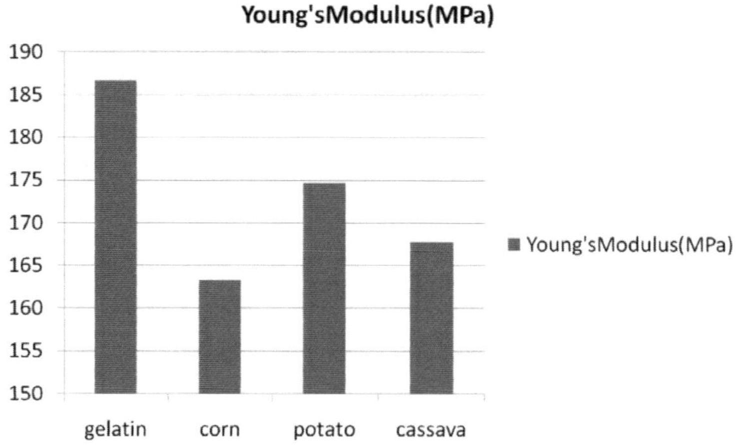

Figure 4: Compares the young's modulus of Gelatin, Corn, Potato and Cassava. Ref (14-17)

As the volume of bioplastic used increased, the percentage of elongation at break increased as well. This is due to the bioplastic's flexibility, which is due to the large level of starch and glycerol it contains. The Figure 5 shows the elongation of gelatin bioplastic and also compares with other bioplastics.

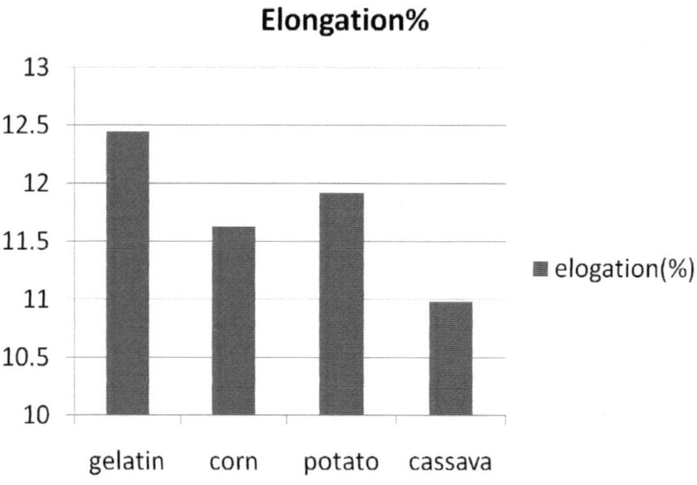

Figure 5: Shows the Elongation of Gelatin, Corn, Potato and Cassava during tensile test. Ref (14-17)

The added tensile force to the specimen is gradually raised. Under the impact of this tensile force, the specimen elongates at the same time. It is obvious that the specimen is being worked on. Up to the point of rupture, the average energy consumed per unit volume of the specimen. At split, the tensile energy is 0.38J.

4.3 Shore hardness test

This indentation test procedure uses a regular indenter to determine the hardness of a rubber specimen. Several rubber hardness measuring scales are referred to by ASTM D2240-00 (A, B, C, D, DO,O, OO, and M).It's used to test the indentation hardness of elastomers, thermoplastic elastomers, vulcanized rubber, cellular, gel-like, and composite materials. [15] The procedure entails indenting the specimen with a hardened steel indenter with precise geometry and force, according to the chosen measurement scale.The indenter tip displacement is used to calculate the material's hardness. The displacement data is converted into a hardness number, which is limited to a range of 0 to 100, using a mathematical relationship. The fig 6 represents the test rig and the two most widely used scales are A and D.

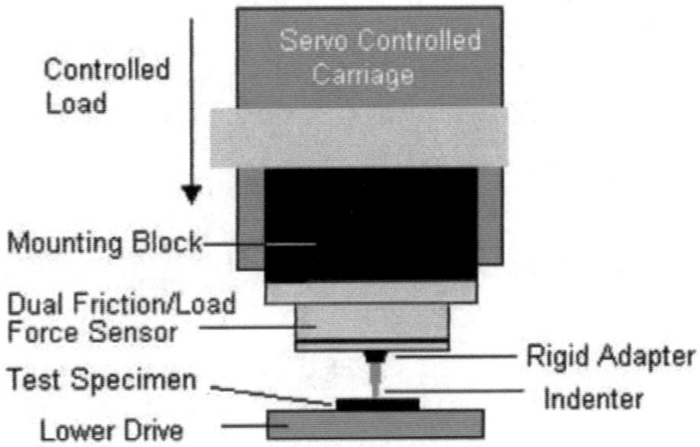

Figure 6: illustrates the experimental setup

This study observed an average shore hardness value of 48.44DM and resulting in very good hardness value. The temperature at which a polymer transforms from a hard glassy material to a soft rubbery material is known as the glass transition temperature. The glass transition temperature does not apply to pure crystalline polymers; the glass transition temperature only applies to amorphous polymers. Amorphous polymers have no melting point; instead, they have a glass transition temperature. Most polymers, on the other hand, have both amorphous and crystalline forms [18]. This ensures that certain polymers have both a melting and a glass transition temperature.The melting temperature is higher than the glass transition temperature. Since different polymers have different glass transition temperatures, certain polymers are better suited for some applications than others. The average glass transition temperature (TG) is 98° degrees Celsius, with a melting point of 120° degrees Celsius.

4.4 Biodegradability tests

To assess the biodegradability of this polymer, conducted Acid test, alkaline test and solubility test and the result are tabulated. Sulphuric acid and Acetic acid were used as the acid test solvents; the Sulphuric acid was a heavy acid, while the Acetic acid was a mild acid. The bioplastic is then put in the beaker with 0.5 M of both acids in 400ml. Check the time it takes for the bioplastics to become completely soluble in strong and weak acids after they appear to dissolve in acids.The aim of the alkaline test was to determine the length of time that the bioplastic is completely soluble in alkaline. Sodium hydroxide was used as the alkaline test solvent, and the NaOH was a solid alkaline. Take 400ml of 0.5 M NaOH, then put the bioplastic in the beaker. As the bioplastic begins to dissolve in alkaline, determine how long it would take for the bioplastic to become completely soluble in NaOH.Water is used as the solubility measure solvent. We take 400 mL deionized water and put the bioplastics in a beaker; the bioplastics begin to dissolve; now verify the time it takes for the bioplastics to completely dissolve in water.

Table.2 The test results of biodegradability.

S.No	Test	Solvent	Duration of time to completely dissolve
1	ACID TEST METHOD	**SULPHURIC ACID** (Strong Acid Test)	70 Minutes
2		**ACETIC ACID** (Weak Acid Test)	175 Minutes
3	ALKALINE TEST	**SODIUM HYDROXIDE**	90 Minutes
4	SOLUBILITY TEST	**DEIONIZED WATER**	490 Minutes

And finally, flammability test is also conducted to estimate the time taken to convert the polymer into ash. Current test was performed only to verify the biodegradability of bioplastics.Bioplastics become ashes if it is biodegradable. It is the easiest and quickest test to find biodegradability of bioplastics.The flame measurement is carried out in the presence of vapor. Bunsen burners are used to burn the bioplastics.Indirect heat is applied, and bioplastic starts burning without catching fire. The average time taken to burnout the sample is 20minutes. The results of the biodegradability test are recorded and tabulated in the Table 2. Further SEM analysis is carried out to observe the surface defects and foreign inclusions in the polymer.

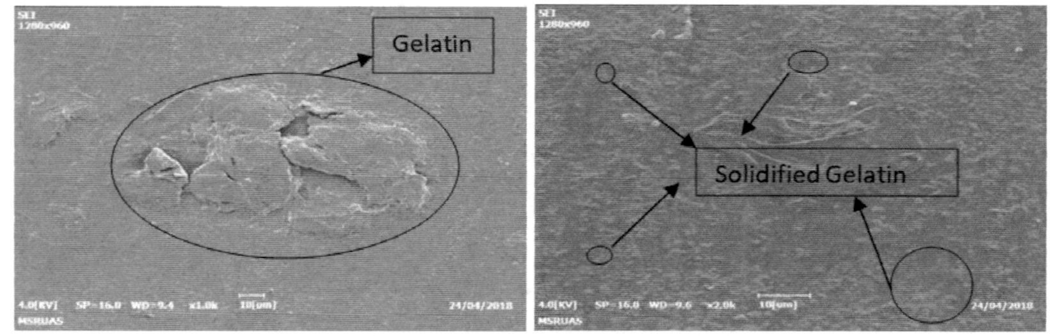

Figure 7: shows the surface of the Gelatin polymer at magnification rate of 1000 (left), 2000(right)

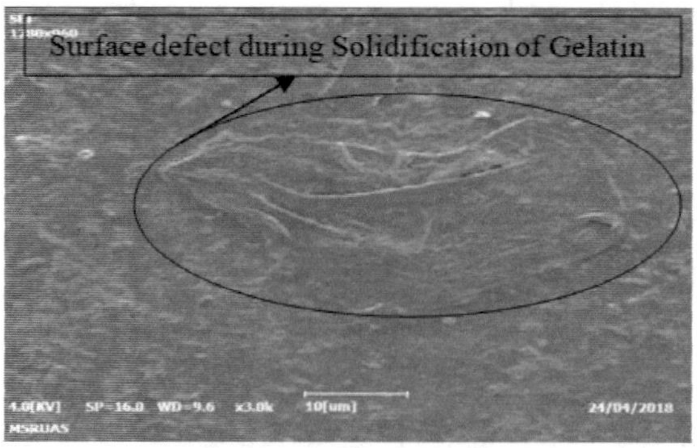

Figure 8: shows the surface of the Gelatin polymer at magnification rate of 3000

In the Figure 7 and 8, the SEM analysis, revealed that the solution prepared is mixed well and no agglomerations are found. Surface is almost plain because of constant cooling rate. Other surface defects like porosity and cavities are not found like in the resin based polymers. Traces of dust inclusions are found on the specimen,might be during the pouring of the solution and curing the solution.

5. Conclusions

1. Gelatin plastic has much less water absorption than other bioplastics.
2. Theyoung's modulusand tensilestrengthofgelatin bioplastic is comparatively better.
3. Some Low density plastics (with Young's modulus between 110MPa to 450MPa) can be replaced with Gelatin.
4. Spoons, Food wrapping covers can be replaced with Gelatin bioplastic.
5. Obtained bioplastic is biodegradable and Non-toxic.

References

[1] Mehdi Borghei, AbdolrezaKarbassi, Shahrzad khoramnejadian1, AbdolrasoulOromiehie and AmirhosseinJavid, Microbial biodegradable potato starch based low density polyethylene, (2010)*African Journal of Biotechnology*, 9(26), pp. 4075-4080

[2] K Wuttisela, S Shobsngob, W TriampoAmylose/amylopectin simple determination in acidhydrolyzedtapioca starch(2008)*Journal of the Chilean Chemical 400 Society*, 3, 1565–1567.

[3] J. Zhou, J. Zhang, Y. Ma, J. Tong, Surface photo-crosslinking of corn starch sheets, (2008)*Jounal of Carbohydr. Polym.* 74 (3) 405e410.

[4] X. Ge, L. Yu, Z. Liu, H. Liu, Y. Chen, L. Chen, Developingacrylatedepoxidized soybean oil coating for improving moisture sensitivity and permeability of starch-based film, (2019) *Int. J. Biol. Macromol.* 125 370e375.

[5] K. Kanimozhi, sethuraman et.al Development of ricehusk ash reinforced bismaleimide toughened epoxy nanocompsites, (2014) *Frontiechemistry* . DOI: 10.3389/fchem.2014.00065

[6] L. Meng, S. Li, W. Yang, R. Simons, L. Yu, H. Liu, L. Chen, Improvement of interfacial interaction between hydrophilic starch film and hydrophobic biodegradable coating, (2019) *ACS Sustain. Chem. Eng.* 7 (10) 9506e9514.

[7] L. Meng, H. Liu, L. Yu, Q. Duan, L. Chen, F. Liu, Z. Shao, K. Shi, X. Lin, How water acting as bothblowing agent and plasticizer affect on starch-based foam, (2019) *Ind. Crops Prod.* 134 43e49.

[8] Q. Duan, L. Meng, H. Liu, L. Yu, K. Lu, S. Khalid, L. Chen, One-step extrusion to minimize thermal decomposition for processing PLA-based composites, (2018) *J.Polym. Environ.* 27 (1) 158e164.

[9] D. Platt, Chapter 3 - overview of biodegradable polymers in biodegradable polymers: market report, (2006),*SmithersRapra Limited, Shropshire, UK*, 26

[10] A.B. Altemimi, Extraction and optimization of Ipomoea batatas starch and its application as a stabilizer in yogurt manufacturing, (2018) *Foods* 7 14–24, https:// doi.org/10.3390/foods7020014.

[11] A.S. Abreu, M. Oliveira, A. de Sa, R.M. Rodrigues, M.A. Cerqueira, A.A. Vicente, A.V. Machado, Antimicrobial nanostructured starch based films for packaging, (2015)*Carbohydr. Polym.*129 127e134.

[12] T. Ladeira, H. Souza, R. Pena, Characterization of the roots and starches of three Manihotesculentacultivars, (2013) *Int. J. Agric. Sci. Res.* 2 12–20.

[13] F. Benhacine, A.s. Hadj-Hamou, A. Habi, Development of long-term antimicrobial poly (ε-caprolactone)/silver exchanged montmorillonite nanocomposite films with silver ion release property for active packaging use, (2015) *Polym. Bull.* 73 1207e1227.

[14] Maulida1, M Siagian1 and P Tarigan1 Production of Starch Based Bioplastic from Cassava Peel Reinforced with Microcrystalline CellluloseAvicel PH101 Using Sorbitol as Plasticizer (2016) *J. Phys.: Conf. Ser.* 710 012012

[15] Bezirhanarikan, h. DuygubilgenProduction of bioplastic from potato peelwaste and investigation of its biodegradability, (2019)*International Advanced Researches and Engineering Journal*, Volume 3 , Issue 2, Pages 93 – 97

[16] Ashok A., R. Abhijith, C. R. Rejeesh, Material characterization of starch derived bio degradable plastics and its mechanical property estimation. (2018)*Materials Today: Proceedings*, 5(1): p.2163–2170.
[17] J. Muller, C. Gonzalez-Martínez, A. Chiralt, Poly(lactic) acid (PLA) and starch bilayer films, containing cinnamaldehyde, obtained by compression moulding, (2017) *Eur.Polym. J.* 95 56e70.
[18] K. Cooksey, Antimicrobial food packaging materials,(2001) *Addit. Polym.* 2001 6e10.

Studies on Stainless Steel 304 Samples Joined by Submerged Arc Welding, MIG Welding and Hybrid Welding

G. B Veeresh Kumar[1], Pragada Sai Ganesh[1], Atul Pandey[1], G. Surya Bharath[1], K. Vinay Kumar[1]

[1]Mechanical Department Engineering, National Institute of Technology – Andhra Pradesh, Tadepalligudem Town, West Godavari Dist, Andhra Pradesh State, India.

* Corresponding author's e-mail: veereshkumargb@nitandhra.ac.in

Abstract. The present paper focuses on the characteristics of Stainless Steel304 welded joints by using Submerged Arc Welding (SAW), Metal Inert Gas (MIG) Welding and Hybrid Welding techniques. MIG welding with a current of 170 A, SAW welding with a current of 105 A, and hybrid welding with currents of 105/170 A are the welding stages used. Hardness, flexural strength, and tensile strength is measured in MIG, SAW, and hybrid SAW-MIG weldments. A SAW-MIG hybrid welded specimens of 105/170A has the highest tensile strength, flexural strength, and percentage elongation of 0.591kN/mm², 15.09KN and 57.143% respectively. SAW – MIG hybrid welding can also be recommended for high-tech industrial applications like nuclear, aerospace, food processing, and the automotive industry, as well as surgical devices, jewelry, food, and beverage manufacturing.

Keywords: Stainless Steel, MIG, SAW, Hybrid welding, Hardness, Tensile, Flexural Strength.

1. Introduction

Stainless steel (SS) is used in situations where the material must resist corrosion. They are divided into different categories depending on their microstructural behaviour at room temperature, such as ferrite, martensitic, precipitation hardening, duplex, and SS [1].Due to its amazing welding characteristics, high corrosion resistance, non-magnetic design, and high formability, austenitic SS is the most common and widely used [2]. Weldability of SS304 is mostly thought to be arc and resistance welding methods. However, owing to the deposition of coarse grains and inter-granular Chromium-rich carbides nucleated along grain boundaries in the Heat Affected Zone (HAZ), which in turn weakens the mechanical properties of the joints being welded, conventional arc welding is often delicate for industrial uses [3].MIG welding is an arc welding technique that uses a welding gun to heat and feed a continuous solid wire electrode into the weld pool [4]. MIG welding can be used to weld a wide variety of materials of different thicknesses, it can even dig deeper, which is particularly useful for industrial applications, and it can achieve faster welds. Alongside advantages, it also has few difficulties like instability of arc, burn-back, manual wire feeding, a greater number of sparks are generated while welding, more smokes and fumes, irregular formation of the weld bead which is uneven in nature. SAW is a kind of arc welding in which an arc is formed between a continuously fed electrode and the work piece (W/P) [5,6].Using SAW one can weld strong weld, minimal fumes, minimal arc light, less distortion; flux can be recovered up to maximum extent [6]. Alongside advantages, it also has few difficulties like welding can perform only on flat surfaces; one must take care of flux handling system as well, it is also limited to high thickness materials as well [7].

By incorporating a hybrid welding process that includes both MIG and SAW, hybrid welding is a revolutionary way to enhance and solve the drawbacks of the above-mentioned issues. For a modern MIG-SAW hybrid welding method, the first pass of welding will be achieved by SAW, and the second pass will be performed by MIG welding, to achieve a better weld with improved mechanical properties as opposed to actual MIG and SAW specimens [3].All the welded samples of MIG, SAW, and Hybrid are later characterized based on hardness, tensile strength, and flexural strength.

2. Experimental Procedure

2.1. Materials and process selection

For the current investigative work SS304 was selected shown in figure 1, as the materials for welding. Before welding, the samples were of size 6mm thick austenitic SS304 plates were prepared by cutting, polishing & chemical cleaning. The sample SS304 plates as per ASTM standards with dimensions 100mm×20mm×6mm (for tensile test),120mm×40mm×6mm (for flexural test), 120mm×40mm×6mm (for harness test) [8] were obtained. Later specimens were welded using all three welding techniques via MIG, SAW and MIG-SAW hybrid welding. While welding a minimum gap of 1mm is maintained between the specimens. This helps filler material to pass in between the specimens held for joining and makes it a strong weld. Chemical composition of SS304 is mentioned in table 1, followed by physical and mechanical properties in the table 2.

Figure 1. SS 304 Sheet [9].

Table 1.Austenitic304SS Chemical Composition in wt.%.

Alloy Element	C	Si	Mn	P	S	Cr	Ni	Fe
Weight Percentage	0.03	0.75	2.00	0.045	0.03	19.50	12	Bal.

Table 2. Mechanical Properties of the 304 austenitic stainless steel.

Property	Melting Point	Density	Modulus of Elasticity	Thermal Conductivity	Tensile Strength (MPA)	Thermal Expansion
Value	1450°C	8.00 g/cm^3	193 GPa	16.2 W/m.K	540 -750	17.2 x 10-6/K

The MIG welding was carried out on MIG welding machine and in the similar way SAW was carried out on SAW machine with carbon di oxide gas as the shielding gas. The diameters of the filler material (Mild steel) used for MIG and SAW welding are 1.2 &2.5mm respectively.

In hybrid welding SAW is the 1st pass that was performed on 304 SS plates through the gap we maintained b/w the plates, from the start of the specimen to the end of the specimen using the filler rod of diameter 2.5mm, after that within 5minutes followed by MIG as Second pass of welding on formerly welded beads by means of a filler diameter of 1.2 mm as shown in the figures 2 and 3 respectively. Further the different welding parameters for MIG, SAW and hybrid welding are presented in the table 3.

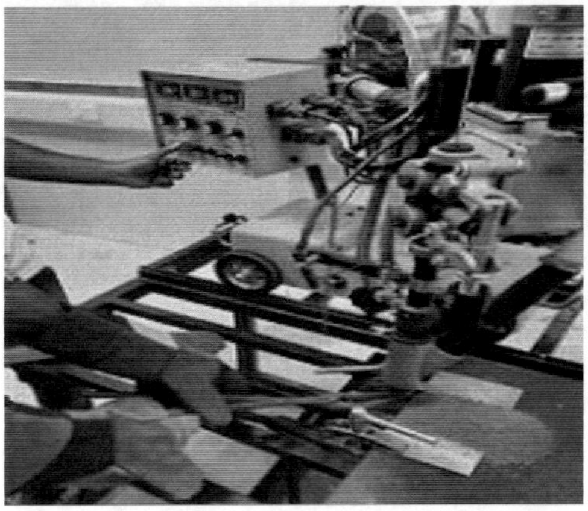
Figure 2. Submerged Arc Welding Setup.

Figure 3. MIG Welding Setup.

Table 3.Parameters for samples prepared using MIG, SAW and hybrid welding [10].

Parameters	MIG Sample	SAW Sample	Hybrid Welding	
			Submerged	MIG
Current (A)	170	105	105	170
Voltage (V)	26	20	20	26
Filler Supply Type	Manual	Automatic	Automatic	Manual
Gas Flow Rate (L/min)	10-15	N/A	N/A	10-15
Filler Material Diameter (mm)	1.2	2.5	2.5	1.2
Shielding Gas Type	Carbon Di Oxide	N/A	N/A	Carbon Di Oxide
Welding Speed (Cm/Min)	N/A	25	25	N/A

3. Results and Discussions

3.1 Test Sample visual Inspection

Figures 4a to d depict the physical appearances of the SS plates prior to welding. The top and bottom views for specimens being welded using hybrid welding technology are shown in figures 4a to d, and the Surface Visual Inspection of welded specimens is shown in table 4.After observing the weldments on both the sides (top and bottom) there are no spatters observed for three different weldments that are carried on austenitic SS 304 samples. The weld shape absorbed was narrow and weld bead width varied in between 3-5mm approximately [3].

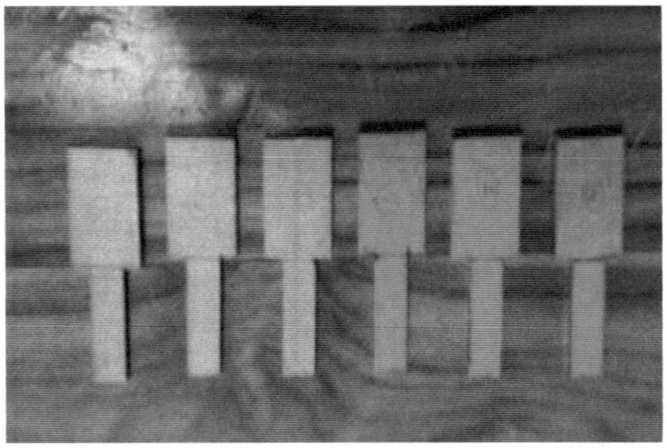

a. Specimens for tensile test

b. Specimens for flexural test

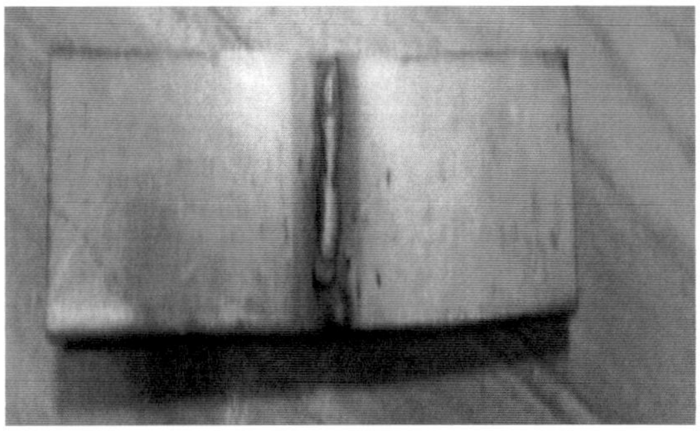

c. Top view of the welded specimen

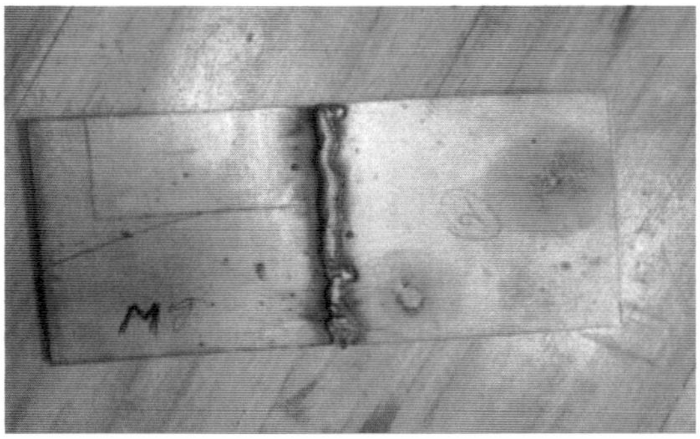

d. Bottom view

Figure 4a – d. SS304 plates before and after welding.

Table 4. Surface Visual Inspection of welded specimens.

Weld Properties	MIG Sample	SAW Sample	Hybrid Sample
Spatters	No Trace	No Trace	No Trace
Cracking	No Trace	No Trace	No Trace
Slag Inclusions	No Trace	No Trace	No Trace
Under Cutting	No Trace	No Trace	No Trace
Excess Penetration	No Trace	No Trace	Little Trace
Incomplete Penetration	No Trace	No Trace	No Trace
Porosity	No Trace	No Trace	No Trace
Uniformity of Weld Bead	Un Uniform	Uniform	Un Uniform

3.2 Tensile Test Outcomes

The tensile strength test is conducted to find out how strong the material joint and helps us to find out how much the material may be stretched before it undergoes into failure. Yield strength, UTS, and ductility, as well as strain hardening properties, Young's modulus, and Poisson's ratio, are all determined using this test process. Figure 5 depicts a cross-sectional image of a specimen for tensile testing in accordance with ASTM E8/E8M-13a.From the SS sheet of 304, three samples of MIG, SAW and hybrid welding were cut for performing tensile test by taking the above-mentioned specimen dimensions [4]. It is also absorbed that; the entire deformed samples are fractured at the weld zone as shown in the figure 6.

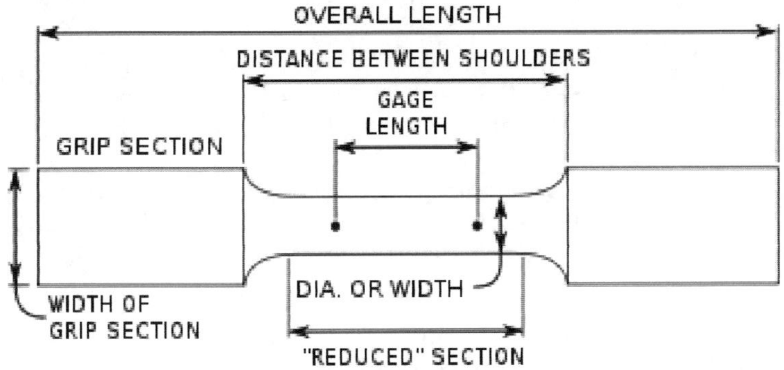

Figure 5. Cross-sectional view of specimen for tensile test as per ASTM E8/E8M-13a standard.

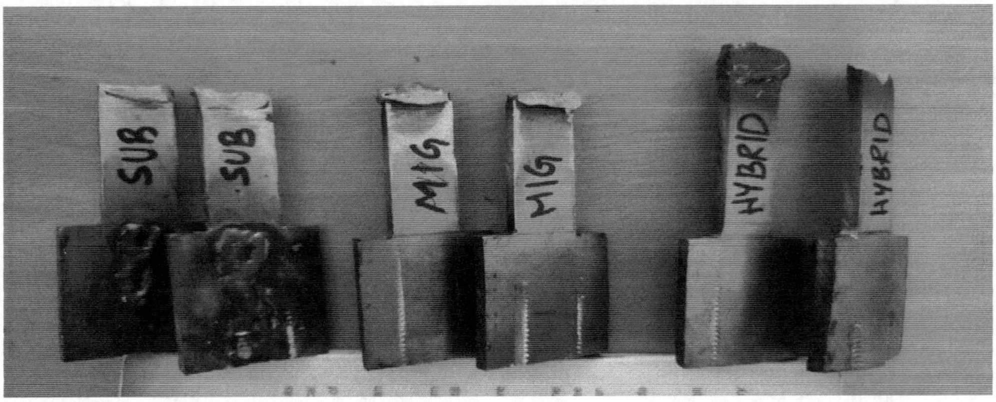

Figure 6. Fractured specimens of MIG, SAW and hybrid welding after tensile test.

Table 5. Tensile strength test outcomes for SAW [1].

1.	Maximum force	24.360KN	6.	Reduction area (z)	20.833%
2.	Displacement at max force	20.200mm	7.	Gauge length	105mm
3.	Max displacement	23.000mm	8.	Width	20mm
4.	Tensile strength	0.203KN/mm²	9.	Area	120mm²
5.	Elongation (%)	23.810%	10.	Thickness	6mm
6.	Final gauge length	130mm	12.	Final area	95mm²

IOP Conf. Series: Materials Science and Engineering 1185 (2021) 012024 doi:10.1088/1757-899X/1185/1/012024

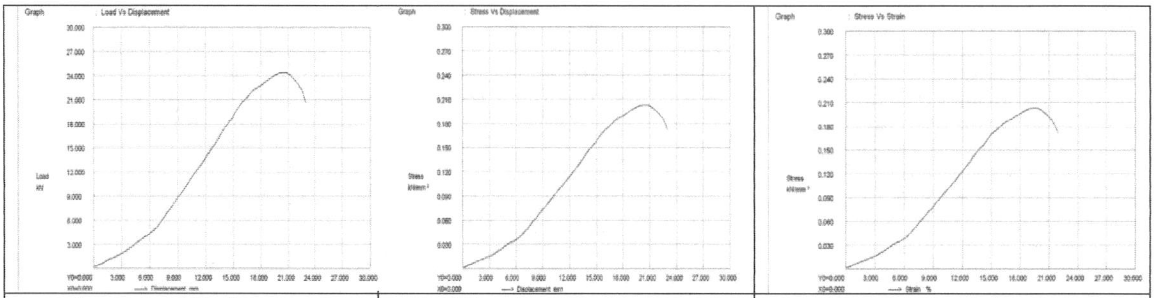

a.Load vs. Displacement | b.Stress vs. Displacement | c. Stress vs. Strain

Figure 7a to c. different graphs of tensile strength tested SAW samples.

Table 6. Tensile strength test outcomes for MIG.

1.	Maximum force	49.050KN	6.	Reduction area (z)	25.000%
2.	Displacement at max force	18.900mm	7.	Gauge length	105mm
3.	Max displacement	19.100mm	8.	Width	20mm
4.	Tensile strength	0.409KN/mm²	9.	Area	120mm²
5.	Elongation (%)	19.048%	10.	Thickness	6mm
6.	Final gauge length	125mm	12.	Final area	90mm²

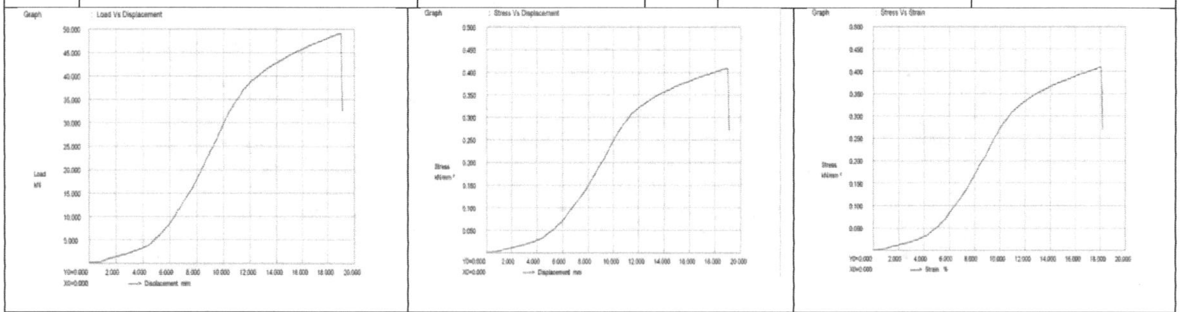

a.Load vs. Displacement | b.Stress vs. Displacement | c. Stress vs. Strain

Figure 8a to c. different graphs of tensile strength tested MIG welded samples.

The UTS of the submerged arc sample is recorded as 0.203 KN/mm² table 5, same for the MIG welded sample is 0.409 KN/mm² table 6 and UTS of the hybrid welding is recorded as 0.591 KN/mm²table 7. The UTS of the welded specimens are shown in the figures 7, 8 and 9 for SAW, MIW and hybrid welding, respectively. The welding parameters shown have a significant effect on the UTS produced. The hybrid sample provided by hybridization of SAW and MIG welds at currents of 105/170 A has the highest UTS of 0.591KN/mm^2 among the three welded samples evaluated in this study. Whereas lowest average UTS value of 0.203 KN/mm² was evaluated for SAW weld sample, which was welded at current of 105 A [3].

Table 7. Tensile strength test outcomes for hybrid welding.

1.	Maximum force	70.890KN	7.	Reduction area (z)	25.000%
2.	Displacement at max force	60.900mm	8.	Gauge length	105mm
3.	Max displacement	61.000mm	9.	Width	20mm
4.	Tensile strength	0.591KN/mm²	10.	Area	120mm²
5.	Elongation (%)	57.143%	11.	Thickness	6mm
6.	Final gauge length	165mm	12.	Final area	90mm²
13.	Yield load	46.650 KN	14.	Yield stress	0.389KN/mm²

With a value of 0.409 KN/mm^2, the UTS of the MIG weld sample drop between the UTS of the other samples. The UTS is critical when choosing materials for structural applications that are likely to experience high overloads and need significant quantities of energy to be absorbed.

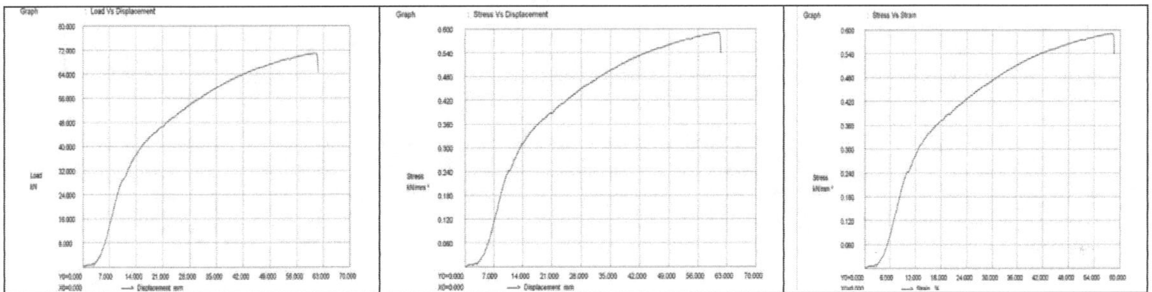

a.Load vs. Displacement | b.Stress vs. Displacement | c. Stress vs. Strain

Figure 9a to c. different graphs of tensile strength tested hybrid welded samples.

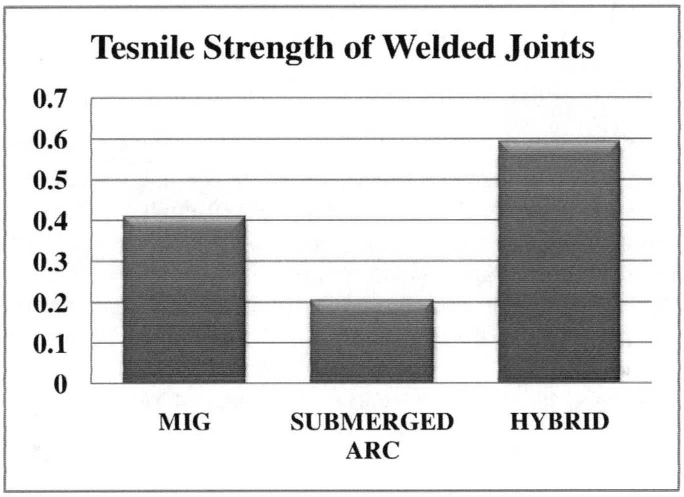

Figure 10. UTS chart for the MIG, SAW and Hybrid welded joints.

The comparisons of the UTS of all the samples are presented in the figure 10. Form the figure 10, the hybrid joint is having highest UTS than the other two joints considered in the current investigation. This result of the hybridised joint is mainly attributed to the deeper penetration.

3.3 Hardness Test Outcomes for Welded Specimens

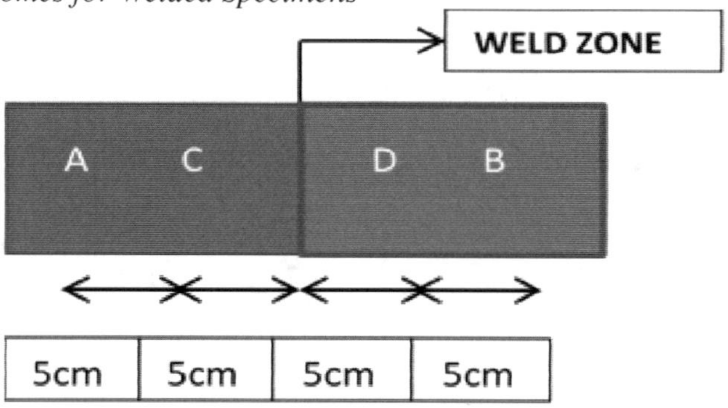

Figure 11. Hardness profile for welded samples.

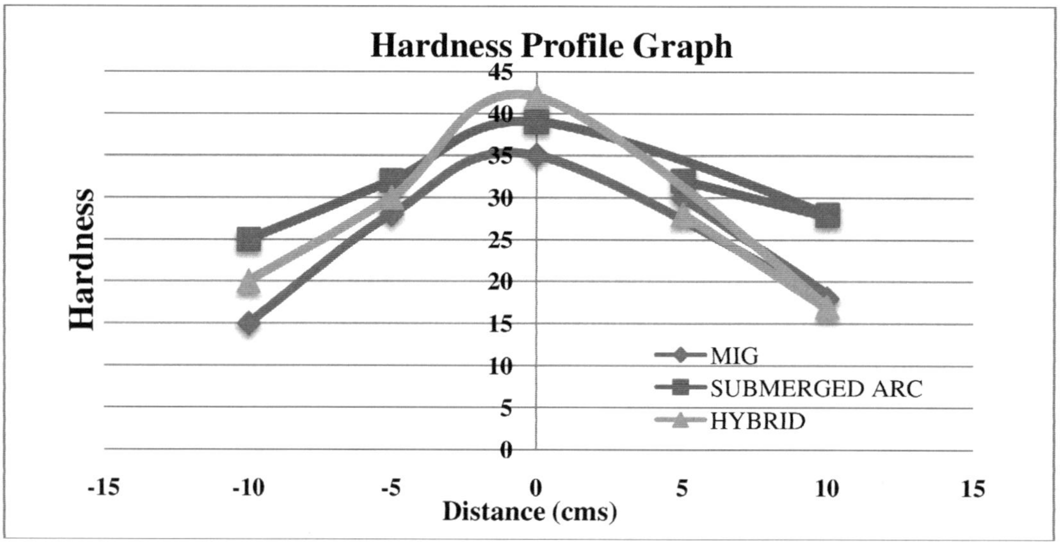

Figure 12.Hardness profile graph for welded samples.

The hardness tests were conducted on all the three weldments with a hardness profile as shown in the figure 11 for welded samples. The hardness values of welded samples were determined by indenting 5cms from left of the weld to centre of the weld and to right of the weld. As tested on HRC scale, the hardness values rise from the left to the middle, then collapse to the right after reaching the peak values in welded samples. Figure 12 also shows that the welded zone has a small rise in hardness as comparison to both the left and right adjacent zones [3]. The repeated thermal cycles encountered during the welding process with the melting and solidification of the filler metal used result in an improvement in the hardness value at the weld zone. Figure 12 shows that nearly all the samples have a higher hardness value in the welding region than in the base material and heat effected zones. At the weld region, the sample of hybrid weld has the highest hardness value of 42HRC, while the sample of MIG has the lowest.

3.4 Flexural Test Outcomes

Table 8. Flexural strength test outcomes for MIG, SAW and hybrid welding.

Parameters	MIG Specimens	SAW Specimens	Hybrid Specimens
Peak load	11.73KN	10.17KN	15.09KN
Displacement yield	19.4mm	14.7mm	48.5mm
Displacement	29.9mm	16.8mm	59.4mm
Break load	5.28KN	4.140KN	7.53KN
Yield load	11.13KN	10.32KN	12.39KN
Display yield	15.5mm	16.2mm	18.7mm

Figure 13.Fractured specimens after flexural test.

The flexural strength of the welded sample is measured by using Blue Star Make Universal Testing Machine UTE-60.The figure 13 is presents the fractured welded specimen after flexural test.From the table 8, it can be deduced that maximum peak load of 15.09KN was achieved on hybrid welded sample and minimum peak load of 10.17KN was achieved on submerged arc welding and peak load of MIG welded sample lies between the remaining two. Same with the case of displacement as well, maximum displacement of 59.4mm was achieved on hybrid weld.

4. Conclusions and future scope

This project attempts to conduct a hybrid welding of SAW-MIG welding on austenitic SS304 grade by using mild steel as filler material, the significant conclusions are as follows:

- Better Bead is obtained on SAW when compared to MIG welding. Observed excess penetration in hybrid welding due to combined welding ofMIG as well SAW.
- Deeper penetration is observed in case of MIG welding when compared to SAW due to higher deposition rate of filler material (which is uncontrollable factor due to manual feeding system of filler in MIG welding)
- Hybrid SAW – MIG welding of SS304 is possible with the excellent weld joint efficiency and with the excellent percentage elongation (57.143%) of the welded sample.

- Sample of Hybrid SAW-MIG Weld which is welded at 105/170A exhibits the highest UTS of 0.591KN/mm², when compared with the other two type weldments.
- Maximum peak load of 15.09KN was achieved on hybrid welded sample and minimum peak load of 10.17KN was achieved on submerged arc welding and peak load of MIG welded sample lies between the remaining two. Same with the case of displacement as well, Maximum displacement of 59.4mm was achieved on hybrid weld.
- SAW-MIG Hybrid sample at the weld region, weld has the highest hardness value of 42HRC, while MIG sample has the lowest.
- Lack of uniformity of welding is observed in MIG welding due to non-uniform feed rate of the filler material.

References

[1] SaurabhKumar Kand Ravi G 2017Comparison of Hardness and Tensile Strength of TIG and MIG Welding Using Stainless Steel 202*International Journal of Scientific & Engineering Research* **4** 264

[2] Arka G, BhattacharyaK, DeshpandeV and Sushovan B 2019 Microstructural andMechanical Properties Analysis of Plasma ARC Welded 316L Austenitic Stainless-SteelJoints without Addition of Filler Material*International Journal of Trend in Scientific Research and Development* **3** 849-65

[3] Emmanuel O, Esther T and Mutiu F 2018 Study on microstructure and mechanical properties of 304 stainless steel joints by TIG–MIG HYBRID WELDING *Surface Review and Letters***25** 245

[4] Labesh Kumar C, Vanaja T, Murti K and PrasadV 2017 Optimization of MIG Welding Process Parameters for Improving Welding Strength of Steel*International Journal of Trends and Technology***50** 21

[5] Prachya P 2017 Investigation of the Effects of Submerged Arc Welding Process Parameterson the Mechanical Properties of Pressure Vessel Steel ASTM A283*Journal of Engineering* **2** 10

[6] Arul Marcel Moshi A, Sundara BharathiR, Rajesh Kumar R and Kumar R 2016 Factors Influencing submerged arc welding on stainless steel: a review*Journal of engineering and Applied sciences***11** 1237-41

[7] Ghosh N, PalK and Nandi G 2016 Parametric optimization of MIG welding on 316L Austenitic stainless steel by Taguchi method *Science Direct (RAEREST 2016)* **25** 1038-48.

[8] Harshal M, Pargaonkar and KulkarniR 2018Process Parameter Optimization for Tensile Strength for Plasma Arc Welding of SS304*International Journal of Engineering Research & Technology* **7** 345.

Fabrication and Tribological Behavior of Al3003-SiC Reinforced MMCs

G B Veeresh Kumar[1], Pinapathina Gowtham[1], P N V N S Sai Ram[1], V Sai Ganesh[1], P Sai Praneeth[1]

[1]Mechanical Engineering Department, National Institute of Technology – Andhra Pradesh,Tadepalligudem, Andhra Pradesh State, India.

*** Corresponding author:** Email: veereshkumargb@nitandhra.ac.in

Abstract. The world is moving in the direction of development of novel materials, there must be an alternative for everything which makes utilization of resources efficiently. In this domain, the significance of Aluminium (Al) Metal Matrix Composites (MMCs)use increased in many industrial segments.This paper mainly focuses on the tribological behavior of Al3003 alloy with is reinforced with Silicon Carbide (SiC) powderof 25μm size. The Al3003-SiCMMCs were fabricated by using liquid metallurgy technique stir casting process. The SiC reinforcements were incorporated 0 to 4.5% in the phases of 1.5% weight percent (wt%). The obtained MMC specimens were subjected to investigative studies related to density, optical andtribological properties. All the studies and experimental work were carried out in accordance to the ASTM standards. The Al3003-SiC composites theoretical and practical evaluated density values are in line with each other and increased with increase in SiC amount in Al3003 matrix material. The optical micrographs of the Al3003-SiC MMCs showed respectable bonding between matrix and reinforcement particle with minimal casting defects. The sliding wear tests have been done on Pin on Disc tribology meter at different loads varying 10 to 40N in the strides of 10Nby maintaining other tribological parameters constant. The wear test outcomes showed that higher resistance to wear was shown by Al3003-SiC composites compared to unreinforced alloy and the composite contain higher SiC amount showed highest resistance to wear.

Keywords: Aluminium, Al3003, Silicon Carbide, Density, Microstructure, Wear.

1. Introduction

As time progress the engineering community is looking to upgrade and develop new type of materials which are durable and highly useful in various industrial applications. So as growing needs are being increased the importance and significance of particulate reinforced MMCs usage and applications are increased. There are many methods to fabricate the particle reinforced MMCs like liquid and powder metallurgy techniques, of which liquid casting metallurgy route is most widely adopted because to its easy to fabricate and economics. The choice of fabrication process depends on its property requirement, cost of production, reliability,and future application [1]. More recently hybrid MMCs with particulate reinforcement are widely adopted in engineering applications with the combination of minimum two types of reinforcement materials with added advantage of their properties to get the better mechanical and physical properties [2]. This will allow a better degree of freedom to design that different material combinations material in any applications. The improvement in the properties of reinforced material or alloy depends on their intrinsic properties of composites like shape, volume fraction, size, orientation, and distribution of reinforcement phase in MMCs [3]. Many scholars or research had shown applicability and usage of hard fibres/whiskers/particles as a ceramic reinforcement in various Al alloy MMCs as a main material in the automobile engines, cylinder liners, marine and aircraft applications etc,there are certain techniques other than traditional casting have been developed by scientists to fabricate MMCs, some of them are stir casting, rheo-casting, die casting, spray forming etc [4].Among these, the stir casting process has its own advantages like less cost of production coupled with economical, better flexibility ability to fabricate various composites, suitable for mass production, we can get good matrix reinforcement compatibility and near net shapes. Al-based MMCs have superior mechanical properties to all other MMCs [5-7]. The wear rate of an Al6061-Titanium Dioxide composite is affected by several factors, including load, rpm, sliding size, and reinforcement [8, 9]. Wear of the modules causes most system breakdowns [10, 11].When Al6061-Al_2O_3 composites were subjected to a dry sliding wear inspection, very different mechanisms were observed [12]. Temperature and load also influence the wear rate of Al6061-Silicon Carbide (SiC) composites. According to reports, Al6061-SiC composites would have a lower wear rate as load increases [13].It was discovered that composites of Al6061 and short fibresaffil fillings had increasing wear resistance when tested [14]. The MMCs with 15% SiC particulates as reinforcement and manufactured using powder metallurgy technique have strong wear tolerance, according to the publications [15].The microstructure, sliding speed, sliding distance, and load applied all have significant impact on wear of composite. It was discovered that as applied load increased to Al2219-SiC composites, wear, and cracking of SiC particles increased [16]. Sliding wear is one of the causes for the composite's maximum wear when exposed to various loads and speeds [17].When dry sliding, there is a transition in wear rate from mild to extreme as temperature of Al6061 alloy rises [18]. The properties of Al6061-Albite and Al6061-Graphite (Gr) were investigated, and it was discovered that as percentage of Gr filler increased, properties like elongation, modulus, and strength increased, while hardness decreased.Increased Albite content in Al6061-Albite composites resulted in increased stiffness and decreased ductility [19]. The key goal of this project is to create composites with Al3003 alloy and SiC micro-powder as reinforcement. It also entails the use of different experiments to investigate the composites' physical, optical, and tribological properties.

2. Fabrication, Matrix and Reinforcement Material Details

2.1 Matrix Al3003 and SiC Reinforcement Material Details

Like other aluminium series alloys,Al3003 also has good corrosion resistance, moderate strength but not heat treatable and generally develops strengthening from mainly cold working.Al3XXX series is an alloy of Al and Manganese (Mn) commonly utilized in sheet metal applications such as kitchen and gas pipeline applications, downspouts, gutters, food processing units, roofing, and siding. In the current studies Al3003 was selected as matrix material and obtained from Mallinath Metal, Mumbai, Maharashtra, in the form of sheets as shown in Figure 1. The reinforcement of SiC laboratory grade of 25μm was procured from Parshwamani Metals Mumbai, Maharashtra, as shown in Figure 2. The composition of Al3003 alloy is shown in table 1, and properties of matrix and reinforcement particulates are shown in table 2.

Table 1. Chemical composition of Al3003 alloy by wt%.

Chemical composition	Si	Fe	Cu	Mn	Zn	Others	Al
Al6061	0.60	0.70	0.10	1.30	0.10	0.1 to 0.5	Bal

2.2 Fabrication of MMCs

The Al3003-SiC composites were made using the stir casting process system of liquid metallurgy techniques. The Al3003 alloy was heated to 710°C in a container of SwamEquip's computerized bottom pouring style stir casting furnace in Chennai, TN.Coverall powder was applied to improve wettability, remove gases from molten metal, and form a protective barrier between molten material and atmosphere after securing molten alloy catalyst such as Magnesium chips and degassing tablets Hexachloroethane.

Table 2. The base matrix and reinforcement materials properties.

Material	Density (g/cc)	Hardness (HB500)	Elastic Modulus (GPa)	Tensile Strength (MPa)
Al3003	2.73	35	70-80	125
SiC	3.1	2800	410	3900*

Figure 1. Sheets of Al3003 Alloy.

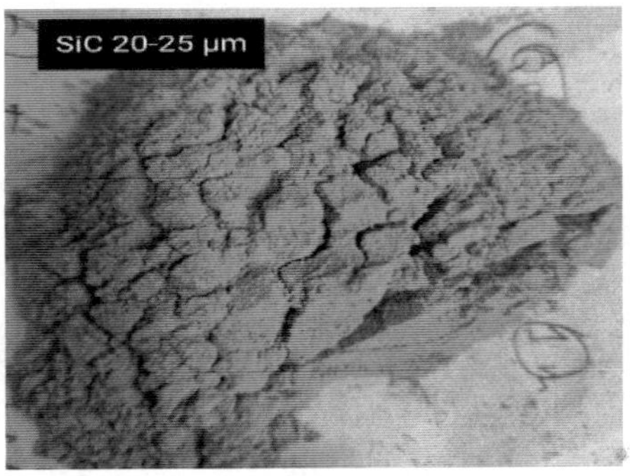

Figure 2. SiC powder of size 25µm.

At a stirrer speed of 400rpm, a steel stirrer coated with graphite was used to stir the molten alloy. The micro particulates of SiC powder were preheated to 350°C using a muffle furnace after sufficient degassing was assured, before being added into the vortex created by stirring.

Figure 3. Cylindrical steel mold box.

Figure 4. Casted Al3003 alloy.

Figure 5. Casted Al3003-1.5% SiC MMC.

Figure 6. Casted Al3003-4.5% SiC MMC.

Stirring was continued for another 10-12 minutes to ensure uniform SiC reinforcement distribution. The molten MMC was then poured into a cylindrical mold box with a length of 180mm and diameter of 30mm. Adding various weight percent (0 percent to 4.5 percent) of SiC micro particulates resulted in four castings.Figure 3 to 4 shows the different photographs of the casting details of the Al3003 and its SiC filled MMCs.

2.3 Experimental details

The stir cast castings were machined to the necessary specifications using CNC lathe machines from Hytech Automation Pune, Maharashtra, in accordance with ASTM standards. After that, properties including microstructure, density, and wear resistance were investigated on the test samples.The density was calculated using the rule-of-mixtures algorithm, and the results were compared to measurements taken in the field. Microstructural photographs of Al3003-SiC MMC specimens were obtained using an NIKHON model 150 ECLIPSE metallurgical microscope from Japan.The wear height loss of the composites was measured using a computerised Pin-on-disc tribometer made by Ducom, Bangalore, under various loads ranging from 10 to 40N. The cylindrical MMC specimens used in this study were 15mm in length and diameter.The Al3003 alloy and Al3003-SiC MMC specimens were made to dry slide against a 60 HRC coated high carbon hardened EN31 steel counter surface disc.Wear height loss in cylinder-shaped Al3003 alloy and Al3003-SiC composites samples was measured every 30 seconds in the test using a 1.0m precision LVDT transducer to draw graphs of wear height loss and sliding speed. The disc was set to spin at 500rpm and had a 3.0km sliding distance.To investigate the form of wear, SEM photographs of worn-out Al3003 alloy and Al3003-SiC composites surfaces were taken.

3. Results and discussions

3.1 Density of Al3003-SiCMMCs

The theoretical density of the four different compositions, namely 0%, 1.5%, 3%, and 4.5% weight percentage compositions, was determined using the law of mixtures, and experimental densities were measured using weight to volume ratios. Then, using the X, Y Scatter graph seen in figure 7, the theoretical and experimental densities were compared.

The density of the composites of varying Wt% of SiC from 0, 1.5, 3, to 4.5 in Al3003 alloy were observed to increasing, it is because the SiC has better density than the Al3003 alloy, so when added with the Al3003 alloy the density will be increased according to rule of mixture [20]. But when it comes practical, we may not see that the same results as in case of theoretical densities. It is due to some other external factors like porosity, casting defects, inclusion of foreign particles (dust), but in both the cases it was observed increase in the densities as weight fraction of SiC percentage increases [21]. This increase in density will lead to increase in the mechanical characteristics.

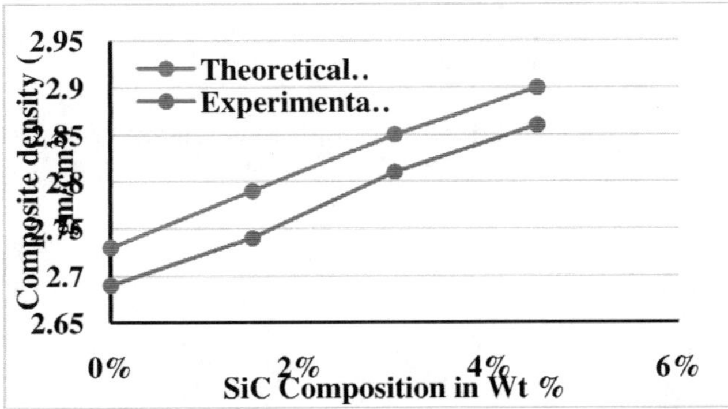

Figure 7. Theoretical and Experimental Densities comparisons of Al3003-SiC MMCs.

3.2 Optical Studies on Al3003-SiCMMCs

The reinforcement grain size is the most important aspect when it comes to mechanical and metallurgical characteristics of MMCs. It plays vary important role in enhancing physical and mechanical characteristicsin Al MMC's, which can be identified by destructive and non-destructive testing like tensile, hardness, compressive and dry sliding wear tribological tests.In the present investigative work, the Al3003-SiC MMCs samples were subjected the after disc polishing under Optical Microscope to view the grain size and distribution of reinforcement particles to compare how its microstructure growth influences the wear properties of the composites for different compositions. After Disc Polishing of the specimens the samples subjected for etching with using Kellar solution for around 30 sec. So that it will give better understanding of grain distribution. For all four compositions samples, images were taken for the 500X Magnification in the 10 μm square area. The image samples for all 4 Compositions are shown in figure 8a to 8d.

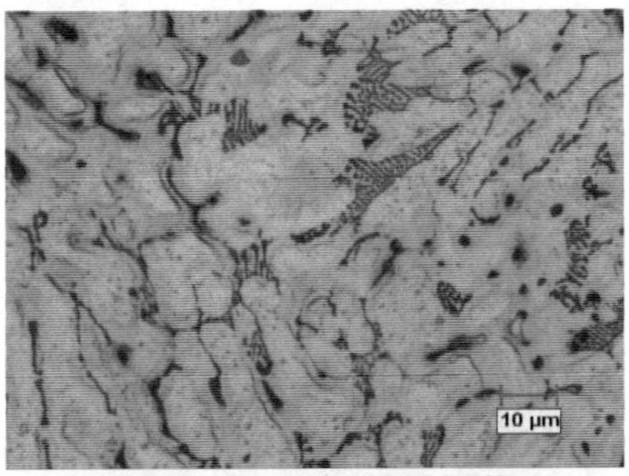

a. OM image of Al3003 alloy.

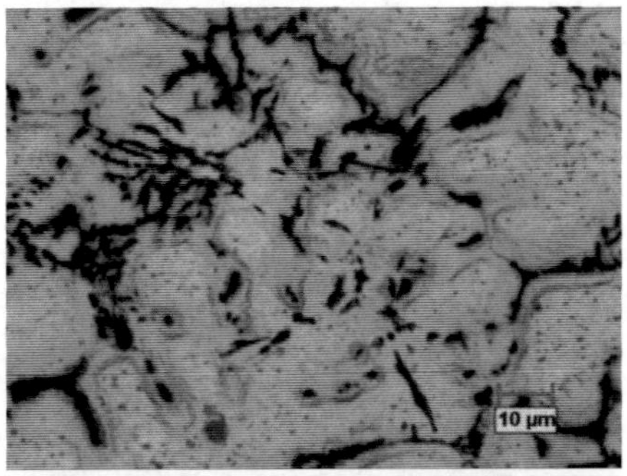

b. OM image of Al3003-1.5%SiC MMCs.

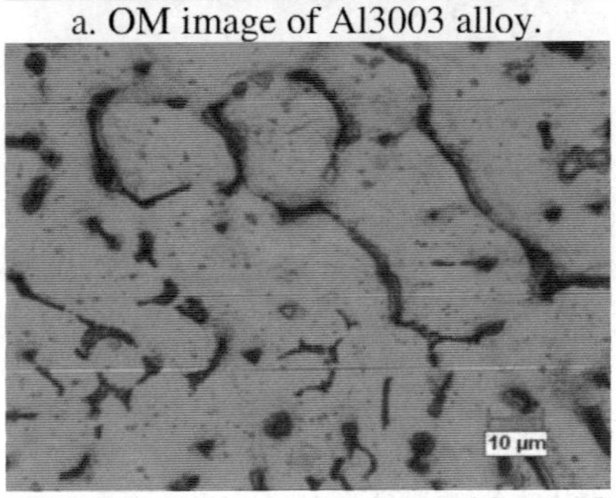

c. OM image of Al3003-3.0%SiC MMCs.

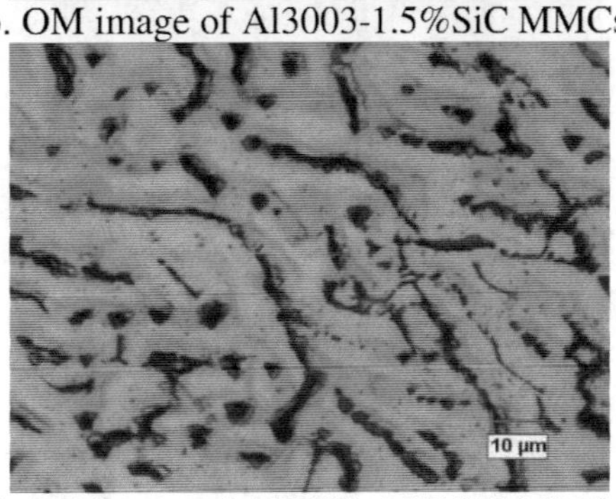

d. OM image of Al3003-4.5%SiC MMCs.

Figure 8a-d. 500X Microphotographs of Al30036061 alloy and SiC reinforced MMCs.

From the OM images of all four samples, we can observe that the grain boundary distribution and size vary as the SiC composition changes which effects the hardness, strength of the material.Infer that the grain borders are in the shape of dendrite regions based on the photographs. These dendrite regions show a uniform distribution of SiC reinforcements with low porosity and good bonding between the Al3003 matrix and the SiC reinforcement materials.The photographs show that the Al3003-4.5 percent SiC composite has extra dendrite regions due to the inclusion of SiC reinforcements in greater quantities.

3.3 Wear studies on al3003-Sic

Wear tests were done on Pin on Disc (POD) machine toanalyserate of wear, volume loss, and friction for all the four different compositions of Al3003-SiC MMC's. The technical parameters of the POD were given in table 3. The specimens of size 15mm both diameter and height in the shape of cylindrical pins were undergone into testing for all samples varying load from 10N, 20N, 30N, and 40N with other parameters like Sliding distance = 3000m,sliding rotation speed = 500rpm are remains same.During the POD test the sliding disc rotates, the pin surface in contact with disc, so when disc rotates in opposite direction there will be a friction generated which results in wear, which gives the height loss against sliding distance values in the digital monitor. The dry sliding wear experiments were executed according ASTM-G99 standards [22]. Further the figure 9 and 10 shows the POD machine set up.

Table 3: Pin on Disc set up parameters and specifications.

S.No	Parameters Name	Specifications of the Setup parameters
1	Disc Dimensions	Range from 8 to 12 mm thickness and up to 160mm dia
2	Dimensions of the Pin	Ranges from 10 to 50 mm length and 3 to 15 mm diameter
3	Wear Track diameter	Ranges from 20 mm to 120 mm
4	Sliding Velocity	Ranges from 0.5 to 26 m/s
5	Load Applied	Maximum 200N load
6	Disc rotational speed	100 to 20000rpm

A pin-on-disc tribometer was used to test the MMCs' wear resistance. In last 25 years, numerous studies on Al-MMCs have been carried out to determine the wear properties of these composites, and numerous reports have been submitted on the topic. The MMCs are subjected to a dry sliding wear test at room temperature in this paper.The sliding velocity is kept constant to 3.14ms^{-1}.

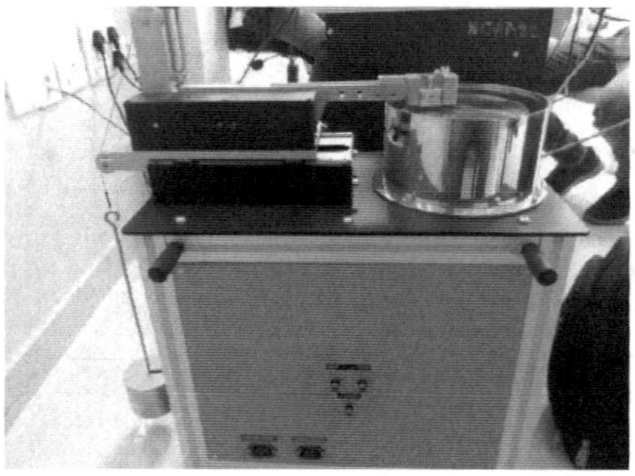

Figure 9. POD experimental set up.

Figure 10. Cylindrical pins.

In general, either the volume loss or wear volume loss may be used for obtaining the wear rate. Using a pin-on-disc tribometer, the wear resistance of MMCs was investigated. In the last 25 years, several studies on Al-MMCs have been carried out to determine the wear properties of these composites, as well as numerous papers. At room temperature, the MMCs are subjected to a dry sliding wear test in this paper.

The figures 11a-d characterizesheight loss at different sliding distances when 04 different loads were applied. It can be deduced from the statistics that wear height increases as sliding distance increases. This is because as the sliding distance rises, the frictional force changes, causing temperature of MMCs sliding surfaces to rise. As the temperature rose, the contact materials weakened, resulting in increased wear loss in height.

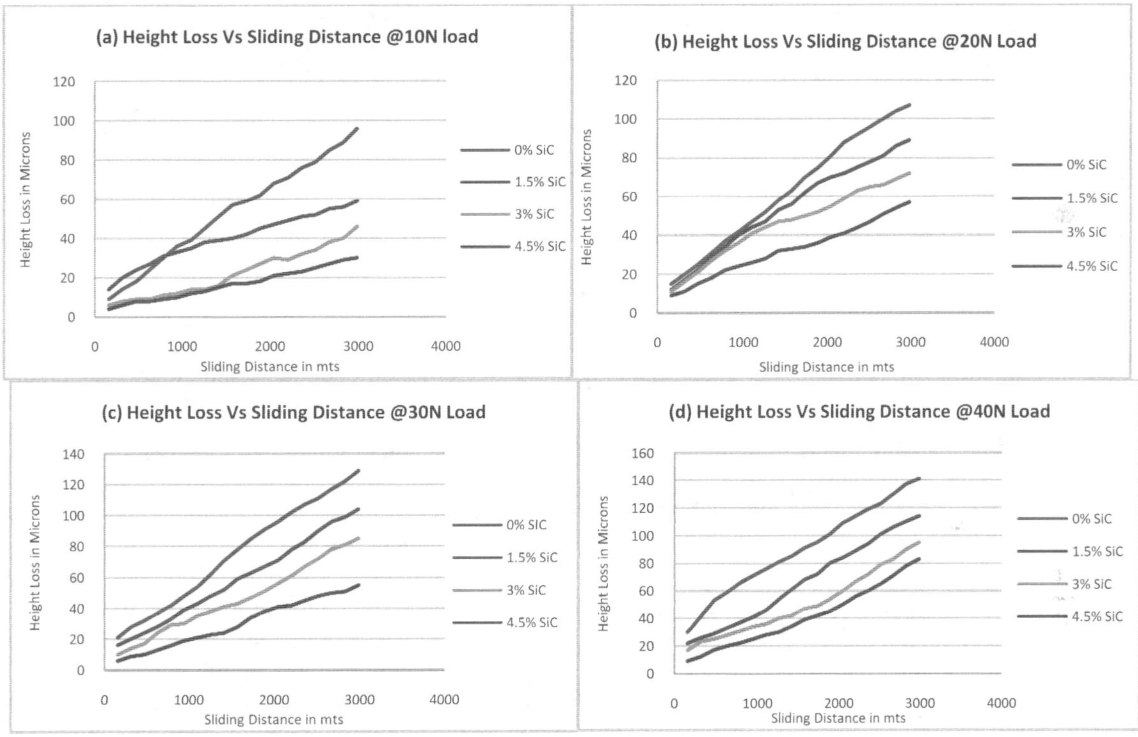

Figure 11a to d. Height Loss Vs Sliding Distance Al3003 and its SiC MMCs.

The wear volumetric loss vs sliding distance for the compositions and 10 to 40N loads is seen in Figure 12a to d. When a 10N load was applied, the volumetric wear loss decreased progressively as the percent sum of SiC reinforcement was increased. This indicates that the stiffness has increased.Similarly, 12b to 12d describe the same for 20, 30, and 40N loads, indicating that as the stiffness of Al3003-SiC MMCs rises, so does the resistance to slipping wear [23]. The load added has a major impact on the wear rate of Al3003 and its SiC composites [24].As four graphs are compared, it can be shown that the wear rate increases as the applied load increases from 10N to 40N. These improved wear properties of Al3003-SiC MMCs could be attributed to the SiC reinforcement's increased stiffness.The sliding dry wear under both composites obtained showed promise with the SiC reinforcement Al3003 composites and demonstrated superior sliding wear resistance [25, 26].

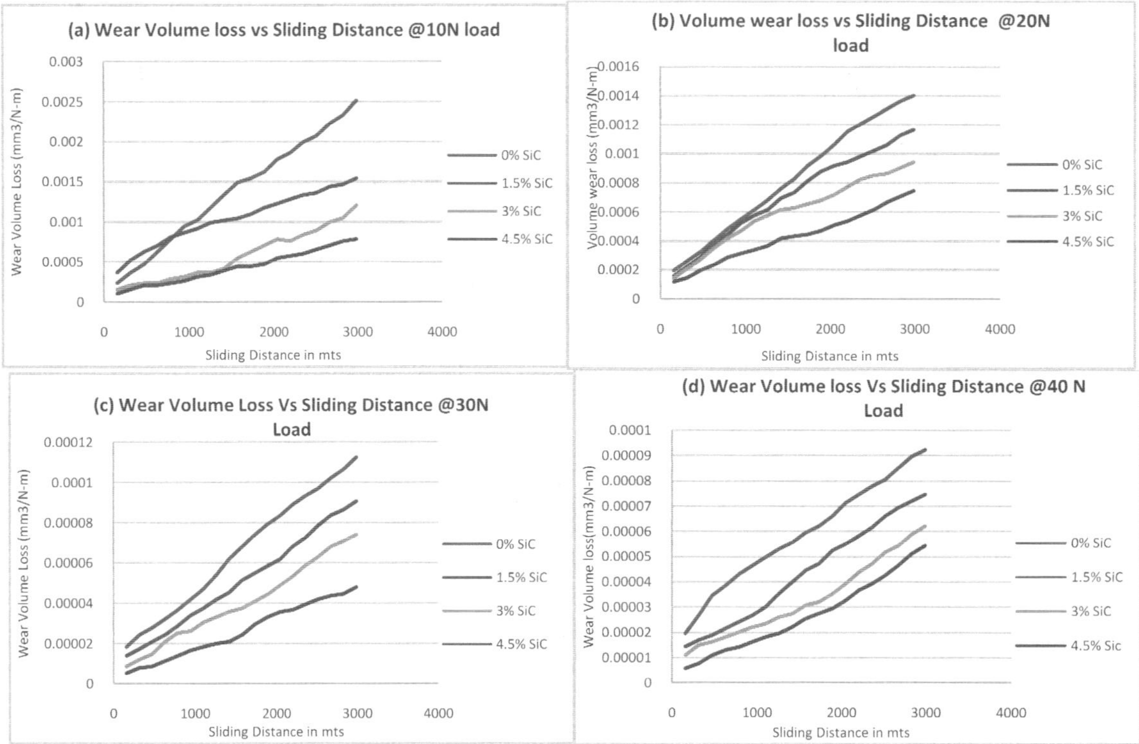

Figure 12a to d. Wear volume loss Vs sliding distance in meters for Al3003-SiC MMCs.

3.4 SEM Analysis of Fractured Surface

The Scanning electron microscope is also known as SEM is used to identify and analyse the wear and its mechanism of SiC particles in matrix Al3003. Through scanning the sample with a directed beam of electrons, SEM creates images of the sample. Electrons communicate with atoms in the sample and provide a variety of signals that provide knowledge about the polished surface structure of the sample. In most cases, the electron beam is scanned in a raster scan pattern, and the direction of the beam is coupled with the observed signal to create an image [27].SEM analysis is used to determine the distribution and orientation of reinforcement particles in MMCs. Following the POD inspection, the samples were subjected to SEM for pictorial wear recognition after slipping at 3km at 40N. At 500x magnification, the worn-out surface of pins for all four combinations of SiC from 0% to 4.5 percent is found, as seen in figures 13a to d.

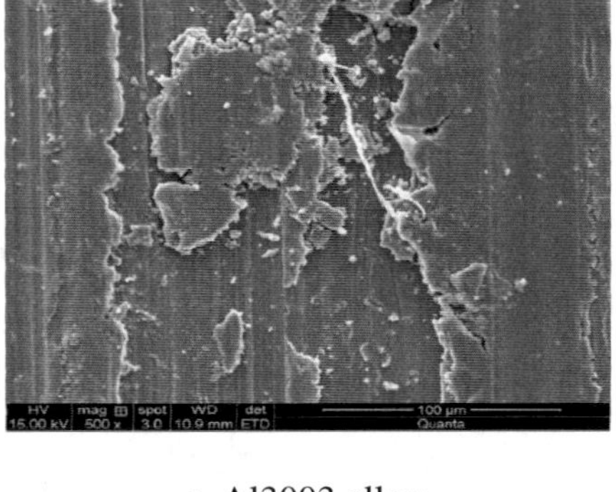

a. Al3003 alloy.

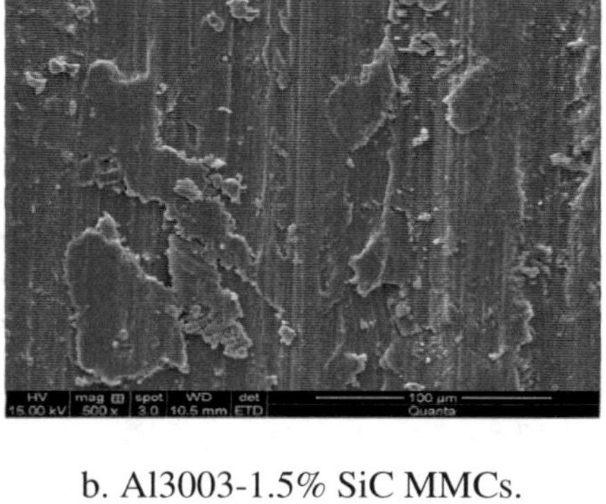

b. Al3003-1.5% SiC MMCs.

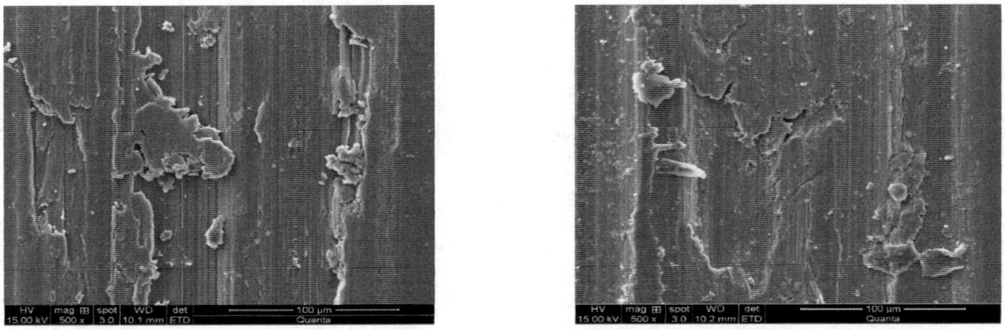

c. Al3003-3.0% SiC MMCs. d. Al3003-4.5% SiC MMCs.

Figure 13a to d. SEM images of worn-out surface of Al3003-SiC MMCs.

The SEM images of MMCs are taken following dry sliding them for distance of 3.0km with 500rpm constant speed. Figure 13a to d, characterizes the SEM micrographs of Al3003, Al3003-1.5% SiC, Al3003-3.0% SiC and Al3003-4.5%SiCcomposites correspondingly. From SEM images it is very much clear that the Al3003-SiC MMCswith 4.5wt.% reinforcement has undergone lesser wear in comparison to other samples in current study. SEM photographs show that the number of grooves and pits is higher in the composite with 1.5wt.% SiC reinforcement and lower in the MMC with 3.0wt.% SiC reinforcement. However, in the SEM picture of the MMC with 4.5 percent SiC reinforcement, there are more tracks with smaller pits and grooves.

The behaviour of wear in reinforced specimens, as well as physical and mechanical characteristics, are the most important considerations in tribology.

Extrinsic to undergoing material interactions at surfaces level like effect wear normal load at the tribo-contact, sliding velocity, sliding wear distance, environment, temperature, reinforcement orientation, counterpart materials and surface finish, factors inherent to material undergoing interactions at the surface level are distribution of the filler, shape, size and reinforcement type, matrix microstructure and volume fraction of reinforcement [28, 29].

4. Conclusions

The project was attempted to fabricate Al3003-SiC MMC's by Stir Casting Process under fixed four different weight percentage of SiC reinforcement and then finding the Microstructure Growth and also evaluated the Tribological Properties of the fabricated composites.

- The Stir Casting technique was successfully implemented for all the four percentage of Silicon Carbide Reinforcement (0%, 1.5%, 3%, 4.5% of SiC) in the Al3003 matrix.
- As the percentage of reinforcement SiC in Al3003 alloy was increased from 0%, 1.5%, 3% up to 4.5% the densities of the fabricated composites were observed increasing because of adding high density ceramic particles.
- Uniform distribution of the reinforcement was observed by using OM images for all the compositions of Al3003-SiC MMCs.
- Wear Volume loss was decreased as the SiC reinforcement percentage increases in the MMC of Al3003-SiC and the composite contain higher SiC filler content showed excellent resistance to wear in comparison to other materials considered in the current studies.
- Further the excellent resistance to wear sample was inferred by the SEM images of the worn-out surfaces of Al3003-4.5% SiC filled MMCs with formation of the lesser pits and grooves.

IOP Conf. Series: Materials Science and Engineering 1185 (2021) 012025 doi:10.1088/1757-899X/1185/1/012025

References

[1] GB Veeresh Kumar,CSP Rao and N Selvaraj 2010 *Journal of Minerals & MaterialsCharacterization& Engineering USA***10** 59-91

[2] Mazumdar S K 2002 *Composites Manufacturing CRC Press New York*

[3] Jaswinder Singh and Amit Chauhan 2017 *Transactions ofNonferrousMetals Society of China***27** 2573-86

[4] Arun Kumar Sharma, Rakesh Bhandari, Amit Aherwar andCameliaPincaBretotean2020*Proc. Materials Today***27**(2) 1608-12.

[5] Ramkumar KR and Natarajan S 2018*Mater. Sci. Eng. A***727**51–60.

[6] KRRamkumar and SNatarajan 2019 *Composites Part B: Eng.***175**107-159.

[7] K Anil Kumar,S Natarajan, MuthukannanDuraiselvam and S Ramachandra 2019*Materials and Manufacturing Processes* **34**(2) 183- 191.

[8] GB Veeresh Kumar, CSP Rao, Bhagyashekar MS and Selvaraj N 2009 *Journal of Minerals &Materials Characterization &Engineering USA***9**(1) 47-59.

[9] GB Veeresh Kumar, P S Shivakumar Gouda, R Pramod and C S P Rao 2017 *Advanced CompositesLetters* **26**(1) 18 – 23.

[10] Veeresh Kumar G B, Pramod R, Shivakumar Gouda P S and CSP Rao2016 Effect of Tungsten Carbide Reinforcement on the Aluminium 6061 *Alloy Journal of Testing and Evaluation*ISSN 0090-3973.

[11] GB Veeresh Kumar, CSPRao, Selvaraj N andBhagyashekar MS 2009 *Journal of Minerals &Materials Characterization &Engineering USA* **9**(1) 47-59.

[12] R Pramod, GB Veeresh Kumar 2014 *Journal of Material Science &Engineering***1**(3) 06-11.

[13] A Martin, J Rodriguez and J Llorca1999 Temperature effects on the wear behavior of particulate reinforced Al-based composites *Wear***229** 615–620.

[14] Szu Ying Yu, Hitoshi Ishii, Keiichiro Tohgo, Young Tae Cho and DongfengDiao 1997.Temperature dependence of sliding wear behavior in SiC whisker or SiC particulate reinforced6061 aluminum alloy composite *Wear***213** 21-28.

[15] HC How and TN Baker 1997 Dry sliding wear behaviour of Saffil-reinforced AA6061 Composites*Wear***210** 263-272.

[16] YN Liang, Z Y Ma, S Z Li, S Li and J Bi 1995Effect of particle size on wear behaviourof SiC particulate-reinforced aluminum alloy composites *Journal of Materials ScienceLetters* **14** 114-16.

[17] S Basavarajappa, G Chandramohan, K Mukund, M Ashwin and M Prabu 2006 Drysliding wear behavior of Al 2219/SiC-Gr hybrid metal matrix composites *Journal of Materials Engineering and Performance***15**668.

[18] S Basavarajappa and G Chandramohan2005Wear studies on metal matrix composites -Taguchi approach *Journal of Mat Sci and Technology***21**(6) 845-50.

[19] J Zhang and A T Alpas 1997Transition between mild and severe wear in aluminium alloys *ActaMateilia***45**(2) 513-528.

[20] A Ramesh, J N Prakash, A S Shiva Shankare Gowda and SonnappaAppaiah2009Comparison of the Mechanical Properties of AL6061/Albite and AL6061/Graphite Metal Matrix *CompositesJournal of Minerals & Materials Characterization & Engineering***8**(2) 93-106.

[21] V Ramakoteswara Rao, N Ramanaiah and MMM Sarcar 2016 Tribological properties of Aluminium Metal Matrix Composites AA7075 Reinforced with Titanium Carbide (TiC) Particles *International Journal of Advanced Science and Technology***88**13-26.

[22] SSivananthan, VRajalaxman Reddy and CSamson Jerold Samuel 2020 Preparation and evaluation of mechanical properties of 6061Al-Al2O3 metal matrix composites by stir casting process*Proc. Materials Today***21**(1) 713-716.

[23] 1999*Annual Book of ASTM Standards* USA: American Society for Testing and Materials.

[24] CS Ramesh, AR Anwar Khan, N Ravikumar and P Savanprabhu 2005 Prediction of wear coefficient of Al6061–TiO2 composites *Wear***259** 602–608.

[25] DP Mondal, S Das, AK Jha and AH Yegneswaran 1998Abrasive wear of Al alloy– Al2O3 particle composite: a study on the combined effect of load and size of abrasive*Wear* **223**131–138.

[26] Veeresh Kumar G, Pramod, R, Shivakumar Gouda, P and C S P Rao 2019 Effect of Tungsten Carbide Reinforcement on the Aluminum 6061 Alloy*Journal of Testing and Evaluation* 47(4) 56-59.

[27] R Pramod, G B Veeresh Kumar, PS Shivakumar Gouda, Arun Tom Mathew 2018A Study on the Al2O3 reinforced Al7075 Metal Matrix Composites Wear behavior using Artificial Neural Networks*Materials Today Proc.***5**(2) 11376-85.

[28] Raghavendra M J, Praveen Kumar K, Arun R and Arjun S 2017 A Study on Microstructure and Characterization of Aluminium 7075 Metal Matrix Reinforced with Silicon Carbide Particles Using Stir Casting method *IJRET: International Journal of Research in Engineering and Technology* eISSN: 2319-1163 | pISSN: 2321-7308.

[29] APSannino, HJRack 1995 Dry sliding wear of discontinuously reinforced Al composites: review and discussion *Wear* **189** 1-19.

[30] RK Uyyuru, MK Surappa, S Brusethaug 2006 Effect of reinforcement volume fraction and size distribution on the tribological behavior of Al-composite/brake pad tribocouple*Wear***260** 1248–55.

Brief Survey of the Microstructural, Mechanical and Corrosion Properties of Duplex Stainless Steels SMAW Weldments.

Ibrahim Momoh-Bello Omiogbemi [1,2*], Rajneesh Kumar Gupta[3], Danjuma Saleh Yawas[2], Matthew Olatunde Afolayan[2], Emmanuel Toi Dauda[4],

[1] Department of Mechanical Engineering, Air Force Institute of Technology, Kaduna-800282, Nigeria
[2] Department of Mechanical Engineering, Ahmadu Bello University, Zaria-10211, Nigeria
[3] Engineering Division, CSIR-National Metallurgical Laboratory, Jamshedpur- 831007, India
[4] Department of Metallurgical and Materials Engineering, Ahmadu Bello University, Zaria-10211, Nigeria

*Corresponding Authors e-mail: omiogbemi1@gmail.com, rkgupta@nmlindia.org

Abstract. Duplex stainless steels (DSSs) structures, especially the American Iron and Steel Institute (AISI)type 2205 and 2507 have greatly attracted the attention of many researchers, engineers, manufacturers as well as the end-users of the productsbecause of its superior engineering properties like strength, good toughness, particular resistance to corrosive environments and to stress corrosion cracking (SCC). Due to its worldwide progressive growth, demand and utilization, this novel steel (DSS) is rising very fast, especially in marine, power plants, chemical process, mining, petrochemical, oil and gas, pharmaceutical and many other related engineering applications. Generally, the joining of DSSs alloys is a big problem due to its susceptibility to sensitization caused by the precipitation of additional phases when heated above 600°C. Conversely, improper selection of welding parameters, imbalance ratio of austenite/ferrite phases can lead to corrosion susceptibility, solidification cracking and susceptibility to plastic deformation. Shielded metal arc welding (SMAW) technique with recommended consumables (E2209 and E2594 or E2595) is used in the fabrication of the multiphase steels with the aim of obtaining the optimum weldment with the desired input welding criteria, optimum mechanical properties with minimum defects from the microstructural perspective and excellent corrosion behaviours. This review is therefore geared towards accentuating the influence of arc welding processes, and SMAW in particular on the microstructure, mechanical and corrosion properties of DSSs weldment.

Keywords SMAW; AISI 2205; AISI 2507; Microstructural properties; Mechanical properties; Corrosion resistance

1 Introduction

Welding can be seen as a fabrication technique of creating a permanent joint resulting from the melting of the metal surfaces of the work-piece to be joined together, with or without the employment of pressure and a filler rod, having a similar composition with the base metal [1].

Generally, the duplex stainless steels (DSSs) structure has been recognized as a well-established type of stainless-steel family for more than 90 years. The earliest mention of DSSs officially was from 1927 [2]. It became obvious during the 20th century that the ferrite phase in DSSs provides excellent protection for the steel against chloride-induced SCC that is a major menace of the counterpart austenitic stainless steels (ASSs). However, the consumption of DSSs in many engineering and other industries increase annually by up to 20% and more [3].

Considering the appropriate compositions, physicochemical and thermo-mechanical processing, the novel DSSs, which display a dual-phase structure of austenite–ferrite, can be realized. The increased yield strength and excellent opposition to SCC are some of the advantages presented by DSSs upon single-phase grades of stainless steel [4-7].

DSSs structure has recorded several outstanding successes in contrast to its counterpart ASSs due to the volatile nature of the price of nickel (Ni) alloy in the general market year over year [8]. One of the major problems of ASSs structures is their susceptibility to several aggressive media [9]. Hence, many industries and engineering sectors are trying to adopt alternative stainless steel with a low percentage of nickel (Ni), having better or comparable performance as that of usual ASSs [10].

Many technical and scientific reports on DSSs delivered in conferences held around the world have aggrandized the great demand for the novel steel structure globally. Some of these conferences are St. Louis in the United State of America, 1982, Den Hauge-Netherland, Beaune-France, Yokohama-Japan, York-United Kingdom, Maastricht-Netherland, Venice-Italy and Grado-Italy in 1986, 1991, 1993, 1994, 1997, 2000, 2007 respectively till the present time. During these conferences, scientific and valuable technical presentations relating to the aspect of metallurgy such as phase precipitations and corrosion behaviour together with the mechanical characterization of dual-phase DSSs based on the experiences in the industry concerning the valuable areas of utilization of lean DSS grade 2304, standard DSS grade 2205 and Super DSS grade 2507. These types of DSS mentioned above were ratified to be essential grades with better performance in service applications experience for more than two decades which resulted in the conclusion made by the aforementioned conferences that there is a high demand for DSSs structure as an alternative of austenitic 300-series counterpart [11-14]. This has further led to the implementation of novel steel (DSSs) in combination with other alloying elements for different operation by several countries like Saudi Arabia, Spain, West Indies, Malta, Sharjah and Cyprus in various sectors such as brackish water reverse osmosis (BWRO), large multi-effect desalination (MED) plants, bleaching plants, multi-stage flash (MSF) plants, and large seawater reverse osmosis (SWRO) plants [15-18].

Amidst the various classifications of stainless-steel grades, the manufacturing of DSSS constitutes less than 200,000 tonnes which are approximately one percentage (1%) of the whole production of stainless steels worldwide, in the year 2007. The conference held in Feinox, 2008 forecasted that by 2020, the manufacturing/production of DSSs will be increased by about 4% as illustrated in Fig. 1. Conversely, the ASSs was 63.1% in demand during the year 2004. It decreased by 1.1% in 2007, as a result of the fluctuation of the price of nickel (Ni) alloy. Therefore, based on this factor, a 53% share of ASSs was forecasted up to 2020 as shown in Fig.1 [19-20].

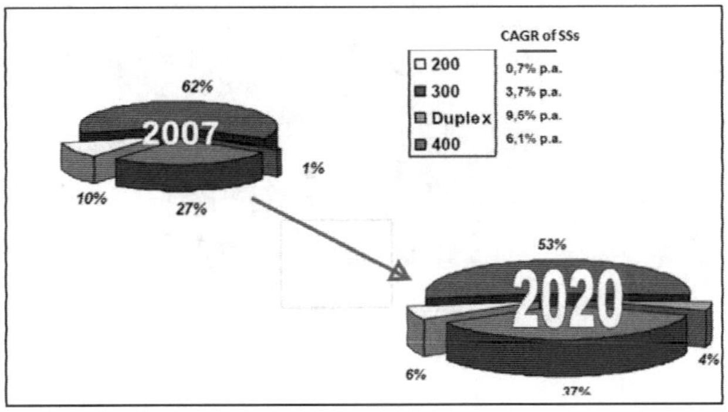

Figure 1 The trend and forecast of stainless steel; associated with Baldo [20].

Furthermore, the volatility of the price of Ni-alloy is unstable which has adversely affected the general stainless steel producers and end-users. Since nickel is the major alloy in most largely used austenitic stainless steels 300-series, its supplies (SSs) to the global market demand has not been met. This resulted in the search for better alternatives with low nickel alloy such as ASS 200-series (Cr-Mn) and the 400-series (ferrite grades) normally applicable in mild aggressive environments and for structural purposes as depicted in Figure 1. Presented in the figure, it is obvious that 10% of austenitic stainless steel 200-series was produced in the year 2007 and about 6% was forecasted up to the year 2020. Looking at the years; 2013 and 2014, 19.8% and 18.9% were produced respectively for the same steel (200-series) which signifies a drop of about 0.9% in the production rate between 2013 and 2014 as illustrated in Figure 2 and Figure 3. While the ferritic grades (400-series) demand and manufacturing are geometrically going up as a result of the low price, better thermal productivity, and easy recyclability. About 27% of 400-series was produced in 2007 and 37% prediction was made up to 2020. However, 25.5% and 25.2% were produced in 2013 and 2014 respectively which indicate a 0.3% decrease in the production between the two years (2013 and 2014). Also, stainless steel products global demand was about 37.7 million tonnes in 2015, there was a little increase from 37.1 million tonnes in 2014 (2% increase). From literature, the annual average growth of stainless steel generally from 2011 to 2014 was 8% while the overall global production of steel in 2015 was about 1.6 billion tonnes and stainless steel was about 2.6% of the total [21-23].

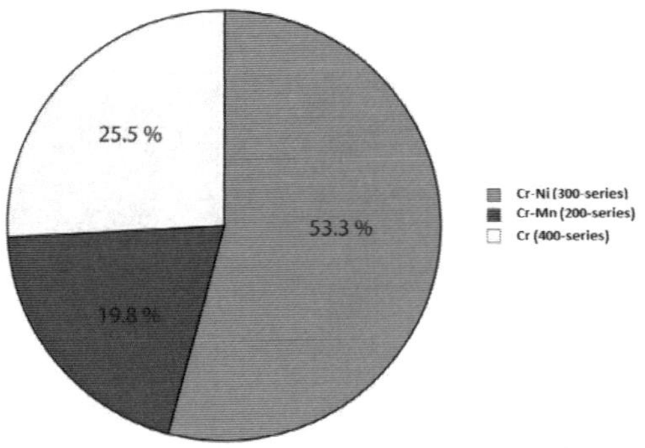

Figure 2 Some grades of stainless-steel production in 2013; Associated with [22].

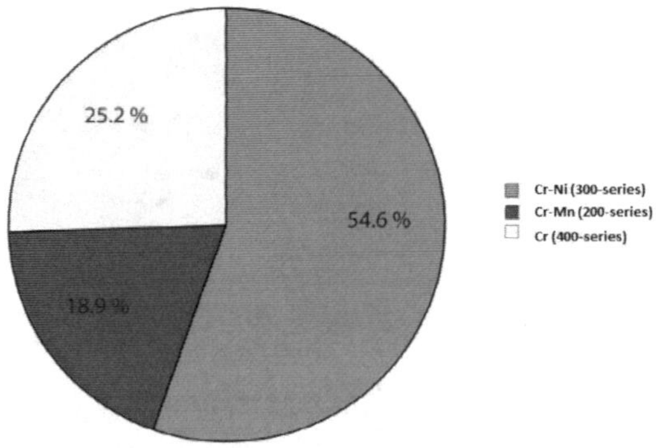

Figure 3 Some grades of stainless-steel production in 2014; Associated with [22].

1.1. Weldability of stainless steels

As posited by some researchers [24] welding happens to be the most essential technique used in the joining of similar and dissimilar metals. To obtain high productivity in welding fabrication, large heat input is needed. The validity of several welding recommendations in comparison with those made available during the period of DSSs conferences of more than two decades ago is still useful. This is because the use of welding fabrication and development is an essential factor in the practical utilization of steel on a large volume since productivity stands as the main factor of any manufacturing industry [25-26]. A paper publication by Nowacki and Rybicki [27] dwells more on the effect of heat input on stainless steel materials but so far, there has not been a proposal on the systematic approach on the influence of high heat input with respect to corrosion properties of DSSs.

Many researchers have immensely contributed to the investigations of welding, mechanical and corrosion characteristics of low nickel alloy stainless steel (SSs) structure in a less corrosive environment [28-30], but for a better choice for highly corrosive media, low nickel alloy of DSSs (AISI 2205 and AISI 2507) is required and this formed the motivation for further investigations of this novel alloy in this review paper.

In the present investigation, an attempt was made to systematically analyse the outcomes of many researcher's findings that are geared towards the influence of welding procedure, especially SMAW on the microstructural, mechanical and corrosion resistance of DSSs grades (AISI 2205 and AISI 2507) neglecting the other classes of DSSs because of the few research done in them. The phase

transformation when DSSs [standard duplex stainless steel (DSS) and super duplex stainless steel (SDSS)] are undergoing solidification during welding together with some selected electrodes for its sound welding was simplified. Finally, the present issues and future focus/perspective of the review were explored.

2 Transformation of austenite- ferrite phases in DSS and SDSS structures

Presented in Table 1are the chemical compositions of standard DSS and SDSS with their respective welding consumables [31, 32]. The respective ferrite content proportion of the parent metals of DSS and SDSS are 46% and between 40-65% measured by a method called image analysis technique [31, 119]. Verma et al. [33] and Dandekar et al. [32] concluded that standard DSS grade 2205 and SDSS grade 2507 can be successfully welded by E2209 and E2594 electrodes respectively having the same composition with its parent metals of 2205 and 2507 grades.

Table 1Chemical analysis (wt. %) of base metals of DSS 2205 and SDSS 2507

	Element	C	S	Mn	Si	Cr	Ni	Mo	P	N	Fe
DSS	Wt. % (2205)	0.053	<0.020	1.420	0.450	21.100	5.800	2.700	<0.030	0.165	Bal.
	Wt. % (E2209)	0.019	<0.025	1.040	0.788	22.880	8.550	3.110	<0.035	0.090	Bal.
SDSS	Wt. % (2507)	0.030	0.020	0.790	0.610	26.860	6.420	3.630	0.030	0.260	Bal.
	Wt. % (E2594)	0.030	0.010	0.800	0.790	24.600	8.600	3.800	0.030	0.250	Bal.

Hindering the formation of secondary phases seems not to be the major problem when subjecting DSSs to solution heat treatment or welding fabrication. An optimum balance of the dual-phase (ferrite-austenite) of the DSSs structureare key factors to obtaining better corrosion resistance and mechanical properties [34-36]. These dual-phases (ferrite-austenite) are the constituents that made up the microstructure of DSSs parent metal as shown in Fig. 4 (a)- (c) having micrographs of the optical micrograph, SEM-micrograph and TEM-micrograph respectively at different magnifications.

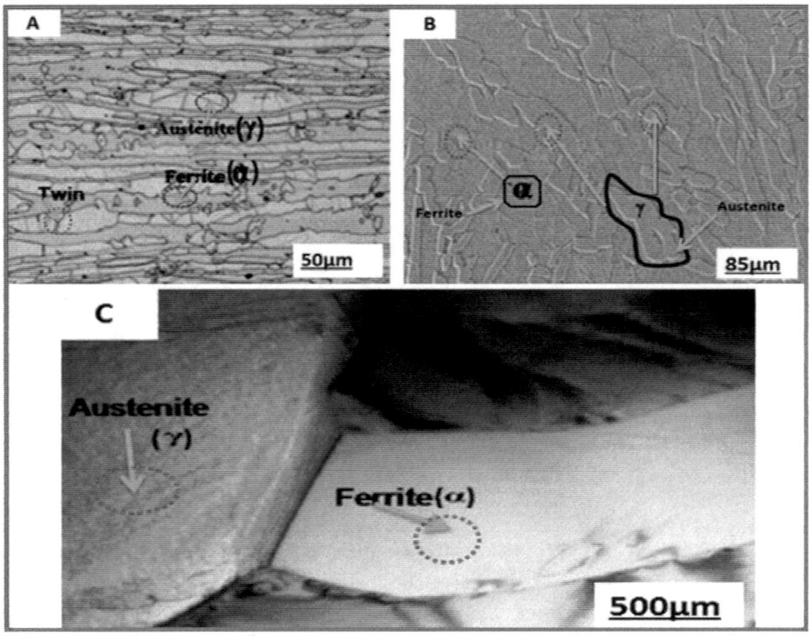

Figure 4 Parent metal of DSS micrographs (a) optical view (b) SEM view and (c) TEM view; Associated with [35].

The transformation of the microstructure of DSSs during cooling can be conveniently expounded. The basic influence of the important alloying elements can be summarized using the current parameters of two equivalents (nickel-Nieq and chromium-Creq) calculation in equations 1 and 2 respectively [38, 39].

$$Ni_{eq} = \%\ Ni + 30\%\ (C) + 0.5\ \%(Mn) + 30\%\ (N) \quad (1)$$

$$Cr_{eq} = \%Cr + \%Mo + 0.5\ \%(Nb) + 1.5\ \%\ (Si) \quad (2)$$

Generally, duplex stainless steels' solidification is completely in the ferrite (δ) mode while the austenite (γ) phase nucleates basically below a specific temperature called the ferritic solvus in the order of $L + \delta \rightarrow \delta + \gamma$, where L is the liquid. To expound more on the mode of solidification in DSSs which is based on the ratio of Cr_{eq}/Ni_{eq}, the Fe–Cr–Ni pseudo-binary diagram, that can be used as a vital instrument for predicting solidification mode [40-44]. Essentially, the constituents of DSSs are within the alpha (α) and gamma (γ) phases. It is noted that in several stainless-steelstructures chemistry, there is an enlargement of phase to conveniently separate both the high and low-temperature ferrite phases. From the pseudo-binary Fe-Cr-Ni diagram, it is observed that there is the existence of ferrite (δ) uninterruptedly from the solidification region to the ambient temperature (room temperature precisely) for the DSSs structure which resulted in the recognition of all ferrite to be classified as ferrite (δ)because of the continuous revolution with the same composition [45]. The low impurity has practically rendered DSSs to be highly resistant to solidification cracking. From the works ofLippold

[46] and Sieurin [47], it is evident that the solidification mode is completely ferritic if Cr_{eq}/Ni_{eq},> 1.95. Hence, for the parent metal of DSSs, the range is between 2.25-3.5, where Cr_{eq}/Ni_{eq} for SDSSs equals 2.25 and that of DSS equals 2.62,and this was confirmed byFourie and Robinson [45] in their attempt to evidentially show the chromium and molybdenum diffusion at the ferrite (δ) region.

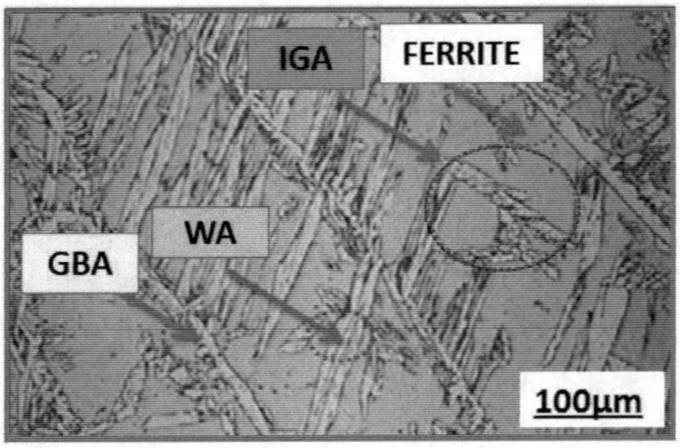

Figure5 Optical micrograph of DSS welded section; Associated with [48].

Subsequently, as presented in Fig. 5, it is evident that during solidification at the weldment (fusion zone FZ), the various morphology of austenite ($\boldsymbol{\gamma}$) phase appears; the first phase that directly nucleates from the matrix of ferrite called the grain boundary austenite (GBA) formed the part of austenite network in the system. At a solidification temperature range of 1350 to 800°C, there exists GBA nucleation. The formation of Widmanstatten austenite (WA) grain from GBA can subsequently be evident but has sufficient nickel (Ni) in comparison with the ferrite region and also, as compared to GBA, it possesses a limited amount of chromium, nitrogen, and molybdenum [48, 49, 50-52]. The cooling rate effect causes the formation of intergranular austenite (IGA) driven by a high propelling force but later precipitates at a moderate temperature which made the DSSs susceptible to pitting corrosion problem as a result of the presence of Cr, Mo, and N that possesses limited alloying element in the system[53].

It was confirmed by Shinohara et al. [54] that the precipitation of σ phase in stainless steel structures possesses a diffusion-regulated phase revolution capability in some certain compositions of chromium and nickel alloys (about 25% Cr to 20% Ni). This is evident in the Cr alloy having an essential role in the rate of precipitation increase of the σ phase of the steel. However, the rate of σ phase precipitation happens to be about 100 times higher in δ-ferrite than in the counterpart γ-austenite phase [55]. This simply implies that the speed of diffusion of the σ phase in $\delta-$ ferrite was higher in contrast to the γ-austenite region. The percentage composition of chromium and molybdenum alloys in the $\delta-$ ferrite depreciated and there was a resultant simultaneous appreciation of nickel alloy at the of σ phase precipitation in the steel. The whole stated mechanism resulted in the secondary austenite phases (γ_2) formation through the δ –ferrite [56].

3. Microstructure, mechanical and corrosion properties of DSS AISI 2205 and SDSS AISI 2507 weldment

Generally, the welding of standard DSS and super DSS to a design specification as recommended is always a difficult task. Though, their weldability by most of the conventional welding techniques such as SMAW is a possibility. The mechanism in the SMAW process is such that it can be efficiently used for positional welding and to perfect fabrication on joints that have restricted accessibility requiring mostly direct current electrode positive (DCEP) that gives excellent welding output. The recommendation of a short arc for the SMAW technique helps to promote stability and decrease the risk of nitrogen pick up capacity during welding since N is an essential factor in the balanced proportion of the dual-phases of the steel structure (δ -ferrite and γ-austenite phases) in the weldment

[57]. Table 2 presents the welding parameters together with the recommended standard electrodes for SMAW of DSSs grade AISI 2205 and AISI 2507 [58, 59-60]

Table 2 Recommended welding parameters and standard electrode for SMAW of DSS and SDSS

Welding technique	Grade of DSS	Electrode diameter (mm)	Voltage (V)	Speed (cm/min)	Current (A)	Electrode type
SMAW	AISI 2205	2.50	20-22	4-6	50-60	E2209
		3.25	23-25	7-9	80-100	E2209
SMAW	AISI 2507	4.00	24-26	15-25	125-135	E2594

3.1 The Behaviour of welding on the mechanical properties of DSS AISI 2205 and SDSS AISI 2507 structures

3.1.1 Effect of heat treatment temperature on the mechanical properties of AISI 2205 and AISI 2507

Essentially, intermetallic phases and compounds have anadverse effect on the toughness of duplex stainless steel. Considering the heat treatment temperature on the Charpy impact energy of duplex stainless steel (AISI 2205) as posited by Badjia et al., [61],it was evident from their findings that the temperature from which the highest amount of sigma phase was noticed was at specimen annealed at 850 °C and this temperature, the lowest toughness of the steel was observed. They also studied the corresponding variations in the tensile strength and elongation at different annealed temperatures (from 800-1050°C). The complicated phase transfiguration observed during the heat treatment (using the annealing method) of duplex stainless steel (AISI 2205) resulted in the variation of its mechanical characteristics. It was also evident from their results that the monumental increase in the tensile properties of the steel (AISI 2205) and the corresponding decrease in its elongation was caused by the application of heat treatment from 800 to 1050 °C. The optimum combination of these parameters (tensile strength and elongation) can be seen at a heat treatment temperature of 1050 °C (σ =775 MPa, A =36.45%), and the gradual reduction in the tensile strength after annealing at a temperature above 1050 °C was as a result of the increment in the volume fraction of ferrite in the steel in accordance to the works of Zucato*et al.* [62] and Sato [63] on the relationships between microstructural phase transformation of DSSs to its mechanical properties.From their fractographical study on the HAZ after annealing at two different temperatures (850 and 1050°C), it can be concluded that the variations in the failure mode mechanisms from an intermediate kind of failure to a dimpled rupture failure were a product of the annealed temperatures observed from 850 to 1050 °C.

Wang et al. [64] subjected some samples for heat treatment mechanism own to the interest of investigating its influence on the microstructure of SDSS AISI 2507. As presented in Figure 6 is the heat treatment temperature of SDSS AISI 2507 weldment with their corresponding σ—phase content and hardness values respectively. There is a consistent trend of increasing and reducing the capacity of the two parameters (σ—phase content and hardness) as the heat treatment temperature increases. It was evident that the maximum σ—phase content (up to 35%) and hardness value (up to 374 HV) were observed at a heat treatment temperature of 850°C. Further increase in the heat treatment resulted in a sharp decrease in these properties (σ—phase content and hardness). Figure 6 (b) shows an appreciable increase of the hardness value of SDSS AISI 2507 structure after heat treatment as seen with the control sample whose hardness value was 264 HV and significantly improve above this range after heat-treated from 800 to 950 °C. Calliari et al. [65], classify σ—phase as brittle with the hard intermetallic compound having a mean hardness value of 800HV approximately. This infers that σ—phase performs a vital role in the hardness properties of the steel (SDSS AISI 2507) structure as it tends to enhance its value and usefulness in-service environment.

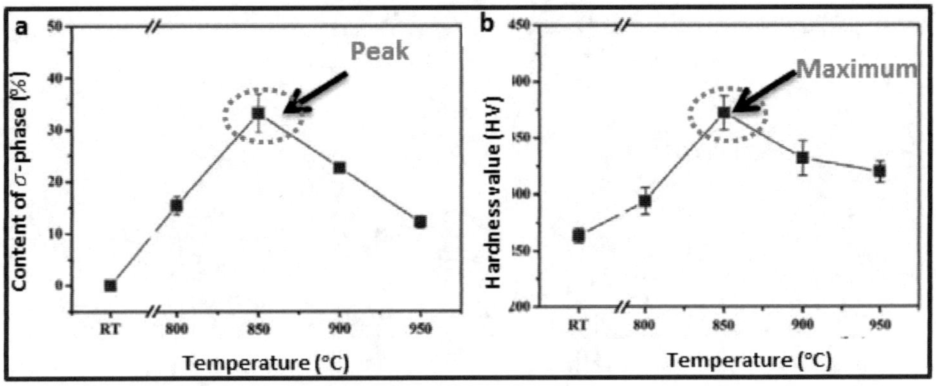

Figure 6 Heat treatment temperature of SDSS AISI 2507 weldment with their corresponding (a) σ—phase content and (b) hardness values; Associated with [64].

3.1.2 Effect of welding on the hardness and tensile properties of AISI 2205 and AISI 2507

Presented in Figure 7 [66], are the micrographs of the various zones (base metal-BM, heat affected zone-HAZ and fusion zone-FZ) of DSS AISI 2205 with their corresponding hardness values. Figure 8presents the comprehensive average hardness results of both DSS AISI 2205 and SDSS AISI 2507 at their respective zones (BM, FZ and HAZ). Figure 8a indicates the welding process used (SMAW) in obtaining the different hardness values while Figure 8b illustrates the DSS electrode type used with their corresponding hardness values in the regions (BM, FZ and HAZ). The results show a strong correlation between the hardness values obtained at the FZ and the HAZ regions of DSS structures. The changes in the hardness values at the microstructure at the different regions of the steel is an indication of fast heating and cooling welding cycle that could have caused a remarkable change in the ferrite-austenite phase transformation [67].

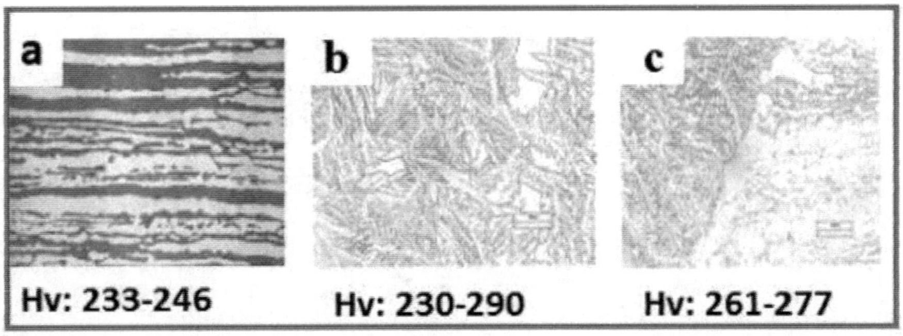

Figure 7 Micrographs and hardness values of duplex stainless steel: (a) BM, (b) FZ and (c) HAZ. Adapted from welding product program; Associated with [66].

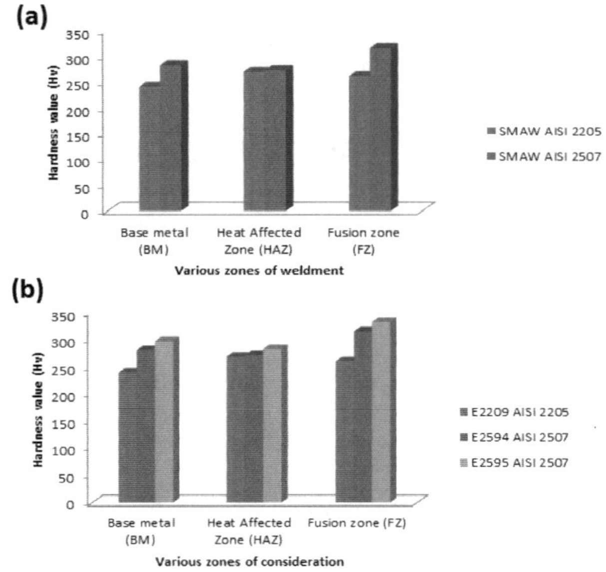

Figure 8 Mean Hardness values (Hv) of DSS AISI 2205 and SDSS AISI 2507 at various zones of welded joints and base metals indicating (a) the welding process used and (b) the electrode type used; Associated with [36, 66].

Dandekar et al. [32], Lima et al. [68] and Belkessa et al. [69] have examined the mechanical properties of DSS AISI 2205 and SDSS AISI 2507 weldments quantitatively with the aid of tensile technique to analyse the tensile strength and yield strength of the structures with their corresponding elongation using the SMAW, GTAW and LBW methods with the recommended electrodes. It was observed that during the tensile analysis, some weldments fractured at the BM with a significant fracture mode of ductility. The ultimate tensile strength of the SDSS AISI 2507 welded with SMAW and LBW were relatively the same for all the weldments (approximately 867.6 MPa and 862.0 MPa) except for GTAW whose ultimate tensile strength is relatively less (approx. 834.0 MPa) as shown in Table 3 and Figure 9.

Table 3 Tensile properties of DSS AISI 2205 and SDSS AISI 2507 welded joints

Welding technique	Grade of DSS	Ultimate tensile strength (MPa)	Elongation (%)	Region of fracture	Electrode/ rod type
SMAW	AISI 2205	768.15	35.81	-	E2209
SMAW	AISI 2507	869.3	14.3	BM	E2594
SMAW	AISI2507	867.6	14.5	BM	E2595
GTAW	AISI 2507	834.0	61.3	BM	-
LBW	AISI 2507	862.0	86.9	BM	-

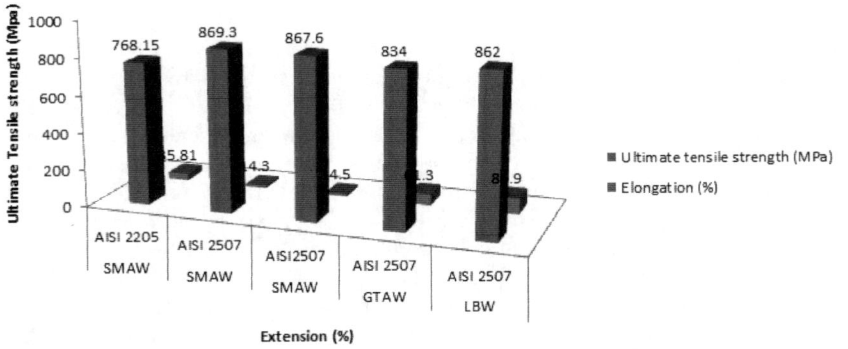

Figure 9. Ultimate tensile strength against elongation for DSS AISI 2205 and SSDSS AISI 2507 welded joint.

The features of the microstructure in welding play an essential role in the fabrication of steel generally and in the effectiveness and integrity of the welded joint of DSSs in particular. Welding fabrication, however, has to be carried out to achieve an optimal ferrite to austenite balance in the weld, at the same time free from dangerous intermetallic phases as well as the nitrides phases (CrN, Cr_2N) [25, 70].It has been well reported [71,72] that the type of fusion welding techniques that enhances the sufficient formation of austenite phase in AISI 2205 and AISI 2507 weldment is the higher arc energy welding techniques such as SMAW, gas metal gas welding (GMAW), gas tungsten gas welding (GTAW) and SAW as a result of the gradual cooling mechanism involved. This is not so with lower energy welding techniques like electron beam welding (EBW) and laser beam welding (LBW) method which supports rapid cooling rate and result in low precipitation of austenite at the weldment.

However, below 25% austenite composition contained in DSSs joint is generally unsatisfactory in many sectors of utilization [73], notwithstanding there is an acceptable recommendation by the Norsok standard 74] that the percentage of the content of austenite should be 30% in the weldment for the inspection of piping structures. Asif et al. [75] pointed that when DSSs undergo a post-weld treatment immediately after welding fabrication for about 15mins at a temperature of 1050°C accompanied fast water quenching mechanism to annul the formation of sigma (σ) as well as other phases like nitrides and intermetallic at the point of cooling which invariably promotes the composition of the austenite (γ) phase. Some mechanical properties of the weldment such as impact toughness are essentially improved. Also, it is evident that post-weld heat treatment promotes the Cr and Mo diffusion in the ferrite (δ) region whereas the austenite stabilizers such Ni and N enhances the austenite phase to achieve an efficiently balanced ratio of ferrite to austenite in the weldment.

Heat input plays a crucial part in the geometry of weld bead, microstructural characterisation, mechanical and corrosion opposition properties of DSSs weldment [76]. However, to achieve satisfactory austenite to ferrite phase balance in DSS AISI 2205, the required heat input is at the range of 0.5-2.5kJ/mm while for SDSS AISI 2507 weldment, the recommended heat input is between 0.3-1.5kJ/mm. This will help to prevent the formation of detrimental phases in the weld [77].

Based on several works of other researchers on DSSs, it is noted that the realisation of precipitation of Cr_2N and CrN in DSS AISI 2205 grade and SDSS AISI 2507 grade is mainly a result of the application of low heat input and rapid cooling mechanism. It is however evident that at high heat input above the recommended heat input above and a faster cooling cycle of the weldment, there is the possibility of Cr_2N phase formation because of the nucleation of N in the ferrite region at a temperature relatively below 900°C while below the temperature of 1100°C, CrN formation is inevitable on thecondition that Cr_2N and austenite phases are subdued. This is following the claim of Hertzman [78], that there is the presence of CrN in DSSs weldment along with some alloying metals like Cr, Fe and Mo.

3.1.3 The effect of welding on the corrosion properties of DSS AISI 2205 structure and SDSS AISI 2507 structures

As evident from the several types of research previously carried out by some researchers, DSSs such as AISI 2205 and AISI 2507, generally demonstrate strong resistance to SCC in chloride-containing environments [79]. Although their susceptibility to SCC is a consequence of different parameters such as microstructures, applied load/stress, alloying elements or the kind of medium of application [79, 80-87]. The selective dissolution of the various components of the steel (DSSs) phases or pitting corrosion may be the cause of the initial initiation of cracks in the steel structures [82, 83].

Tsai [88] used the potentiodynamic polarization method to assess the susceptibility of SCC of AISI 2205 in nitrogen (N_2) deaerated sodium chloride (NaCl) medium of about 1-26 wt.% of a 6.0 pH at a specified temperature range of 25 to 90^{0}C. Using a specially designed saturated calomel (SCE) as the reference electrode (kept at room temperature) whiles the platinum mesh was used as the counter electrode. The slow strain rate testing (SSRT) technique [82] was utilized to determine the SCC susceptibility of AISI 2205 weldment in N_2 deaerated NaCl medium (1-26 wt%) of 6.0 pH value at a range of temperature between 25 to 90^{0}C using austenitic stainless steel of type AISI 316 as a comparison. From the study,a vast rate range of passive part was observed with a current surge seen at -160mV which corresponds to the potential of pitting corrosion. Validation of these pits was done via SEM analysis on the surface of the experimental sample.

Bhattacharya [89] investigated the corrosion susceptibility of AISI 2205 in a sulfide-containing caustic environment where the steel (AISI 2205 was exposed to the media of 150g/L of NaOH in combination with 50g/L of Na_2S for the corrosion analysis. Presented in Figure 10 is the overview of the findings of the experimentation. Conversely, the findings are quite distinct from those obtained from the chloride-containing solution where the general and localized corrosion susceptibility of DSSs is significantly caused by the influence of microstructure [90]. This characteristic of DSS is contrary to that in the acidic chloride containing media, where the condensation of intermetallic phases like σ-phase and chi phases happen to cause the initiation of the corrosion via selective dissolution of Cr and Mo reduced parts within the parameters of precipitations [91].

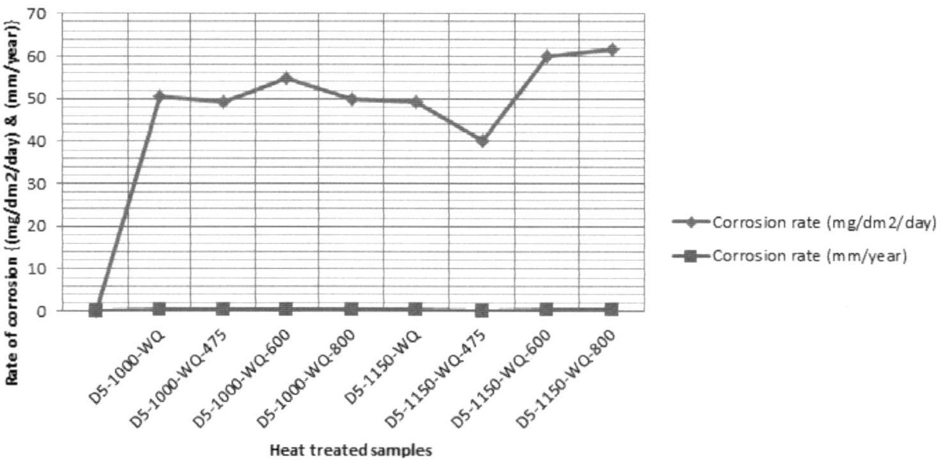

Figure 10 Corrosion rates of DSS AISI 2205 in sulfide-containing caustic solution; Associated with [89].

Paranthaman- et al. [92] investigated the electrochemical properties of gas tungsten arc welding (GTAW) of AISI 2507 SDSS weldment and compared it with its base metal (BM) through a technique of potentiodynamic polarization test. The analysis was carried out in various concentrations of sodium chloride compositions of 3.5M of NaCl, 4.5M of NaCl and 5.5M of NaCl to establish a basement for the findings. It was evident from their results that the samples analysed in 3.5M of NaCl medium for corrosion susceptibility of the AISI 2507 SDSS weldment showed minimal deterioration current density and migrate to the positive direction of corrosion potentials in comparison with the specimens analysed in the various concentrations of the chloride media (4.5M NaCl and 5.5M NaCl). From the potentiodynamic polarizationanalysis, it was noticed that the γ-austenite phase present in the weldment in a relative amount resulted in excellent electrochemical properties of the AISI 2507 SDSS specimens welded with GTAW. This shows a higher percentage of γ-austenite phases in the FZ than at the BM with a less stable δ-ferrite [93].

These findings are in agreement with that of Alsarraf [34] who carried out a potentiodynamic polarization on SDSS AISI 2507 weldment to comprehend the significant function of the configuration as well as the microstructure on the corrosion and passivation properties of SDSS AISI 2507 immersed in aqueous NaCl (3.5%) salt at ambient temperature.

3.1.4 Effect of process parameters on the properties of DSS AISI 2205 and SDSS AISI 2507 weldment

Different welding techniques have been deployed over the years for the fabrication of AISI 2205 and AISI 2507 structures. Reviewed here are some of these techniques:

Yurtisik*et al.* [94] researched on the characterization of DSS weld metals using a hybridized welding process of plasma-metal inert gas welding method. The focus was to achieve duo purposes of deep penetration properties in keyholes area and metal deposition capability of plasma and metal inert gas welding processes respectively. The hybridization of these two welding techniques helped to achieve the following; a less heat input as compared to the conventional individual welding processes as illustrated in Table 8 and suppression of the precipitation capability of the secondary phases that serve as great barriers to the deterioration and toughness properties of the novel steel at the weldment and heat-affected zones [95]. This resulted in the achievement of efficient cooling time balance austenite to ferrite phases.

Hu and Xue [96] researchedon the influence of process parameters on the weld quality during double-pulsed GMAW of AISI 2205 duplex stainless steel and considered the influence of the number of both strong and weak pulses alongside the welding speed as illustrated in Figure 11. The mechanical and microstructural characterisations were conducted on the welded samples of DSS AISI 2205 also. The findings showed that amongst the three factors considered (welding speed, number of strong pulses and number of weak pulses), welding speed had the greatest influence on the quality of weldment of the steel (DSS AISI 2205) and then the number of weak pulses and lastly the number of strong pulses as seen in their average ranges R as 6.25, 5.50 and 5.25 respectively and it is in agreement with the method adopted by Xie et al. [97] in analysing the weld quality of aluminium alloy double pulsed GMAW. The systematic order mathematically is $S > N_2 > N_1$

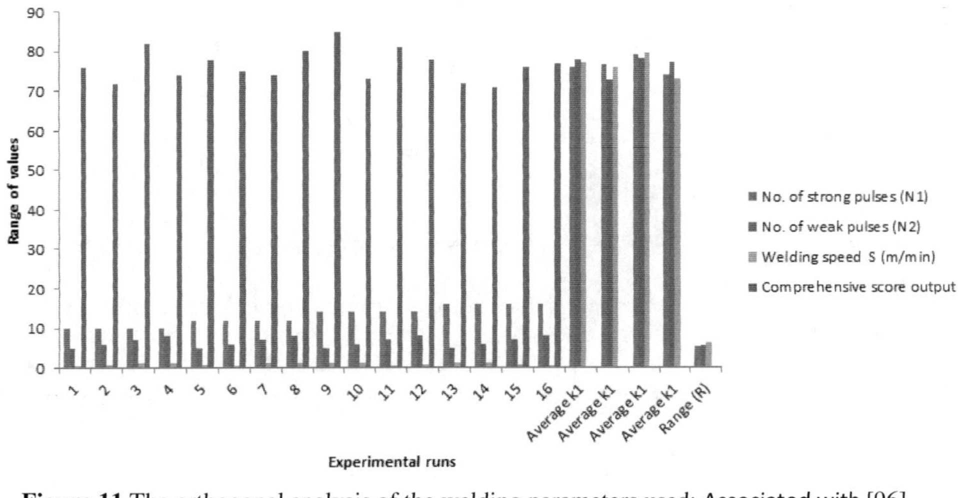

Figure 11.The orthogonal analysis of the welding parameters used; Associated with [96].

Taban and Kaluc [98] have studied the welding behaviour of DSSs and SDSSs using laser and plasma arc welding methods. Mechanical and microstructural examinations were carried out on the weldment together with the ferrite content measurement. It was found that at controlled heat input, the ferrite-austenite ratio at the weldment was balanced which resulted in an excellent toughness of DSS and SDSS welded with plasma arc welding technique as compared to the counterpart laser welding method. Subsequently, there was variation in the content of ferrite in the weldment in an acceptable range which shows its importance in realizing sound mechanical and microstructural properties [99-101]. Subsequently, Urena *et al.* [35] researched on the best welding conditions (such as welding intensity and travel speed) determination for welded butt joints of DSS structure of grade 2205 with the aid of plasma-arc welding. The conduction mode and the keyhole mode as types of welding modes were utilized for the evaluations of the metallurgical weldability and the possible net energy input for adequate application. Investigations on the parametric effect of welding for individual modes on the dimensional geometry of the weldment as well as their ferrite content analysis were carried out successfully [102]. It was noticedthat the best heat input was at the ranges 2.5kJ/cm and 3.2 kJ/cm adopting the keyhole mode. Also, a substantial increase in the content of the ferrite up to about 20% was evident in the microstructuralweldzone [103].

Paulraj and Garg [104] investigated the effect of welding parameters on the deterioration characteristics of GTAW of DSS and SDSS weldments. The influence of heat input on the mechanical properties (Table 7), and cooling rate, inter-pass, shielding gas on the corrosion resistance of the weldments were observed after welding DSS AISI 2205 and SDSS AISI 2507 pipes with the GTAW technique. Figure 12 present the summary of the influence of heat input on the mechanical properties of welded DSS AISI 2205 and SDSS AISI 2507 pipes employing the GTAW technique. Samples No.1 to 4 are samples of DSS AISI 2205 welded under low PREN condition and samples No. 5 to 8 are samples of DSS AISI 2205 welded under high PREN condition with their corresponding heat inputs. The same tread was applied for samples No. 9 to 12 and samples No.13 to 16 for SDSS AISI 2507 welded under low and high PREN conditions respectively with their unity standard deviation error.

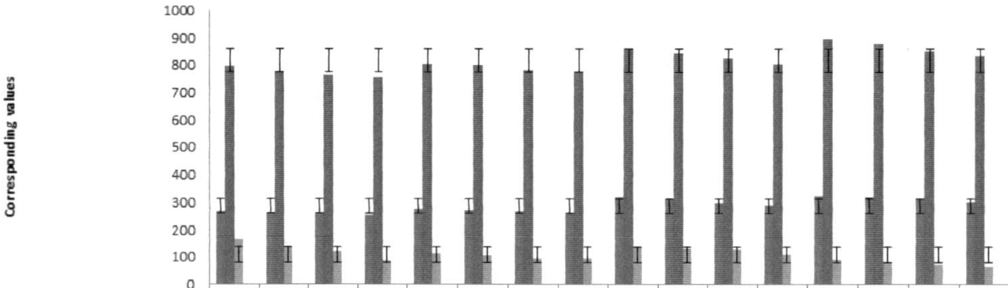

	1	2	3	4	5	6	7	8	9	10	11	12	13	14	15	16
Hardness (VHN)	266	264	259	255	275	270	266	261	319	312	298	288	326	320	315	302
Tensile strength (MPa)	795	780	767	755	805	802	785	777	862	844	827	805	896	880	855	835
Impact analysis @ -46 oC (J)	168	140	123	85	115	104	97	94	140	130	126	107	90	81	72	64
Heat input (kJ/mmin)	1.05	1.1	1.15	1.2	1	1.05	1.1	1.15	0.95	1.05	1.15	1.25	0.75	1	1.1	1.2

Figure12. Summary of the influence of heat input on the mechanical properties of welded DSS AISI 2205 and SDSS AISI 2507 under low PREN and high PREN conditions; Associated with [10].

Microstructural characteristics, as well as the determination of the ferrite content of the materials, were carried out using an appropriate standard. The microstructures of the weld metal were contrasted with the corrosion findings. The various phases of the weldment with respect to pitting resistance were investigated along with the influence of austenite secondary phase on the degradation of the steels (AISI 2205 AND AISI 2507) which can result in the loss of resistance of the passivation of the steel [105].

Luo *et al.* [31] researched on the effect of post-weld heat treatment (PWHT) on the welded joint of DSS grade 2205 using the submerged arc welding (SAW) method with the specified welding parameters. The sigma (σ) and delta (δ) phases of the DSS weldment are increased as the amount of gamma (γ) phase decreases after PWHT. Since the microstructure of the HAZ can be enhanced by PWHT, the micro-hardness at this zone (HAZ) was observed to be reduced by PWHT. The fine dendrite structure of the microstructure of base metals changes gradationally to coarse columnar grains from the boundary of fusion to the centre weld zone. The successive segregation of the compositions of sigma (σ) and delta (δ) phases was as a result of the formation of the columnar grain at the weld zone of DSS weldment. Yuan *et al.* [106] studied that if by any means the Ni composition in the welding rod of DSS is high, the resultant effect will be higher austenite formation when the SAW process is employed in the weldment. Hence, the mechanical properties and corrosion resistance of weldment are excellently enhanced. Using the same SAW method, Sieurin and Sandstrom [48] have studied that the quota of austenite transformation in the HAZ of DSS grade 2205 hinders the formation of the sigma phase. It was found that in SAW of DSS grade 2205, the mean ferrite content in the weldment increases at the heat-affected zone of 150°C during interpass temperature [27, 107]. As a result of the heat treatment (HT), there were many formations of the sigma phase elements at the fusion zone which invariably led to the reduction in the toughness of the welded joint (FZ) and thereby increases the microhardness [108-110].

Due to its flexibility, portability, simplicity, and applicability, the shielded metal arc welding (SMAW) technique seems to be one of the most extensively utilized welding methods in several engineering fields of applications [46, 111]. SMAW method serves as a better alternative for preventing nitrogen loss for nitrogen alloy steel structure through the provision of self-protecting slag as a cover [112] Srinivasan *et al.* [113] studied the microstructure and corrosion properties of DSS grade 2205 and low alloy steel weldment using SMAW method. It was evident that the DSS electrode grade E2209 and ASS electrode grade E309 can be used to join DSS 2205 to a low alloy steel structure with the aid of the SMAW method. They both possess good general corrosion resistance but their pitting corrosion resistance was observed to be inferior when compared with the base metal of DSS grade 2205. There was also an observance of the carbon depletion region near the fusion boundary at the side of the low alloy steel structure.

Verma *et al.* [33] investigated the microstructure, mechanical and intergranular corrosion behaviour of dissimilar steels of grade 220 (DSS and grade 316L ASS welded by SMAW method using electrodes with a similar configuration to the base metal of the DSS (E2209 electrode). 0.45 to 0.60kJ/mm heat inputs were considered by the authors. The optical emission microscope (OES) and scanning electrode microscopy (SEM) were utilized in the characterization of the microstructures while energy dispersive spectroscopy (EDS) attached to the SEM was used in obtaining the localized chemical information. The intergranular corrosion analysis was carried out based on the level of sensitization using the double loop electrochemical potentiokinetic reactivation technique. Paying close attention to the evaluation of the effect of weld dilution on mechanical properties and the measurement of ferrite content of the weldment using ferritoscope. It was noticed that the weld joint accomplished the needed ferrite content for the different heat inputs mentioned above. The high cooling rate resulted in higher ferrite content which invariably increased the hardness and tensile properties of the steel at the lower heat input in comparison with higher heat input. It was also evident that the impact energy observed using the E2209 electrode for the welding was high without any reasonable defect.

Wang *et al.* [114] have investigated that the Tungsten inert gas welding (TIG) technique using filler materials of grade ER2209 seems to have better suitability for the joining of dissimilar metals of DSS grade 2205 and 16MnR steel as compared to using the SMAW method. There was no observance of precipitation of intermetallic phases in the two welding techniques. Susceptibility to pitting corrosion in chloride environment (3.5% sodium chloride) of the weldment produced by the shielded metal arc welding method was higher than the joint produced by the TIG welding technique due to the finer grains in the TIG welding method.

Gupta et al. [118] investigated the effect of heat input on microstructure and corrosion behaviour of SDSS 2507 welded by SMAW technique with E2595 electrode adopting duo heat inputs mechanism low (0.54kJ/mm) and high (1.10kJ/mm). The calibration of the ferritscope used in determining the ferrite content of the weldment at the BM, FZ and HAZ was done following AWS A4.2 standards in collaboration with the World Institute, United Kingdom and the preparation of tensile samples for experiment on the SDSS AISI 2507 weldment for the two levels of heat input (low and high), using ASTM E8M-04 standard [115-117]. The results of the mechanical properties of the weldment deduce the similarity among the parameters of interest; tensilestrength, hardness and impact toughness using the two heat inputs. It is also evident from the corrosion studies using 3.5% concentration of common salt (NaCl) solution that the steel (SDSS 2507) indicated about less a percentage (1%) level of sensitization which indistinguishably possesses pitting possibility for the used heat inputs (low and high), resulting in the recommendation of low heat input in the fabrication of SDSS 2507 structures [118].

From the above brief literature survey, the results of other authors (though relatively scarce) with respect to their researches on the influence of welding, particularly the SMAW process on the microstructural, mechanical and corrosion properties of DSSs weldment (AISI 2205 and AISI 2507 grades) were concisely analysed.

3 Conclusions

4

Having gone through some scientific research works from other researchers on general DSS (AISI 2205 and AISI 2507) in particular using various welding techniques but specific attention on SMAW, the following conclusions can be drawn: SMAW technique can be successfully used to fabricate DSS AISI 2205 and SDSS AISI 2507 when all relevant welding conditions are maintained during welding. Improper control of welding parameters to achieve a balanced ferrite-austenite phase during welding of DSS AISI 2205 and SDSS AISI 2507 structures can adversely affect its service application. Painstaking control of the ferrite content in the weldment of the steel structures can effectively promote better mechanical properties and excellent corrosion resistance of both DSSs (AISI 2205 and AISI 2507). The consequential effect on the precipitation quantum of σ −phase, and its subsequent influence on the hardenability of the two distinct plates of steel weldment could be engineered by heat treatment temperature. Adequate attention must be given to the weld cycle solidification to avoid the precipitation of the sigma phase in DSSs which is usually between 600°C-1000°C. ASTM standard electrodes for effective fabrication of AISI 2205 and AISI 2507 structures are E2209 and E2594 electrodes respectively due to their relatively similar compositions with the base metal chemistry. It was evident that microstructures have a great influence on the weldment of AISI 2205 and AISI 2507 in respects to their corrosion-resistant performance in chloride-containing environments. Low heat input is relatively recommendable for the fabrication of AISI 2507 owe to its insignificant influence on the weld integrity.

Funding and Acknowledgments

The great support received from the CSIR-TWAS funding in CSIR-National Metallurgical Laboratory, Jamshedpur is highly acknowledged (Award No.: 22/FF/CSIR-TWAS/2018).The authors also would like to thank Dr S. C. Ghosh, Dr A. Das and Mr Roshan from CSIR- National Metallurgical Laboratory, Jamshedpur for creating a smooth environment for the final draft of this review.

Compliance with ethical standards

The authors declare that there is no conflict of interest in the course of the review

References

1. Ghusoon R.M., *et al.,*Effects of Heat Input on Microstructure,Corrosion and Mechanical Characteristics of WeldedAustenitic and Duplex Stainless Steels: A Review. *Metals2017. 7(39); DOI: 10.3390/*met7020039
2. Bain E.C.and W. E. Griffith, An introduction to the iron-chromium-nickel alloy. *Trans. AIME. 75: p. 166-213.*
3. Smuk O. (2004) Microstructure and Properties of Modern Powder Metallurgy of Super Duplex Stainless Steels. *PhD thesis presented to the Department of Materials Science and Engineering, Division of Ceramics, Royal Institute of Technology, Stockholm, Sweden.*
4. Lo K. H.,*et al.,*Recent developments in stainless steels. *Materials Science Engineering. R Rep.2009.**65***: p.39–104.

5. Bonollo F., et al., Welding processes, microstructural evolution and final properties of duplex and super duplex stainless steels. *In Duplex Stainless Steels; John Wiley & Sons, Inc.: Hoboken, NJ, USA; 2013. p. 141–159.*
6. Kisasoz A., et al., Effect of annealing time and cooling rate on precipitation processes in duplex corrosion-resistant steel. *Metal Sci. Heat Treat, 2016.**57***: p.544–547.
7. Srikanth S., et al., Development of Lean Duplex Stainless Steels (LDSS) with Superior Mechanical and Corrosion Properties on Laboratory Scale. *Adv. Mater. Res. 2013.**794***:p. 714–730.
8. Oshima T., et al., Efforts to save nickel in austenitic stainless steels. *ISIJ Int. 2007.**47**: p. 359–64.*
9. Malik A.U., et al., Relevance of corrosion research in the material selection for desalination plants. *In: Proceeding of 2nd scientific symposium on maintenance planning and operations.1993. P. 885–900.*
10. Boillot P. and J.Peultier, Use of stainless steels in the industry: recent and future developments. *Procedia Engineering, 2014. **83**: p. 309–321.*
11. Charles J. and S.Bernhardson, Materials to meet your needs. *In: Proceeding conference on duplex stainless* steel *1991. p. 3–48.*
12. Charles J., Why and where duplex stainless steels. *In: Proceeding of 5th international conference on duplex stainless steel 97, stainless steel world. KCI Publishing, 1997.* p. 29.
13. Charles J., A review after DSS '07*". In: Proceeding conference, stainless steel world*, 2007
14. Olsson J, Snis M (2007) Duplex—a new generation of stainless steels for desalination plants. *Desalination, 205: 104–13.*
15. Olsson J. and H.Groth, A new approach to reduced costs. *In: Proceeding IDA and WRPC world conference on desalination and water treatment, evaporators made of duplex stainless steel, 1993.*
16. Olsson J. and K. Cosic, Stainless steels for SWRO plants high-pressure piping. *In: Proceeding IDA world conference on desalination and water re-use, 2003.*
17. Olsson J.,*et al.*, MSF chambers of solid duplex stainless steel. *In: Proceeding IDA World conference on desalination and water re-use, 2003.*
18. Peultier J., et al., New trends in selection of metallic material for desalination industry. *In: Proceeding IDA world congress, 2009.*
19. Charles J., Past, present and future of the duplex stainless steels. *In: Proceeding of international conference and expo on duplex stainless steels, 2007.*
20. Baldo S., Innovative steels for structural and corrosion resistance applications. *PhD Thesis presented to the Department of Engineering's Chemical Processes (DPCI), University of Padua, Italy, 2010.*
21. Outokumpu Annual report 2015, (Accessed on 6th April 2020 from www.outokumpu.com//Outokumpu Annual report, 2015) pp. 1–116
22. What prospects for stainless steel in 2016? Stainless Steel World January/February magazine, (Accessed on 6th April 2020 from www.stainless-steelworld.net/pdf/What prospects for stainless steel in 2016.pdf) p. 1–5
23. The ferritic solution. *International stainless steel forum published in April 2007.(Accessed on 6th April 2020 from www.worldstainless.org)*
24. Yang Y., *et al.*,The effect of large heat input on the microstructure and corrosion behaviour of simulated heat-affected zone in 2205 duplex stainless steel. *Corrosion Science, 2011.**53**:p.3756–3763.*
25. Van-Nassau L., *et al.*,Welding duplex and super duplex stainless steel, *Doc. IIW-1165, Welding in the World, 1999.**31** (5):*p. 323-343.
26. Jacques C. and B.Sven, Duplex Stainless Steels. *Proceedings Conference Duplex Stainless Steels'91, Beaune, France, 1991. www.tib.eu/en/search/id/TIBKAT%3A124397204/Duplex-stainless-steels-91-28-3Octoberre-1991-Beaune/*
27. Nowacki J. and P. Rybicki, The influence of welding heat input on submerged arc welded duplex steel joints imperfection. *Journal of Materials Processing Technology, 2005.**164-165**: p. 1082–1088.*

28. Vashishtha H., *et al.*,Welding behaviour of low nickel chrome-manganese stainless steel. ISInternationalal, 2014. **54**: p. 1361–1367.
29. Taiwade R. V.,*et al.,* Studies on welding and sensitization of chrome manganese austenitic stainless steel. *Trans Indian Inst Met. 2011.**64**: p. 513–8.*
30. Taiwade R. V.,*et al.,* Effect of welding passes on heat-affected zone and tensile properties of AISI 304 stainless steel and chrome-manganese austenitic stainless steel. *ISIJ International, 2013.**53:** p. 102–109.*
31. Luo J., *et al.,* Microstructure of 2205 duplex stainless steel joint in submerged arc welding post-welded heat treatment. *Journal of Manufacturing Processes*, 2014. **16**(1): p. 144-148.
32. Dandekar T. R.,*et al.,* Shielded metal arc welding of UNS S32750 steel: Microstructure, Mechanical Properties and corrosion behaviour". *Materials Research Express,2018.* **5**(10);IOP Publishing Ltd, *DOI: 10.1088/2053-1591/aad99a*
33. Verma J., *et al.,* Microstructure, Mechanical and Intergranular Corrosion Behavior of Dissimilar DSS 2205 and ASS 316L Shielded Metal Arc Welds. *The Indian Institute of Metals, (IIM),2016. DOI 10.1007/s12666-016-0878-8*
34. Alsarraf J., Hydrogen Embrittlement Susceptibility of Super Duplex Stainless steel. *PhD Thesis submitted to School of Applied Sciences, Cranfield University UK, 2010.*
35. Urena A.,*et al.,* Weldability of a 2205 duplex stainless steel using plasma arc welding. *Journal of Materials Processing Technology, 2007. **184**:p. 624–631*
36. Zou N. D.,*et al.,* Deformation characteristic and prediction of flow stress for as-cast 21Cr economical duplex stainless steel under hot compression. *Mater. Des., 2013. **51**: p. 975–982.*
37. Bettahar K., *et al.,* Microstructure and mechanical behaviouror in dissimilar 13Cr/2205 stainless steel welded pipes. *Mater. Des., 2015. **85**: p. 221–229.*
38. Alvarez-Armas I. and S. Degallaix-Moreuil, Duplex Stainless Steels". *Published by ISTE Ltd. and John Wiley & Sons,*2009. P. 141-155.
39. Vinoth J. A.,*et al.,* Weldability, machinability and surfacing of commercial duplex Stainless Steel AISI 2205 for marine applications – A recent review. *Journal of Advanced Research, 2017.DOI:10.1016/j.jare.2017.01.002*
40. Sun Z., *et al.,* Effect of dual torch technique on duplex stainless steel welds. *Materials Science and Engineering A, 2003. **356**:p. 274–282.*
41. Leone G. L. and H. W. Kerr, The ferrite to austenite transformation in stainless steels. *Welding Journal, 1982.**61**:p.13s–22s.*
42. Glownia J.,*et al.,* Delta ferrite predictions for cast duplex steels with high nitrogen content. *Mater Charact, 2001.**47**: p. 149–155.*
43. Jayachitra R., *et al.,* Characterization of duplex stainless steel heat-treated at 1300 °C.*Int J Sci Res Publ., 2012. **2**: p. 1–6.*
44. Di X., *et al.,* Microstructural evolution of transition zone of clad X70 with duplex. *Mater Des., 2016.**95**: p. 231–236.*
45. Fourie J. W. and F.P.A. Robinson, Literature review on the influence of weld-heat inputs on the mechanical and corrosion properties of duplex stainless steels". *J S Afr Inst Min Metall, 1990.**90**: p. 59–65.*
46. Lippold J. C. and D.J. Kotecki, Welding Metallurgy and Weldability of Stainless Steels. *WileyHoboken. Interscience, New Jersey, 2005.*
47. Sieurin H., Fracture toughness properties of duplex stainless steels. *Dissertation. Royal Institute of Technology, 2006.*
48. Sieurin H. and R. Sandstrom, Austenite Reformation in the Heat-Affected Zone of Duplex Stainless Steel 2205. *Materials Science and Engineering, A,2006. **418**: p. 250–256*
49. Kacar R., Effect of solidification mode and morphology of microstructure on the hydrogen content of duplex stainless steel weld metal. *Mater Des., 2004. **25**: p. 1–9*
50. Wang S. H., *et al.,* Gamma phase transformation in pulsed GTAW weld metal of duplex stainless steel. *Mater Sci Eng A, 2006. **420**: p. 26–33.*
51. Liu H. and X.Jin, Secondary austenite morphologies in fusion zone of welded joint after post-weld heat treatment with a continuous wave laser. *J Mater Sci Technol., 2012.**28**(3): p. 249–54.*
52. Geng S.,*et al.,* Evolution of microstructure and corrosion behaviour on 2205 duplex stainless steel GTA-welding joint. *J Manuf Process, 2015.**19**: p. 32–37.*

53. Nilsson J. O., *et al.,* Secondary austenite formation and its relation to pitting corrosion in duplex stainless steel weld metal. *Mater Sci Technol.*, 1995.**11**: p. 276–283.
54. Shinohara K., *et al.,* Recrystallization and sigma phase formation as concurrent and interacting phenomena in 25%Cr-20%Ni steel. *Materials Transactions, 1979.**20**(12): p. 713-723.*
55. Baerlecken E. and H. Fabritius, Umwandlungskinetik der sigmaphase in einer Eisen-Chrom-Legierungmit 48% Chrom," Arch Eisenh"uttenwes, 1995.**26**: p. 679-686.
56. Garz C. M. and A. J. Ramirez, Growth kinetics of secondary austenite in the welding microstructure of a UNS S32304 duplex stainless steel". *Acta Materialia, 2006.**54**(12): p. 3321-3331*
57. Goswami P. and R. Bapat, Duplex stainless steel: Metallurgy, engineering codes &welding practices Part 2. *Stainless Steel World, 2014. p.1-5. www.stainless-steel-world.net*
58. Karlsson L., Welding duplex stainless steels—a review of current recommendations. *Weld World, 2012.**56**: p. 65–76.*
59. Verma J. and R. V.Taiwade, Effect of welding processes and conditions on the microstructure, mechanical properties and corrosion resistance of duplex stainless steel weldments—A review. *Journal of Manufacturing Processes, 2017.**25**: p. 134–152 DOI:* http://dx.doi.org/10.1016/j.jmapro.2016.11.003
60. Messer B., *et al.,* Duplex stainless steel welding: best practices. *Fluor Canada, Stainless Steel World, 2007.*
61. Badjia R., *et al.,* Phase transformation and mechanical behaviour in annealed 2205 duplex stainless steel weld. *Materials Characterization, 2008.**59**:p. 447-453*
62. Zucato I., *et al.,* Microstructural characterization and the effect of phase transformations on toughness of the UNS 31803 duplex stainless steel aged treated at 850 °C. *Mater Res., 2002.**5** (3): p. 385 –389.*
63. Sato Y. S. and H. Kokawa, Preferential precipitation of sigma phase in duplex stainless steel welds metal. *Scr Mater., 1999.**24**: p. 659 –663.*
64. Mengjiao W., *et al.,* Effect of Heat Treatment Temperature and Lubricating Conditions on the Fretting Wear Behavior of SAF 2507 Super Duplex Stainless Steel. *Journal of Tribolog, Transactions of the ASME, 2019.**141**: p. 101601-101608*
65. Calliari I., *et al.,* Influence of Isothermal Aging on Secondary Phases Precipitation and Toughness of a Duplex Stainless Steel SAF 2205, *J. Mater. Sci., 2006.**41**(22): p. 7643 –7649*
66. Welding Product Programme: Duplex and Super Duplex stainless steel, Selectarc Welding. *Accessed on 15th April, 2020 from www.fsh-welding.com*
67. Zhang W., *et al.,* Modelling of ferrite formation in a duplex stainless steel weld considering non-uniform starting microstructure. *Acta Mater., 2005.**53**: p. 4441 –453.*
68. Lima M.S.F.,*et al.,*Mechanical and Corrosion Properties of a Duplex Steel Welded using Micro-arc or Laser". *Materials Research*, 2015. ***18**(4)*: p. 723-731. DOI: http://dx.doi.org/10.1590/1516-1439.007115
69. Belkessa B., *et al.,* Microstructure and Mechanical Behavior in Dissimilar SAF 2205/API X52 Welded Pipes. *Acta Metall. Sin. (Engl. Lett., 2016.**29**(7): p. 674–682.* DOI 10.1007/s40195-016-0428-8
70. EN 1011-3, Welding – Recommendations for welding of metallic materials, Part 3: Arc welding of stainless steels.
71. Kannan T. and N.Murugan, Effect of flux-cored arc welding process parameters on duplex stainless steel clad quality *J Mater Process Technol., 2006.**176**: p. 230–239.*
72. Nowacki J., Ferritic-austenitic steel and its weldability in large size constructions. *J Achiev Mater Manuf Eng., 2009.**32**(2): p. 115–141.*
73. Hsieh R. I., *et al.,* Effects of cooling time and alloying elements on the microstructure of the gleeble-simulated heat-affected zone of 22% Cr duplex stainless Steel. *J Mater Eng Perform, 2001. **10**(5); p. 526–536.*
74. Norsok Standard M601, Welding and inspection of piping. *Lysaker, Norway: Standards Norway-2004.*
75. Asif M. M.,*et al.,* Effects of post-weld heat treatment on friction welded duplex stainless steel joints. *J Manufac Process, 2016.**21**: p. 196–200.*

76. Giridharan P. K. and N. Murugan, Optimization of pulsed GTA welding process parameters for the welding of AISI 304L stainless steel sheets. *Int J Adv Manuf Technol., 2009.**40**: p. 478–489.*
77. Sadeghian M., *et al.,* Effect of heat input on microstructure and mechanical properties of dissimilar joints between super duplex stainless steel and high strength low alloy steel". *Mater Des., 2014.**60**: p. 678–684.*
78. Hertzman S., The influence of nitrogen on microstructure and properties of highly alloyed stainless steel welds. *ISIJ Int., 2001.**4** (6): p. 580–589.*
79. Nilsson *J. O., Overview*: Super duplex stainless steels.*Materials Science and Technology, 1992.**8**: p. 685-700*
80. Kim K.Y., *et al.,* Electrochemical and Stress Corrosion Properties of Duplex Stainless Steels Modified with Tungsten Addition. *Corrosion., 1998.**54**(11): p. 910-921. DOI: 10.5006/1.3284810*
81. Kudo T., *et al.,* Stress Corrosion Cracking Resistance of 22%Cr Duplex Stainless Steel in Simulated Sour Environments. *Corrosion., 1989.**45**(10): p. 831-838.DOI: 10.5006/1.3584990*
82. Laitinen A. and H. Hänninen, Chloride-Induced Stress Corrosion Cracking of Powder Metallurgy Duplex Stainless Steels. *Corrosion, 1996.**52**(4): p. 295-306. DOI: 10.5006/1.3293641*
83. Miyasaka A., *et al.,* Critical Stress for Stress Corrosion Cracking of Duplex Stainless Steel in Sour Environments. *Corrosion, 1996.**52**(8): p. 592-599. DOI: 10.5006/1.3292149*
84. Mark-Wilhelm S. and R. D. Kane, Effect of Heat Treatment and Microstructure on the Corrosion and SCC of Duplex Stainless Steels in H2S/Cl− Environments". *Corrosion, 1984. **40**(8): p. 431-439. DOI: 10.5006/1.3593951*
85. Magnin T. and J. M. Lardon, Cyclic deformation mechanisms of a two-phase stainless steel in various environmental conditions**.** *Materials Science and Engineering A, 1988.**104**: p. 21-28, DOI: 10.1016/0025-5416(88)90402-8*
86. Van-Gelder K., *et al.,* The stress corrosion cracking of duplex stainless steel in $H_2S/CO_2/Cl^-$ environments. *Corrosion Science, 1987. **27**(10-11): p. 1271- 1279, DOI: 10.1016/0010-938X(87)90114-4*
87. Jargelius R. F. A., *et al.,*in: Charles J, and Bernhardsson S, (Eds.), *Duplex Stainless Steel `91, vol. I, Les Editions de Physique, 1992. p. 211*
88. Tsai W. T. and M. S. Chen, Stress corrosion cracking behaviours of 2205 duplex stainless steel in concentrated NaCl solution. *Corrosion science, 2000.**42**:p. 545-559. PII: S0010-938X (99) 00105-5*
89. Bhattacharya A., Stress Corrosion Cracking of Duplex StainlessSteels in Caustic Solutions. *A PhD Dissertation Submitted to the Academic Faculty, Georgia Institute of Technology Atlanta, U.S.A., 2008.*
90. Bhattacharya A. and P. M. Singh, Role of microstructure on the corrosion susceptibility of UNSS32101 Duplex Stainless Steel. *Corrosion, 2008.**64**(6): p. 532 540; DOI: 10.5006/1.3278489*
91. Magnabosco R. and N.Alonso-Falleiros, Pit morphology and its relation to microstructure of 850 ºC aged duplex stainless steel. *Corrosion, 2005.**61** (2): p. 130-136 DOI: 10.5006/1.3278167*
92. Paranthaman V., *et al.,*Analysis of structure property relationship of super duplex stainless steel AISI 2507weldments for severe corrosive environments. *Materials Research Express, 2018.**5**(9); DOI:https://doi.org/10.1088/2053-1591/aad855*
93. Rousset N. B. S.,*et al.,*Role of surface finishing on pittingcorrosion of a duplex stainless steel in sea water. *Journal of Materials Engineering and Performance, 1996.**5**(2): p. 225-231.*
94. Yurtisik K., *et al.,* Characterization of duplex stainless steel weld metals obtained by hybrid plasma-gas metal arc welding.*Soldag. insp., 2013. **18**(3) São Paulo July/Sept. 2013, Technical PapersArtigosTécnicosDOI: 10.1590/S0104-92242013000300003*
95. Nilsson, J. O. and A. Wilson Influence of isothermal phase transformations on toughness and pitting corrosion of super duplex stainless steel SAF 2507. *Materials Science and Technology, 1993.**9**: p. 545–554. DOI: 10.1179/mst.1993.9.7.545*

96. Hu Y. and J. Xue, Effects of Process Parameters on the Weld Quality During Double-Pulsed Gas Metal Arc Welding of 2205 Duplex Stainless Steel. *Transactions on Intelligent Welding Manufacturing, 2018. © Springer Nature Singapore Pte Ltd, https://doi.org/10.1007/978-981-10-8330-3_7*
97. Xie H., *et al.,* Analysis of strong and weak pulse ratio on weld quality of AA6061 aluminium alloy double pulsed gas metal arc welding. *Trans China Weld Inst., 2015.**36** (12): p. 77–80*
98. Taban E. and E. Kaluc, Welding Behaviour of Duplex and Super-DuplexStainless Steels using Laser and PlasmaArc Welding Processes.*Welding in the World Peer-reviewed Section, 2011**55**(8): p. 48-57.*
99. Forgas J. A., *et al.,* Ferrite quantification methodologies for duplex stainless steel. *Journal of Aerospace Technology and Management, 2016.**8**(3): p. 357–362. DOI: 10.5028/jatm.v8i3.653*
100. Bermejo V. A. M., Predictive and measurement methods for delta ferrite determination in stainless steels. *Weld J, 2012.**91**(4): p. 113-s to 121-s*
101. Farrar M. C. J., The measurement of ferrite number (FN) in real weldments-Final Report. *Weld in the World, 2005.**49**(5–6): p. 13–2. DOI:*10.1007/BF03263405.
102. Klug H. P. and L. E. Alexander, X-ray Diffraction Procedures for Poly Crystalline and Amorphous Materials, *2nd edition, John Wiley & Sons Inc., New York, 1974.*
103. Zhang M. and S. B. Zhang, Observation of the keyhole during plasma arc welding. *Welding Research Supplement*, 1999.p. 53-s-58-s.
104. Paulraj P. and R. Garg, Effect of welding parameters on pitting behaviour of GTAW of DSS and super DSS weldments. *Engineering Science and Technology, an International Journal.* Karabuk University. Publishing services by Elsevier B.V., *2016. DOI:10.1016/j.jestch.2016.01.013*
105. Yuko D. K. and S. Wolynec, Evaluation of the corrosion-resistant phase formed during the sigma phase precipitation in duplex stainless steels, *Materials Research, 1999.**2**: p. 239–247. DOI: 10.1590/S1516-14391999000400002*
106. Yuan X., *et al.,* Microstructure and XRD Analysis of Brazing Joint for Duplex Stainless Steel Using a Ni-Si-B Filler Metal. *Material Characterization, 2009.**60**(9):923-931*
107. McPherson NA, Li Y, Baker TN (2000) Microstructure and Properties of as Welded Duplex Stainless Steel. *Science and Technology of Welding and Joining, 5(4): p. 235-244.*
108. Luo J., *et al.,* Double-Sided Single-Pass Submerged Arc Welding for 2205 Duplex Stainless Steel. *JMEPEG, 2013.**22**: p. 2477–2486. DOI: 10.1007/s11665-013-0529-8*
109. Deng B., *et al.,* Effect of Thermal Cycles on the Corrosion and Mechanical Properties of UNSS31803 Duplex Stainless Steel. *Corrosion Sciencece, 2009.**51**(12):2969-2975.*
110. Fonseca G. S., *et al.,*Sigma Phase in Super duplex Stainless Steel: Formation, Kinetics and Microstructural Path. *Materials Research, 2017.*20 (1):249-255. DOI: 10.1590/1980-5373-MR-2016-0436
111. Sidhu G. S. and S. S. Chatha,Role of shielded metal arc welding consumables on pipe weld joint". *International Journal of Emerging Technology Advanced Engineering, 2012.***2**(12): p. 746-750.
112. Gunn R. N., Duplex Stainless Steels, Microstructure Properties and Applications, *Abington Publishing, Cambridge, England, 1997.*
113. Srinivasan P. B., *et al.,* Microstructure and corrosion behaviour of shielded metal arc-welded dissimilar joints comprising duplex stainless steel and low alloy steel. *Journal of Materials Engineering and Performance, 2006.**15**: p. 758-764. DOI: 10.1361/105994906X150902*
114. Wang S., *et al.,*Characterization of microstructure, mechanical properties and corrosion resistance of dissimilar welded joint between 2205 duplex stainless steel and 16MnR. *Materials and Design, 2011. **32**: p. 831-837. DOI: 10.1016/j.matdes.2010.07.012*
115. Practical guidelines for the fabrication of duplex stainless steel. 3rd Edition, IMOA 1999-2014, *published by International Molybdenum Association (IMOA), London, UK. www.imoa.info*
116. Standard Test Method for Tensile Testing of Metallic Materials. *E8-04, ASTM, PA, USA, 2004. DOI:10.1520/E0008-04, www.astm.org*
117. Standard Procedures for Calibrating Magnetic Instruments to Measure the Delta Ferrite Content of Austenitic and Duplex Ferritic-Austenitic Stainless Steel Weld Metal, *AWS A4.2M (ISO 8249:2000 MOD), 2006.*

118. Gupta A., *et al.,* Effect of Heat Input on Microstructure and Corrosion Behavior of Duplex Stainless Steel Shielded Metal Arc Welds. *The Indian Institute of Metals, 2018. https://doi.org/10.1007/s12666-018-1294-z*
119. Hosseini V. A.,*et al.,* Ferrite content measurement in super duplex stainless steel welds".*Welding in the World, 2019.*

A Review on Electrical Discharge Coating (EDC) and its Multi-Optimization Techniques

JagadeeswaraRao Maddu[1], Buschaiah K[1], Riyaaz Uddien Shaik[2]

[1]Department of Mechanical Engineering, University College of Engineering, Osmania University, Hyderabad, India

[2]Department of Astronautical, Energy and Environment, University of Rome "La Sapienza", Rome, Italy.

E-mail: jaganmuddu@osmania.ac.in

Abstract. This paper, primarily, enlightens the properties of the Electrical Discharge Coatings in the view of hardness. Coatings that withstand extremely high temperature, chemical corrosion, stress and other hostile environments are required in industries as functional surfaces. This type of coatings has already applied for purposes such as biomedical, aerospace, automobile etc. Then, secondarily, brings out the different types of optimization techniques and machine learning models used in the field of electrical discharge machining/coating.
Keywords: Coatings; CVD; EDC; Electro-Plating, Optimization Techniques; PVD.

1. Introduction

Improvisation and modification of material surface or workpieces with respect to mechanical, physical and biochemical properties could be carried out by hard coatings [1]–[4]. Surface modification with hard coatings have enhanced the resistance to corrosion, erosion and abrasive wear. In the field of biomedical, enhancement of biomechanical and morphological compatibilities between the already available and new receiving body tissues for promoting osteointegration, surface modification plays a vital role [2], [5]. Hard coatings have strong applications in dental and orthopaedics and previously, this was carried by carburising, electroplating and plasma spraying. Figure 1 explains the different types of coating procedure viz., PVD, CVD, electroplating, thermal spraying and EDC.

Recently, Electro-Discharge Coatings (EDC) have led to renewed interest which modifies the surfaces by electrical discharge energy. In EDC, an extremely high temperature of around 10000 degrees is used to melt the electrode material and deposit it on the surface of workpiece. This process will also be carried out on the Electrical Discharge Machine (EDM) as shown in Figure 2. In the Figure 2, three types of EDM's, normal EDM, powder mixed in dielectric medium EDM and powder metallurgy electrode EDM are shown.

Like EDM, is used for brittle conductive and hard materials since it can melt any conductive materials regardless of its hardness without being in contact, EDC can also coat on the workpiece. Within a constant spark in minute gap between tool electrode and workpiece material, in a short span of time, a series of electrical sparks or discharges occur [27]. In this paper, EDM to EDC, Types of Electrical Discharge Machines used for EDC and techniques involved in achieving the higher thickness and list of optimization techniques are covered.

IOP Conf. Series: Materials Science and Engineering 1185 (2021) 012027 doi:10.1088/1757-899X/1185/1/012027

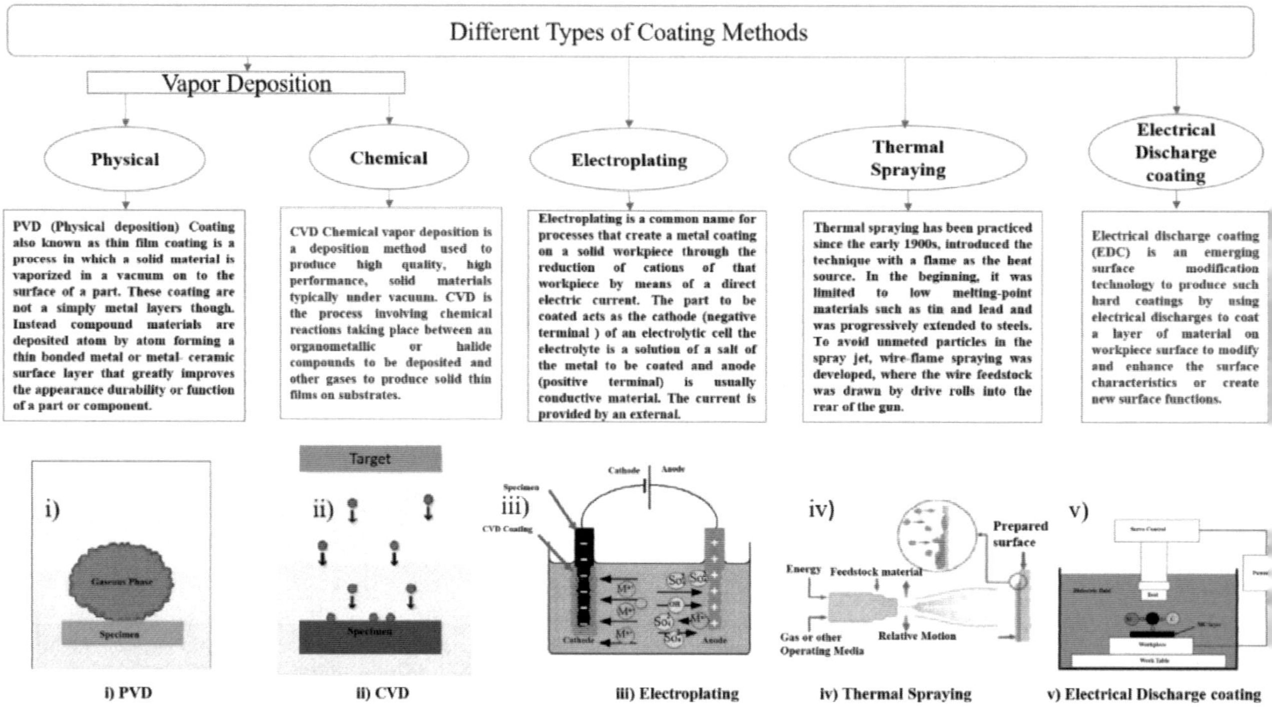

Figure 1. Different Types of Coating Methods [1,6,7,8,9].

2. Review of EDC

Figure 3 shows the processes involved in ED Machining and ED Coating. In the same, material removal and material deposition can be carried out by fixing the suitable parameters. For EDC, Reverse polarity (electrode positive terminal) is more suitable than straight polarity (electrode negative terminal). Surface roughness, layer thickness and Material Deposition Rate can be controlled by peak current and pulse-on time [28], [29].

Different Types of Electrical discharge machine (EDM) for Electrical discharge coating (EDC). Figure 4 shows the different types of EDM machines, necessary process parameters and performance parameters.

But very few are only used for coating which are explained in Figure 5.

3. Case Studies

In the first case, Titanium alloy (Ti6Al4V) is being used a workpiece and brass as electrode. In the dielectric medium of de-ionised water, molybdenum disulphide (MoS2) powder of micron level is mixed and coated with micro-EDC[30], [31]. SEM and EDS image are shown in Figure 6. Dielectric fluid mixed powder can enhance the coating quality and reduce defect such as micro-holes. Significant effect can be observed by increase in concentration and particle size of the powder used and it can reduce environmental pollution problems by reducing dielectric fluid.

In the second case, Ti6Al4V workpiece is being used to test the micro-electro discharge coating using WS2 and Brass PM electrode. Figure 7 shows the SEM and EDS of the coating obtained. Layer thickness of the coating and tool wear rate depends upon the compact load applied while manufacturing PM electrode.

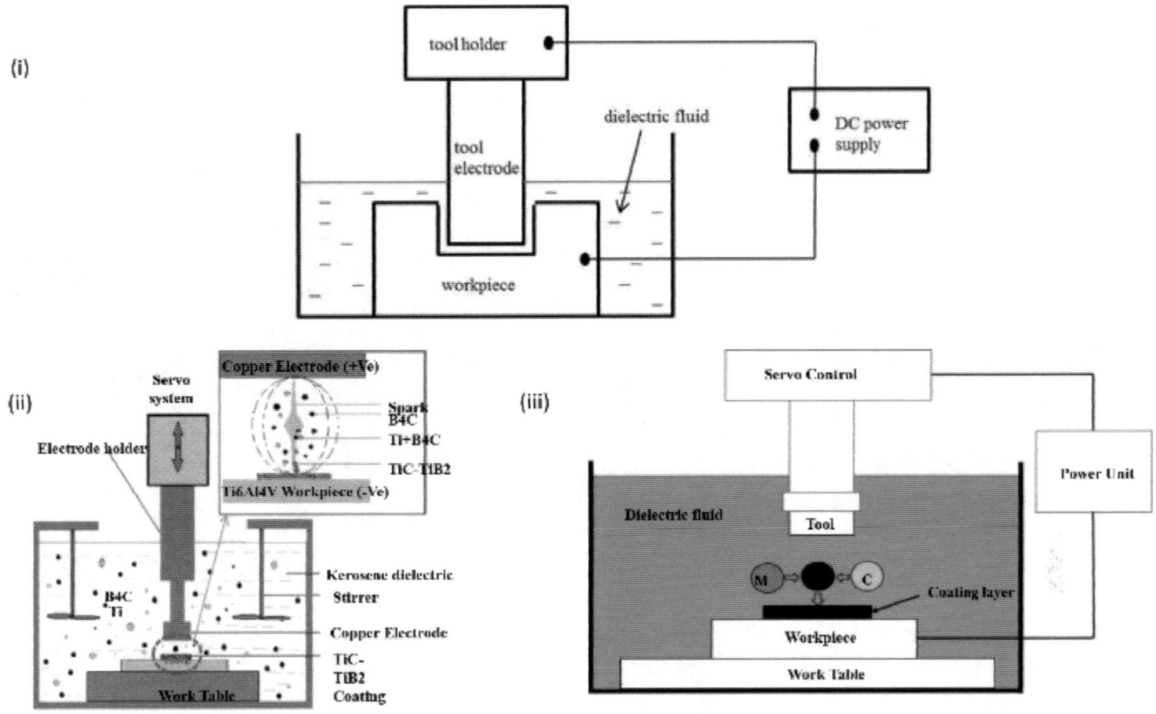

Figure 2. Different Types of EDM, (i) EDM, (ii) powder mixed EDC and, (iii) powder metallurgy electrode EDC [10, 11, 19, 20, 21, 22, 23, 24, 26].

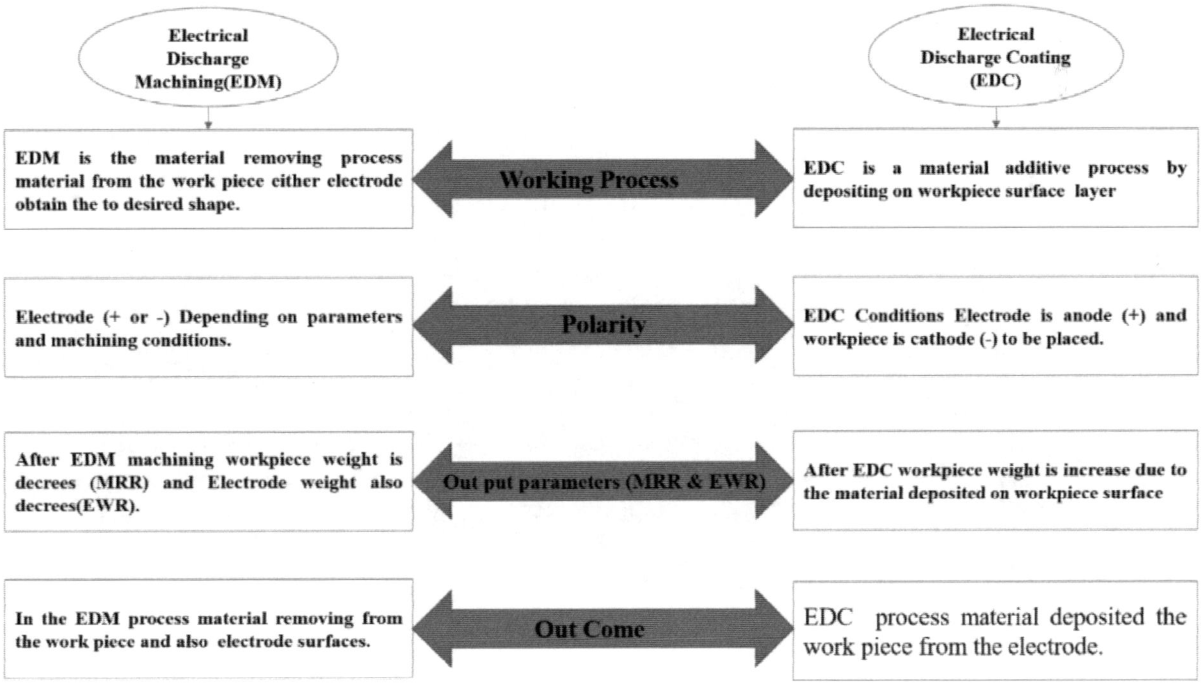

Figure 3. EDM process Vs EDC process.

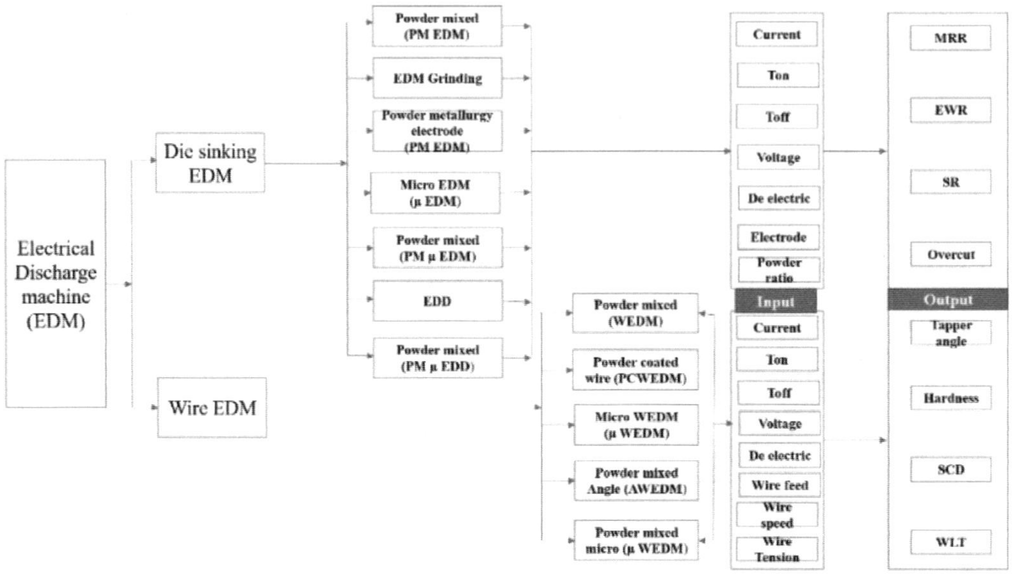

Figure 4. Procedure and process parameters involved in EDMs.

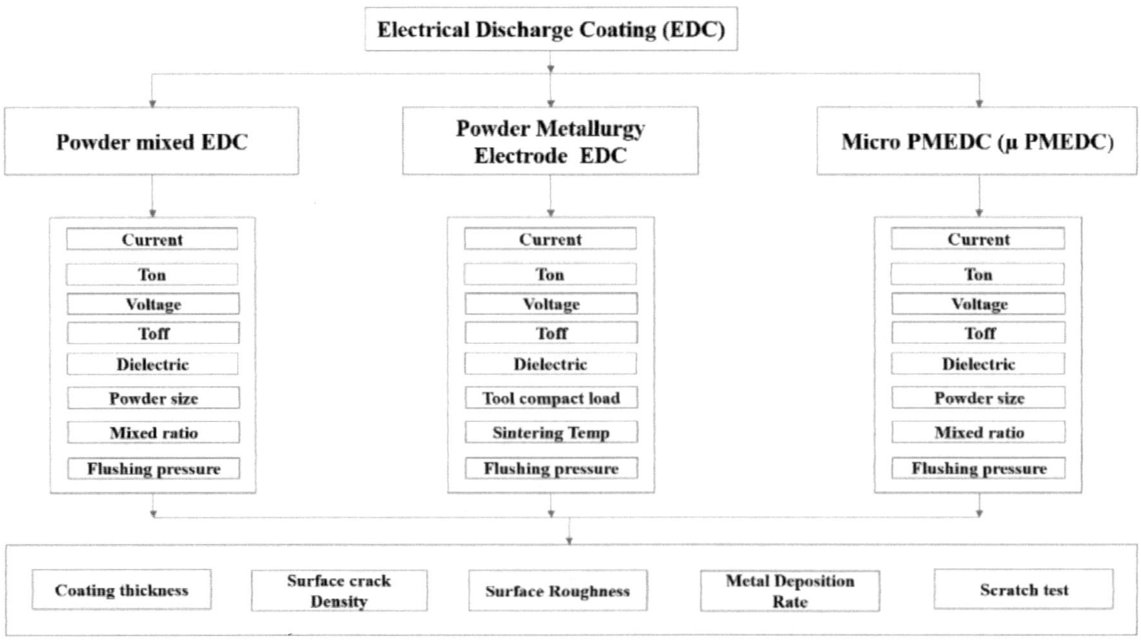

Figure 5. Types of EDC being in use.

IOP Conf. Series: Materials Science and Engineering 1185 (2021) 012027 doi:10.1088/1757-899X/1185/1/012027

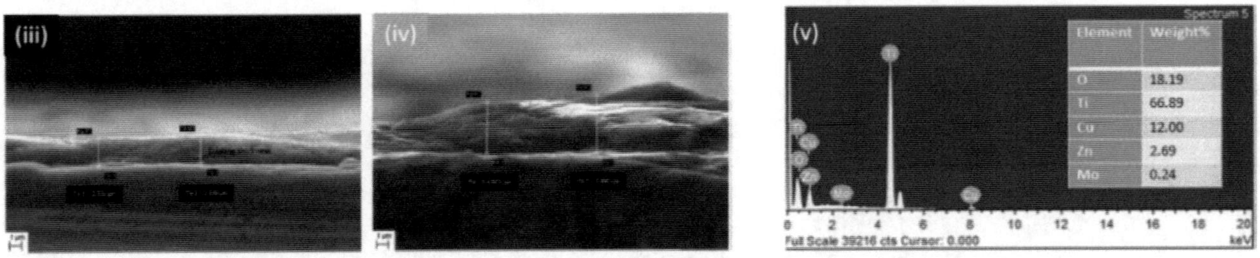

SEM and EDS image showing elemental composition Ti6Al4V workpiece at 60 V, 70% duty factor and 8 g/L MoS2 powder concentration

Figure 6. SEM and EDX of Ti6Al4V workpiece in powder mixed dielectric medium [30].

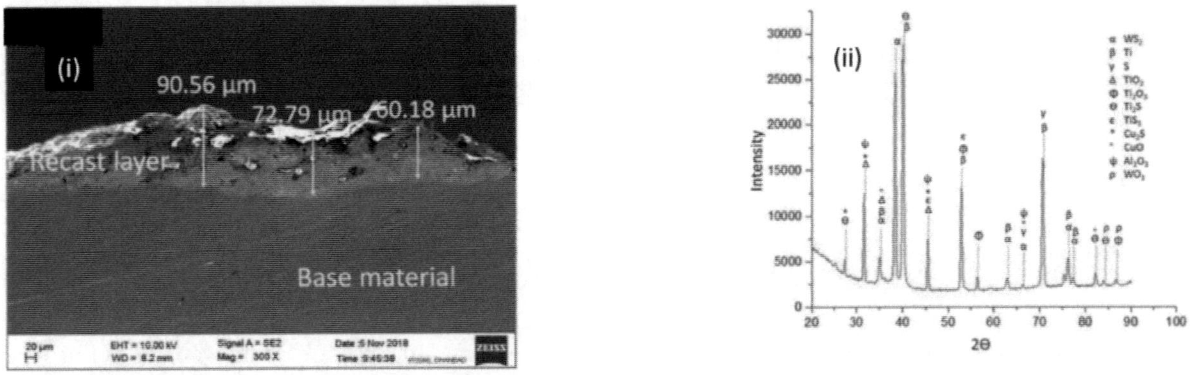

SEM and XRD analysis of Titanium alloy coated sample and PM electrode at 60:40 mixing ratio (WS2: Brass), 40V and 50% duty factor.

Figure 7. SEM and XRD of Ti6Al4V with PM electrode [32].

4. Optimization Techniques

Techniques used by the researchers in this field for optimizing the parameters of Electrical Discharge Machines is been explained below. Techniques summarized could be useful for the researchers working in this field of EDC.

T1. Taguchi Experimental Optimization [33,34,35,36]
T2. Response Surface Methodology [30,37,38]
T3. Grey Relational Analysis [38,39]
T4. TOPSIS [38,40]
T5. Neural Network [38,41]
T6. Genetic Algorithm [42,43]
T7. Artificila Bee Colony Optimization [44]
T8. Finite Element Method [47,48]
T9. Simulated Annealing [45,46]
T10. Ant Colony Optimization Algorithm [49,50]
T11. Particle Swarm Optimization [50,51]
T12. Fuzzy-Optimization [51,52]
T13. Sheep Flock Heridity Algorithm [42,45,53]

Former 1 to 4 technics are most often used by the researchers because of its simplicity so are they not considered here for explanation. Among the remaining techniques, procedures of few are explained in the figure as flowcharts. For in-depth knowledge about the above-mentioned techniques, refer to the citations.

4.1. Neural Network (NN)
NN works by learning the relation between input and output parameters. It has two types of propagations: feed forward and backward propagation methods in artificial neural networks. But only disadvantage of using neural network is it completely depends upon the amount of data. In order to have better accuracy, large amount of data is required. In the field of EDM, around 100 experiments were performed to apply neural networks.

4.2. Genetic Algorithm (GA)
GA developed on the foundation of probability that searches in a parallel and random way via operations of crossover, mutation and reproduction. 'Survival of the fittest' method is being used to preserve and control a population of solutions and implement them later for answers. It has the ability to solve both linear and non-linear problems and basically it was inspired by the Darwin's concept of evolution.

4.3. ABC Optimization
ABC optimization is encouraged by the forage behaviour of honey bees. It works by two groups: un-looker bees and employed bees. In which, skilled bees will have an idea about the food source (nectar position) and food standards (nectar quantity). When unemployed bees receive information about food source, they turn into employed by luring the food from employed. So, always, unemployed bees are proportional to the employed bees. This concept is being used to solve continuous optimization problems.

4.4. Finite Element Method (FEM)
FEM is one of the most used techniques in the field of mechanical design to solve engineering problems. In this area of EDM/C, it could be used to solve structural issues occurring the process of material removal or deposition. ANSYS is the most used software for FEM and the accuracy of the solution depends upon the mesh shape and size. Meshing tactics are important to obtain high accuracy.

4.5. Simulated Annealing (SA)
As the name suggests, it works based on the cooling phenomenon of molten metal and it has evolutionary algorithms for searching the global optimum.

4.6. Fuzzy Logic (FL)
Any reasoning method of human within the linguistic terms helps to enable any user to model any reasoning method. Within the fuzzy set, the fuzzy values are set by the elementary functions that outline the degrees of elements of a selected object. Collection of the semantic statements characterizes the fuzzy model made with fuzzy rules to build up the link between inputs and outputs. Specific fuzzy system contains different number of fuzzy standards and fuzzy rules involved with every input variable.IF-THEN rules are accustomed for making the fuzzy logic. Single scalar terms, de-fuzzification method converts scalar quantity from the fuzzy quantity for output process [24].

4.7. Sheep Flock Heridity Algorithm (SFHA)

As the name suggests, sheep live within their own flock under the manage of shepherds. Hence, the inheritance of genetics develops within the flock only. Certain characteristics will grow only within the flock by heredity and the sheep with large fitness characteristics to their surroundings will breed in the flock. Sheep in neighbouring flock can be inherited to the sheep in additional flock by mixing of flocks. After the mixing of flocks, the one with superior characteristics to the field surroundings, breeds most. In contrast with this process is natural evolution which correspondence to genetic operations. This again has two types: (i) sub-chromosome level genetic operation and (ii) chromosome (global) level genetic operation.

5. Conclusions

Electrical Discharge Coating (EDC) is the recent development on the Electric Discharge machine for surface modification. In this paper, evolution of EDC from the EDM is explained along with the different types of ED machines available. Methodologies, parameters and characteristics of various EDC processes were discussed. Explanation of EDC in comparison with other available hard coating procedures is given. Various types of optimization techniques used in the field of EDM for optimizing input parameters w.r.t to required output performance parameters are listed and discussed.

References

[1] P. J. Liew, C. Y. Yap, J. Wang, T. Zhou, and J. Yan, *Surface modification and functionalization by electrical discharge coating: a comprehensive review,* Int. J. Extrem. Manuf., vol. **2**, no. 1, p. 012004, Feb. 2020, doi: 10.1088/2631-7990/ab7332.

[2] Ching Yee Yap Pay Jun Liew Jiwang Yan *Surface modification of tungsten carbide cobalt by electrical discharge coating with quarry dust suspension.* Int J Adv Manuf Technol **111,** 2105–2116 (2020). https://doi.org/10.1007/s00170-020-06268-9

[3] N. Ahmed et al., *Formation of thick electrical discharge coatings,* J. Mater. Process. Technol., vol.**285,** Nov. 2020, doi: 10.1016/j.jmatprotec.2020.116801.

[4] A. F. Mansor, A. I. Azmi, M. Z. M. Zain, and R. Jamaluddin, *Parametric evaluation of electrical discharge coatings on nickel-titanium shape memory alloy in deionized water,* Heliyon, vol. f6, no. 8, Aug. 2020, doi: 10.1016/j.heliyon.2020.e04812.

[5] B. E. J. Lee, S. Ho, G. Mestres, M. Karlsson, P. Koshy, and K. Grand, *Dual-topography electrical discharge machining of titanium to improve biocompatibility,*Surface & Coatings Technology , vol. **296,** pp. 149–156, 2016, doi: 10.1016/j.surfcoat.2016.04.024.

[6] S. J. Algodi, J. W. Murray, M. W. Fay, A. T. Clare, and P. D. Brown, *Electrical discharge coating of nanostructured TiC-Fe cermets on 304 stainless steel,* Surf. Coatings Technol., vol. **307,** pp. 639–649, Dec. 2016, doi: 10.1016/j.surfcoat.2016.09.062.

[7] R. Tyagi, K. Darmalingam, V. S. Patel, A. K. Das, and A. Mandal, *Deposition of WS 2 and Cu nanopowder coating using EDC process and its analysis,* materialstoday proceedings., Volume **18,** Part 7, Pages 2201-5564 (2019)

[8] V. D. Bui, J. W. Mwangi, A. K. Meinshausen, A. J. Mueller, J. Bertrand, and A. Schubert, *Antibacterial coating of Ti-6Al-4V surfaces using silver nano-powder mixed electrical discharge machining,* Surf. Coatings Technol., vol.**383,** Feb. 2020, doi: 10.1016/j.surfcoat.2019.125254.

[9] P. Janmanee and A. Muttamara, *Surface modification of tungsten carbide by electrical discharge coating (EDC) using a titanium powder suspension,* Appl. Surf. Sci., vol. **258,** no. 19, pp. 7255–7265, Jul. 2012, doi: 10.1016/j.apsusc.2012.03.054.

[10] A. A. Bruzzone, *of EDM,* vol. **48,** pp. 123–126, 1999.

[11] E. Uhlmann and D. Carlos, *Dressing of graphite electrodes for EDM of seal slots in nickel-base alloy,* Procedia CIRP, vol.**42,** no. Isem Xviii, pp. 328–333, 2016, doi: 10.1016/j.procir.2016.02.176.

[12] X. Yue and X. Yang, Molecular dynamics simulation of material removal process and mechanism of EDM using a two-temperature model, Appl. Surf. Sci., p. **147009,** 2020, doi: 10.1016/j.apsusc.2020.147009.

[13] Rahul, Mishra, D.K., Datta, S. et al.*Effects of Tool Electrode on EDM Performance of Ti-6Al-4V.* Silicon **10,** 2263–2277 (2018). https://doi.org/10.1007/s12633-018-9760-0

[14] H. K. Dave, S. Kumar, V. J. Mathai, K. P. Desai, and H. K. Raval, *Application of Artificial Neural Network for predicting multiple responses during EDM process* 3 rd International Conference on Production

and Industrial Engineering**CPIE-2013** Application of Artificial Neural Network for predicting multiple responses durin,no. March, 2013.

[15] U. Maradia, M. Boccadoro, J. Stirnimann, I. Beltrami, F. Kuster, and K. Wegener, *Die-sink EDM in meso-micro machining,* vol. **1,** pp. 166–171, 2012, doi: 10.1016/j.procir.2012.04.029.

[16] M Jagadeeswara Rao, Riyaaz Uddien Shaik, K Buschaiah *Electrical Discharge Machining: A Comparative Surface Integrity Study for Incoloy-800* Materials Today: Proceedings Volume**22** Pages 3286-3296

[17] A. A. A. Alduroobi, A. M. Ubaid, M. A. Tawfiq, and R. R. Elias, *Wire EDM process optimization for machining AISI 1045 steel by use of Taguchi method, artificial neural network and analysis of variances,* Int. J. Syst. Assur. Eng. Manag., 2020, doi: 10.1007/s13198-020-00990-z.

[18] R. Nadda, C. K. Nirala, and P. Saha, *Tool Wear Compensation in Micro-EDM.* Springer Singapore, 2019.

[19] S. Narayan, S. Kumar, and N. Chandra, *ScienceDirect Multi-objective Optimization of EDM Process with Performance Appraisal of GA based Algorithms in Neural Network Environment,* Mater. Today Proc., vol. **18,** pp. 3982–3997, 2019, doi: 10.1016/j.matpr.2019.07.340.

[20] M. Bhaumik and K. Maity, *Effect of Electrode Materials on Different EDM Aspects of Titanium Alloy,* Silicon https://doi.org/10.1007/s12633-018-9844-x pp. **187–196,** 2019.

[21] A. Kumar, S. Siba, and S. Mahapatra, *Surface Characteristics of EDMed Titanium Alloy and AISI 1040 Steel Workpieces Using Rapid Tool Electrode,*"Arab. J. Sci. Eng., vol.**45,** no. 2, pp. 699–718, 2020, doi: 10.1007/s13369-019-04144-7.

[22] N. A. Mufti, M. Rafaqat, N. Ahmed, M. Q. Saleem, A. Hussain, and A. M. Al-Ahamri, *Improving the performance of EDM through relief-angled tool designs,*" Appl. Sci., vol. **10,** no. 7, 2020, doi: 10.3390/app10072432.

[23] J. E. A. Qudeiri, A. Zaiout, A. H. I. Mourad, M. H. Abidi, and A. Elkaseer, *Principles and Characteristics of Different EDM Processes in Machining Tool and Die Steels,* Appl. Sci., vol. **10,** no. 6, pp. 1–46, 2020, doi: 10.3390/app10062082.

[24] M. Al-Amin, A. M. Abdul Rani, A. A. Abdu Aliyu, M. A. Abdul Razak, S. Hastuty, and M. G. Bryant, *Powder mixed-EDM for potential biomedical applications: A critical review,* Mater. Manuf. Process., vol. **00,** no. 00, pp. 1–23, 2020, doi: 10.1080/10426914.2020.1779939.

[25] P. D. Machkale and B. M. Dabade, *Experimental investigation of tungsten and copper carbide coating on AISI 1020 steel using electro discharge coating process,* in Materials Today: Proceedings, 2019, vol. **26,** pp. 2915–2920, doi: 10.1016/j.matpr.2020.02.601.

[26] N. Beri, S. Maheshwari, C. Sharma, and A. Kumar, *Technological advancement in electrical discharge machining with powder metallurgy processed electrodes: A review," Mater. Manuf. Process.,* vol. **25,** no. 10, pp. 1186–1197, 2010, doi: 10.1080/10426914.2010.512647.

[27] A. Descoeudres, *Characterization of Electrical Discharge Machining Plasmas,*" vol. **3542,** p. 137, 2006, doi: 10.5075/epfl-thesis-3542.

[28] A. Banu and M. Y. Ali, *Electrical Discharge Machining (EDM): A Review Electrical Discharge Machining (EDM): A Review,* no. November, pp.**2–10,** 2016, doi: 10.26776/ijemm.01.01.2016.02.

[29] S. K. Choudhary and R. S. Jadoun, *Latest Research Trend of optimization Techniques in Electric Discharge Machining (EDM): Review Article,*" no. March 2014, 2016.

[30] S. Mohanty, B. Bhushan, A. Kumar Das, and A. Rai Dixit,*A study on parametric optimization of Micro-electrical discharge coating process using response surface methodology,* Mater. Today Proc., Aug. 2020, doi: 10.1016/j.matpr.2020.07.415.

[31] D. Sharma, S. Mohanty, and A. K. Das, *Surface modification of titanium alloy using hBN powder mixed dielectric through micro-electric discharge machining,* Surf. Coatings Technol., vol. **381,** Jan. 2020, doi: 10.1016/j.surfcoat.2019.125157.

[32] A. R. Siddique, S. Mohanty, and A. K. Das, *Micro-electrical discharge coating of Titanium alloy using WS2 and Brass P/M electrode,* Mater. Manuf. Process., vol. **34,** no. 15, pp. 1761–1774, Nov. 2019, doi: 10.1080/10426914.2019.1666988.

[33] S. Varma, M. Jagadeeswararao, and S. R. Uddien, *Optimization and Regression Analysis of Process Parameters for wire-EDMed HCHCr steel using Taguchi ' s Technique.*

[34] E. Osmania, OPTIMIZATION OF MACHINING PARAMETER FOR EDM USING TAGUCHI METHOD ON INCOLOY-800, vol. **6,** no. 11, pp. 119–133, 2016.

[35] H. A. Hegab, M. H. Gadallah, and A. K. Esawi, *Modeling and optimization of Electrical Discharge Machining (EDM) using statistical design,* 2015, doi: 10.1051/mfreview/2015023.

[36] U. Ashok and P. Laxminarayana, *ScienceDirect Optimization of Electrode Tool Wear in micro holes machining by Die Sinker EDM using Taguchi Approach,* Mater. Today Proc., vol. **5,** no. 1, pp. 1824–1831, 2018, doi: 10.1016/j.matpr.2017.11.281.

[37] A. Kumar, V. Kumar, and J. Kumar, *Microstructure analysis and material transformation of pure titanium and tool wear surface after wire electric discharge machining process,* Mach. Sci. Technol., vol. **18,** no. 1,

pp. 47–77, 2014, doi: 10.1080/10910344.2014.863632.

[38] S. M. Towhidul Islam Nayim, Muhammed Zahid Hasan, Anbesh Jamwal, Sunil Thakur, and Sumit Gupta, *Recent trends & developments in optimization and modelling of electro-discharge machining using modern techniques: A review,* AIP Conference Proceedings **2148,** 030051 (2019); https://doi.org/10.1063/1.5123973

[39] L. Raju and S. S. Hiremath, *A State-of-the-art Review on Micro Electro-Discharge Machining,* Procedia Technol., vol. **25,** no. Raerest, pp. 1281–1288, 2016, doi: 10.1016/j.protcy.2016.08.222.

[40] A. Y. Joshi and A. Y. Joshi, *A systematic review on powder mixed electrical discharge machining,* Heliyon, vol.**5,** no. 12, p. e02963, 2019, doi: 10.1016/j.heliyon.2019.e02963.

[41] S. Velpula, K. Eswaraiah, and S. Chandramouli, *ScienceDirect Prediction Of Electric Discharge Machining Process Parameters Using Artificial Neural Network,*Mater. Today Proc., vol. **18,** pp. 2909–2916, 2019, doi: 10.1016/j.matpr.2019.07.160.

[42] P. Dharmin, M. R. Mihir, S. P. B. C. Khatri, and P. Janak, *Optimization Techniques used in Electric Discharge Machining – A Technical Optimization Techniques used in Electric Discharge Machining – A Technical Review,* no. November, 2015, doi: 10.13140/RG.2.1.2245.0644.

[43] Zubair Butt1, Riffat Asim Pasha, Faisal Qayyum, Zeeshan Anjum, Nasir Ahmad, Hassan Elahi,*Generation of electrical energy using lead zirconate titanate (PZT-5A) piezoelectric material: Analytical, numerical and experimental verifications* Journal of Mechanical Science and Technology vol. **30(8)**:3553-3558 DOI: 10.1007s12206-016-0715-3

[44] N. Kumar, N. Mandal, and A. K. Das,*Micro-machining through electrochemical discharge processes: a review,* Materials and Manufacturing Processes, vol.**35,** no. 4. Taylor and Francis Inc., pp. 363–404, Mar. 11, 2020, doi: 10.1080/10426914.2020.1711922.

[45] G. Ugrasen, H. V Ravindra, G. V. N. Prakash, and R. Keshavamurthy,*Process optimization and estimation of machining performances using artificial neural network in wire EDM,* MSPRO, vol.**6**, no. Icmpc, pp. 1752–1760, 2014, doi: 10.1016/j.mspro.2014.07.205.

[46] Ali Safaran, Masoud Azadi Moghaddam, Farhad Kolahan *Optimization of backpropagation neural network-based models in EDM process using particle swarm optimization and simulatedannealing algorithms,*Journal of the Brazilian Society of Mechanical Sciences and Engineering (2020) **42:73** https://doi.org/10.1007/s40430-019-2149-1

[47] A. Gholipoor, M. R. Shabgard, and M. Mohammadpourfard, *A novel approach to plasma channel radius determination and numerical modeling of electrical discharge machining process,* J. Brazilian Soc. Mech. Sci. Eng., vol.**42,** no. 4, pp. 1–10, 2020, doi: 10.1007/s40430-020-2244-3.

[48] H. Wu, J. Ma, Q. Meng, M. P. Jahan, and F. Alavi,*Numerical modeling of electrical discharge machining of Ti-6Al-4V,* Procedia Manuf., vol. **26,** pp. 359–371, 2018, doi: 10.1016/j.promfg.2018.07.044.

[49] R. Musa, J. Arnaout, and H. Jung, *Computers & Industrial Engineering Ant colony optimization algorithm to solve for the transportation problem of cross-docking network,* Comput. Ind. Eng., vol.**59,** no. 1, pp. 85–92, 2010, doi: 10.1016/j.cie.2010.03.002.

[50] R. V. Rao and V. D. Kalyankar, *Optimization of modern machining processes using advanced optimization techniques: A review,* Int. J. Adv. Manuf. Technol., vol. **73,** no. 5–8, pp. 1159–1188, 2014, doi: 10.1007/s00170-014-5894-4.

[51] Fan Liu and Meng Joo Er (2010). *Development of Fuzzy Neural Networks: Current Framework and Trends,* New Trends in Technologies: Control, Management, Computational Intelligence and Network Systems, Meng Joo Er (Ed.), ISBN: 978-953-307-213-5, InTech,

[52] H. Fazlollahtabar and H. Gholizadeh, *Fuzzy possibility regression integrated with fuzzy adaptive neural network for predicting and optimizing electrical discharge machining parameters,* Comput. Ind. Eng., vol.**140,** Feb. 2020, doi: 10.1016/j.cie.2019.106225.

[53] P. Dharmin, M. R. Mihir, S. P. B. C. Khatri, and P. Janak, *Optimization Techniques used in Electric Discharge Machining – A Technical Review,*" vol. **3,** no. 02, pp. 424–428, 2015

Investigation on StructuralCharacteristics of Bamboo-Laminates

Veeresh Kumar G B[1]*,ChellaNanda Kishore[2],Krishna Kanth V[2,3], Pramod R[4]

[1]Mechanical Engineering Department, National Institute of Technology-Andhra Pradesh, Tadepalligudem, West Godavari district, Andhra Pradesh State, India.
[2]Research Scholar, Mechanical Engineering Department, National Institute of Technology-Andhra Pradesh,Tadepalligudem, West Godavari district, Andhra Pradesh State, India.
[3]Mechanical Engineering Department, Raghu Institute of Technology, Viskhapatnam, Andhra Pradesh, India.
[4]Mechanical Engineering Department, Amrita School of Engineering, Bengaluru, Amrita Vishwa Vidyapeetham, India.

*Corresponding author'se-mail: veereshkumargb@nitandhra.ac.in

Abstract.Composite materials are structural and functional materialsformedartificially or synthetically by combining two or more dissimilar materials having different characteristics (physical or chemical) which are in macroscopic level and are essentially not soluble in each other. Our present venture is focused on polymer composites due to their light weight, high damping factor, and high stress concentration factor, anti-corrosiveness and anti-chemical properties.The need for strengthened composites (FRP) for renewable fibers has never been as predominant as it is now. The Natural fiber provides both densities decrease and cost savings in comparison with glass fibres. Nevertheless, natural fiber resistance is not as good as glass, the specific characteristics are similar.In the current work bamboo fibre woven as cloth is used to manufacture a polymer composite through and lay process which,includes the application of uniform pressure using rollers to impregnate the fibres with the adhesive resin. The Natural fiber-epoxy composite is then subjected for mechanical properties tests like Brinell's hardness, flexural, tensile and compression strength tests. Further also observes the laminate orientation and observes the deformation and von-mises stress under the load conditions using ANSYS. The composite shows greater promises to be a better natural fibre composite with strengths comparably better than other natural fibre composites like that of jute, linen, kenaf.

Keywords: Bamboo, Polymer Composite, Mechanical Properties, Hand Layup, fibre orientation.

1. Introduction

Many scientists have researched widely the outstanding efficiency of composite materials relative to monolithic materials[1]. Because of their outstanding mechanical and thermal characteristics, strengthened polymer composites are commonly used in many sectors such as automotive, aviation and submarine among polymer composite materials FIBER.Nevertheless, it is unknown and non-biodegradable to discard synthetic and polymer products at the end of life. The use and manufacturing technique of these products will therefore harm nature. Natural fibers, on the other hand, are environmentally friendly and renewable materials; they have reduced prices, densities and mechanical characteristics that are satisfactory.Many scientists and researchers are therefore interested in replacing them with synthetic products for environmental preservation. Natural fibers are split into three distinct classifications based on their sources: crop fibers (hemp, flax, sisal, bamboo, etc.), parts of livestock including protein (hair, silk, wool, etc.) and minerals.Hemicellulose, cellulose, lignin, and pectin are the main components of plant fibres. Cellulose fibers kept together in the framework of plant fibers hemicellulose and lignin matrix. As the amount of cellulose increases, the mechanical characteristics will improve[2, 3]. Cellulose fiber has been used in automotive, building and bridge construction as polymer-reinforced composites.Zakikhani P R. et.al.,[4]described that the bamboo fiber reinforced with epoxy resin has more advantage over the chemical and mechanical properties when compared with glass fibers. In the event of replacing synthetic fibers many researchers [5 -7] have reported that the utilization of natural fibersinexhibiting high tensile modulus, high tensile strength, low breakage elongation and in-sensitive to chemicals and appropriatefor low load applications.

In the present investigations an attempt was made to manufacture Bamboo woven fabric reinforced with suitable epoxy-based resin was used for the various mechanical tests to characterize the potential applications of semi bio-based bamboo fiber composite system.

2. Materials

The raw materials used are Bamboo Fiber woven fabric, Araldite Epoxy LY556 and Araldite HY951 hardener. The Bamboo fabric was obtained from weavers in Erode, for which the yarn was supplied by Aditya Birla Yarns and knit supplies Pvt.Ltd.and it contains-starch about 2-6%,deoxidized saccharide about 2%, 2-4% fat and protein to an extent of 0.8-6%. The physically theBamboofibers appears as shown in Figure 1.Resin and the hardener were obtained from Huntsman Resins.

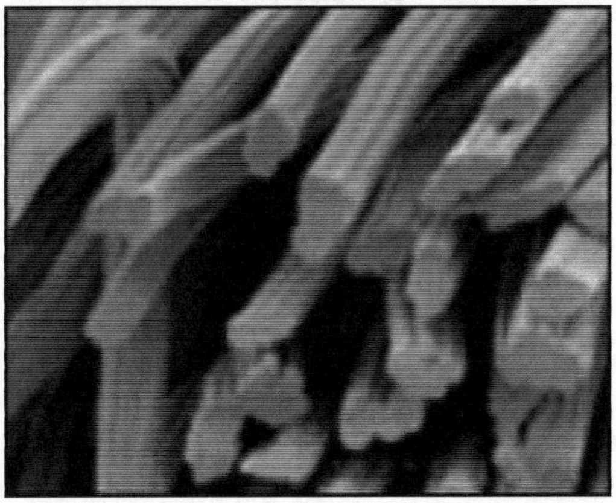

Figure 1. Physical appearance of Bamboo fibers.

3. Composite manufacturing

Plywood is trimmed to 350mm*350 mm in size. A square of 310* 310 mm was labelled on the plate and wooden beadings were stuck along the labelled rows shown in Figure 2. Based on the resin percentage literature fiber was selected as 55:45. Mold releasing agent was applied by means of thin steel blades throughout the prepared mould box and on the inner sides of the beadings.Using a blow warm air drier, the mould box was subsequently dried and held aside for about 24 hours. The hardener and resin were carefully blended in the 9:1 ratio and slender steel blades were used to apply the standardized blend to the mould box. Alternatively, the resin and fibers are placed one by one, followed by a prearranged order of stacking. Pressure of approximately 5kg / sq Ft., after full laying of laminates., applied to the laminates and left for 2 days to curing. The laminates were removed from the mould box after appropriate curing by cracking the mould box.Final thickness of Bamboo fiber composite, resin, hardner and its composition are summarized in theTable 1.

Figure 2.Mould preparation for Bamboo composite laminate

Table 1Bamboo fiber composite composition and properties

Bamboo Fibre		**Resin withHardener AralditeLY556**	
Density	1.25 g/cm^3	Density	1.18 g/cm^3
Thickness	0.474 mm	Shear Modulus	6.21GPa
Poisson's Ratio	0.38	Poisson's Ratio	0.35
Young's Modulus	20GPa	Young's Modulus	25.53GPa

3.1 Mechanical Testing

To know the physical behavior of the Bamboo fiber reinforced composite laminates various strength tests have been carried out. The detailed methodology followed for the tests as discussed under following sections.

3.1.1 Tensile Test

The thin flat strip of laminates with a steady rectangular cross-section region was installed in the mechanical testing machine grips and exposed to tension loading while noting down the force applied. The material's Ultimate Tensile Strength (UTS) was determined from the material's maximum strength prior to failure.If the coupon strain has been monitored with displacement or strain transducers then the material's stress and strain response can be derived from which the elasticity tensile module, ultimate tensile strain, transition strain, and Poisson's ratio can be derived.Grips at each head of the testing device will carry one grip to hold the test sampling so that the direction of force applied to the specimen coincided with the sampling longitudinal axis. The grips apply enough lateral pressure to prevent slipping between the top of the cover and the coupon.Itwasextremely desirable to use the grips that are rotational and self-aligning todiminish bending stress in the coupon.

Tensile properties were gotten by conducting uniaxial tensile strength test using Universal Testing Machine (UTM)as per ASTM D3039. The dimensions of the specimen were as shown in Figure 3.

IOP Conf. Series: Materials Science and Engineering 1185 (2021) 012028 doi:10.1088/1757-899X/1185/1/012028

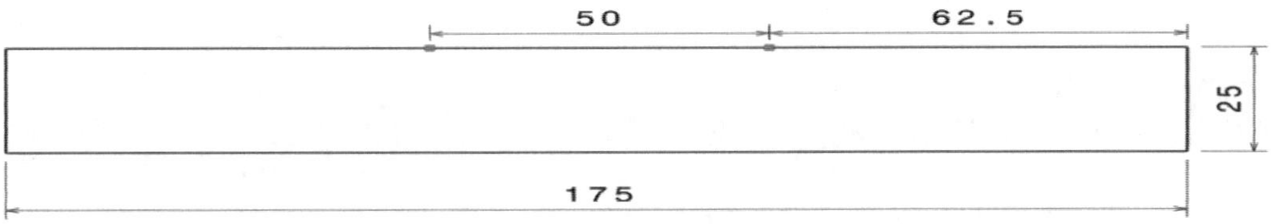

Figure 3.Dimensions(in mm) of Tensile Test specimen.

3.1.2 Compression Test

Composite strengths against compression wascarried-outby conducting uniaxial compression test using UTM as per ASTM D3039. The dimensions of the specimen are as shown in Figure 4.

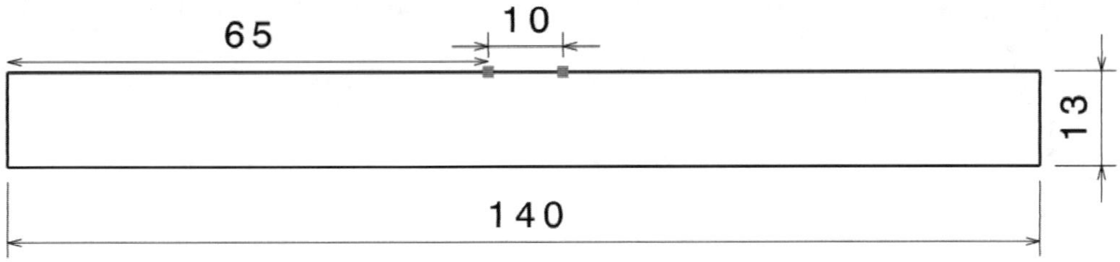

Figure 4.Dimensions(in mm) of Compression Test specimen.

3.1.3 Flexural Test

The flexural strength determination of the fabricated sample is asignificant characterization for any structural composites. It gives the information on the ability of material to sustain the bending before attainmentof the breaking point. Conservatively a three-point loading test was conducted for the determination of flexural properties of the fabricated composites using computer controlled UTM. A sample length of 80 mm and thickness of 4mm (Figure 5) under low strain rates of 6 × 10 were taken and the speed of cross head was continued at 4 mm/min at room temperature. At the immediate failure, the bending resistance of a material was demonstrated as the stress on a bent sample specimen's outermost or farthest fiber. The test was conducted at room temperature in accordance with ASTM D790.

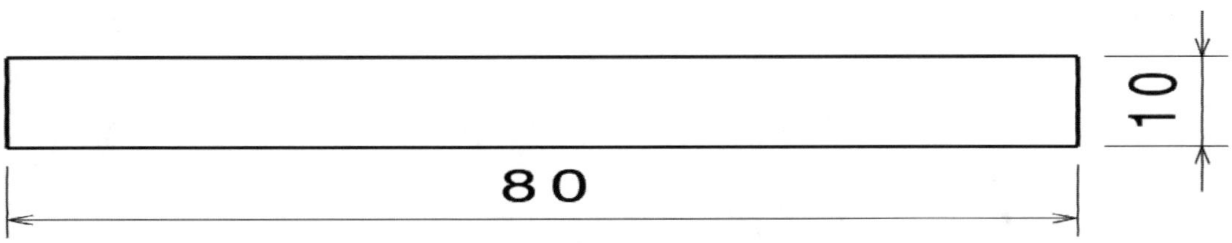

Figure 5.Dimensions (in mm) of Flexural Test specimen.

3.1.4 Hardness test

The measurement of hardness testing was performed using Rockwell hardness testing devices as per ASTM D785-08.The hardness of the specimen tested isindicated directly on the dial gauge of the instrument.Rockwell hardness number was always quoted with a scale symbol in lieu of the indenter size, dial scale used and load. EachRockwell scale division signifies 0.002-mm [0.00008-in.] vertical movementof the indenter.

IOP Conf. Series: Materials Science and Engineering 1185 (2021) 012028 doi:10.1088/1757-899X/1185/1/012028

3.1.5 Effect on orientation of laminates

The analysis of laminate orientations will result in improving further strength in the laminate composite. The change in orientation of laminates can be analyzed by using the Ansys APDL software[8]. First it starts with selection of element type by selecting shell element of 8 node 281 type. Applied the laminate properties as investigated from the experiments. The layer creation tools will help us to design the layers in different orientations. In this work three orientations are considered to study the laminate behaviour under loads, following table 1 will illustrates the details of layer orientations. The positions and orientations of the laminated layers are showed in the figure 6 below.

Table 2 List of experiments and details

S.no	Specimen Name	Angles orientation from 1 to 8 layers
1	Symmetry Smaple-1	0,30,45,-45,-45,45,30,0
2	Symmetry Smaple-2	0,30,-45,45,45,-45,30,0
3	Symmetry Smaple-3	0,-3,45,-45,-45,45,-30,0

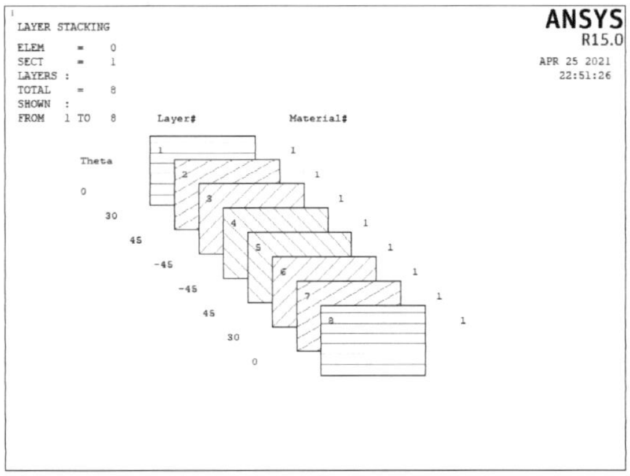

(a)

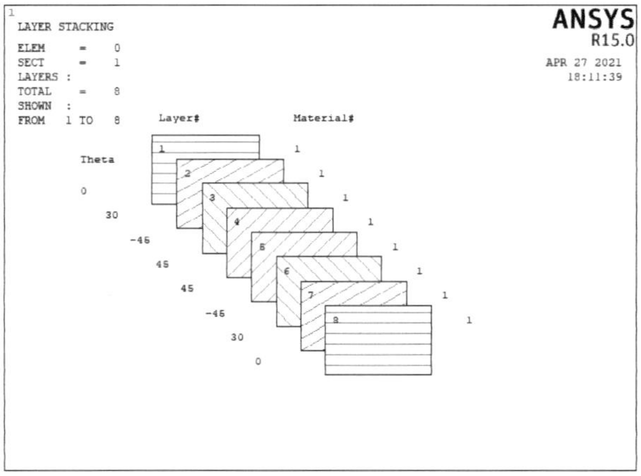

(b)

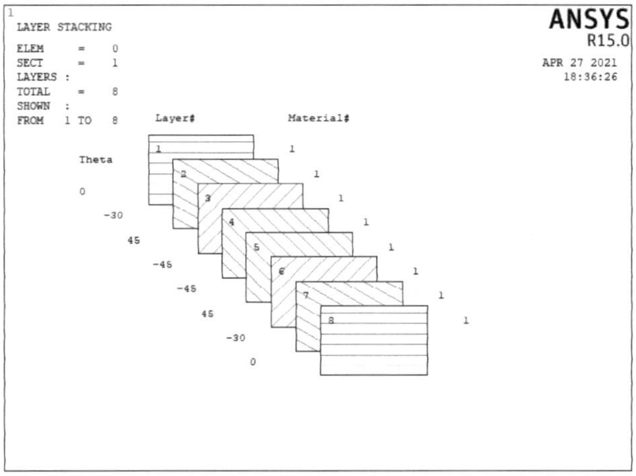

(c)

Figure 6 (a, b and c). Shows the three types of orientations that are considered.

IOP Conf. Series: Materials Science and Engineering 1185 (2021) 012028 doi:10.1088/1757-899X/1185/1/012028

The laminate composite is created in a form of plate 2m x 2m and global sets are created. An uniform mesh is generated. The below figure 7 shows the meshed plate. Boundary conditions are applied as all the sides of the plate are fixed by making the degree of freedom is zero. And a load of 1000N is applied at the centre of the plate in order to estimate the load carrying capacity. The below figure 8 shows the applied boundary conditions.

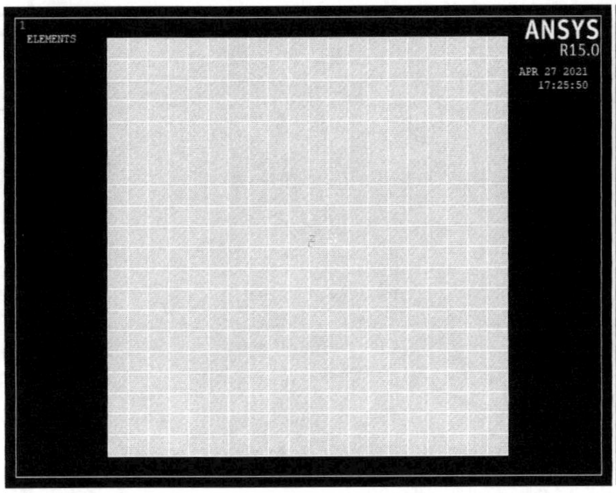

Figure 7. shows the details of mesh on the plate.

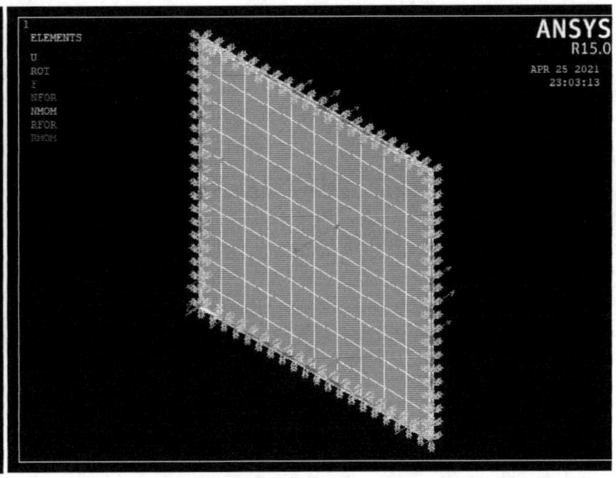

Figure 8. Shows the applied boundary conditions.

All the three different oriented laminate plates are designed as per above procedure and applied boundary conditions. Form the results minimum and maximum deformations of the plate will be recorded and minimum and maximum von-mises stresses are recorded for all the three laminated plates.

4. Results and discussion

To characterize thecandidate Bamboo composite laminates, experiments have been carried out as per the ASTM standards. Physical behaviours under tensile, compression, flexuralandhardness using various specimen configurations has been done by conducting mechanical tests. In the following parts, the examination of the outcomes and the inspiration of different parameters on the different characteristics were summarized.

Table 3:Summary of mechanical properties

Composite	**Tensile Strength (MPa)**	**Flexural strength (MPa)**	**Compressive Strength (MPa)**	**Hardness**
Bamboo composite laminate	27.59	2.45	10.30	87

IOP Conf. Series: Materials Science and Engineering 1185 (2021) 012028 doi:10.1088/1757-899X/1185/1/012028

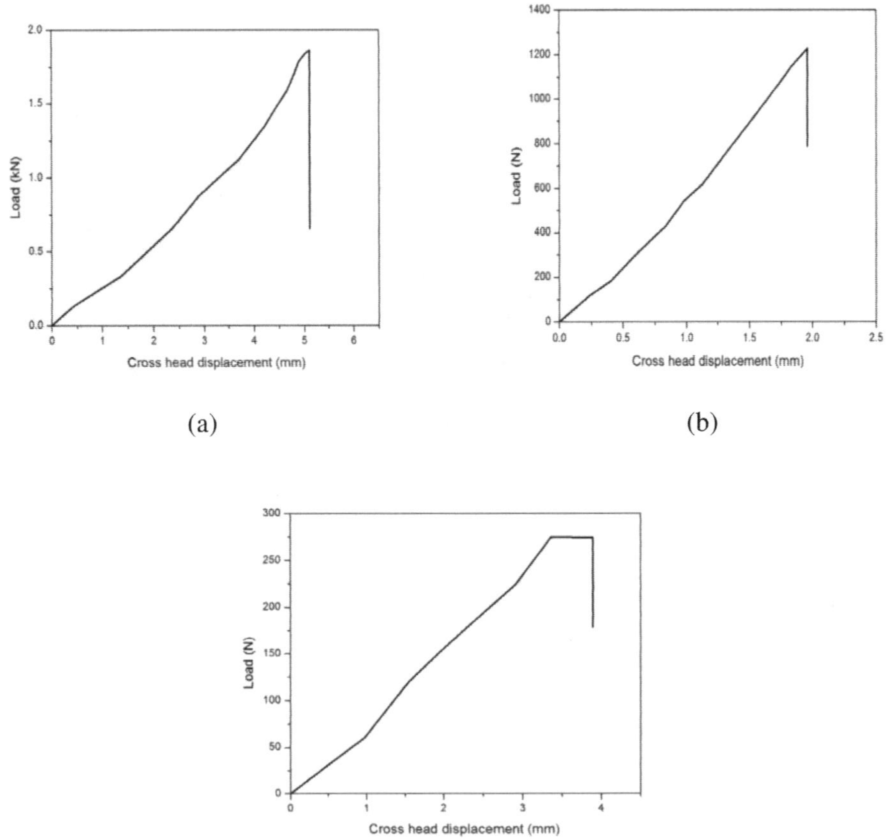

(a) (b)

(c)

Figure 9. Load versus displacement plots for (a) tensile, (b) compression and (c) flexural tests

Form the results the laminated properties are obtained and applied to the laminated plate which is created in ansys. The deformation of the plates and the von-mises stresses distribution on the three plates are showed in the figure 10, 11 and the results are plotted in the figures 12 and 13.

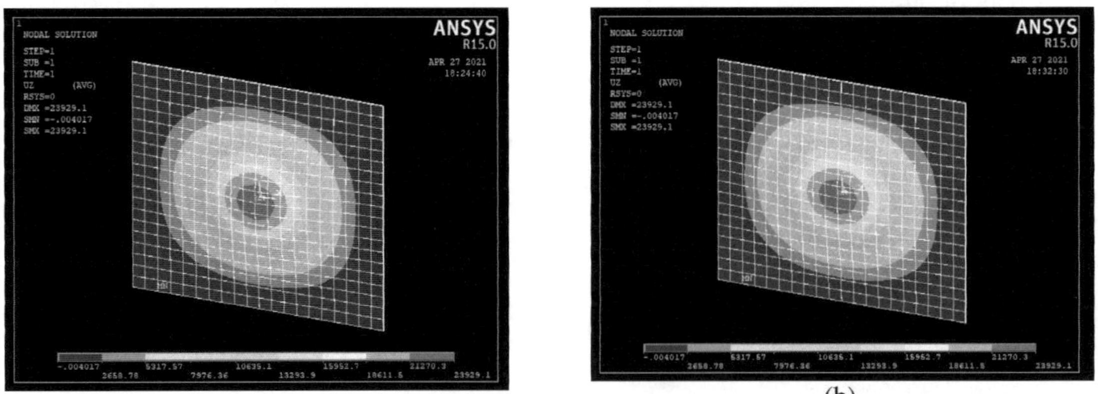

(a) (b)

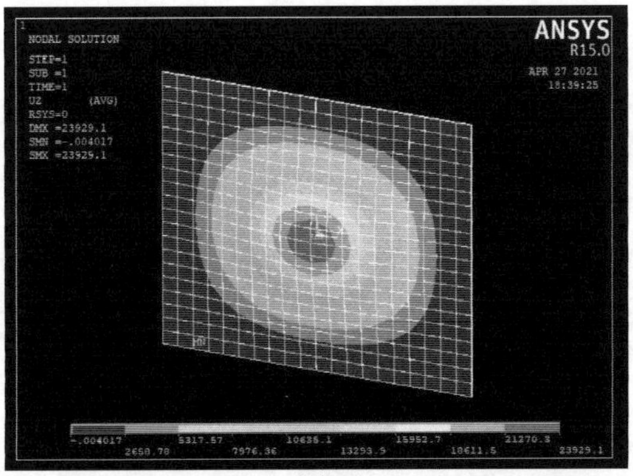

(c)

Figure 10 Shows the plotted results for deformation in the plate in sample 1 (a), sample 2 (b), sample 3 (c).

The results show the change in orientation of each layer effects the behavior of the plate for the boundary conditions applied. The sample 3 (c) in the figure 10, with the oriented angles of 0,-30,45,-45,-45,45,-30,0 is an symmetry orientation which shows the better results in all the three conditions. The sample three (c) is less deformed than sample 1(a) and 2(b).

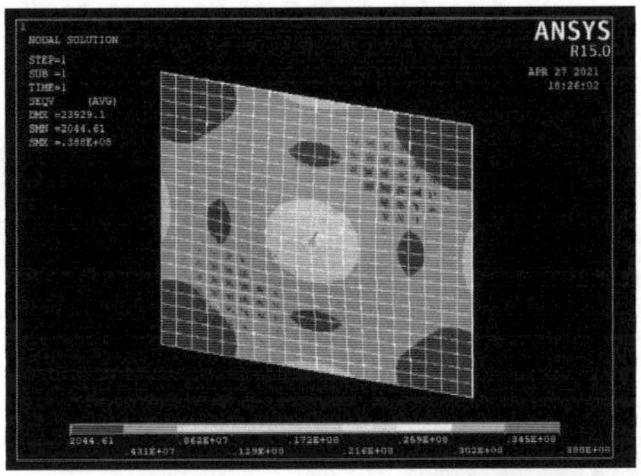

(a)

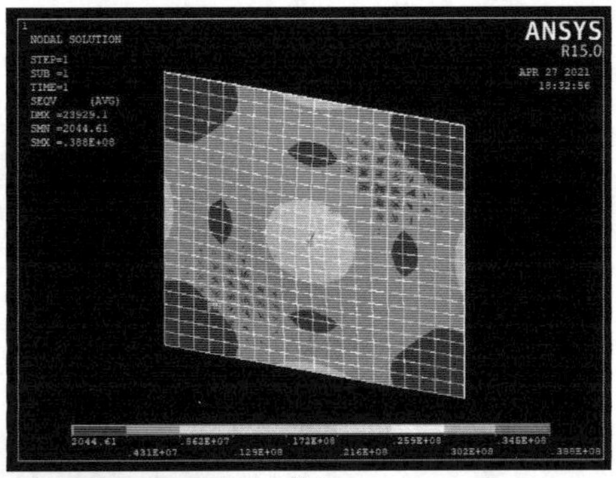

(b)

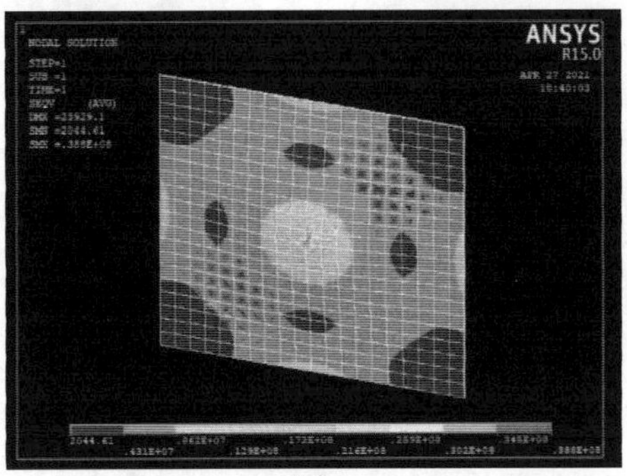

(c)

Figure 11 Shows the plotted results for von-mises stress in the plate in sample 1 (a), sample 2 (b), sample 3 (c).

The graphs are plotted below (figure 12,13) for the both deformation and von-mises stress in all the plates. The von-mises stress is also low in sample 3 (c) when compared with other two conditions.

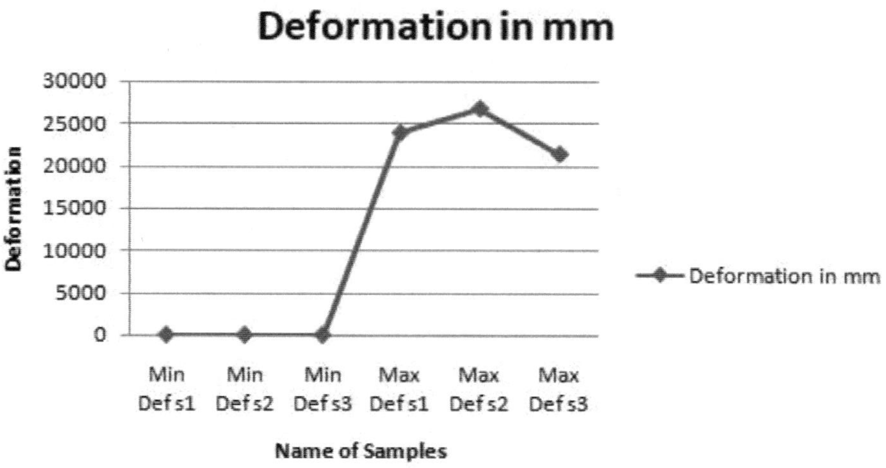

Figure 12 Shows the maximum (max) and minimum (min) deformation in the 3 samples

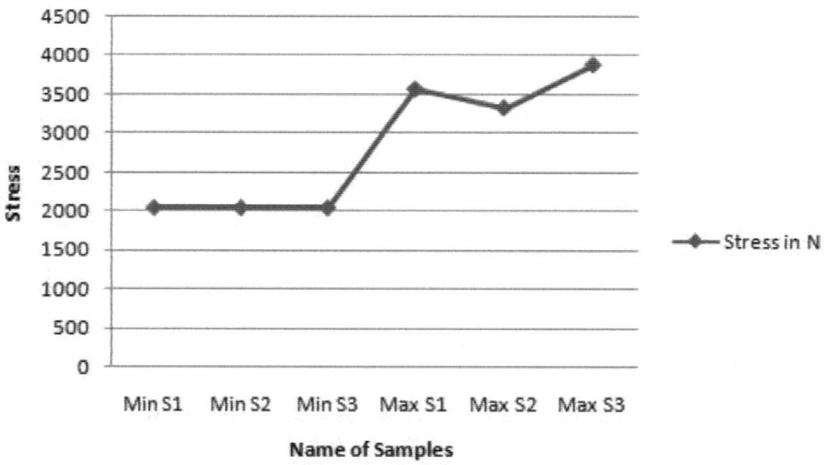

Figure 13 Shows the maximum (max) and minimum (min) stress values in the 3 samples.

The results show that bamboo laminate composites are comparatively better than other composites [9]. The study of laminate orientation in this work figured out some potential results and this study also improves the better understanding of the laminates.

5. Conclusion

- Laminates with a greater volume proportion of fiber have found their ultimate tensile strength to have a greater tensile strength. Consequently, the conclusion of the findings is that the Fiber provides more support for greater tensile strength. By also moving the weave pattern and pattern setup, tensile strengths can be further improved.
- In comparison with many other natural fiber composites, the compressive strength of laminates is improved. By giving extra filler or reinforcement content, the compressive resistance can be enhanced.
- Laminate flexural strength with a lower resin proportion showed an improvement in flexural strength.

IOP Conf. Series: Materials Science and Engineering 1185 (2021) 012028 doi:10.1088/1757-899X/1185/1/012028

- The results shows that layer orientations in the laminated composite plays an important role in improving the load baring capacity. Sample 3 orientation can make the composite bit high strength and can be suggestible while applying.
- Bamboo laminates are showing better results when compared with others.
- All characteristics can be further enhanced if it is possible to add additional reinforcements such as E-glass or S-glass alongside natural fiber.

References

[1] Veeresh Kumar, G.B. and Pramod, R., 2015. Influence of WC particulate reinforcement on the mechanical properties and sliding wear of Al6061 alloys. In *Applied Mechanics and Materials* (Vol. 813, pp. 67-73). Trans Tech Publications Ltd.

[2] Uppin, V.S., Huddar, D.S., Kodancha, K.G., Sridhar, I. and Gouda, P.S., 2016, September. Investigation on pseudo-ductility to improve mechanical behavior in glass-cellulose epoxy composites. In *IOP Conference Series: Materials Science and Engineering* (Vol. 149, No. 1, p. 012112). IOP Publishing.

[3] Uppin, V.S., Sridhar, I. and Gouda, P.S., 2016, September. Interlaminar fracture toughness in glass-cellulose reinforced epoxy hybrid composites. In *IOP Conference Series: Materials Science and Engineering* (Vol. 149, No. 1, p. 012113). IOP Publishing.

[4] Zakikhani, P., Zahari, R., Sultan, M.T.H. and Majid, D.L., 2014. Bamboo fibre extraction and its reinforced polymer composite material. *Int. J. Chem. Biomol. Metall. Mater. Sci. Eng*, *8*(4), pp.271-274.

[5] Nunna, S., Chandra, P.R., Shrivastava, S. and Jalan, A.K., 2012. A review on mechanical behavior of natural fiber based hybrid composites. *Journal of Reinforced Plastics and Composites*, *31*(11), pp.759-769.

[6] Bharath, K.N., Sanjay, M.R., Jawaid, M., Harisha, Basavarajappa, S. and Siengchin, S., 2019. Effect of stacking sequence on properties of coconut leaf sheath/jute/E-glass reinforced phenol formaldehyde hybrid composites. *Journal of Industrial Textiles*, *49*(1), pp.3-32.

[7] Kesavarao, Y., Ramakrishna, C. and Arji, A., 2015. Stress Analysis of Laminated Graphite/Epoxy Composite Plate Using FEM. *International Journal of Mechanical Engineering (IJME)*, *4*, p.5.

[8] Saineel KamalArji, Y Kesavarao, et al.,2015. Design And Simulation Of Laminated Graphite/EpoxyComposite Using Numerical Method And FEM.*BEST: International Journal of Management, Information Technology and Engineering* Vol. 3, Issue 7.

[9] Bharath, K.N. and Manjunatha, G.B., 2018. Investigating the contribution of geometrical parameters and immersion time on fracture toughness of jute fabric composites using statistical techniques. *FratturaedIntegritàStrutturale*, *12*(46), pp.14-24.

Evaluation of Mechanical and Wear Roperties of AA2024 Through Co_2 Casting Using Silica Sand

K. Ajay[1*], B Sateesh[2]

[1]P. G Scholar, Department of Mechanical Engineering, Vignan'sInstitute of Information Technology (A), Visakhapatnam, India - 530049.
[2] Professor, Department of Mechanical Engineering, Vignan's Institute of Information Technology (A), Visakhapatnam, India – 530049.

* Corresponding author e-mail: ajay.kilari15@gmail.com

Abstract: Silica sand is generally used moulding sand in the casting industries and majority of the castings produced worldwide because of its easy availability and binding properties. In the present study, silica sand is taken as moulding material and studies the effect of different shapes on the mechanical and wear properties, micro hardnessand wear tests. For this casting the selected material (AA2024) into two different shapes (cube and cylinder) using silica sand as the moulding material and study the effect of split piece pattern on the properties and wear of the casted components. Co_2 casting is chosen as the industrial process and study mainly focused on to produce the defect free surfaces on AA2024 with split piece pattern which is difficult to control. Destructive tests like hardness, surface roughness are conducted and wear properties of the components is also observed. It is observed from the results that cylinder component has higher hardness due to the faster cooling rate, lower surface roughness and lower wear rate. Analysis concluded that Co_2 process is better working for the cylinder type of objects.

Keywords: Silica, AA2024, Co_2 process, cylinder.

1. Introduction

Aluminum alloys are used widely for various applications due to their lightweight and specific strength. In particular, transportation and automotive industries most often offer the Aluminum due to the greenhouse effect. Therefore, modifying the properties is an alternative way to promote the alloys for using them effectively. Adding various elements to the substrate resulted in improving mechanical properties, thereby microstructure characteristics. The casting process is one of the most suitable ways to make substrates changes with low cost and high quality. Most of the new alloys were investigated to determine their behavior during the casting and following treatment. A study was previously performed on Al and 4.5 Cu alloys using Co_2 casting method [1]. The casted parts were compared to the mechanical and wear properties [2].

In general, the casting process involves the formation of several undesirable effects while making the new alloys. Most of them were related to solidification defects due to insufficient amount and improper mixing of elements into substrates. Therefore, the final product may for the presence of micro cracks and the crystalline structure. Greensand mold casting leading the manufacturing units most effectively compared to other methods. Some researchers were studied on AA6061 by varying the sand permeability and pouring temperature [5]. Also, Co_2 casting was also making its way to use it in the casting industries in now a day because of its ease of making the mould. It contains mainly of pure silica sand mixed with sodium silicate and by passing the Co_2 gas through holes made in the mould which makes it stronger to hold the pressure made by the molten metal. Almost every manufacturing unit using this method raised its usage in the casting industries. This sand's primary sources are from the coastal lines and every foundry industry importing in a considerable range. Huge quantities of silica sand preferred to be used in the workshop applications as a molding material to produce ferrous and non-ferrous castings.

The alloys AA2024 suffer severely from various casting defects on both surface and subsurface of the components. The cause of these defects is due to multiple sources such as shape of the element, design, type of material and process parameters. The quality of castings drastically improves only through the sand's quality and properties, such as molding sand's physical and chemical properties. These properties vary with the physical properties like size, shape, and density; and also chemical properties of chemical composition [6]. The output of the silica sand casting results was highlighted with the casting parts' remarkable properties and the reduction of the defects. Srinivas et al. [7] reported that the LM24 Al alloy exhibited the properties through the silica sand casting method excellent than the other casting methods. It was also said that the Silica sand has an influence on the improvement of grains and changes its shape into a spherical shape with a combination of fine and coarse sizes [8].

Therefore, the present study aims to use the silica sand moulds for the AA2024 alloy. Co_2 casting was used to produce the components which were of different shapes namely Cube, Gear, Flat plate and Cylinder. Here the different shapes of the components were taken in which two were prepared by split piece pattern (Cube and cylinder). The main objective of the study is to produce the defect free components with best suitable component by silica sand as moulding sand through Co_2 casting process out of these components. To study the best various tests were performed on the components like Micro hardness, surface roughness and wear properties of the components.

Additionally, heat transfer rate through the sand also performed to study the flow rate of heat through the silica sand particles to know the cooling rate of the components which makes the components to stronger or weaker by forming the coarse and fine grains. Co_2 casting is performed with 6% of sodium silicate as a binder by passing the Co_2 gas for 15 sec. Two components of same volume were taken for the comparison study between it. With this study, for AA2024 alloy which pattern shape of component was best suitable to use through Co_2 casting can be done.

2. Experimental Procedures

Aluminum alloy AA2024 is used for the present study. The alloy was received in wrought form and is cut into small pieces for the ease of performing the casting and the chemical composition of Mg 1.1%, Cu 4.3 %, Cr 0.55%, Fe 0.4%, Si 0.5%, Mn 0.60%, Ti 0.10%, Zn 0.16%. Silica sand is mostly used sand in the present casting industries with nearly 96% of silica grains with other impurities of 4 %. The sand was than cleaned thoroughly in the sieve to remove all the foreign particles and dust and is used for the casting Cube and Cylinder.

The silica sand is weighed in the weighing machine to add the sodium silicate, for every 100 % of weight 92% of silica sand and 8 % of sodium silicate was added to get the molded silica sand. The silica sand was weighed 92% in the weighing machine and it was kept in a tray and 8 % of the sodium silicate

IOP Conf. Series: Materials Science and Engineering 1185 (2021) 012029 doi:10.1088/1757-899X/1185/1/012029

weighed again in the weighing machine and added to the tray and it was mixed with hands thoroughly until it completely gets mixed. The prepared sand was used for filling all the moulds. For every casting 14 kg of silica sand was used for filling the mould. The silica sand used for the casting was shown in Figure 2.0

The composition of the silica sand was given in Table 1. Before the usage of the silica sand on to the casting firstly preheating is done to remove the moisture content.

Table 1 Chemical composition of Silica sand

Element	SiO_2	CaO	FeO	Al_2O_3	MgO
Silica Sand	96.62	0.57	1.02	1.54	0.57

2.1 Casting process

Pattern was the replica of required casting and for the present study it is made of the wood as shown in Fig. Wood is chosen because of its easy availability, low weight, easily shaped and most importantly cheaper. Two patterns are prepared in total of the same volume and they were cube and cylinder.

Mould is prepared for all the components with by taking the silica sand of 94 % and 6% of sodium silicate for every 1 kg of the moulding sand. Silica sand and sodium silicate is mixed thoroughly and is filed in the drag part by placing the patterns. Ramming was applied gently on the mould after filling the mixed sand. After that on the cope part sprue and riser is placed and again filled it with the sand, similarly it is also gently rammed to make the mould stronger enough to withstand the pressure exerted by the molten metal. For cube and cylinder the pattern was placed in both the drag and cope part as they are split piece patterns. Mould preparation is shown in the Figure 1.

Figure 1 Mould Preparation for the casting

After the mould preparation is done the aluminum cut pieces are properly cleaned and placed in the graphite crucible. Crucible furnace is used for the melting of AA2024 alloy. It is set to 700°C as it is the pouring temperature as shown in Figure 2. Graphite crucible is placed into the furnace slowly with the help of pair of tongs. Soon after the furnace reaches to the pouring temperature, the AA2024 started melting and after it reaches to molten state it is taken out carefully and poured in the cavity. After that the temperature was measured at 8 points for 50 minutes thereafter it is left for cooling down to room temperature. For two components similar process was adopted.

Zwick Vickers's hardness tester is used for measuring the hardness of casting components at three different spots and taken as an average of it. The load of 0.3 kg applied for 15 seconds. Tally surf used to measure the surface roughness of the all the components to study the surface quality.

The equipment used for the wear test is pin on disc tester model TR-20 LE, which is used for knowing the wear rate of all the casting components. The wear test was conducted at two rotational speeds which were at 600 rpm and 780 rpm at a load of 2 kg was used for the study for all the components. The sliding distance of 600 m was taken to study the wear characteristics of the components.

Figure 2. Melting Furnace used for melting

3. Results & Discussions

3.1 Hardness Properties

Hardness is a measure of resistance offered by a material to localized plastic deformation induced by either mechanical indentation or abrasion. Hardness of the different shaped components like cube and cylinder was calculated using Vickers hardness test using a diamond shaped indenter. The Vickers hardness test is used as an alternative to the Brinell method to measure the hardness of materials as it is easier to use than other hardness tests. It has widest scales among hardness tests and can be used for all metals. From the hardness values obtained graph is plotted for the two shapes and from the graph, the following results are obtained.

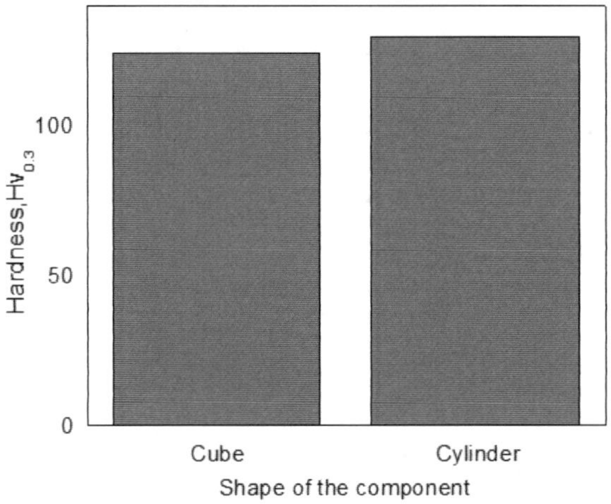

Figure 3. Comparison of Hardness values vs Specimen Shape

The hardness profile of the silica sand for different shapes of casting components were shown in Figure 3. The microhardness profile showing that the higher microhardness shown at cylinder component i.e., 137.6Hv compared with the other and lowest microhardness observed at cube component with 122.1Hv. It is because of the grain size of the cylinder was finer than the remaining components and also the cooling rates were higher in the cylinder compared to the other component. It is evident that also the thermal conductivity plays a vital role in the conduction of heat from the component to all the area. The surface contact area of the cylinder was more compared to the cube as a result cooling rate in the cylinder was more which results in higher hardness.

3.2 Surface Roughness Analysis

Surface roughness is a surface texture of a material as it undergoes to erosion and friction. It was measured by the deviations from the real surface to textured surface. Rough surface was formed with large deviation occurs and if the deviation is less then smooth surface was formed. In general, higher wear rate takes place when the material was rough and smooth surfaces forms when the coefficient of friction was higher. Surface Roughness was also used to predict the performance of any component as irregularities on the surface of the material forms nucleation sites for cracks.

In this study, the surface roughness was measured from the average deviation from the regular surface. It was the average value of ordinates from mean line. The surface roughness for the two components is measured by a talysurf. The Talysurf was used to measure the surface roughness and surface texture of the material quickly and accurately. Talysurf shows the measured roughness (R_z) as well as the mean roughness value (R_a) in micro meters or microns (μm).

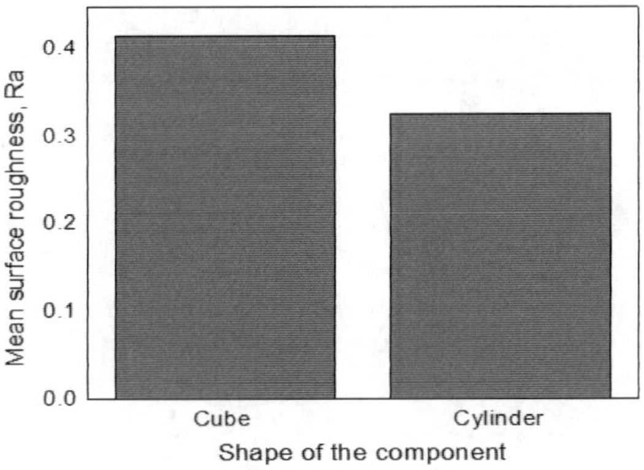

Figure 4. Mean Surface Roughness of casted shapes

In this study the surface roughness for the two components is measured by using portable surface roughness tester and the obtained values are plotted in graph as shown in Figure 4. The surface roughness graph showing the higher surface roughness observed at cube component with 0.415 Ra and lower surface roughness observed in cylinder with 0.325 Ra. At cube component the cooling rate was lower compared with the other component as a result it was prone to the coarse grains results in the softer surface. While the machining was done at higher cutting speeds and higher depth of cuts the surface of the cube component was becomes softer resulting high indentation on the surface with the deeper scratches resulting in the higher surface roughness. At the cylinder component it was showing the lower surface roughness as higher cooling rates were observed in the component, due to that finer grain structure was observed and resulting the lower surface roughness.

3.3 Wear Test

Wear test is a measure of the material erosion with time. It was used to predict the wear mechanism, wear property and amount of wear rate of any material. The major reason to perform the wear test was that whether the given material can withstand to resistance to wear.

From the Figure 5 it was clear that in both components cube and cylinder 600 rpm showing the higher wear rate and also cylinder component was showing the higher wear rate than the cube in both the rotation speeds of 600 rpm and 780 rpm. Here the interesting fact was, as the speed increases the wear rate should increase but interestingly lower rotational speeds showing the higher wear rate than the higher speeds the reason behind that was as the speed increases the friction generates in between the work piece and disc as a result heat generates which softens the material at the higher speeds which was lowering the wear rate of the component.

In both the components at both rotational speeds the wear rate was decreasing with the time. At starting after 100 seconds it was showing the peak wear ate after that as the time increases gradual decrease in the wear rate was observed. There are some fluctuations was observed in the wear rate as in few cases there was linear decrease and a non linear decrease in the wear rate can be observed in few cases. It was due to as the time increases the contact between the work piece with disc fluctuates as the work piece gets soften with the time and the softened material acts like a lubricant in between them due to that the contact between them fluctuates and which reduces the wear rate as the time increases.

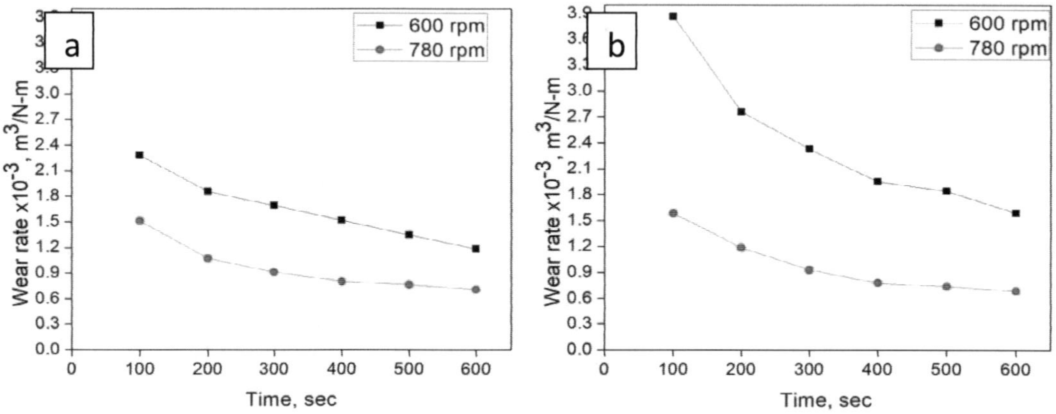

Figure 5. Wear rate graph of (a) Cube,(b) Cylinder

4. Conclusions

This study is focused on the properties of AA2024 casted through silica sand. This study also focused on which shape yields better properties to the casting when silica sand is used as the moulding material. The following conclusions are drawn through this study:

1. Casting of both the components were successfully prepared without any defect and shape deviation through CO_2 casting process.
2. At cylinder component higher hardness was observed due to the faster cooling rate and lower surface roughness.
3. Wear test was conducted on the castings shows that cylinder was showing the lower wear rate and at 780 rpm rotational speed in both the cases showing the lower wear rate.
4. Out of those two components Co_2 process is better working for the cylinder type of objects.

References

[1] Kulkarni S N and Radhakrishna K 2005 Evaluation of metal–mould interfacial heat transfer during the solidification of aluminium –4.5% copper alloy castings cast in CO_2–sand moulds*Materials Science-Poland*Vol. **23** No. 3 200.

[2] HalilKarakoc, Ismail Ovalı, SibelDündar and Ramazan, Cıtak 2019 Wear and mechanical properties of Al6061/SiC/B4C hybrid composites produced with powder metallurgy*journal of material research and technology*. 2019 8(6):5348–5361.

[3] Yang L J 2003 The effect of casting temperature on the properties of squeeze cast aluminium and zinc alloys*Journal of Materials Processing Technology* 140 (2003) 391–396.

[4]Akhlaghi F, Lajevardi A and Maghanaki 2004 Effects of casting temperature on the microstructure and wear resistance of compocast A356/SiCp composites: a comparison between SS and SL routes*Journal of Materials Processing Technology* 155–156 (2004) 1874–1880.

[5] OlawaleOlarewajuAjibola, Daniel ToyinOloruntoba, and Benjamin OAdewuyi 2015 Effects of Moulding Sand Permeability and Pouring Temperatures on Properties of Cast 6061 Aluminium Alloy*International Journal of Metals* 2015 Article ID 632021 Page 13.

[6] Narasimha Murthy I and BabuRao J 2016 Investigations on Physical and Chemical Properties of High Silica Sand, Fe-Cr Slag and Blast Furnace Slag for Foundry Applications*Procedia Environmental Sciences* 35 (2016) 583 – 596.
[7] KarthickSrinivasan, Siddharth C S K, Arun L VKaarthic and Thenarasu M2018 Evaluation of Mechanical Properties Economic and Environmental Benefits of Partially Replacing Silica Sand with Biomass Ash forAluminium Casting *Materials Today Proceedings* 5 (2018) 12984–12992.
[8] Narasimha Murthy I and BabuRao J 2016 Investigations on Physical and Chemical Properties of High Silica Sand, Fe-Cr Slag and Blast Furnace Slag for Foundry Applications*Procedia Environmental Sciences* 35 (2016) 583 – 596.
[9] Narasimha Murthy I ArunBabu N and BabuRao J 2018 Microstructure and Mechanical properties of A356 alloy Castings made in Sand and Granulated Blast Furnace Slag Moulds*Materials TodayProceedings* 5 (2018) 161–167.
[10] Narasimha Murthy I and BabuRao J 2018 Non Destructive Evaluation of A356 alloy Castings made in Sand and Granulated Blast Furnace Slag Moulds*Materials TodayProceedings* 5 (2018) 168–174.
[11] Skočilasová B and Skočilas J 2016 Effect of mold material on temperature distribution in alloy cast *AIP Conference Proceedings* 1768 020017 (2016).
[12] WasiuAjibolaAyoola, Samson OluropoAdeosun, Olujide Samuel Sanni and AkinlabiAyetunji 2012 Effect of casting mould on mechanical properties of 6063 aluminum alloy Journal of Engineering Science and Technology Vol7 No. 1 (2012) 89 – 96.
[13] Narayan Prabhu K and Suresha K M 2004 Effect of Superheat, Mold, and Casting Materials on the Metal/Mold Interfacial Heat Transfer During Solidification in Graphite-Lined Permanent Molds*journal of materials engineering and performance* (2004) 13:619-626.
[14] Feifan Wang, Qingxian Ma and Wen Meng and Zhiqiang 2017 Ha Experimental study on the heat transfer behavior and contact pressure at the casting-mold interface in squeeze casting of aluminum alloy*International Journal of Heat and Mass Transfer* 112 (2017) 1032–1043.
[15] Sung B S and Kim I S 2008 The molding analysis of automobile parts using the die-casting system *Journal of materials processing technology* 201 (2008) 635–639.

IOP Conf. Series: Materials Science and Engineering 1185 (2021) 012030 doi:10.1088/1757-899X/1185/1/012030

Microstructural and Corrosion Characteristics of a Heat Treatable Stainless-SteelWeldments

Adegoke Adesanmi[a], Ibrahim Momoh-Bello Omiogbemi[b,c*], Nicholas Sylvester Gukop[a], Anand Kumar Gummadi[d],SemirMohammadibrahim Negash[e], Ishaya Musa Dagwa[f]

[a] Mechanical Engineering Department, University of Jos, 930003, Nigeria
[b] Mechanical Engineering Department, Air Force Institute of Technology, Kaduna 800282, Nigeria
[c] Mechanical Engineering Department, Ahmadu Bello University, Zaria 810107, Nigeria
[d] Engineering Division, CSIR-National Metallurgical Laboratory, Jamshedpur-831007, India
[e] Mechanicaland Industrial Engineering Department, Addis Ababa University /AAiT, P. O. Box: 1176, Addis Ababa, Ethiopia
[f] Mechanical Engineering Department, University of Abuja, Gwagwalada, Abuja 900001, Nigeria

*Corresponding Authors e-mail: omiogbemi1@gmail.com, professorgok@gmail.com

Abstract. In this research, AISI 316 stainless steel samples were welded and subjected to post-weld heat treatment (PWHT). Thereafter, theywere evaluated for microstructural andcorrosion behaviours. PWHT was carried out on some of the welded samples at 400°C for 12 minutes in the muffle furnace. Investigation of microstructure and chemical characterization of the samples was performed using scanning electron microscope, whilst also corrosion rate was evaluated by weight loss method in a controlled environment The As-welded and post-welded samples were immersed in 0.5M of tetraoxosulphate (IV) acid at room temperature and then studied for 15days with each sample removed at 3days interval. It was evident that the post-weld heat-treated samples under controlled heating condition showed higher resistance to corrosion than the as-welded sample which is attributed to the fact that PWHT increases steel weldment resistance to stress corrosion cracking ultimately. Therefore, corrosion susceptibility of stainless steel in H_2SO_4 medium during exposure time could be a pointer to the aggressive sulphide ion (SO_4^-) which continuously breaks down the protective film on the metal sample.Two major carbide precipitates of Cr23M7 and CR7C3 were evident after the simulation of the research material, which could pose a detrimental effect on the general performance of the steel in engineering and other industrial applications.

Keywords: AISI 316; Arc Welding; Post weld heat treatment, H_2SO_4**;** Corrosion; SEM.

1. Introduction

Stainless steel was discovered in the early stages of the 20th century. Its classification was based on the constituent microstructure of the individual stainless-steel members. There are austenitic, ferritic, martensitic and precipitation-hardenable (PH) stainless steels with the first two being the most common and abundant of the family. Austenitic stainless steels form the largest stainless-steel family with reference to alloys and usage in the steel industry. The basic alloying element in stainless steel is said to be the chromium compound. Others include nickel, molybdenum, nitrogen and carbon at different proportions. Applications at elevated temperature and aggressive media/ environment in chemical industries made a new dimension and significant step on the discovery of stainless steel. Corrosion and oxidation resistant materials were introduced into the engineering sectors [1, 2]. A combination of the austenitic and ferritic stainless-steel species gives a much stronger and better variety of stainless steel. It is called duplex stainless steel (DSS). The modern development of the DSS grade was introduced which had improved welding properties strictly due to the nitrogen application in the alloy matrix [3].

The corrosion process has been right from the time a basic issue of concern in the metallic and engineering world. Its prevention and control as well have posed a substantial degree of retardation in the technological development of metallic compounds and materials [4]. The constant degradation of a material's properties (both chemical and physical) due to interaction with the environment is a simple definition of corrosion [5]. A reduction in a system's Gibb's energy is the driving force for corrosion to occur i.e. the natural tendency for materials and metals to return to their natural state.

Heat treatment is defined as a sequence of heating and cooling that is designed to obtain some desired combination of properties on the material been heated [6]. Phase transformations as well as structural changes like Chromium rich carbide formation and dendritic grain formation which do take place around the welded joint thereby making the weldment susceptible to corrosion attack [7]. But post-weld heat treatments are the causes or reasons for changes in the properties of the material after treatment. The "principles of heat treatment" are factors that determine and control these changes. A substantial amount of residual stress occurs on weldment which is basically due to restraint by the parent metals of the steel component during weld solidification which can become a source of stress corrosion while in service [8].

Additionally, the presence of asymmetrical grains, element segregation, and thermal stresses lead to lower ductility in the material, and poor performance [9, 10]. An estimated US$4 trillion in global annual loss due to corrosion and corrosion protection expenses includes corrosion as well as everything related to its protection [11, 12]. Pitting during the corrosion in stainless steels is one of the primary failure mechanisms [13-16]. The heat treatment here was selected to eliminate the internal stress and get homogenized the microstructure after the welding processes [17]. This study is therefore geared towards validating the metallurgical and corrosion characteristics of AISI 316 weldment with a view to ascertaining specific scientific measures in curbing the possible menace welding could pose. All these stated challenges necessitated this research focus.

2. Experimental procedure

2.1. Materials and equipment

Materials used for the research work include the following; stainless steel sheet, acetone, distilled water, paper grits (80, 120, 220, 320, 400, 600, 800 and 1000 grit sizes), stainless steel electrodes (AWS A5.4-E316), tetraoxosulphate IV acid (H_2SO_4) solution while equipment include; Miller blue thunder series 443

SMAW Arc welding machine (Model: SKU MIL-029016236), JSM-F100 Schottky field emission scanning microscope (SEM), Electric furnace (Model: CHO-3,5.5.3.5/12,5), desiccators, digital weight balance and beakers.

2.2 Microstructure and chemical analysis
Both as-welded and post-welded samples were cut out and carefully examined for compositional characteristics in weight percentage as shown in Table 1.

Table 1 Chemical composition (wt. %) of the research material of parent metal, welded metal and post welded metal

Element (wt.%)	C	Mn	Si	S	P	Cr	Ni	Ti	Al	Fe
Parent Metal	0.540	1.730	0.480	0.008	0.008	18.530	8.4900	0.008	0.006	Bal.
Welded Metal	0.680	1.280	0.750	0.002	0.001	17.310	7.2400	0.012	0.004	Bal.
Post Welded Metal	0.730	1.260	0.650	0.001	0.001	17.430	7.6000	0.010	0.003	Bal.

2.3 Welding and sample preparation
A size of 40mm x 30mm x 1.6mm specimens were welded in pairs to obtain ten different variations. The electric arc welding process was used to join the cut pieces together. The voltage and current passing through the welding machines were set at 22V and 100A respectively. A butt joint was utilized in the welding process. Stainless steel electrode was used for electric arc welding due to the thickness of the specimen.

2.4 Post-weld heat treatment

Post-weld heat treatment was done on five welded samples using controlled heating at 400^{0}C for 12 minutes in a furnace. Thereafter, water was employed as the quenching medium to cooling.

2.5 Determination of corrosion behaviour
The weight loss system of corrosion studies was conducted as per the ASTM methods to determine the corrosion characteristics of the samples. Each of the prepared specimens was weighed before and after the immersion in order to determine the corrosion behaviour. The digital weight balance machine which measures a weight of 0.001g was used. The welded portion (the weldment and the HAZ) was cut out from the unaffected zones. This was then weighed and the result taken for all five samples. The samples were then wholly immersed in 0.5M of H_2SO_4 solution at ambient temperature. The post-welded samples were also immersed in the same solution for corrosion investigation.

3. Results and discussion

3.1. Microstructure study
Scanning Electronic Microscopy (SEM) analysis was performed to study the surface morphology of the samples after immersion in the corrosive medium. The SEM micrographs of the specimen are shown in Figure 1 (a)-(g). A passive layer is formed due to corrosion in all samples that is; As-received sample

(Figure 1 (g)), As-welded sample (Figure 1 (a)-(c)), and PWHT sample (Figure 1 (d)-(f)). But the thickness of the layer is differing from sample to sample. There is more aggressive behaviour in the As-welded sample due to the effect of heat from the weld pool. This region is called the heat-affected zone (HAZ); it implies that the microstructure is greatly affected by the heat generated during welding as posited by Clark [18]. Many researchers have investigated the HAZ of the AISI 316 welding and came to a conclusion that there exist some features within the confined of HAZ region which are: the coarse grain (CGHAZ) which is fully austenite phase at a temperature range of 1000°C-1250°C, fine grain (FGHAZ) boundary layer which is the austenite phase and the inter-critical zone (ICHAZ) which is at 50% of austenite phase transformation. Due to these disturbances in phases of AISI 316 welded structure, precipitation is bound to occur which plays a significant role in the aggressive corrosive nature in the regions [19-23].

Also, presented in Figure 2 (a)-(d) are the optical micrographs of the tested parent metal studied for corrosion characteristics for 3 days and 7 days in 0.5M H_2SO_4 medium. Essentially, the parent metal composed of largely coarse austenite grains with tiny amounts of ferrite. Annealed twins were observed crossing the grain boundaries. The results in figure 2 (d) show significant damage occurred on the surface of the material after 7 days of immersion and a craggy-like terrain is observed. Evident from Figure 2 (b) and (d) are the pitting profiles of the steel which indicate the pattern of pitting on the material of experimentation. These observations are consistent with other researchers' view on corrosion of AISI 316 stainless steel [24, 25].

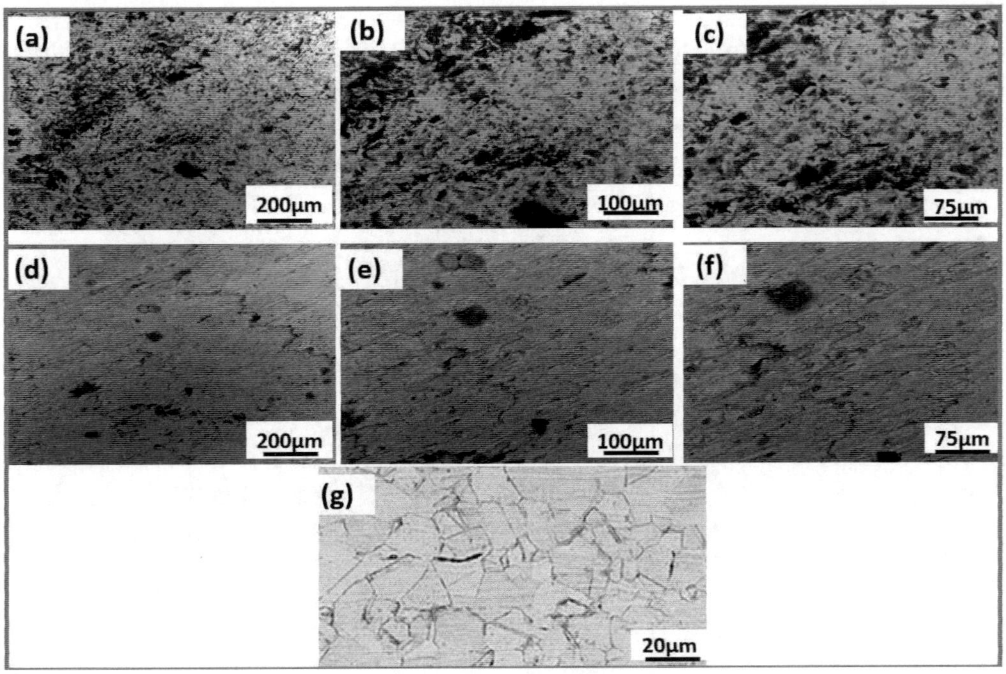

Figure 1 SEM Microstructure of: (a) As-welded stainless-steel joint at 500x, (b) As-welded stainless-steel joint at 1000x, (c) As-welded stainless-steel joint at 1500x, (d) Post-welded stainless-steel joint at 500x, (e) Post-welded stainless-steel joint at 1000x, (f) Post-welded stainless-steel joint at 1500x and (g) parent metal at 4000x magnifications

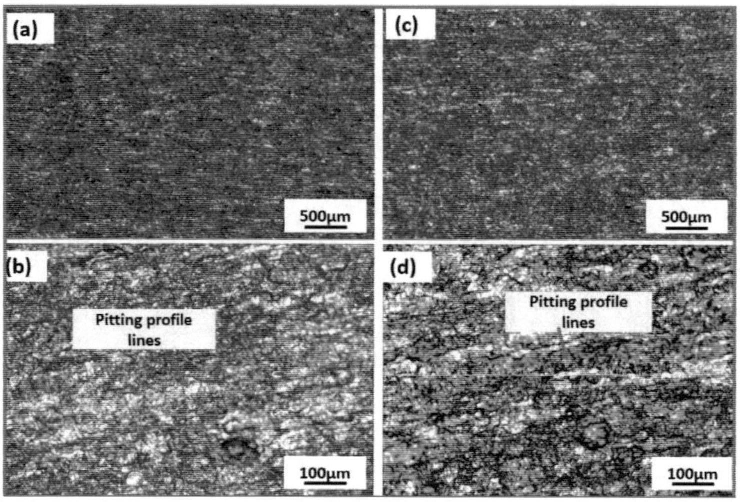

Figure 2 Optical Microscopic images of the corroded samples of AISI 316 parent metal examined in 0.5M H_2SO4 at Magnifications of: (a) 50x for 3 days (b) 200x for 3 days, (c) 50x for 7 days and (d) 200x for 7 days

It is observed that the post-weld heat treatment gives the low corrosive pitting than the as-received and welded samples with respect to pitting and surface roughness, pits distribution and size with a depth of cavities as shown in Figures 3 and 4.

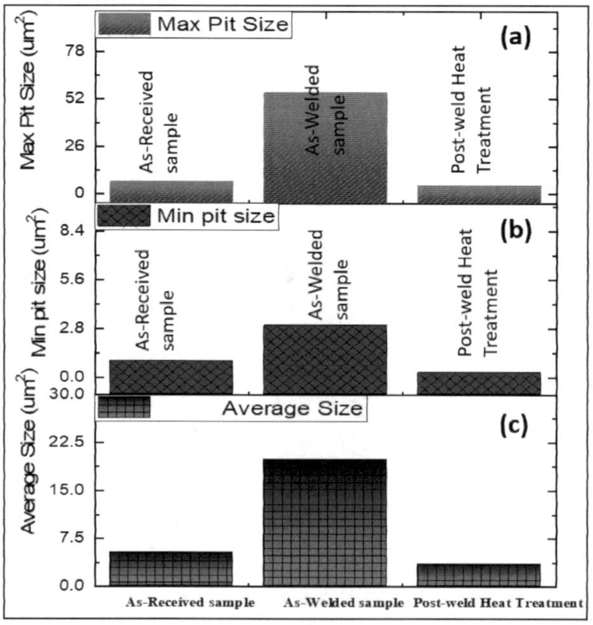

Figure 3 Graphs of As-received samples, As-Welded samples and Post-weld Heat treatment for: (a) Maximum pit size, (b) Minimum pit size and (c) Average pit size of austenitic stainless steel

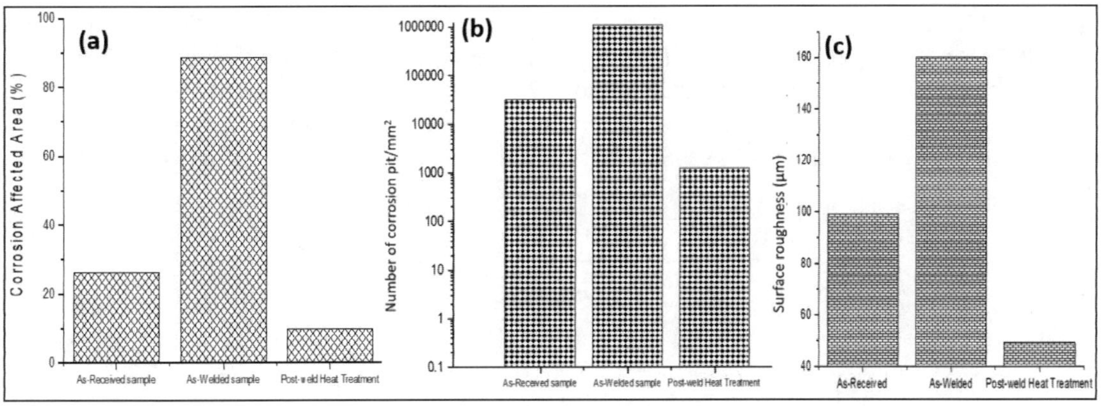

Figure 4 Mechanism of corrosion of As-received samples, As-Welded samples and Post-weld Heat treatment: (a) corrosion affected area (%), (b) number of corrosion pit/mm^2 and (c) surface roughness (µm)

3.2. Weight loss Analysis

Weight loss experiments were conducted in accordance with the ASTM methods described previously. Gravimetric measurements were carried out in an electrolysis cell equipped with a thermostat-cooling condenser. The carbon steel specimens used have a rectangular form of 40mm x 30mm x 1.6mm. After the immersion period, the specimens were cleaned according to ASTM G-81 and reweighed in order to determine the corrosion rate [26]. Duplicate experiments were performed in each case, and the mean value of the weight loss was reported. The weight loss mechanism gives room for the calculation of the mean corrosion rate as expressed in (mg.cm^{-2}.h-1). It can be determined either by chemical analysis of dissolved metal in solution or by gravimetric method measuring the weight of specimen before and after exposure in the aggressive solution applying the following equation:

$$W(corrosion) = \frac{Mi-Fi}{st} \quad (1)$$

Where Mi, Fi, s and t denote initial weight, final weight, the surface of specimen and immersion time, respectively.

3.3. Corrosion behaviour

The computed values for resistance to corrosion of both samples after 15 days at 3 days interval using weight loss method is tabulated below.

Figure 5 shows the corrosion rate (mm/yr) for AISI 316 steel specimen immersed in 0.5M H_2SO_4 for 15 days at 3days intervals for each of the five samples. These five samples are functions of the time of their exposure to the corrosive medium at room temperature which ranges between 26 -33^{0}C. Results obtained show an indication that the rate of corrosion as the days increased from 3 to 15 days for uniform concentration (0.5M) of H_2SO_4 acid, decreased moderately for treated samples and significantly decreased for untreated samples. This is in agreement with the works of Arwati*et al.,* [27] and Iliyasu*et al.,* [28]. This unique behaviour of the medium is very much attributed to the corrosive or acidic condition of the acid at the onset of the procedure (first 3days). As time goes on, acidity tends to decrease due to its reaction with the steel specimen. The graphs indicate a curve tending towards a horizontal position to be constant no matter how long it will take.

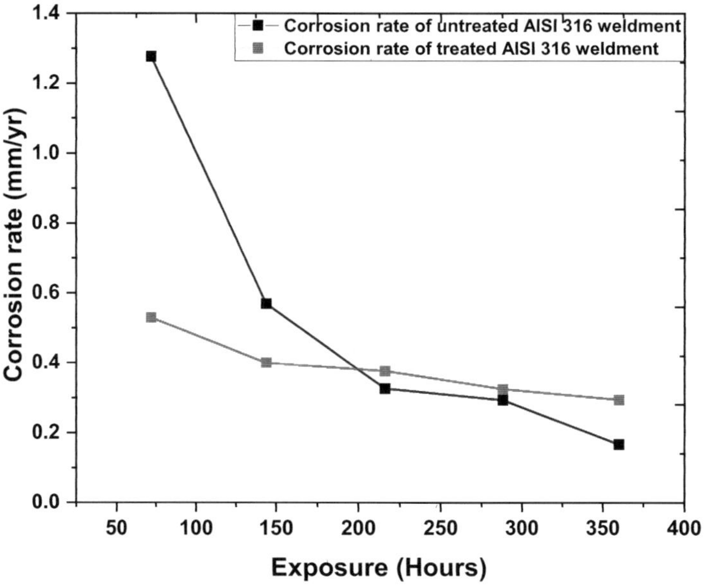

Figure 5 Corrosion rate (mm/yr) for AISI 316 steel specimens of untreated and treated weldment immersed in 0.5M H_2SO_4 for 15 days at 3days intervals for each of the five samples.

3.4.*Phase Study of research material.*

Figures 6 and 7 are the graphs generated from the phase property diagram simulation using the JmatPRO software. From the simulated data, it can be seen that the major phases of ferrite are present in the steel generally and a negligible amount of cementite is existing. The austenite commences from around 750^{0}C, this is the Ac1 temperature and the 100% transformation of austenite at 900^{0}C, that is Ac3 temperature. Therefore, according to Kou [19], the CGHAZ is generated at 1300^{0}c (Ac3<<<); FGHAZ is estimated at 900^{0}C (Ac3) and the ICHAZ is estimated at 850^{0}C (<Ac1>). The Liquid phase is starting from around 1500^{0}C. At 1400^{0}C the austenite dissolved and transforms into delta ferrite (DF) as a result of the gamma alpha transformation, which is dangerous and leads to poor mechanical properties of the steel [29-30].From the phase transformation study, it can be noted that precipitations can also be generated in these steels such as M23C6 and M7C3. The M23C6 precipitated carbides is the Cr23C6 with a mixture of molybdenum richened carbides and the M7C6 precipitates are the Cr7C3 with a rich iron in the carbides.

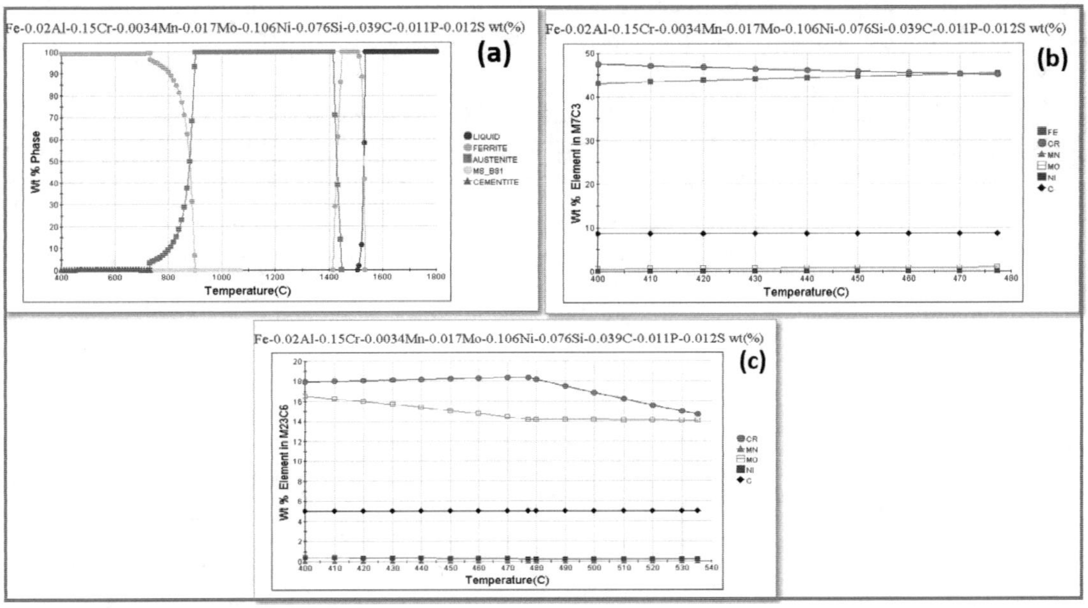

Figure 6 Phase diagrams of: (a) the simulated steel, (b) and (c) M7C3 carbides chemical composition

3.5. Carbon and chromium content

Presented in figure 7 (a) and (b) are the percentage of carbon and chromium present in the steel phases. Carbon is an interstitial alloying element in steel that can easily diffuse speedily through the structure and focus on the grain boundaries. The most challenging issues with carbon are the precipitation of chromium carbide that tends to deplete the grain boundaries of chromium, thereby posing a high risk to the steel in terms of its resistance-ability to corrosion. While the chromium content plays the role of toughens the steel and subsequently improves its resistance to corrosion at elevated temperature. These views are in agreement with the works of John and Tverberg [31]. The concentration of chromium and carbon is richer in the precipitated carbide of Cr23M7 and CR7C3. These precipitates are the main reason for creep and impact toughness enhancement improve the creep and impact toughness by blocking and creating the peening effect to stop/ slow down the dislocation movements [32-35].

Looking at this from a different perspective, as the welding progresses, the microstructure tends to distort invarious morphology with respect to their thermal effect that resulted in the HAZ formation. Precipitation of MC, Cr7C3 and Cr23C6 carbides are inevitable in this zone. Precisely, there is segregation of carbon and chromium from the ferrite and austenite matrix, giving rise to a more stable bond with chromium carbides. It was evident that there is a reduction in the percentage of chromium and carbon elements in the welded sample of experimentation than the As-received sample, which resulted in corrosion aggression on the welded samples as shown in Figure 3 and 4. The corrosion aggression in the PWHT sample possesses less susceptibility than As-received samples. This due to the available time of heat treatment at 400oC to dissolve all precipitates in the matrix. It will be of great recommendation for PWHT for welded AISI 316 to enhance its corrosion resistance-ability properties.

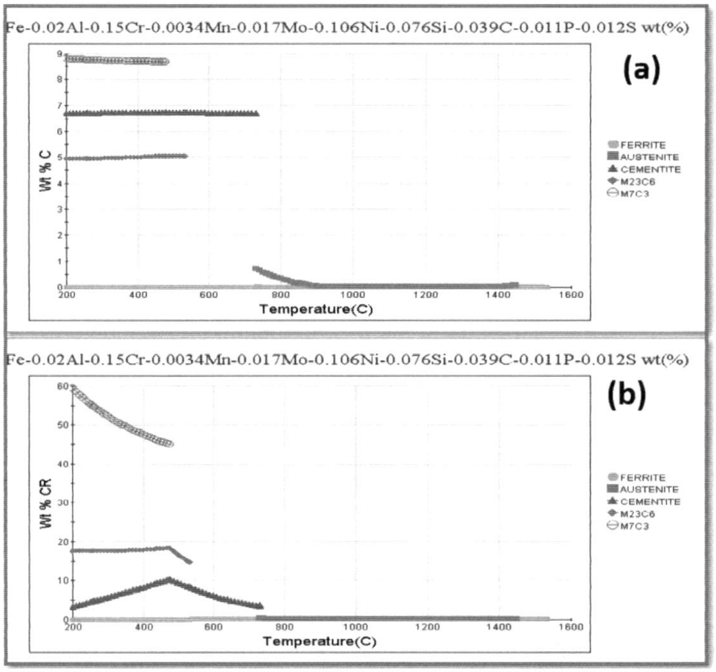

Figure 7 The percentage of: (a) Carbon present in the steel phases and (b) the percentage of Chromium present in the steel phases.

4. Conclusion

The following conclusion can be drawn from this study;

i. The corrosion resistance of post-weld heat-treated samples of AISI 316 stainless steel welded metal is higher than that of its parent metal and As-received samples.
ii. Chromium content in AISI 316 seems to be a strong determinant of corrosion resistance of the steelweldment.
iii. The general resistance to corrosion due to its low ability to form insoluble (precipitates) compounds in the acid medium was not significantly affected by Mn additions.
iv. Additionally, the formation of insoluble molybdenum, Carbon, Chromium has increased the corrosion performance of the steel.

v. Controlled post-weld heat treatment improves corrosion behaviour of an as-welded sample.
vi. Corrosion susceptibility of stainless steel in H_2SO_4 medium during exposure period could be attributed to the aggressive sulphide ion (SO_4^-) which continuously breaks down the protective film on the metal sample.
vii. Two major carbide precipitates of Cr23M7 and CR7C3 were observed after the simulation of the research material, which could pose a detrimental effect on the general performance of the steel in engineering and other industrial applications.

Acknowledgements
The authors would want to acknowledge the Mechanical Engineering Department, University of Jos, Nigeria for providing a conducive environment for the research work.

Compliance with ethical standards

The authors declare that there is no conflict of interest in the course of this research article.

References

[1] Roberge, P. R., 2019 *Handbook of Corrosion Engineering.* 3rd edition, Ed S. Scholl and D. M. Martone (New York: McGraw-Hill Education) p 2-5.
[2] Omiogbemi, I.M.B., et al., Analysis of Corrosion Behaviour of Welded Austenitic Stainless Steel Immersed in Aqueous Alkaline Medium, *British Journal of Applied Science & Technology, 2016.* ***15**(1): p. 1-8, DOI: 10.9734/BJAST/2016/23837*
[3] Sunil, K., et al., Effect of Tempering Treatments on Microstructure and Intergranular Corrosion of 13 wt% Cr Martensitic Stainless Steel. *Corrosion, 2017.* ***73**(4): p. 362-378.*
[4] Loto, C. A., et al., Inhibition Effect of Extracts of Carica papaya and Camellia sinensis Leaves on the Corrosion of Duplex (α β) Brass in 1M Nitric acid. *International journal of electrochemical Science, 2011. 6: p. 4900-4914.*
[5] Barbara, S., and K. Robert, What is corrosion? *The electrochemical society interface, 2006.* ***6**: p. 24-26.*
[6] Ahmad, E., et al., Effect of martensite morphology on tensile deformation of dual-phase steel. *Journal of materials engineering and performance, 2012.* ***21**(3): p. 382-387.*
[7] Abioye, T. E., et al. *"Effects of post-weld heat treatments on the microstructure, mechanical and corrosion properties of gas metal arc welded 304 stainless steel."* World Journal of Engineering, 2020.***7**(1): p. 87-96. https://doi.org/10.1108/WJE-11-2019-0323.*
[8] Kumar, M. V, et al. "*Stress corrosion cracking behaviour of gas tungsten arc welded super austenitic stainless-steel joints." Defence Technology, 2015.* ***11**(3): p. 282-291.*
[9] Liu, Y.J., et al., Microstructure, defects and mechanical behaviour of beta-type titanium porous structures manufactured by electron beam melting and selective laser melting, *Acta Mater. 2016.* ***113**: p. 56–67.https://doi.org/10.1016/j.actamat.2016.04.029*
[10] Dong, C.F., et al., Effects of hydrogen-charging on the susceptibility of X100 pipeline steel to hydrogen-induced cracking, *Int. J. Hydrogen Energy, 2009.* ***34**: p. 9879–9884.https://doi.org/10.1016/j.ijhydene.2009.09.090*
[11] Hou, B., et al., The cost of corrosion in China, *NPJ Mater.Degrad., 2017.* ***1**(4): p. 4–14.*
[12] Li, X.G., et al., Materials science: Share corrosion data, *Nature, 2015.* ***527**: p. 441–442.*

[13] Wei, X., et al., Electrochemical measurements and atomistic simulations of Cl^--induced passivity breakdown on a Cu_2O film, *Corros. Sci., 2018.* ***136***: *p. 119–128.https://doi.org/10.1016/j.corsci.2018.02.057*
[14] Dong, C.F., et al., Passivity Breakdown on Copper: Influence of Temperature, Macdonald, *J. Electrochem. Soc. 2016.* ***163***: *C707–C717.https://doi.org/10.1149/2.0401613jes*
[15] Gao, S. J., et al., Scanning electrochemical microscopy study on the electrochemical behavior of CrN film formed on 304 stainless steel by magnetron sputtering, *Electrochim. Acta, 2013.* ***114***: *p. 233–241.https://doi.org/10.1016/j.electacta.2013.10.009*
[16] Cheng, X., et al., Interaction between austenite-ferrite phases on passive performance of 2205 duplex stainless steel*J. Mater. Sci. Technol. 2018.* ***33***: *p. 2140–2148.https://doi.org/10.1016/j.jmst.2018.02.020*
[17] Yan, F., et al., Characterization of nano-scale oxides in austenitic stainless steel processed by powder bed fusion,*Scripta Materialia, 2018.* ***155***: *p. 104–108.*
[18] Clark, N.J.,Manual metal arc weld modelling: Part 2 Treatment of multi-pass weld, *Materials Science and Technology, 1985.* ***1*** *(12): p. 1069-1080. https://doi.org/10.1179/mst.1985.1.12.1081*
[19] Celi, R., et al., A Comparison of As-Welded and Simulated Heat Affected Zone (HAZ) Microstructures, *Mtaec9, 2016.* ***50***(*3): p. 1580-2949. doi:10.17222/mit.2016.006*
[20] Lolla, T., et al., Understanding the Role of Initial Microstructure on Intercritically Reheated Heat-Affected Zone Microstructures and Properties of Micro-alloyed Steels, *Proc. Of the 9th Inter. Conf., June 4–8, 2012 Chicago, p. 34-42.*
[21] Kou, S., 2003 *Welding metallurgy*, 2nd ed., (New York: Wiley–Interscience)p. 406
[22] Adonyi, Y., Heat-Affected Zone Characterization by Physical Simulations, *Welding Journal, 2006.* ***10:*** ***p.*** *42-47*
[23] Welding – Recommendations for welding of metallic materials – EN 1011-2 Part 2: Arc welding of ferritic steels, Accessed 20th March, 2021 from http://www.energomsw.ru/images/EN-1011-2.pdf
[24] Shieu, S. F., et al., Microstructure and Corrosion Resistance of a Type 316L Stainless Steel, *Corrosion Science,* 1998. **39**(7): p. 1267-1279
[25] Kumar, S. S., et al., Effect of friction stir welding on mechanical and microstructural properties of AISI 316L stainless steel butt joints, *Welding in the World,* 2018. https://doi.org/10.1007/s40194-018-0621-7
[26] Veisi, H., et al., Convenient One-Pot Synthesis of 2,4,5-Triaryl-1H-imidazoles from Arylaldehydes, Benzyl Alcohols, or Benzyl Halides with HMDS in the Presence of Molecular Iodine, Bull. *Korean Chem. Soc.* 2012. **33**: p. 1231.
[27] Arwati, G.I., et al., Temperature Effects on Stainless Steel 316L Corrosion in the Environment of Sulphuric Acid (H_2SO_4), *IOP Conf. Series: Materials Science and Engineering,* 2018. **343**: P. 012016 doi:10.1088/1757-899X/343/1/012016.
[28] Iliyasu et al., Corrosion Behavior of Austenitic Stainless Steel in Sulphuric Acid at Various Concentrations, *Advances in Applied Science Research,* 2012. **3** (6): p. 3909-3915
[29] Schäfer, L., Influence of delta ferrite and dendritic carbides on the impact and tensile properties of a martensitic chromium steel:*Journal of Nuclear Materials,* 1998. **258–263** (2): p. 1336-1339.
[30] Wang, L., et al., Effect of delta ferrite on impact properties of low carbon 13Cr–4Ni martensitic stainless steel:*Materials Science and Engineering: A,* 2010. **27**(13-14): p. 3210-3216
[31] John C. and Tverberg, P.E., The Role of Alloying Elements on the Fabricability of Austenitic Stainless Steel. *Metals and Materials Consulting Engineers*(Mukwonago: Wisconsin), 2018. p. 1-5
[32] Maribel L., et al., Precipitation Characterization and Creep Strength at 600°C for Creep Resistant Cr–Mo Steel,*ISIJ International*,2020. **60**(9): p. 2059-2067 ttps://doi.org/10.2355/isijinternational.ISIJINT-2019-682

[33] Fujita, T., et al., Creep Rupture Strength and Microstructure of Low C-10Cr-2Mo Heat-Resisting Steels with V and Nb. *Metall Mater Trans A,* 1981. **12**: p. 1071–1079
[34] Kumai. N. and Masuyama, F., Microstructural Degradation of the HAZ in 11Cr-0.4Mo-2W-V-Nb-Cu Steel (P122) during Creep, *SIJ International,* 2002. **42**(12): p. 1364-1370.https://doi.org/10.2355/isijinternational.42.1364
[35] Pandey, C., and Mahapatra, M.M., Effect of Heat Treatment on Microstructure and Hot Impact Toughness of Various Zones of P91 Welded Pipes. *J. of MateriEng and Perform, 2016.***25**, 2195–2210.

Harmonic Analysis of Flex Seal of Rocket Nozzle

Chinta Jagadish[*1], Ch. Siva Rama Krishna[2] and Pendyala Abhishek[3]

[1*]P. G. Scholar, Department of Mechanical Engineering, Vignan's Institute of Information Technology (A)

[2]Associate Professor, Department of Mechanical Engineering, Vignan's Institute of Information Technology (A)

[3]P. G. Scholar, Department of Mechanical Engineering, Vignan's Institute of Information Technology (A)

- Corresponding author e-mail: chintajagadish99@gmail.com

Abstract: A flex seal in a rocket nozzle provides directional change of the rocket and acts as a pressure tight seal. In the present analysis, ground test conditions and forced vibration test are simulated using 3D finite element method to check the effect of number of pads and shims on the flex seal, to study the effect of isotropic material and composite material on the flex seal for shims. The problem is modeled in ANSYS Workbench 2020 R2 student version. 3D analysis is performed for flex seal with varying design parameters (shims and pads) to perform the harmonic analysis. The constraints and loading are given as per the ground test conditions followed in industry. A maximum extended operating pressure of 9MPa is applied. Proper mesh refinement is done for reasonably acceptable finite element solution. From the design parametric study, model 5 with 9 shims and 10 pads shows better deformations and better stress values. The remaining models (1, 2, 3 and 4) are showing maximum deformations and peak stresses which are going to failed in the resonance conditions. Model 5 under harmonic analysis does not allow the structure to fail in the resonance conditions.

Keywords: Finite Element Method, Flex Seal , Rocket Nozzle, Flex Nozzle, Pivot point, Actuator, Flex seal sub assembly, ANSYS and Finite Element Method.

1 Introduction

The Flex Nozzle System mainly consists of

- Flex seal
- Pivot point
- Actuator

The detailed structure of above mentioned components are as shown below in the Fig 1.

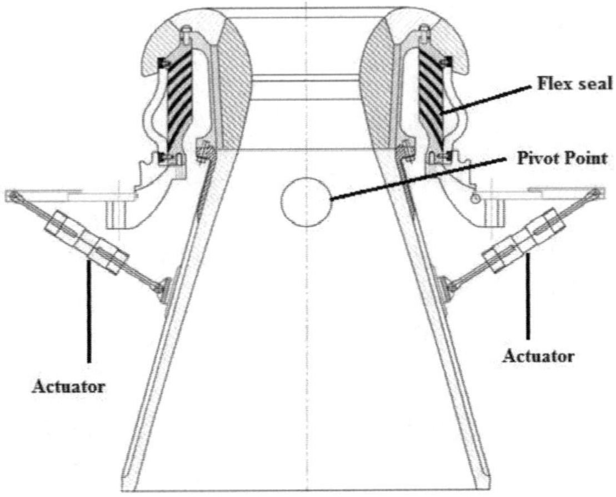

Figure 1. Flex Nozzle System (FNS)

The Flex Nozzle System consists of 4 sub assemblies i.e., Convergent sub assembly, Divergent sub assembly, Intermediate dome sub assembly and Flex seal sub assembly. John T. Herbert et al [1] invented the flexible joint. This flexible joint is invented for coupling of shafts, conduits, and more particularly to an annular joint means comprising of alternate layers of rigid reinforcing andresilient material. E.Y.Wong [2] in this a solid rocket motor with an inner diameter of 260 inch is designed, fabrication, and test of a large diameter, flight weight, and omnidirectional flexible seal meeting their requirements are presented. Form this simplified flexible seal was designed which consists of uniform, conical shapes of rubber and steel layers. Donat, James R [3] the solid rocket motors, direction are controlled by controlling the thrust vector from the nozzle. In this the flex seal core geometry is estimated. The core is analyzed for performance characteristics such as stress, weight, and the torque required vectoring the core based on the requirements/constraints. Fredrick R. et al [4]. This invention related to liquid propellant reaction engines for controlling the direction of rocket and amount of thrust required. Vector control of the craft, pitch and yaw changes, has been obtained by deflecting the exhaust stream in a desired direction by installing vanes or jetavators in the exhaust stream. Louis E. Miltenberger [5] invented a system providing a rocket powered vehicle with complete thrust vectoring capabilities including a convergent-divergent nozzle attached to a motor casing by a flexible joint member. Gerald F. Kobalter [6] told that a flexible nozzle consists of conically shaped body with alternative layers of pads and shims bonded together in a fluid tight manner. The studies regarding misalignment and thermal growth in the system due to redirect flow of applied forces due to pressure integrity are elaborated. Stanley P. Desjardins, et al [7]. The present invention represents a further advance in the art by incorporating the flexible bearing not as a separate component, but as an integral, annular portion of the divergent cone of the thrust nozzle. Gary M. Wendel [8] a thrust nozzle of the hollow-cone type fixed to a stationary shell having external insulation on the nozzle is invented. The extended part with lining of a thin light weight insulation made of carbon-epoxy liner provides more efficient energy conversion. John T. Herbert [9] the swivelled type rocket motor joint and the purpose of its exhaust guidance are presented in this paper. James R. Johnston, et al [10] in this performance of Rocket nozzle throat insert materials were investigated using three small-scale solid-propellant rocket engines. The graphite nozzles performed well with the least oxidizing 5600'F propellant but generally eroded severely with the other propellants. Fibre reinforced plastic nozzles as a class was the least erosion resistant materials.

IOP Conf. Series: Materials Science and Engineering 1185 (2021) 012031 doi:10.1088/1757-899X/1185/1/012031

2 Problem Description

2.1 Problem Objective & Scope of the work

To simulate the structure of the flex seal of a rocket nozzle under harmonic loads using finite element analysis with reference to the ground test conditions by varying number of shims and elastomeric pads. The scope of the work is to develop and verify 3D finite element models in ANSYS software and performing non-linear static analysis for studying the normal stresses (Axial & Radial) and directional deformations (Axial & Radial).

2.2 Geometry

The geometrical dimensions of the flex seal of the rocket nozzle [2] for the present analysis which are considered are shown in Figure 2.

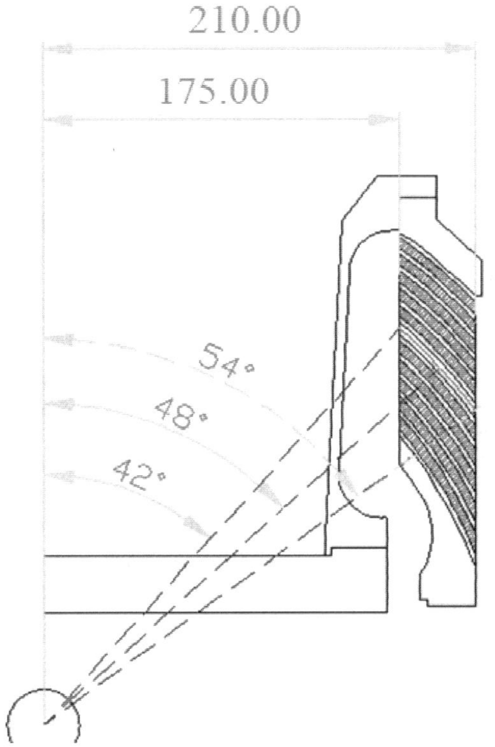

Figure 2. Flex seal Geometry

Diameter (Inner): 350 mm	Diameter (Outside): 420 mm	Full height : 190 mm
Joint angle (Inner): 42 degrees	Joint angle (Outer): 54degrees	Joint angle(Mean): 48 degrees
Radius of pivot: 248 mm	Diameter(Throat): 160 mm	Geometry of Shim: Spherical

The cross sectional (2D) view of the Flex Seal (5 Shims and 6 Pads) with parts and 3D view are shown in the below Figure 3.The list of the flex seal models used for the harmonic analysis and their respective quantity of shims and pads and their thickness are as shown in Table 1.

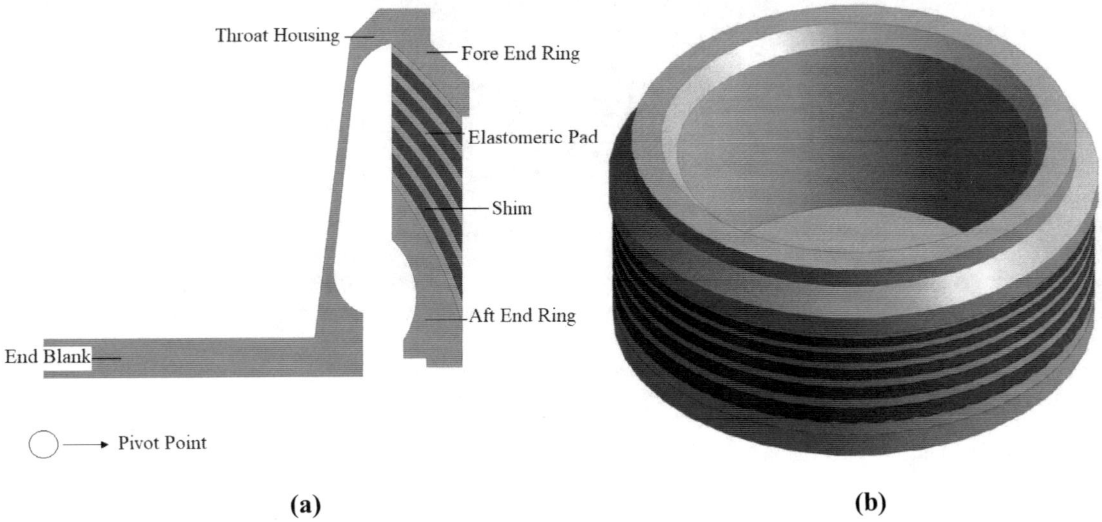

Figure 3. (a) 2D and (b) 3D Geometry of Flex Seal

Table 1 Geometric Model Details

Model Number	Quantity		Thickness (mm)		Weight (Kg)
	Shims	Elastomeric Pad	Shim	Elastomeric Pad	
Model 1	5	6	11	6.67	64.958
Model 2	6	7	10	5	66.199
Model 3	7	8	9	4	66.945
Model 4	8	9	8	3.45	66.958
Model 5	9	10	7	3.2	67.197

2.3. Meshing

Meshing is the process of converting the geometrical entities to finite element entities, such as nodes and elements. Here two types of elements are used for the analysis i.e., solid 186 and solid 187.Solid 186 is used for the flat portions and Solid 187 is used for the curved portions of the flex seal of the rocket nozzle.The three dimensional mesh model of the flex seal of the rocket nozzle having 5 shims and 6 pads are shown in the Figure 4.

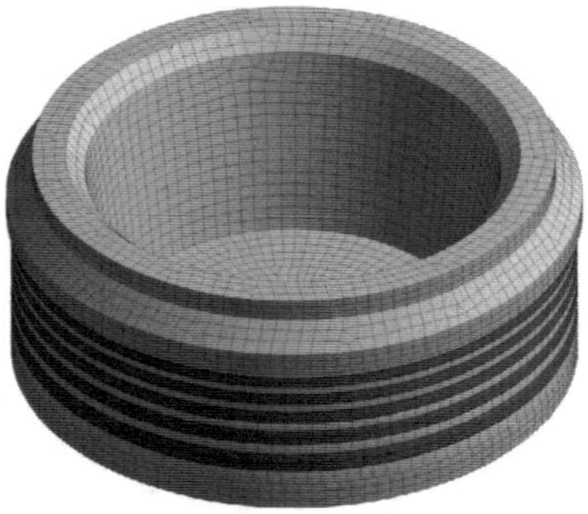

Figure 4.Flex Seal with mesh

A cylindrical coordinate system is assigned to the geometry of the flex seal with all models. X-Axis is assigned to the radial direction, Y – Axis is assigned to the circumferential direction and Z-Axis is assigned to the longitudinal direction.

2.4 Material Properties

Steel and mooney rivlin 2 parameters rubber material are used for harmonic analysis. Fore end ring, aft end ring, throat housing & Shims are assigned with steel material, Elastomeric pads are assigned with mooney rivlin two parameters and shims are assigned with the steel. Steel material has density of 5.68 kg/m^3, Young's Modulus of 2.1e5, poisson's ratio of 0.3 and yield strength of 1000 MPa and Mooney rivlin parametersare C01 = 1.0 MPa, C10 = 1.85 MPa with density of 0.8 kg/m^3.

2.5 Loads and Boundary Conditions

A normal pressure of 9 MPa [1] is applied on external surface of the flex seal of the rocket nozzle and bottom surface of the Aft end ring is fixed in all degree of freedoms, therefore a fixed support constraint is given to the bottom surface of the aft end ring which are shown in the below Figure 5.

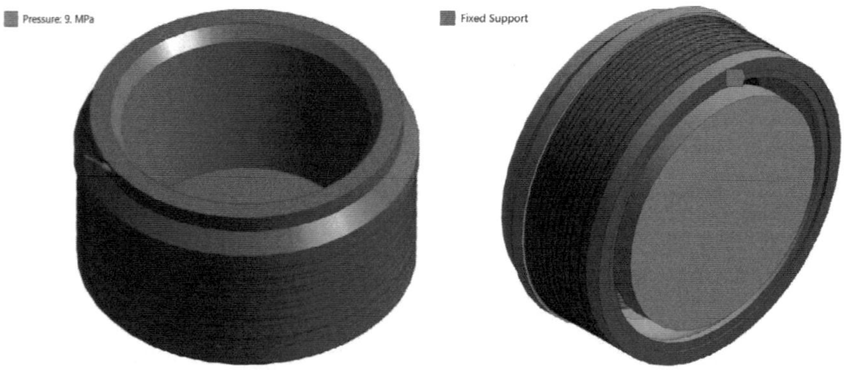

Fig 5. Load & Boundary Conditions.

2.6 Connection

The individual components of the flex seal are glued together to form an assembly. So, in this analysis bonded contacts are given to the individual parts of the flex seal.As discussed in the previous session elastomeric pads contain material non linearity properties. So solution type of nonlinear analysis is to be performed on the flex seal.

3. Harmonic Analysis

In this paper we perform Harmonic analysis on the flex seal of the rocket nozzle [Model1] with a harmonic load (Pressure) of 9 MPa with in a frequency of range 0 to 1000 Hz. For harmonic analysis, a modal analysis is required to perform to find the natural frequencies of the flex seal within the range of 0 to 1500 Hz and natural frequencies are listed in Table 2.

Table 2. Natural Frequencies of the Flex Seal [Model 1]

Mode No	Frequency (Hz)	Mode No	Frequency (Hz)
1	126.45	11	767
2	126.55	12	963.67
3	148.05	13	963.81
4	315.61	14	985.91
5	339.33	15	1014.5
6	339.65	16	1015
7	535.87	17	1186.2
8	536.03	18	1224.4
9	546.88	19	1224.8
10	765.63	20	1303.4

From the above table, it is shown that a total of 20 natural frequencies are obtained between the ranges of 0 to 1500 Hz. The directional deformation of the flex seal [Model 1] shown in the fig 6.0. which subjected to harmonic load

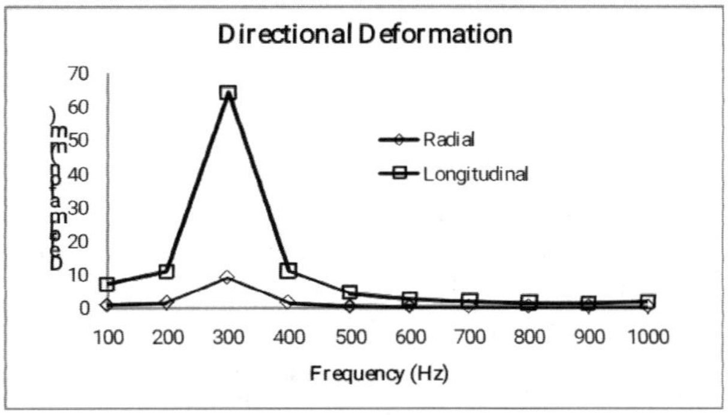

Figure 6. Directional Deformation of the flex seal [Model 1]

From the above Figure 6, it is shown that the directional deformations of the flex seal in longitudinal direction are more than the deformations of the flex seal in the radial direction. The maximum deformations are occurred at the 300 Hz frequency which is very close to the fourth natural frequency of the structure. The normal stresses of the flex seal [Model 1] shown in the Figure 7 which subjected to harmonic load

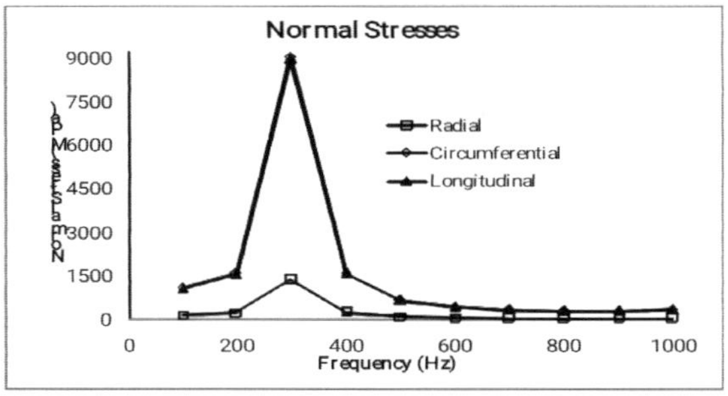

Figure 7. Normal Stresses of the flex seal [Model 1]

From the above figure it is shown that the normal stresses of the flex seal in circumferential direction are more than the stresses of the flex seal in the radial and longitudinal directions. The maximum deformations are occurred at the 300 Hz frequency which is very close to the fourth natural frequency of the structure. Harmonic analysis is performed on the flex seal of the rocket nozzle [Model 2] and natural frequencies are listed in Table 3.

Table 3. Natural Frequencies of the Flex Seal [Model 2]

Mode No	Frequency (Hz)	Mode No	Frequency (Hz)
1	134.06	11	771.35
2	134.1	12	1015.3
3	156.94	13	1015.8
4	307.92	14	1015.9
5	332.74	15	1016.1
6	332.95	16	1042.8
7	560.35	17	1165.7
8	560.42	18	1249.9
9	574.78	19	1250.1
10	771.09	20	1273.8

From the above table, it is shown that a total of 20 natural frequencies are obtained between the ranges of 0 to 1500 Hz.The directional deformation of the flex seal [Model 2] shown in the fig 8. which subjected to harmonic load

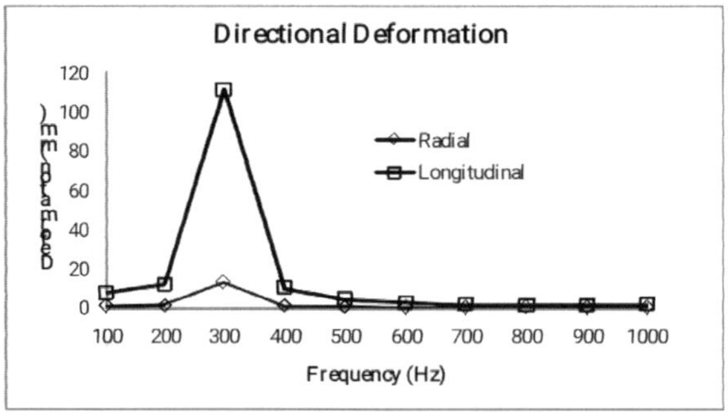

Figure 8. Directional Deformation of the flex seal [Model 2]

From the above figure it is shown that the directional deformations of the flex seal in longitudinal direction are more than the deformations of the flex seal in the radial direction. The maximum deformations are occurred at the 300 Hz frequency which is very close to the fourth natural frequency of the structure. The normal stresses of the flex seal [Model 2] shown in the fig 9. which subjected to harmonic load

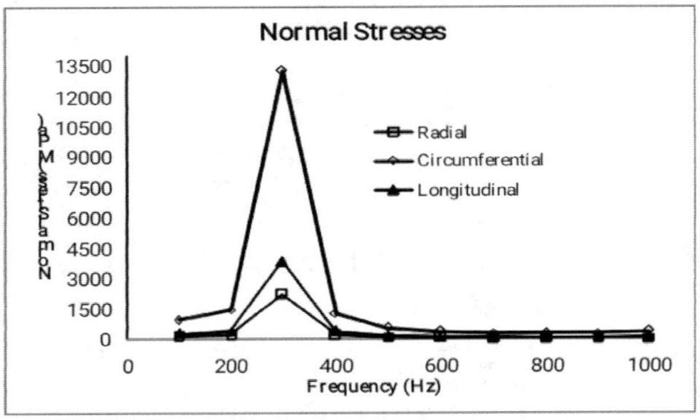

Fig 9. Normal Stresses of the flex seal [Model 2]

From the above figure it is shown that the normal stresses of the flex seal in circumferential direction are more than the stresses of the flex seal in the radial and longitudinal directions. The maximum deformations are occurred at the 300 Hz frequency which is very close to the fourth natural frequency of the structure. Harmonic analysis is performed on the flex seal of the rocket nozzle [Model 3] and natural frequencies are listed in table 4.

Table 4. Natural Frequencies of the Flex Seal [Model 3]

Mode No	Frequency (Hz)	Mode No	Frequency (Hz)
1	175.26	8	881.78
2	175.84	9	899.87
3	202.55	10	954.37
4	407.4	11	954.5
5	414.67	12	1182.5
6	414.95	13	1182.6
7	881.55		

From the above table, it is shown that a total of 13 natural frequencies are obtained between the ranges of 0 to 1500 Hz. The directional deformation of the flex seal [Model 3] shown in the fig 10. which subjected to harmonic load

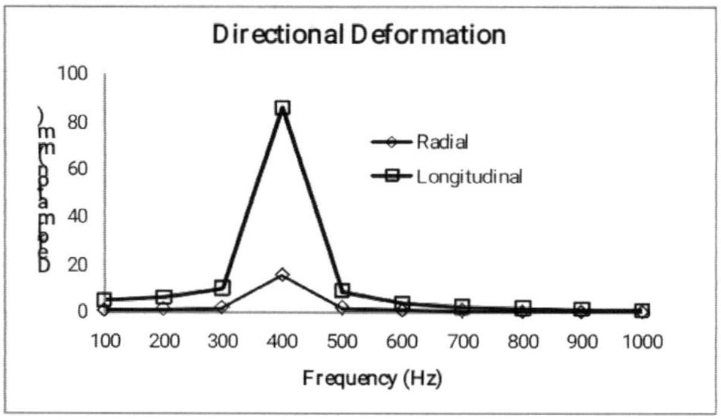

Fig 10. Directional Deformation of the flex seal [Model 3]

From the above Fig10 ,it is shown that the directional deformations of the flex seal in longitudinal direction are more than the deformations of the flex seal in the radial direction. The maximum deformations are occurred at the 400 Hz frequency which is very close to the fourth natural frequency of the structure. The normal stresses of the flex seal [Model 3] shown in the fig 11. which subjected to harmonic load.

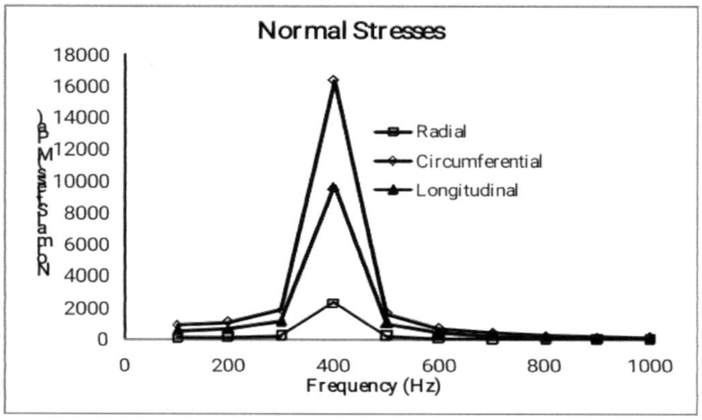

Fig 11. Normal Stresses of the flex seal [Model 3]

From the above figure it is shown that the normal stresses of the flex seal in circumferential direction are more than the stresses of the flex seal in the radial and longitudinal directions. The maximum deformations are occurred at the 400 Hz frequency which is very close to the fourth natural frequency of the structure. Harmonic analysis is performed on the flex seal of the rocket nozzle [Model 4] and natural frequencies are listed in table 5.

Table 5. Natural Frequencies of the Flex Seal [Model 4]

Mode No	Frequency (Hz)	Mode No	Frequency (Hz)
1	287.36	7	1038.1
2	288.02	8	1038.2
3	338.27	9	1308.3
4	534.33	10	1308.5
5	534.61	11	1326.9

6	534.84

From the above table, it is shown that a total of 11 natural frequencies are obtained between the ranges of 0 to 1500 Hz.The directional deformation of the flex seal [Model 4] shown in the fig 12. which subjected to harmonic load,

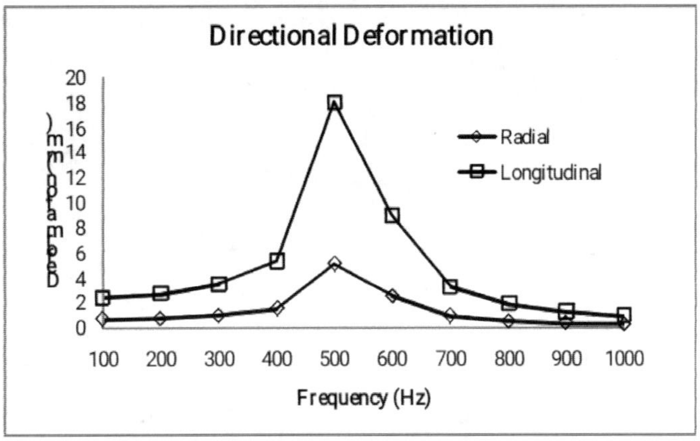

Fig 12. Directional Deformation of the flex seal [Model 4]

From the above figure it is shown that the directional deformations of the flex seal in longitudinal direction are more than the deformations of the flex seal in the radial direction. The maximum deformations are occurred at the 500 Hz frequency which is very close to the fourth natural frequency of the structure. The normal stresses of the flex seal [Model 4] shown in the fig 13. which subjected to harmonic load

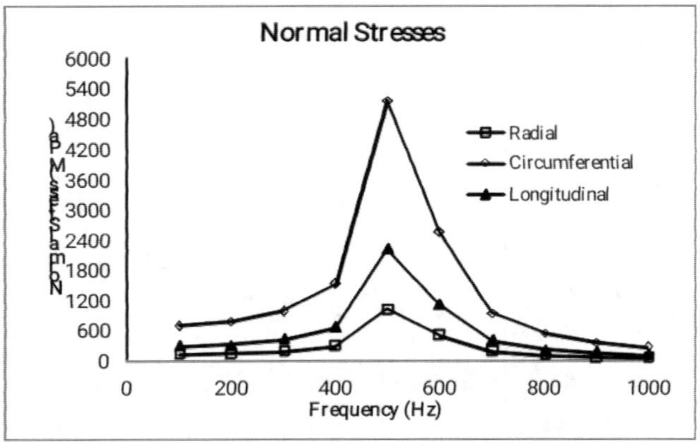

Fig 13. Normal Stresses of the flex seal [Model 4]

From the above figure it is shown that the normal stresses of the flex seal in circumferential direction are more than the stresses of the flex seal in the radial and longitudinal directions. The maximum deformations are occurred at the 500 Hz frequency which is very close to the fourth natural frequency of the structure. Harmonic analysis is performed on the flex seal of the rocket nozzle [Model 5] and natural frequencies are listed in table 6.

Table 6. Natural Frequencies of the Flex Seal [Model 5]

Mode No	Frequency (Hz)
1	893.37
2	894.41
3	1378.8

From the above table, it is shown that a total of 3 natural frequencies are obtained between the ranges of 0 to 1500 Hz. The directional deformation of the flex seal [Model 5] shown in the fig 14 which subjected to harmonic load,

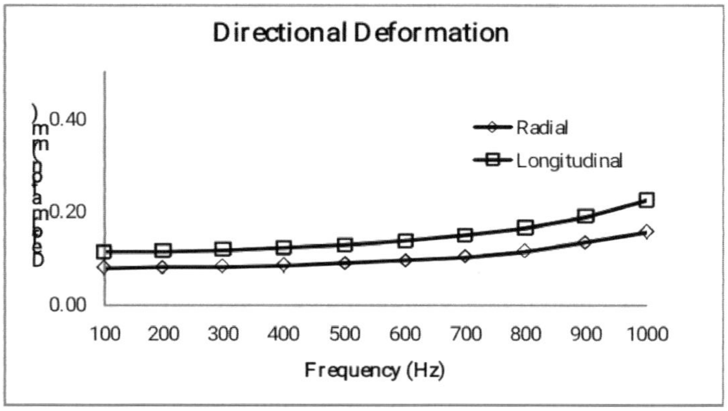

Fig 14. Directional Deformation of the flex seal [Model 5]

From the above figure it is shown that the directional deformations of the flex seal in longitudinal direction are more than the deformations of the flex seal in the radial direction. The maximum deformations are occurred at the 1000 Hz frequency. The normal stresses of the flex seal [Model 5] shown in the fig 15. which subjected to harmonic load

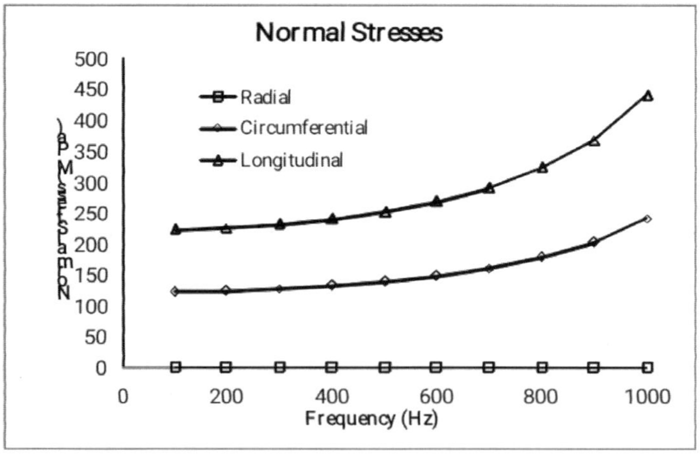

Fig 15. Normal Stresses of the flex seal [Model 5]

From the above fig15. , it is shown that the normal stresses of the flex seal in circumferential direction are more than the stresses of the flex seal in the radial and longitudinal directions. The maximum deformations are occurred at the 1000 Hz frequency.

4. Conclusions

From the present work, harmonic analysis is performed on the flex seal of the rocket nozzle and the results of directional deformations in radial & longitudinal directions and normal stresses in radial, circumferential & longitudinal directions are calculated. The results shows that the directional deformations are higher in the longitudinal direction compared to the radial direction, normal stresses in circumferential directions are higher than the radial & longitudinal directions. The design model 1, 2, 3 & 4 are going to fail in strength point of view under harmonic loads in circumferential direction due to resonance. The design model 5 is in safe position under strength point of view and exhibits better stiffness performance and does not undergoes resonance condition.

References:

[1] Eswara Kumar A., Somanadha Sastry K. Manideep K. and Priyanka M., " Dynamic Analysis of Flex Seal of Solid Rocket Motor Nozzle", *5th International Conference of Materials Processing and Characterization (ICMPC 2016).*

[2] Prakash D, Balakrishna Murthy V., Chandra MohanR., and Karteek N.., " Effect of Material Mismatch on Static behavior of Flex Seal", *5th International Conference of Materials Processing and Characterization (ICMPC 2016).*

[3] Eswara Kumar, Balakrishna Murthy V. and Chandra MohanR, " Effect of Shim Percentage on Non-Linear Static Behavior of Flex Seal of Rocket Nozzle", *4th International Conference of Materials Processing and Characterization (ICMPC 2016).*

[4] Eswara Kumar, Balakrishna Murthy V.,Chandra Mohan R, and Prakash D. " Study on Non – Linear Static Behavior of flex seal of rocket nozzle by varying number of shims", *4th International Conference of Materials Processing and Characterization (ICMPC 2016).*

[5] John Herbert, Redlands, Frank J. Kovitch, Jr, River side, and Max McCorkle."Flexible Joint Means". *United States Patent no. 3,390,899*. July 2, 1968.

[6] E.Y. Wong. "Design, Fabrication, And Test Of Omnidirectional Flexible Seals For Thrust Vector Control Of Large Solid Rocket Motors". *NASA CR 72889, final report*, Aero jet Solid Propulsion Company, Advanced Technical Operations, Sacramento. California. 10 June 1971.

[7] Donat, James R. "Solid Rocket Motor Nozzle Flex seal Design Sensitivity". *AIAA, AHS, and ASEE, Aerospace Design Conference*. Irvine, CA. page no.16-19, Feb 1993.

[8] Fredrick R. Hickerson, Newton, N.J."Variable Thrust Rocket Engine Incorporating Thrust Vector Control", *United States Patent no. 3,192,714*, July 6, 1965.

[9] Louis E. Miltenberger. "Thrust Vectoring Control System for Rocket Nozzles", *United States Patent no. 3,726,480*, Apr. 10, 1973.

[10] Gerald F. Kobalter. "Flexible Nozzle For A Thrust Vector Control System", *United States Patent no. 3,860,134*, Jan. 14, 1975.

[11] Stanley P. Desjardins, Jonathan W, Wilson, Alan R.Canifield. "Movable Thrust Nozzle for rockets", *United States Patent no. 3,696,999*, Oct. 10, 1972.

[12] Gary M.Wendel. "Thrust Nozzle With Insulation", *United States Patent no.4, 649,701*, Mar. 17, 1987.

[13] John T.Herbert. "Swiveled Rocket Motor Nozzle", *United States Patent no. 3,302,885*, Feb. 7, 1967.

IOP Conf. Series: Materials Science and Engineering 1185 (2021) 012031 doi:10.1088/1757-899X/1185/1/012031

[14] James R. Jobnston, Robert A. Signorelli, and John C. Freche. "Performance Of Rocket Nozzle Materials With Several Solid Propellants", *NASA TN-D 3428*Lewis Research center, Cleveland, Ohio, National Aeronautics And Space Administration, Washington, D.C., May 1966,.
[15] Ebrahim Harirbafan, Mohsen Mohseni Shakib, AliAkbar Karimi, Mojtaba Heydari. "Analytical Solution and Simulation of Flexible Joint in Thrust Vector Control System under Bending", Pawtucket Island, 2006.
[16] Woodberry. "Solid Rocket Thrust Vector Control", *NASA SP-8114*,December, 1974
[17] Woodberry. "Flexible joints for thrust vector controls", *AIAA-75- 1221*, October, 1975
[18] Gajbir Singh and Rao, G.V. "New formulas for design of flex bearing", *Journal of Spacecraft*, vol.30, No 6, 1991.
[19] Wood berry, R.F.H. "Rapid design equations for elasto-metallic joints". *Memo to J.E Mcneer, Ref. No. Misc./6/50 – 1113*, Hercules Inc./Bacchus, Nov.30, 1973
[20] Wilson, J. W. "Final report, Poseidon C3 joint venture joint". *Rep. TWR-2486*, Thiokol chemical corp. /Hercules Inc., July 1967
[21] Galford, J.E. "unstable joint analysis – Poseidon investigation task II-A". *Hercules Inc. /magna*, February 1971

Bioactive titanium composites for bone implant applications

Emani Siva Krishna [1], *, G. Suresh [1], B. Ratna Sunil [2], Ravikumar Dumpala [3]

[1] Department of Mechanical Engineering,
Vignan's Foundation for Science, Technology & Research, Vadlamudi 522213, India
[2] Department of Mechanical Engineering,
Bapatla Engineering College, Bapatla 522101, India
[3] Department of Mechanical Engineering,
Visvesvaraya National Institute of Technology, Nagpur 440010, India

* Corresponding author e-mail: sivakrishna.emani@gmail.com, bratnasunil@gmail.com

Abstract: Development of titanium (Ti) based implants has revolutionized the application of materials engineering to address several challenges associated with other metallic implants. Ti exhibit excellent biocompatibility, nontoxicity and lowers the stress shielding effect. However, bio-inert nature of titanium is a limitation and therefore, several surface engineering strategies are adopted to promote bioactivity. Higher bioactivity helps to improve the healing rate by accelerating the implant tissue interactions. On the other hand, developing Ti based composites by incorporating bioactive ceramic phases is another promising route to develop bioactive Ti implants. In the present review, the role of addition of bioactive ceramic phases on tailoring the properties of Ti for bone implant applications is presented. Different processing routes to incorporate bioceramic particles into Ti are also discussed with a brief discussion on future scope of Ti based composites for biomedical implant applications.

Keywords: FSP, MMCs, hybrid composites, machining, hardness.

1. Introduction

Metal are the best candidates to produce bio-implants particularly for load bearing applications in orthopaedics and allied fields. Among the available metallic implants, Titanium (Ti) has emerged as a potential load bearing material to manufacture bone fixing implants and joint replacement implants [1]. Titanium possesses excellent biocompatibility, nontixicity and nancarcinogenicity properties [2, 3]. Titanium also exhibits lower stress shielding effect compared with steels and Co based implants when used in load bearing orthopaedic applications. Stress shielding is a phenomena in which the bone that is supported by a metallic implant is subjected to lower level of bone remodelling due to the lower stresses developed in the bone compared with the implant. This phenomenon is profound when the difference between the young's modulus of the implant material and the bone is higher. Among the available metallic implants, Ti shows mechanical properties close to that of natural bone and hence the mismatch in the mechanical properties of bone and the implant is lower which reduces the stress shielding effect [4]. Fig 1 compares the young's modulus of different metals used in bio-implant applications [4].

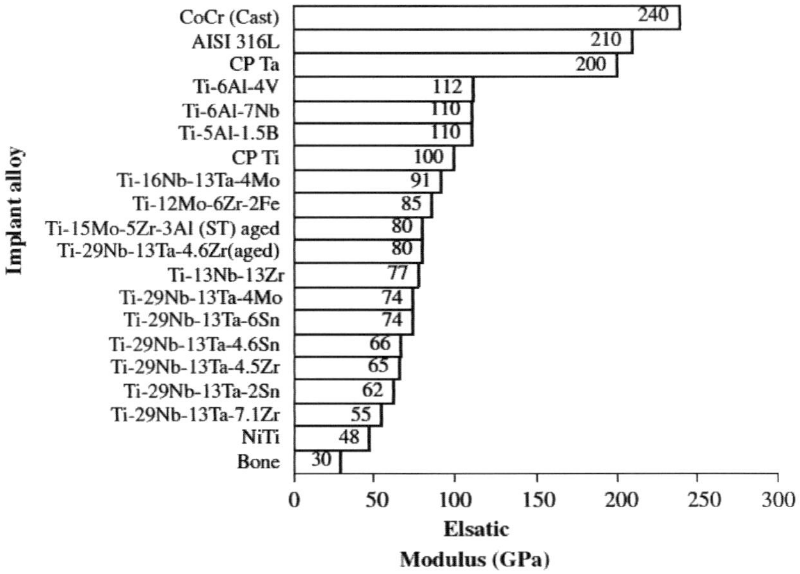

Fig 1 Comparison of Young's modulus of different metals used in biomedical applications [Source: Geetha et al., [4]]

Compared with Co-alloys and steels, pure Ti exhibits nearly 50% reduction in the elastic modulus value. Further by alloying certain elements such as Ni, Ta and Nb, the developed Ti alloys demonstrate elastic modulus values close to that natural bone. Hence, from the mechanical properties point of view, Ti based implants decreases stress shielding in load baring implant applications. However, Ti suffers from lack of bioactivity which influences the interactions between the tissue and the implant surface. There are several strategies such as developing nanostructured surfaces, providing surface coatings, introducing bioactive phases into the surface reported in the literature to improve the bioactivity of Ti [5, 6, 7, 8, 9]. On the other hand, several other reports also demonstrated the superior bioactivity resulted from the addition of ceramic particles as dispersing phases in Ti [10, 11, 12, 13]. The present review provides a brief summary of the state of the art in developing Ti based composites with enhanced bioactivity for bone fixing implant applications in biomedical engineering.

2. Titanium based composites as bio-implants

Interest on Ti as suitable candidate to manufacture biomedical implants arose from its excellent biocompatibility and corrosion resistance in the highly corroding environment. The formation of a thin oxide layer on the surface of Ti implant provides corrosion resistance against the aggressive chemical events in the biological environment. Additionally, the oxide layer that is developed on the Ti surface helps to enhance the bonding ability of the tissue with the implant surface by which, bioactivity of the surface is accelerated. Among the available Ti alloys, commercial grade -2 pure Ti (medical grade) and Ti-6Al-4V alloys are used to manufacture medical implants. Ti based implants are mainly used to support the hard tissue or to replace the hard tissue in bone implant applications. Furthermore, Ti based materials are used in total joint replacement, dental and maxillofacial implants. Composites exhibit hybrid properties and offer more flexibility to choose different composition to achieve desired properties in the end product. The dispersing fibre in the form of particles or fine powder imparts specific properties to the bulk composite. The main objective in developing Ti based composites targeted for biomedical applications is to introduce additional properties which are inferior in Ti such as bioactivity, bio-mineralization and ability to develop strong bonding with the local host tissue.

Therefore, the selection of reinforcing particles to develop the composite is crucial. It is well established fact that the dispersing particle size, shape and chemical composition influence the bulk properties of the composite materials. When it comes to biomedical applications, the dispersing phase that is used to develop the composite also must have the necessary properties to be used as a biomaterial. Hence, it is always better to select approved materials to be used in biomedical applications. Calcium based ceramic phase are well accepted and widely used in the biomedical applications. Among the available bioceramic phases, apatite is the most promising material due to its close resemblance with the natural calcium based phase that is available in the bone. By using apatite as a dispersing phase into Ti, Additional properties such as bioactivity and improved osseointegration can be achieve for the implant. Other ceramic phases such as TiN, TiB, TiC, Nb_2O_5 TiO_2, SiO_2, ZrO_2 and SrO were also used to develop Ti based composites for biomedical applications.

Thirugnanam et al. [10] developed Ti based composites by using nano-hydroxyapatite (nHA) as the dispersing phase. The powders were mixed in appropriate compositions and subjected to high energy ball milling followed by sintering. The crystallite size after ball milling was found to be influenced by milling speed (rpm), milling medium and time, ball to power ratio etc. From the XRD studies, broader peaks were identified which is an indication of decreased crystallite size. The authors have studied the role of added nHA on bioactivity by conducting immersion studies in simulated body fluid (SBF) for one week. The studies demonstrated the promising effect of addition of nHA on enhancing the deposition of apatite from the SBF which is an indication of enhanced bioactivity. From the cell adhesion studies by using human osteoblast cells showed excellent cell adhesion and growth on the surface of Ti-nHA composites compared with commercial pure Ti.

Calin et al., [14] added Nb in $Ti_{75}Zr_{10}Si_{15}$ and $Ti_{60}Nb_{15}Zr_{10}Si_{15}$ alloys and studied the role of addition of Nb on the bioactivity by conducting immersion studies in Ringer's solution. From the observations, they have reported superior bioactivity for the Ti based metallic glasses compared with commercial pure Ti. Furthermore, pitting corrosion was found to be insignificance on the produced metallic glasses compared with the commercial pure Ti. Zhang et al. [15] developed Ti-TiC functionally graded composites with different amount of TiC addition. It was observed that the change in yield strength and ultimate tensile strength was insignificant with the increased amount of TiC. On the other hand, ductility was significantly measured as decreased with the increased TiC content.

Developing *in-situ* composites by promoting chemical reaction within the material is another way of producing Ti based composites. There are several reports in producing Ti based *in-situ* composite. Developing in –situ composites offer additional benefits compared with the conventional processes to develop the composites where externally reinforcements are added. Higher level of strong bonding can be achieved in the Ti based composites produced by in-situ route. Furthermore, Thermodynamic stability is also increased with better interfacial bonding between the secondary phase particles and the matrix. The uniform distribution of the particles within the matrix resulted from *in-situ* reaction helps to achieve better mechanical properties and other bulk properties in the Ti composites [16, 17]. Gu et al., [18] developed *in-situ* Ti composite that has Ti_5Si_3-TiN dispersing phases. From the investigations, They found that the microhardness of the developed in-situ composite is 4 times higher than the base Ti due to the strengthening effect introduced by the intermetallic produced during the in-situ reaction. Attar et al., [19, 20, 21, 22] have done extensive studies on developing in-situ Ti based composites and various studies have been carried out such as the effect of powder morphology, mechanical behaviour, wear properties and demonstrated the superior behaviour of these composites compared with their counter parts.

Balla et al. [23] developed Ti-TiO2 functionally graded composites by LENS techniques and the efficacy of these composites as biomaterials was studied. From the results, higher wettability, and enhanced tissue – implant interactions and better wear resistance were demonstrated. On the other hand, Bandyopadhyay et al. [24] incorporated calcium phosphate (CaP) mineral phase into titanium by

LENS with an aim to produce bioactive implants. From their observations, it was clearly shown that the presence of Ca-P decreases the rate of material loss due to wear by developing a Ca-P tribofilm on the surface of the implant. Additionally, the incorporated Ca-P has also enhanced the yield strength (compressive) and ultimate strength (compressive) and also hardness. The wear test data clearly demonstrate the superior wear resistance for the composites. Studies were also reported in the literature to incorporate hydroxyapatiet (HA) into Ti and Ti6Al7Nb to promote osseointegration and higher bioactivity at the surface [25, 26].

Recently, Li et al., [27] developed Ti- Nb_2O_5 composite by powder metallurgy route. Increased strength was measured for the composites compared with Ti. When the composites are exposed to human osteoblast like cells, excellent tissue implant interactions and cell adhesion was observed for the composites. Improved mechanical properties and bioactivity for the composites were attributed to the presence of Nb_2O_5.

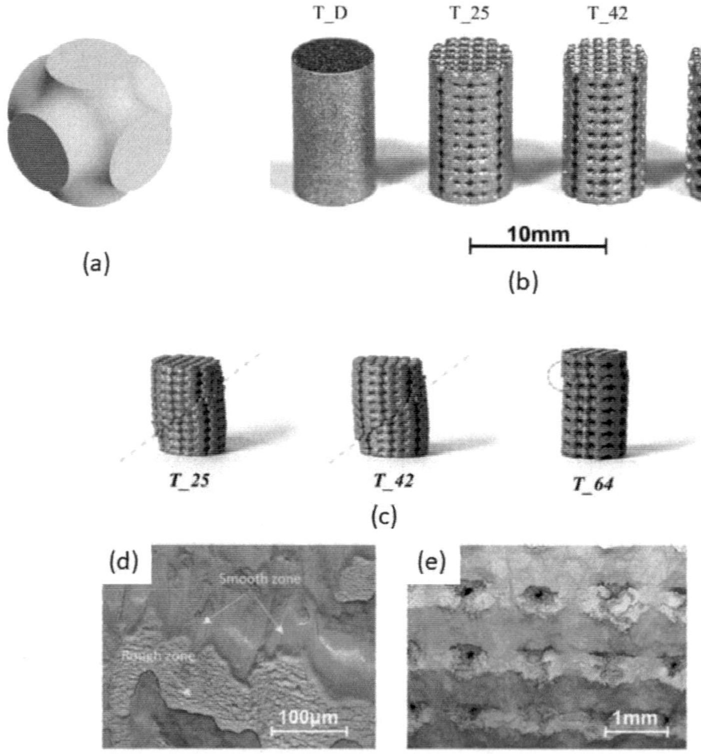

Fig 2 a) 3D modal of the network connectors, b) scaffolds with different % of porosity, c) Scaffolds subjected to compressive loads, d) SEM image of Ti with 0% porosity and e) SEM image of Ti with 25% porosity (Source: Soro et a., [28]).

Several successful attempts were also made to develop porous scaffold structures of Ti based composites as demonstrated in the literature. Soro et al., [28] studied the promising benefits of porous Ti scaffolds for biomedical applications by selecting three levels of porosity (25 %, 42% and 64%). Three dimensional model was created as shown in Fig 2 (a) and the same has been used to produce 3D porous structures as shown in Fig 2 (b). Then the porous scaffolds were produced by selective laser melting (SLM) technique and subjected to compression (Fig 2 (c)). The scaffolds after compression test were observed by using SEM as shown in Fig 2 (d) and (e). Similarly, Xiang Li et al. [29] also developed a composite scaffold contains a silk fibroin (SF) sponge on functionally graded titanium (FG-Ti). With the same level of porosity, FG-Ti exhibited better radial gradation and enhanced mechanical properties compared with homogeneous porous Ti. The composite structure was observed as similar to that of bone structure. The bio-mimicking nature of the developed composite scaffold improved the bioactivity as observed from the cell adhesion studies. Fig 3 shows comparison of rat osteoblast cells cultured on the surface of homogeneous porous Ti and FG-Ti. It can be observed that the cell spreading was higher in the FG-Ti.

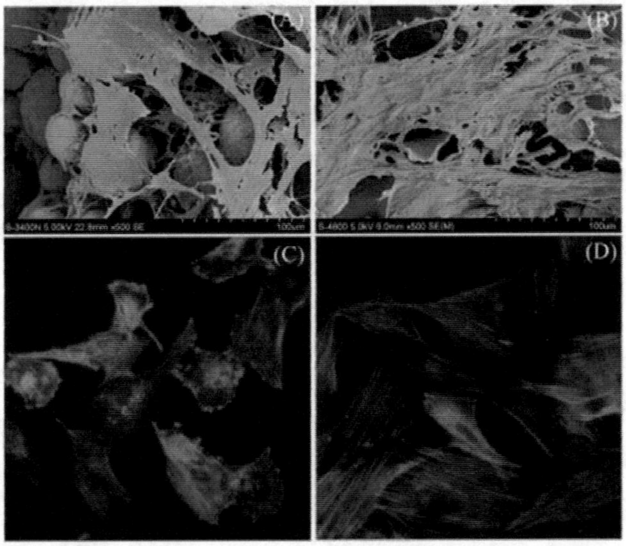

Fig 3 Rat osteoblast cells on the surface of the samples: a) SEM image of Ti, b) SEM image of FG-Ti composite scaffold, c) Confocal laser scanning microscopy of Ti and d) FG-Ti composite scaffold (Source: Li et al., [29]).

3. Challenges and future perspectives

The available literature demonstrates the promising benefits which can be obtained by incorporating appropriate ceramic phases in the form of particles and discontinuous fibres. The selection of reinforcing phase is crucial to impart specific properties to the developed Ti composite. Several ceramic phases are available as reinforcing materials to develop conventional composites. However, limited number of reinforcements only can be used as dispersing phases due to the restriction from the biocompatibility and toxicity issues as the intended use of these composites is for medical applications. Furthermore, the concentration and the distribution of the dispersing phases in Ti are also crucial as the higher amounts of reinforcements may alter the bulk properties of the composite. On the other hand, if liquid processing routes are selected to develop Ti composites, stability of the ceramic phases that are reinforced into the liquid Ti is crucial to study. For example, hydroxyapatie (HA) is a calcium phosphate ceramic phase that can be used as dispersing phase into Ti to improve the bioactivity and osseointegration. However, the stability of HA at higher temperatures is an important concern as the HA is unstable if heated above 1000°C, particularly if Ca fraction is decreased in the HA phase. On the other hand, solid state processing routes such as powder metallurgy can help to complete the manufacturing of Ti based composites without affecting the phase stability of reinforcements. Developing porous structures can be an advantage with powder metallurgy as implants with porosity can perform well in bone fixing applications to promote higher healing rate. However, shape and size of the produced composite are the limitations with powder metallurgy routes. Developing Ti based composites by allowing in-situ chemical reactions is another promising strategy

IOP Conf. Series: Materials Science and Engineering 1185 (2021) 012032 doi:10.1088/1757-899X/1185/1/012032

which helps to obtain stronger bond between the resulted secondary phase and Ti. However, the limitation with in-situ process is from the restrictions of number of phases that can be produced by chemical reactions. Certain bio-ceramic phases which are attractive to alter the Ti matrix properties cannot be produced by in-situ reactions such as calcium-phosphorous based phases. Very recently, additive manufacturing (AM) techniques are widely adopted to produce highly porous, network structures with different cell (open or close) morphologies suitable for bone implant applications [30]. Adopting AM techniques help to produce functionally graded composites with multi phases to impart combination of properties into the Ti composite. By adopting AM techniques, complex shapes of Ti based implants can be manufactured. Even though the AM techniques have demonstrated excellent results, the developments are gradual compared with developments in conventional Ti alloy studies.

4. Conclusions

Developing Ti based composites addresses several issues associated with the bio-inert nature of Ti implants in the biomedical industry. Introducing secondary phases into Ti matrix and achieving phase stability is challenging if the processing routes involve heating of the materials to higher temperatures. In-situ chemical reaction helps to develop matrix compatible intermetallic phases and excellent matrix-fibre bonding can be achieved. However, the number of phase combinations is limited. On the other hand, powder metallurgy route and advanced manufacturing methods such as additive manufacturing facilitate to produce novel Ti based composites by using wide range of secondary phases. From the available information in the literature, addition of reinforcements improves the performance of Ti as biomaterial and future developments certainly help the engineers to produce new age Ti based implants with hybrid properties with higher healing rate for the biomedical applications particularly for load bearing bone implant applications.

References

[1] Park J, Lakes R S 2007 *Biomaterials an Introduction* (Springer - New York, USA).
[2] Jemat A, Ghazali M J, Razali M, Otsuka Y 2015 *BioMed Research Intl*, 2015, 791725.
[3] Cochran D L, Schenk R K, Lussi A, Higginbottom F L, Buser 1998 *J Biomed Mater Research* **40** 1–11.
[4] Geetha M, Singh A K, Asokamani R, Gogi A K 2009 *Prog Mater Sci* **54** 397–425.
[5] C. Wen 2015 *Surface coating and modification of metallic biomaterials* (Elsevier - Cambridge, UK)
[6] Truong V K, Stuart R, Lapovok R, Estrin Y, Wang J Y, Berndt C C, Barnes D G, Fluke C J, Crawford R J, Ivanova E P2009 *Appl Microbiol Biotechnol* **83** 925–937.
[7] Khang D, Lu J, Yao C, Haberstroh K M, Webster T J 2008 *Biomaterials* **29** 970–983.
[8] Zhang L, Webster T J 2009 *Nano Today* **4** 66–80.
[9] Ratna Sunil B, Thirugnanam A, Uday Chakkingal, Sampath Kumar T S 2016 *Mate Technol* **31(13)** 743-755.
[10] Thirugnanam A, Kumar and T S, Uday Chakkingal 2010 *Trends Biomater. Artif. Organs,* **24** **(3)** 146–154.
[11] Mehmet Topuz, Burak Dikici, Mehmet Gavgali, 2021 *J Mech Behav Biomed Mater*, **118**, 104480.
[12] Kenneth Kanayo Alaneme, Oluwaseun P. Adu, Samuel Ranti Oke, Oluwasegun Eso Falodun, Peter Apata Olubambi 2020 *J King Saud Univ - Eng Sci*, **22** (in press),
[13] Wen Shi, Akira Kamiya, Jun Zhu, Akira Watazu 2002 *Mater Sci Eng: A* **337(1–2)**, 104-109.
[14] Mariana Calin, Annett Gebert, Andree a CosminaGhinea, Petre Flaviu Gostin, Somayeh Abdi, Christine Mickel, Jürgen Eckert 2013 *Mater Sci Eng: C* **33** 875-883.

[15] Zhang Y, Wei Z, Shi L, Xi M, *J Mater Process Technol* **206** 438-444.
[16] Gu D, Hong C, Meng G 2012 *Metall Mater Trans* **43** 697-708.
[17] Tjong S C, Ma Z 2000 *Mater Sci Eng R Rep* **29** 49-113.
[18] Gu D, Hagedorn Y C, Meiners W, Wissenbach K, Poprawe R 2011 *Surf Coating Technol* **205** 3285-3292.
[19] Attar H, Bonisch M, Calin M, Zhang L-C, Scudino S, Eckert J 2014, *Acta Mater* **76** 13-22
[20] Attar H, Bonisch M, Calin M, Zhang L C, Zhuravleva K, Funk A, Scudino S, Yang C, Eckert J 2014 *J Mater Res* **29** 1941-1950
[21] Attar H, Prashanth K G, Zhang L-C, Calin M, Okulov I V, Scudino S, Yang C, Eckert J 2015 *J Mater Sci Technol* **31** 1001-1005.
[22] Attar H, Ehtemam-Haghighi S, Kent D, Okulov I, Wendrock H, Bönisch M, Volegov A, Calin M, Eckert J, Dargusch M 2017, *Mater Sci Eng A* **688** 20-26.
[23] Balla V K, De Vas Con Cellos P D, Xue W, Bose S, Bandyopadhyay A 2009 *Acta Biomater* **5** 1831-1837.
[24] Bandyopadhyay A, Dittrick S, Gualtieri T, Wu J, Bose S 2016 *J Mech Behav Biomed Mater* **57** 280-288.
[25] Han C, Wang Q, Song B, Li W, Wei Q, Wen S, Liu J, Shi Y 2017 *J Mech Behav Biomed Mater* **71** 85-94.
[26] Marcu T, Todea M, Maines L, Leordean D, Berce P, Popa C 2012 *Powder Metall* **55** 309-314.
[27] Yuncang Li, Khurram S. Munir, Jixing Lin, Cuie Wen 2016 Bioact Mater **1** 127- 131.
[28] Nicolas Soro, Hooyar Attar, Xinhua Wu and Matthew S. Dargusch 2019 Mater Sci Eng A, **745** 195-202
[29] Xiang Li , Yin-Ze Xiong, Hang Zhang, Rui-Ning Gao 2021, Mater Let **282** 128670.
[30] Hooyar Attar, Shima Ehtemam-Haghighi, Nicolas Soro , Damon Kent, Matthew S. Dargusch 2020, *J Alloy Compound* **827** 154263

Estimation of johnson-cook constitutive model constants from sheppard and jockson model

V V Kondaiah [1], D Sameer Kumar [1], Vasantha Rao N [2], Santhosh Kumar P [2], Jagu S Rao[2*]

[1]Department of Mechanical Engineering, Bapatla Engineering College, Bapatla, Andhra Pradesh, India
[2]Department of Mechanical Engineering, Rajiv Gandhi University of Knowledge Technologies, Nuzvid, Andhra Pradesh, India.

* Corresponding author e-mail: jagusrao@rguktn.ac.in

Abstract. This work introduces a novel approach for determining the material constants of the Johnson Cook constitutive model, which is used to estimate flow stress when a material is open to higher strain rates, stresses, and temperatures. In this study the material characteristics/constants of the Johnson-Cook constitutive model were derived by using another flow stress model which was introduced by Sheppard and Jackson. The input data was collected from published hot compression test results. This method can potentially reduce the experimental efforts required for evaluating the data of the Johnson-Cook model. In this work aluminum alloys 7075-T651 and Ti-6Al-4V have been considered for analysis. These materials have their mechanical properties at high temperatures and are of excellent significance for new manufacturing procedures that use deformation to produce a bonding in the solid state, such as friction stir welding and magnetic impulse welding. The flow stress variation of the two models and similar characteristics have been observed at some points over the temperature range. The strain rate constant (c) and temperature softening coefficient (m) for each material have been determined and results were compared with experimental data which were found in the literature. By adopting this method it is possible to eliminate torsion test to know the strain rate effect on a material. These data could also be used to simulate possibly the best processes as forging , rolling at higher temperatures, and creep.

Keywords. JohnsonCook constitutive model, Zener -Hollomon parameter, Sheppard and Jackson model, Aluminum alloys, material constants, friction stir welding

1. Introduction

A typical model consisting of the plastic behavior in the current generation manufacturing processes, such as friction stir welding (FSW, magnetic impulse welding and other well recognized processes, such as hot rolling, forging to estimate reliable material properties, strains, and strain rates at high temperatures are needed. There is always a need for the newer techniques as the current techniques have lack of experimental measurements at high temperatures, high strain rates, and high strains. The problems involved in measuring mechanical properties at high temperatures are also high.

In the case of friction stir welding (FSW), reproducing the conditions anticipated in the shear layer experimentally is exceptionally hard, and the community agrees that one of the most difficult challenges to bypass the creation of effective and reliable FSW simulations is a shortage of available mechanical properties data. Atomistic simulations can be used to obtain mechanical behaviour of monolithic metals and alloys, however simulations of alloys with grain sizes higher than 1 μm are extremely difficult.

In the absence of real measurements at high temperatures, strain rates, and strains, realistic extrapolations based on trustworthy prepared models are the next best choice. For hot deformation of materials, the Johnson-Cook material model and the constitutive model presented by Sheppard and Jackson are the most extensively used.

For materials subjected to huge strains, high strain rates, and high temperatures, Johnson and Cook (1983) established a constitutive model . This model requires data from torsion testing over a wide range of strain rates, static tensile tests, and dynamic Hopkinson bar tests at elevated temperatures for the material constants. Maheswari et al., (2013) established a dimensionless constitutive modelling containing the relationship between stress, strain rate and temperature is the prominent input for assessing thermo mechanical deformation processing results. The proposed model uses the results of hot compression test. Huang et.al. (2012) proposed a paradigm to determine material parameters Johnson-Cook constitutive model. The direct method and optimization method identifying constitutive model parameters are proposed.

Tello et al. (2010) presented newer constants for the Sellars and Tegart constitutive model for hot metalworking with various metals, which had previously been unpublished. Dorogoy et al. (2009) used the shear compression specimen (SCS) to determine the Johnson Cook material parameters, which included the recognition of the thermal softening effect in quasi-static and dynamic loading, including the strain rate hardening effect in dynamic loading. Slais et al. (2012) proposed a method for determining five parameters of Johnson-Cook constitutive equation for the Ti-6Al-4V titanium alloy.

Mousavi et al. (2010) studied the hot compression behaviour of a medium carbon low alloy steel, analyzing the impact of Zener-Hollomon parameters, strain, and strain rate on the flow stress and estimating the material's deformation activation energy. Yang Hui et al. (2006) used hot compression experiments to investigate the warm-hot deformation behaviour of 20CrMnTi steel. The influence of the Zener–Hollomon parameter on microstructures and mechanical characteristics in pure Cu were explored by Li et al. (2008) using a method including varied plastic deformation with multiple strain rates and at different temperatures. On a Gleeble-1500 mechanical testing machine, Ou, L., Nie, Y., and Zheng, Z. (2014) investigated isothermal hot compressive deformation in the temperature range of 350-500 $^{\circ}$C at strain rates ranging from 0.01-10/s up to a real strain of 0.9. The flow stress rapidly climbed to a maximum value. With increase in deformation temperature and decreasing strain rate, the peak stress was decreased. A Zener-Hollomon parameter with an Arrhenius term can be used to model the effects of strain rate and temperature on hot deformation behaviour. Ning et al. (2018) [11] predicted machining forces under various cutting conditions using the identified J-C constants and then compared them to the corresponding experimental forces. There was a high level of agreement between predicted and experimental forces. Song et al. (2019)[12] adapted the Johnson–Cook constitutive model to define the deformation behavior of the material when it was exposed to high temperatures and strain rates. Ballistic impact testing was used to verify the correctness of the improved model. The hot deformation behaviour of aluminum matrix composites was investigated by Neelima et.al., (2010) [13] developing constitutive equations using Johnson–Cook (JC), modified JC (m-JC), Arrhenius, and modified Zerilli and Armstrong (m-ZA) models, and the capacity of these models was estimated and compared using average absolute error.

In this work a comparison of both Johnson-Cook constitutive model and Sheppard-Jackson model was carried out by using MATLAB software for two materials Al7075-T651 and Ti-4Al- 6V. With these observations, the material constants strain rate constant (c) and temperature softening coefficient (m) of Johnson-Cook constitutive model have been determined and the results were compared with known values which were found in the literature.

1.1. Johnson-Cook Constitutive Model

Gordon R. Johnson and William H. Cook proposed this model in 1983 for materials subjected to significant strains, high strain rates, and high temperatures. The model's basic form is as shown

$$\sigma(\dot{\varepsilon}p, \varepsilon p, T) = [A + B\varepsilon p^n]\left[1 + c\frac{\dot{\varepsilon}p}{\dot{\varepsilon}o}\right]\left[1 - \left[\frac{T - Troom}{Tmelt - Troom}\right]^m\right] \quad (1)$$

Where

σ is the flow stres

εp is the equivalent plastic strain

$\frac{\dot{\varepsilon}p}{\dot{\varepsilon}o}$ is the dimensionless plastic strain rate for $\dot{\varepsilon}o = 1.0s^{-1}$

$\frac{T - Troom}{Tmelt - Troom}$ is the homologous temperature.

A, B, n, C, m are the material constants.

In this model the data for the material constants is collected from torsion tests over a broad spectrum of strain rates, static tensile tests, dynamic Hopkinson bar tensile tests and Hopkinson bar tests at elevated temperatures. The following sections provide detailed information about these tests.

1.1.1. Torsion Test

Torsion tests across a wide variety of strain rates are the primary source of Johnson-Cook material data. The state of the test specimen is well defined, high strains may be reached without forming geometric instabilities, and small to big range of strain can be obtained with the same testing technique, are some important features.

1.1.2. Dynamic Hopkinson Bar Tensile Test

This is also a important test in getting the data of the Johnson-Cook material model. This is done over a range of temperatures. The higher temperatures can be achieved by surrounding the in-place test specimen in an oven such that the temperatures are gradually given for several minutes before to testing. It is also possible to test materials to larger strains; the Hopkinson bar tensile test cannot be accurately evaluated after necking begins to occur in the tensile specimens. In addition, the effects of adiabatic heating at significant strains might make the results even more complicated. The materials are softening as a result of the higher temperatures.

1.1.3Tensile Test

Static tensile tests should also be used in order to obtain test data of the Johnson-Cook model. The stress for the tension test data is based on the neck current area, and the strain is referred as $\ln(A_0/A)$, where A_0 and A represent the initial and areas of the neck.
Static tensile testing should also be used to get test results for the Johnson-Cook model. Because of the presence of hydrostatic tension, this is the case. The Bridgman correction factor is used to approximate the equivalent tensile flow stress obtained from the tension measurements. Usually both tensile and torsion tests have to be conducted such that the disagreement between two modes of deformation can be recognized and rewarded.

1.2 Sheppard-Jackson Constitutive Model

Constitutive equation introduced by sheppard and Jackson

$$\sigma_1(T,\dot{\varepsilon}) = \frac{1}{\eta} ln \left\{ \left(\frac{Z(T,\dot{\varepsilon})}{A}\right)^{1/n} + \left[1 + \left(\frac{Z(T,\dot{\varepsilon})}{A}\right)^{2/n}\right]^{1/2} \right\} \quad (2)$$

Where Z is the Zener-Hollomon parameter:

$$Z(T,\dot{\varepsilon}) = \dot{\varepsilon}\ exp\left(\frac{Q}{RT}\right) \quad (3)$$

Where

T is temperature

$\dot{\varepsilon}$ is the strain rate

Q is the activation energy

R is gas constant

A,n, η are the material constants.

The Zener-Hollomon parameter is used to relate the flow stress to the deformation strain-rate and temperature. It can be used to describe the material properties during solid state joining, such as friction stir welding, cold spray and magnetic impulse welding.

2. Study of Constitutive Models

2.1. Material Constants

Material constants are used to describe the constitutive model for a particular material. Experimental work has been done by many researchers to derive the data of the constitutive models. Material constants for few materials was given in Table 1.

Table 1. Material constants for Different materials [9,10,11]

Model	Material	A s^{-1}	N	Q KJ mol^{-1}	η Pa^{-1}	R	Melting point,K
Sheppard and Jackson	Ti-6Al-4V	$3.92x10^6$	3.6	176	$2.13x10^{-8}$	8.314	1877
Sheppard and Jackson	Al7075-T651	$1.03x10^9$	5.41	129	$1.41x10^{-8}$	8.314	750
Johnson and Cook	Ti-6Al-4V	802	995	0.01	0.5	0.6	1877
Johnson and Cook	Al7075-T651	527	575	0.017	0.72	1.61	750

2.2. Flow Stress Variation

When a material is subjected to large strains, high strain rates and high temperatures its flow stress varies. Many models were proposed for this flow stress variation with temperature and strain rates. Johnson-Cook constitutive model and Sheppard-Jackson model were the most useful in estimating this flow at particular temperature and strain rate. The flow stress variation for Al7075-T651 using both models at different strain rates is given in Figure 1.

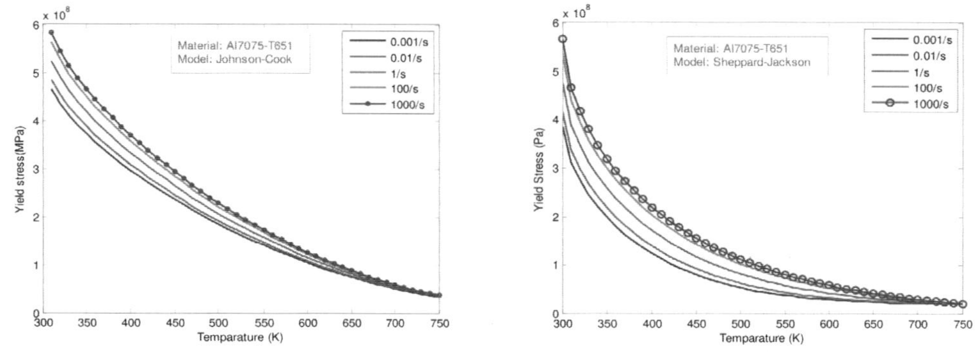

Figure 1. Yield stress vs Temperature for AL7075-T651 by both models at different strain rates

The flow stress variation for Ti-4Al-6V using Johnson-Cook model and Sheppard-Jackson model at different strain rates are given in Figure 2.

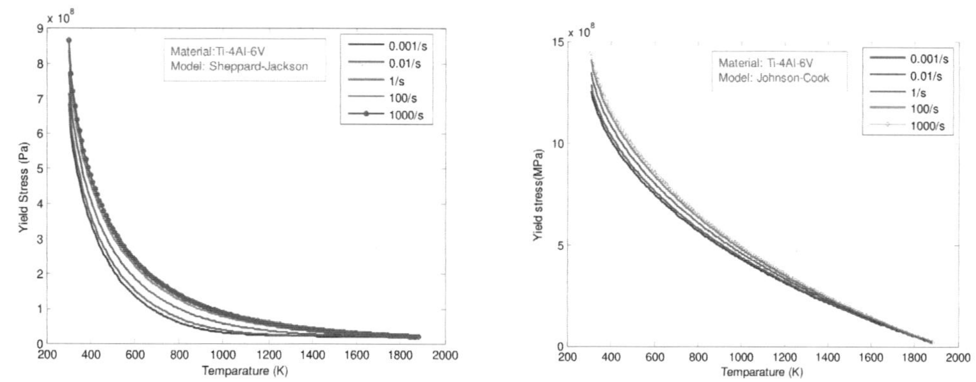

Figure 2. Yield stress vs Temperature for Ti-4 Al-6V by both models at different strain rates

2.3 Comparison of Flow Stress

From the above graphs (Figure 3.) it can be say that the value flow stress is increasing with strain rate at a temperature in both models. We have increased strain rate value from $1s^{-1}$ to $100s^{-1}$ and the corresponding increment in flow stress in both models has been drawn in flowing figures.

Constitutive equation given by sheppard and Jackson is

$$\sigma_1(T,\dot{\varepsilon}) = \frac{1}{\eta} ln\left\{\left(\frac{Z(T,\dot{\varepsilon})}{A}\right)^{1/n} + \left[1 + \left(\frac{Z(T,\dot{\varepsilon})}{A}\right)^{2/n}\right]^{1/2}\right\} \tag{4}$$

First derivative of flow stress with respect to temperature is given as

$$\frac{\partial \sigma_{Sf}}{\partial T} = -\frac{1}{\eta}\frac{1}{\left(\frac{Z}{A}\right)^{1/n} + \left[1+\left(\frac{Z}{A}\right)^{2/n}\right]^{1/2}}\left[\frac{1}{n}\left(\frac{Z}{A}\right)^{\frac{1-n}{n}} + \frac{1}{2\sqrt{1+\left(\frac{Z}{A}\right)^{2/n}}}\frac{2}{n}\left(\frac{Z}{A}\right)^{\frac{2}{n}-1}\right]\dot{\varepsilon}\, e^{\frac{Q}{RT}}\frac{Q}{RT^2}\frac{1}{A} \tag{5}$$

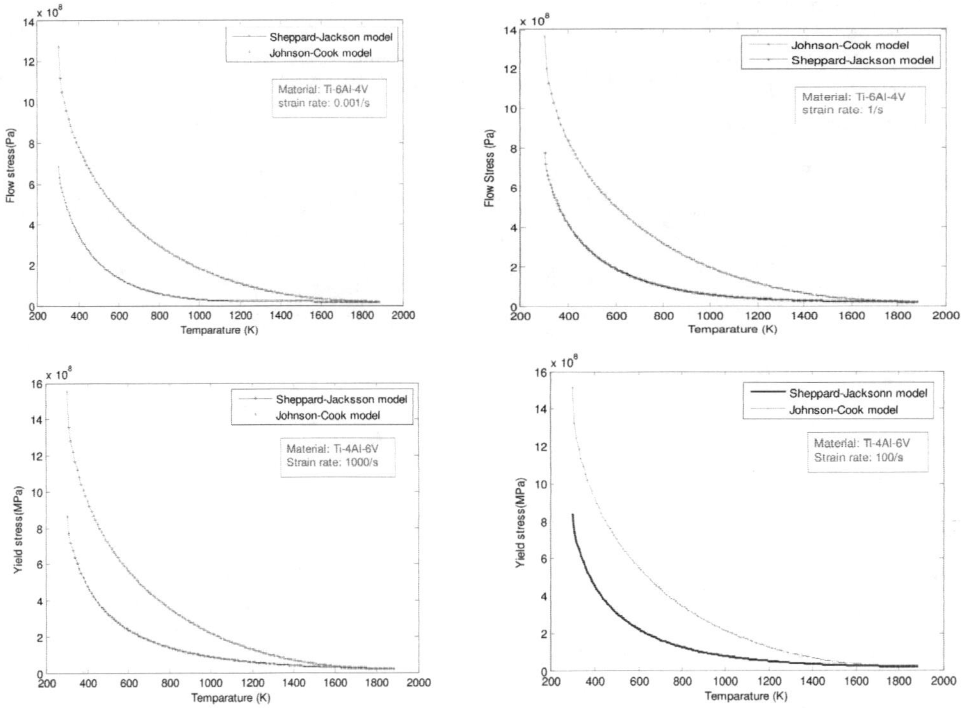

Figure 3. Comparison of Flow stress vs Temperature for Ti-6Al-4V material for both the models at different strain rates

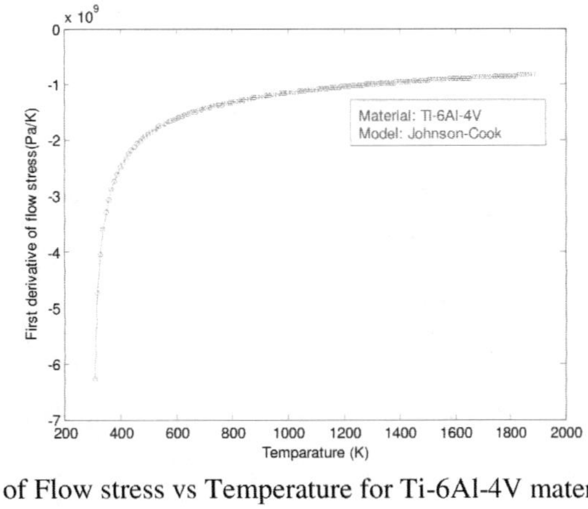

Figure 4. First derivative of Flow stress vs Temperature for Ti-6Al-4V material by Johnson-Cook model

Johnson-Cook Constitutive model is given by

$$\sigma_{JC}(\dot{\varepsilon}p, \varepsilon p, T) = [A + B\varepsilon p^n]\left[1 + c\ ln\frac{\dot{\varepsilon}p}{\dot{\varepsilon}o}\right]\left[1 - \left[\frac{T - Troom}{Tmelt - Troom}\right]^m\right] \quad (6)$$

By taking the partial derivative of first order with respect to temperature

$$\frac{\partial \sigma_{JC}}{\partial T} = -m\left(A + B\varepsilon_p^n\right)\left(1 + C\ ln\frac{\dot{\varepsilon}}{\dot{\varepsilon}_o}\right)\left(\frac{T - T_{room}}{T_m - T_{room}}\right)^{m-1}\frac{1}{T_m - T_{room}} \quad (7)$$

The variation of the its first derivative over temperature is given in the following Figure 5

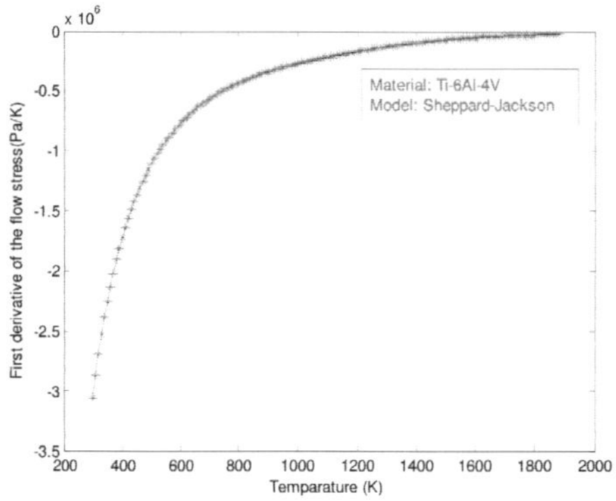

Figure 5. First derivative of Flow stress vs Temperature for Ti-6Al-4V material by Sheppard-Jackson model

From the above two figures (Figure 4 & Figure 5) it can be said that in Johnson-Cook model the flow stress is decreasing very rapidly with temperature and in Sheppard-Jackson model flow stress is decreasing gradually over the temperature.

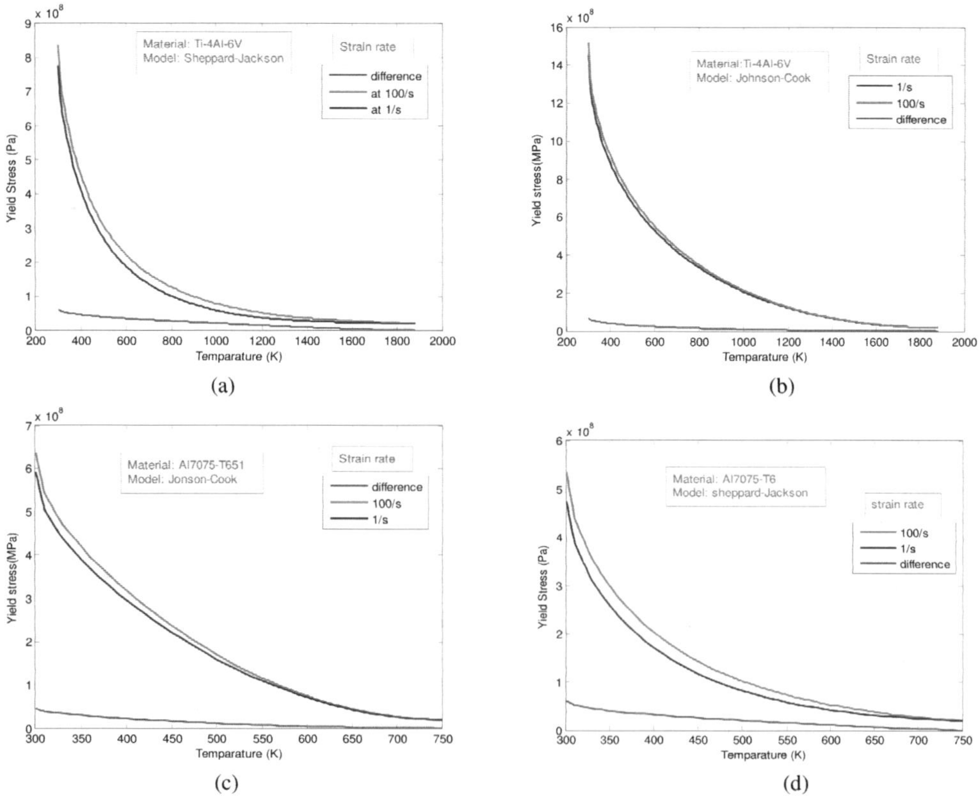

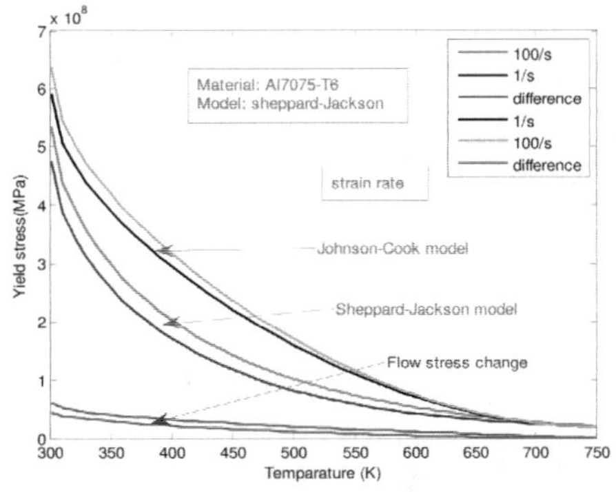

(e)

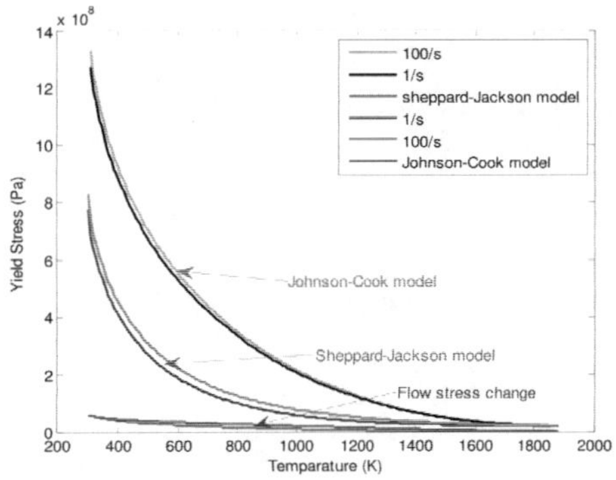

(f)

Figure 6. Comparison of flow stress for the two models

The change in flow stress value with increasing strain rate from $1s^{-1}$ to $100s^{-1}$ in both models was shown in above Figures (Figure 6 (a-f)). By using these observations the detailed procedure to determine C and m was explained in the following section.

3. Determination of Material Constants

By observing above graphs, at a particular temperature, as we increase the strain rate, flow stress is also increasing. And this increment in flow stress is proportionally equal. This characteristic of these two models can be used to predict the value of strain rate constant of the Johnson-Cook constitutive model by using data of the sheppard-Jackson model.

3.1 Determination of Strian Rate Constant C

At a particular temperature, as we increase the strain rate, flow stress is also increasing. And this increment in flow stress is proportionally equal in both the models from Fig. 3.6. with these observations the increment in flow stress at room temperature considered equal for both models and the material constants strain rate constant C and temperature softening coefficient of Johnson-Cook constitutive model have been determined as follows.

For Ti-4Al-6V using Sheppard-Jackson model
At $\dot{\varepsilon}p = 1s^{-1}$, $T=T_{room}$
Flow stress value = 7.7479×10^8 pa

Johnson-Cook constitutive model becomes

$$\sigma_{JC}(\dot{\varepsilon}p, \varepsilon p, T) = \left(A + B\varepsilon_p^n\right) \quad (8)$$

And at $\dot{\varepsilon}p = 100s^{-1}$, $T=T_{room}$

Flow stress value = 8.3285×10^8 pa

$$\sigma_{JC}(\dot{\varepsilon}p, \varepsilon p, T) = \left(A + B\varepsilon_p^n\right)(1 + C\ ln100) \quad (9)$$

From above graphs for Ti-4Al-6V subtractive equation 1 from 2

$$\left(A + B\varepsilon_p^n\right)(C\ ln100) = (8.3285\text{-}7.7479) \times 10^8 \quad (10)$$

From simple quasi-static tensile test
A=802MPa, B=9.95MPa, n=0.5, equivalent plastic strain =0.4
By substituting above values in equation 10
We get **C=0.0091**
For Al7075-T651 using Sheppard-Jackson model

At $\dot{\varepsilon}p = 1s^{-1}$, $T=T_{room}$
Johnson-Cook constitutive model becomes

$$\sigma_{JC}(\dot{\varepsilon}p, \varepsilon p, T) = \left(A + B\varepsilon_p^n\right) \tag{11}$$

And at $\dot{\varepsilon}p = 100s^{-1}$, $T=T_{room}$

$$\sigma_{JC}(\dot{\varepsilon}p, \varepsilon p, T) = \left(A + B\varepsilon_p^n\right)(1 + C\ ln100) \tag{12}$$

From above graphs for Al7075-T651 subtractive equation 12 from 13

$$\left(A + B\varepsilon_p^n\right)(C\ ln100) = (5.3549\text{-}4.7512)\text{x}10^8 \tag{13}$$

From simple quasi-static tensile test
A=527MPa, B=575MPa, n=0.72, equivalent plastic strain =0.03
By substituting above values in equation 13
We get **C=0.021**

4. Results
5.

The change in value of strain rate constant C effects flow stress variation with temperature at a strain rate. The flow stress variation and error in flow stress from experimental value is shown in the below Figure 7.

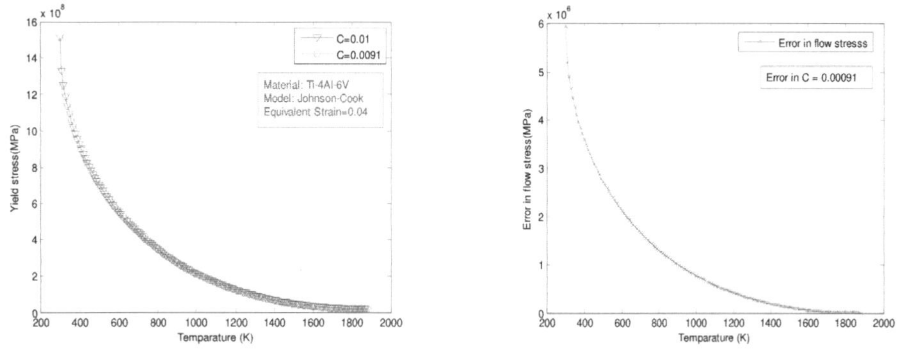

Figure 7. Flow stress variation with C=0.017 and error in flow stress for Al7075-T651

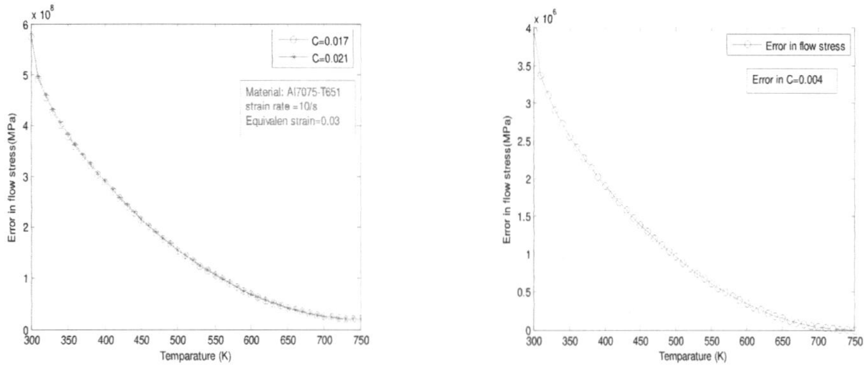

Figure 8. Flow stress variation with C=0.0091 and error in flow stress for Ti-4Al-6V

IOP Conf. Series: Materials Science and Engineering 1185 (2021) 012033 doi:10.1088/1757-899X/1185/1/012033

The above observations and calculations were summarized in the following table 2

Table 2. Material constants comparison with Experimental values

Material	Experimental	Using sheppard Jackson model	Error in value of C (%)	Maximum Error in flow stress(MPa)
Ti-6Al-4V	0.01	0.0091	9	6
Al7075-T651	0.017	0.021	23.5	4

5 Conclusions

From the above results and observations it is concluded that

- The Johnson-Cook constitutive model always gives more flow stress than Sheppard-Jackson model at a particular temperature and strain rate for any material over temperature range. From the graphs of first derivative of the flow stress
- Flow stress of the material using Johnson-Cook model is decreasing with temperature rapidly. But it was decreasing gradually if use Sheppard and Jackson model. Flow stress is increasing with strain rate at a particular temperature in both models and it was observed that the increment in flow stress is proportionally equal in the both models.
- The values of C and m for the materials have been calculated from the Sheppard and Jackson model. The results were well approximated with the know values which were found in the literature. The value of C for Ti-4Al-6V is found to be 0.0091 and for Al7075-T651 is 0.021and the error from the experimental value is found to be low.

Generally to determine material constants of Johnson-Cook constitutive model three tests namely torsion test, Split Hopkinson bar test and quasi-static tensile test have to be conducted. This work can eliminate torsion test which is used to find strain rate constant “C”. If we are able to find thermal softening coefficient “m” in future by any method with doing experiment, it is also possible to avoid split Hopkinson bar test.

These results can also be used to model the plastic behavior of the material in FSW joining processes and also in well established process such as Creep , hot rolling, and forging.

References

[1] Johnson G. R. and Cook W. H. (1983) A constitutive model and data for metals subjected to large strains, high strain rates and high temperatures. *Proceedings of the Seventh International Symposium on Ballistics*, The Hague, pp. 541–547.

[2] Maheswari, A K., Naraya S. P., and Gaur N. K., (2013) Constitutive modelling for flow stress determination using dimensionless proportionality relationship. *International Journal of Advances in Materials Science and Engineering*, Vol. 2, No.2.

[3] Haung X., Zhu J., and Hu W., (2012) Identification of Johnson – Cook constitutive model parameters for high strain rate deformation. *15th International Conference on experimental mechanics*. Port/Portugal, 22-27 July 2012.

[4] Tello K. E., Gerlich A. P., and Mendez P. F., (2010) Constants for hot deformation constitutive models for recent experimental data. *Science and Technology of welding and joining* Vol.15, No.3, pp. 260.

[5] Dorogoy A., and Rittel D., (2009) Determination of the Johnson – Cook material parameters using the SCS specimen, *Experimental mechanics*, Vol. 49, No 6, pp.881-885

[6] Slais, M., Dohnal, I., and Forejt M., (2012) Determination of Johnson – Cook Equation parameters. *Acta Metallurgica Slovaca*, Vol. 18, No. 2-3.

IOP Conf. Series: Materials Science and Engineering 1185 (2021) 012033 doi:10.1088/1757-899X/1185/1/012033

[7] Mousavi A., and Meisami M., (2010) Investigation on Zener – Hollomon parameter of a medium carbon low alloy of 1Cr-1Mn-1.5Ni-1Si-0.1V under hot compression test, Roznov pod Radhostem, Czech Republic, EU, Vol 18.

[8] Hui Y., Hong L. Z., and Liang Z. Z., (2006) Investigation on Zener – Hollomon parameter in the warm – hot deformation behaviour of 20CrMnTi. Yang et al. / J Zhejiang Univ SCIENCE A, 1453-1460, 2006.

[9] Li Y. S., Zhang Y., Tao N. R., and Lu K., (2009) Effect of the Zener–Hollomon parameter on the microstructures and mechanical properties of Cu subjected to plastic deformation. *Acta Materialia* 57, 761 – 772

[10] Ou, L., Nie, Y., & Zheng, Z. (2014). Strain compensation of the constitutive equation for high temperature flow stress of a Al-Cu-Li alloy. *Journal of materials engineering and performance*, *23*(1), 25-30.

[11] Ning, J., Nguyen, V., Huang, Y., Hartwig, K. T., & Liang, S. Y. (2018). Inverse determination of Johnson–Cook model constants of ultra-fine-grained titanium based on chip formation model and iterative gradient search. *The international journal of advanced manufacturing technology*, *99*(5), 1131-1140.

[12] Song, P., Li, W., Wang, X., & Xu, W. (2019). Study on mechanical properties and constitutive model of 5052 aluminium alloy. *Materials Science and Technology*, *35*(8), 916-924.

[13] Neelima, P., Murty, S. N., & Chakravarthy, P. (2019). Comparison of Prediction Capabilities of Flow Stress by Various Constitutive Equation Models for Hot Deformation of Aluminum Matrix Composites. *Materials Performance and Characterization*, *9*(2).

Fatigue life of composite leaf spring assembly

Abhishek Pendyala, Ch Siva Rama Krishna, Chinta Jagadish

Department of Mechanical Engineering, Vignan's Institute of Information Technology, Visakhapatnam 530049, India

Corresponding author e-mail: pendyalaabhishek@gmail.com

ABSTRACT: Suspension system in automobiles plays a major role in observing the shock loadswhich are result from the rough road, collision and vibrations. Leaf springs are one of the major suspension systems in heavy vehicles which are responsible for observing the shock loads and heavy loads. Most of the times under dynamic loads, these leaf springs experience fatigue failure with low life cycles. It is important to study the behaviour of leaf springs under dynamic conditions with respect to static loads. In the present study, a heavy vehicle leaf spring is designed and to be analyzed under dynamic loads i.e., fully reversed loads, zero based loads and fluctuating loads [ratio of -0.5 and 0.5] using composite material (Carbon Fiber reinforced polymer composite). A total of3 models of leaf springs with varying leaf lengths and leaf thickness are modeled in FEA software i.e., ANSYS Workbench. From the static structural analysis, it is observed that Model 3 having fewer stresses compared to the remaining models. The results with Leaf spring with 15mm thickness shows that better strength and stiffness properties when compared to the Leaf Spring with 12 mm thickness. From the fatigue life hand calculations, it is observed that Model 3 having the highest life cycles stresses compared to the remaining models.

Keywords: Finite Element Method, Constant Amplitude, Leaf Spring, Fully ReversedLoad, Zero Based Load, Fluctuating Loads, Static Structural Analysis.

1. Introduction

Fatigue is a failure that comes with dynamic loading conditions. Fatigue failure is mainly 2 types based on loading and life cycles. There are two types of fatigue life calculation approaches. They are high cycle fatigue and low cycle fatigue. If the structure fails above the 1000 cycles, this type of failure comes under High cycle fatigue.For high cycle fatigue, SN curve is used for the life calculations. Structural Life, Aerospace structures and bridges are such common examples of High cycle Fatigue. The high cycle fatigue mainly depends on the SN curve whereas stress values are noted in the curve versus the number of life cycles.If the life of the structure falls in between 1 to 1000 cycles, this type of fatigue failure iscalled the Low Cycle Fatigue. The Low cycle fatigue mainly depends on the cyclic stress strain curve for respective materials. For the Low cycle fatigue, cyclic stress strain curve is used for life calculations.

If the load varies between +P to –P is called a fully reversed load, if the load varies between 0 to P is called Zero based load and if the load varies between +P/2 to +P or –P/2 to -P or –P/2 to -P are called fluctuating loads which are shown in Figure 1.0.

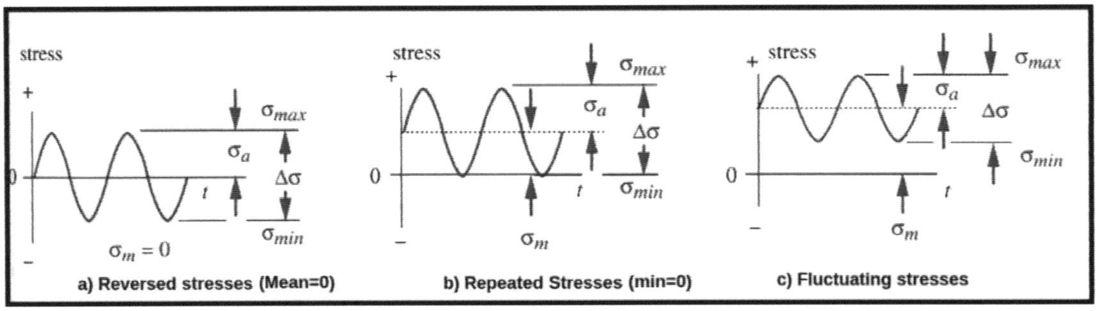

Figure 1 Dynamic Loads

There are 5 types of Leaf Springs in the industry; they are elliptic, semi elliptic, three Quarter elliptic, quarter elliptic and transverse.T G Loganathan [1] performed work on leaf springs and kept focus of study in terms of materialchange from the conventional SAE 5160 steel to CFRP. The present work provides the results of flexural fatiguelife and damage incurred for both the material by FE Analysis. Navuri Karteek [2] performed fatigue analysis on the different shapes of alloy wheel under constant amplitude loading. The material parametric study is also performed on the different cases of the spokes. Theresults show that the carbon steel has shown better results when compared to remaining materials in view of life, damage and factor of safety.Navuri Karteek [3] performed different types of static structural analysis i.e., Radial load case, impact load case, bending load case, torsion load case and maximum deflection case on four different alloy wheels. The induced stresses and deformations are within the limits. M. Saran Theja [4] presented a research article onStructural and Fatigue Analysis of Two Wheeler Lighter Weight Alloy Wheel.The present work attempts to analyse the safe load of the alloy wheel, which will indicate the safe drive is possible. Sahil Bandral [5] performed impact analysis on the four wheeler alloy wheel and calculated the stresses by varying the number of spokes. He concluded 3 spoke alloy wheel shows better results. Anusha R [6] performed static and impact analysis on the four wheeler alloy wheel and calculated the stresses and deformations. Later performed weight optimization and reduced the weight by 20%.

2. Problem Description

2.1 Problem Statement:

To perform Static structural analysis on the leaf spring assembly with realistic loads and boundary conditions using 3D FEA and to verify the design of leaf spring components in strength point of view and to calculate the fatigue life under static loads.

2.2 Geometry:

The 3D model of the leaf spring which are used for the present analysis is taken from the Heavy Vehicle Bus (Ashok Leyland- Ms355)and it is shown in the below figure. The dimensions of the leaf spring are also shown in the below figures. The main full-lengthleaf is 1480 and Leaf Spring [Model 1] is shown in the below figure2.0.

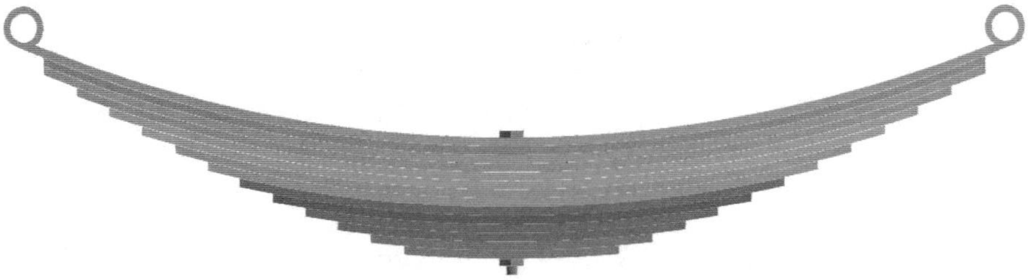

Figure 2. Leaf Spring [Model 1]

IOP Conf. Series: Materials Science and Engineering 1185 (2021) 012034 doi:10.1088/1757-899X/1185/1/012034

Note: The width and thickness of the leaf spring are75 mm and 15mm.
The Graduated decreased leaf's or supporting leafs are used for the leaf spring system to providing extra support for the main leaf springs and the dimensions are listed below.
L1 = 1480 mm, L2 = 1380 mm, L3 = 1380 mm, L4 =1280 mm, L5= 1180 mm, L6 = 1080 mm, L7 =980 mm, L8 = 860 mm, L9 = 780 mm, L10 =680 mm, L11 = 580 mm, L12 =480 mm, L13= 380 mm and L14 =280 mm.
The leaf spring assembly [Model 2] comprises of 14 leaf springs, as same as the previous master model 1. The geometry of the leaf spring assembly [model 2] is shown in the below figure3.0.

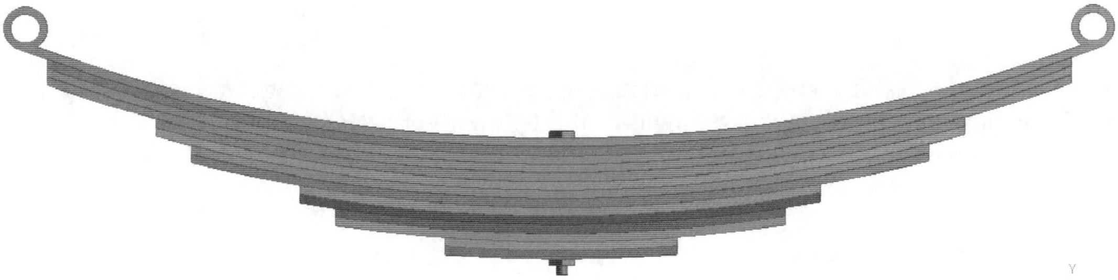

Figure 3 Leaf Spring[Model 2]

The Graduated decreased leaf's or supporting leafs are used for the leaf spring system to providing extra support for the main leaf springs and the dimensions are listed below.
L1= 1480 mm, L2= 1380 mm, L3 = 1380 mm, L4 =1380 mm, L5 = 1080 mm, L6 = 1080 mm, L7= 980 mm, L8 = 980 mm, L9 = 680 mm, L10= 680 mm, L11= 580 mm, L12 = 580 mm, L13= 280 mm and L14= 280 mm.
The leaf spring assembly [Model 3] comprises of 14 leaf springs, as same as the previous model 1 and 2. The geometry of the leaf spring assembly [model 3] is shown in the below figure 4.0.

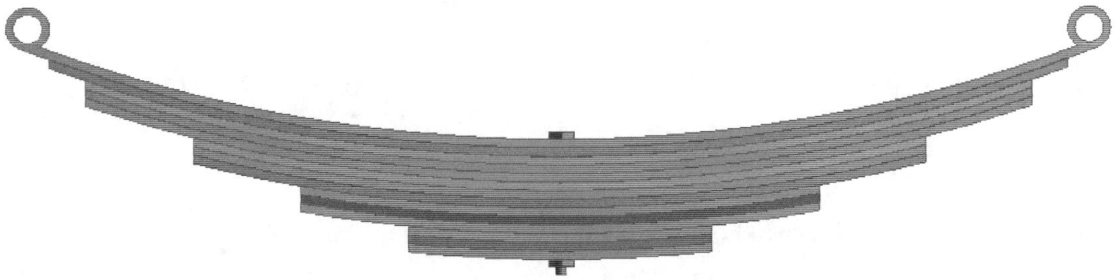

Figure 4 Leaf Spring [Model 3]

The Graduated decreased leaf's or supporting leafs are used for the leaf spring system to providing extra support for the main leaf springs and the dimensions are listed below.
L1 = 1480 mm, L2 = 1380 mm, L3 = 1280 mm, L4=1280 mm, L5 = 1280mm, L6 = 980 mm, L7 = 980 mm, L8 = 980 mm, L9 = 680 mm, L10 = 680 mm, L11 = 680 mm, L12 = 380 mm, L13 = 380 mm and L14 =380 mm.

2.3 Connections:
Here two types of contacts are used for this analysis i.e., Bonded contact and frictional contact with material coefficient of friction is assumed to be as 0.2. The connection details are listed in the below Table 1.

Table 1 Connections details

S. No	Contact	Target	Type
1	Leaf	Leaf	Frictional Contact
2	Leaf	Bolt/Nut	Frictional Contact
3	Bolt	Nut	Bonded Contact

2.4 Mesh:

Meshing is the process of converting geometry entities in to finite elements and nodes. The meshed model of the leaf spring assembly is shown in the below figure. Hexahedron element [Solid 186]is used for the supporting leafs where there is no complicity in nature. Tetrahedron element [Solid 187]is used for the main leaf spring due to curved profile.

2.5 Loads & Boundary Conditions:

Here the total mass of the vehicle (Curb) = 4000 kg, additional passengers' mass = 3000 kg.Therefore, the final force which is acting on the bottom leaf is

$$F = m.a = 7000 \times 9.81 = 68670 \text{ N}$$

$$\text{For Each Spring, } F = 68670/4 \text{ N} = 17167.5 \text{ N}$$

Figure 5. Force acting on the Leaf Spring Assembly

$$\text{Bolt Pretension, } P = \frac{T}{D} = \frac{50}{0.2 \times 0.0015} = 16666 \text{ N}$$

A bolt pretension (P) of magnitude 16666 N is applied on the bolt outer surface. The eyes of the main full length leaf are arrested with respect to the realistic working conditions. The figure of the leaf spring with 2 remote displacements are shown in the below figure. The left eye of the main fulllength leaf is arrested in translation in X, Y, Z direction and rotation also arrested about X and Y except rotation about Z is free. The right eye of the main full length leaf is arrested translation in Y, Z direction and rotation about X and Y. Translation in X direction and Rotation about Z direction are free. The displacement of the leaf springs (Full length and supporting leaf's) and bolt are arrested width direction.

IOP Conf. Series: Materials Science and Engineering 1185 (2021) 012034 doi:10.1088/1757-899X/1185/1/012034

3. Static Structural Analysis

The assembly of leaf spring assembly is assigned with the carbon fiber reinforced polymer composite. The properties of the composite material are taken from the article [1]. The isotropic properties of the carbon fiber and epoxy resin are listed in the below table 2.

Table 2. Isotropic Properties of Carbon Fiber and Epoxy Resin

Properties	Carbon Fiber	Epoxy Resin
Density	1800 Kg/m3	1400 Kg/m3
Young's Modulus	23000 MPa	10500 MPa
Poisson's Ratio	0.4	0.3
Tensile Strength	300 MPa	850 MPa
Bulk Modulus	500 MPa	333 MPa
Shear Modulus	900 MPa	125 MPa
Volume Fraction	0.5	0.5

The carbon fiber and epoxy resin are mixed with together to form a lamina and that lamina are stacked together with different orientations to form a 3D laminate which having equal properties in X, Y and Z directions. The lamina thickness and number of lamina's with orientations are selected depend upon the research article [1].The 3D laminate properties of the carbon epoxy with volume fraction 50-50 is listed in the below table 3.

Table 3. Laminate Properties of Carbon Epoxy

Property	Value	Property	Value	Property	Value
E_{11}	23000 MPa	μ_{12}	0.2	G_{xy}	9000 MPa
E_{12}	23000 MPa	μ_{23}	0.4	G_{yz}	8214MPa
E_{13}	23000 MPa	μ_{31}	0.2	G_{xz}	9000 MPa
ρ	1800 Kg/m3	σ_{yt}	300 MPa	σ_{ut}	600 MPa
		σ_{yc}	285 MPa	σ_{uc}	300 MPa

The fatigue properties of the 3D CFRP composite under fully reversed rotating beam fatigue test are listed in the below table 4.

Table 4 SN Properties of Carbon Epoxy Laminate

Stress (S)	Number of cycles (N)
800	200
700	3200
600	25,000
500	1.227e5
400	4.509e5
300	1.3548e6
200	3.5155e6
100	8.1435e6

Note: For static structural analysis of the composite leaf spring assembly, cylindrial co-ordinate system is assigned to the leaf spring for element orientation.Longitudinaldirection is assigned with width direction, circumferential direction is assigned with curved Length direction and radial direction is assigned with thickness direction. The calculated von-misses stresses and von-misses strains of the leaf spring [Model 1] are shown in the below figure 6.

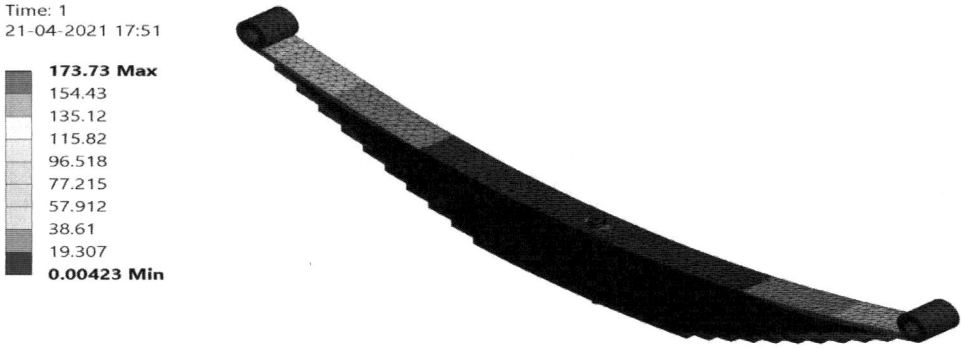

Figure 6. Von-misses stress of the CFRP Composite Leaf Spring Assembly

The calculated values of the von-misses stresses and von-misses strain of the leaf spring with CFRP Composite for Model 1, 2 and 3 are listed in the below table 5.

Table 5. Static Results of Leaf Spring [Model 1, 2 and 3]

Component	Model 1	Model 2	Model 3
Von-misses Stress (MPa)	176.71	173.73	149.35
Von-misses Strain (mm)	0.0080	0.0079	0.0045

From the above table, it is shown that the model 1 is having higher von-misses stresses and von-misses strains followed by the Model 2 and Model 3. It shows that the model 3 showing high strength and hign stiffness. Fatigue life calculation are discussed below.

4. Fatigue Life:

4.1 Fully Reversed Loads

If the load varies between +P to -P, then the loading type is called fully reversed load.

For Model 1, $\sigma_{Von} = 176.71$ MPa

$\sigma_m = 0$

$\sigma_a = 176.71$

$\sigma_{yt}= 300$MPa

$\sigma_e= 176.71$MPa

Log-Log interpolationis $\text{Log}\ \sigma = A \log N + B$

From the SN curve, with log-log interpolation, the below A and B values are obtained.

$A = -0.1958$

$B = 3.3534$

Therefore N = 4082300 Cycles

The fatigue life of the leaf spring [Model] under fully reversed load is 4082300 cycles. Respectively, the fatigue life of the leaf spring [Model 2] and Model 3 are 4181400 Cycles and 4991900 Cycles.

4.2 Zero Based Loads

For Model 1, $\sigma_{Von} = 176.71$ MPa

$\sigma_{max} = 176.71$ MPa

$\sigma_{min} = 0$ MPa

$\sigma_m = 88.355$

$\sigma_a = 88.355$

$\sigma_{yt} = 300$MPa

For Mean stress correction, Soderbergmean stress theory is used

$$\frac{\sigma_m}{\sigma_{yt}} + \frac{\sigma_a}{\sigma_e} = 1$$

$\sigma_e = 113.9140$MPa

N = 6194400 Cycles

The fatigue life of the leaf spring [Model] under Zero based loads is 6194400 cycles. Respectively, the fatigue life of the leaf spring [Model 2] and Model 3 are 6407900 Cycles and 8143500 Cycles.The infinite life of the structure is 8143500 cycles.

4.3 Fluctuating Loads or Alternating Loads (R Ratio) = 0.5

$\sigma_{max} = 176.71$ MPa

$\sigma_{min} = 88.355$ MPa

$$\frac{\sigma_m}{\sigma_{yt}} + \frac{\sigma_a}{\sigma_e} = 1$$

$\sigma_m = 132.5325$

$\sigma_a = 44.1775$

$\sigma_{yt} = 300$MPa

$\sigma_e = 79.14$MPa

N = 8143500 Cycles

The fatigue life of the leaf spring [Model] under fluctuating loads [0.5] is 8143500 cycles. Respectively, the fatigue life of the leaf spring [Model 2] and Model 3 are 8143500 Cycles and 8143500 Cycles.

The infinite life of the structure is 8143500 cycles.

4.4 Fluctuating loads or alternating Loads(R Ratio) = -0.5

$\sigma_{max} = 176.71$ MPa

$\sigma_{min} = -88.355$ MPa

$$\frac{\sigma_m}{\sigma_{yt}} + \frac{\sigma_a}{\sigma_e} = 1$$

$\sigma_m = 44.1775$

$\sigma_a = 132.5325$

$\sigma_{yt} = 300$MPa

$\sigma_e = 155.427$MPa

N = 4769100 Cycles

The fatigue life of the leaf spring [Model] under fluctuating loads [-0.5] is 4769100 cycles. Respectively, the fatigue life of the leaf spring with Model 2 and Model 3 are 4905000 Cycles and 6019000 Cycles.

The comparison study of the fatigue life of the composite leaf spring under fully reversed load, zero based load and fluctuating loads (tensile and compressive) are listed in the below table 6.

Table 6. Fatigue life of the composite leaf spring

	Fully Reversed Loads (1)	Zero Based Loads	Fluctuating Loads	
			Tensile (0.5)	**Compressive(-0.5)**
Model 1	4082300	6194400	8143500	4769100
Model 2	4181400	6407900	8143500	4905000
Model 3	4991900	8143500	8143500	6019000

From the above table, it is noticed that the fatigue life of the composite leaf spring assembly under fully reversed loads is less followed by the fluctuating loads (compressive -0.5) and fatigue life of the composite leaf spring is having infinite life cycles under zero based load and fluctuating load (tensile).

5. Conclusions

The leaf spring of the heavy vehicle is designed in the FEA Software i.e., ANSYS Workbench. The dimensions which are taken for the Leaf spring is taken from the real existing vehicle. Proper meshing, contacts and loading boundary conditions are given to the Leaf Spring Model. A total of 3models are designed by changing the lengths of the leaves with keeping same volumetric ratios. From the present the following conclusions are drawn.From the static structural analysis, it is observed that Model 3 having less stresses and strains compared to the remaining models using composite materials. From the fatigue life hand calculations, it is observed that Model 3 having highest life cycles stresses compared to the remaining models. The weight of the leaf spring assembly is reduced up to the 25% of the whole weight by using composite material compared to the carbon steel material, and its shows less stress values, strain values and high fatigue life cycles.The life of the leaf spring assembly using composite material under fully reversed loads is less compared to the other loading conditions.The life of the leaf spring assembly using metal and composite material under zero based load and fluctuating load (-0.5) shows infinite life cycles.So it is recommended to check the structure of the leaf spring under fully reversed loads. It is recommended to use Model 3 for all type of applications for better strength, stiffness and fatigue life in heavy load applications.

References

[1] Loganathan, T.G., Kumar, K.V. and Madhu, S., 2020. Flexural and fatigue of a composite leaf spring using finite element analysis. *Materials Today: Proceedings*, *22*, pp.1014-1019.

[2] Karteek, N., Krishna, B.S., Babu, B.K. and Paleti, B.M., Fatigue Analysis of Motor Cycle Alloy Wheel.*IJITEE,* Volume 8, Issue 9, ISSN 2278-3075 .

[3] Theja, M.S. and Krishna, M.V., 2013. Structural and fatigue analysis of two wheeler lighter weight alloy wheel. *IOSR Journal of Mechanical and Civil Engineering (IOSRJMCE) e-ISSN*, pp.2278-1684.

[4] Jadhav, S. and Roy, G., Structural Analysis of Spoke Wheel Rim Assembly of Motorcycle. *Altair Technology Conference, India, 2015*

[5] SahilBandral and satnam Singh 2018 "Impat Analysis of car Alloy Wheel Rim using Finite Element Analysis", *International Journal of Engineering research &Technology, Special Issue 2018, Volume* **6**, *Issue 16*

[6] Anusha.R and Tara Sasanka. C, July 2017 "Analysis and optimization of alloy wheel", *International Journal of Innovative Research in science, Engineering and Technology, Volume* **8**, *Issue 7, pp 15328-15336*

IOP Conf. Series: Materials Science and Engineering 1185 (2021) 012035 doi:10.1088/1757-899X/1185/1/012035

Random vibration analysis of composite plate

Yalamanchili Swapna, K Sri Harsha

Department of Mechanical Engineering, Vignan's Institute of Information Technology, Visakhapatnam 530049, India

* Corresponding author e- mail: swapnayalamanchili99@gmail.com

Abstract: Random vibration analysis is one of the major type of dynamic load which are having multiple amplitudes with multiple frequencies at a given particular time. The aerospace structure which contains Printed circuit board [PCB] fails frequently due to the high accelerations, high Stresses levels and high deformations under dynamic loads. So it is very important to analyse the PCB (Composite Plate) under dynamic load conditions. In the present research work, rectangular plate with composite materials which acts as PCB is analysed under random vibration load with different boundary conditions 2 faces fixed using Copper foil and Epoxy E Glass UD materials. The geometry of the rectangular plate is modelled in the FEA Software ANSYS. Proper fine mesh is done for the rectangular plate. Plate with [1/1], [2/2], [2/4] and [4/4] layups are designed in the Ansys ACP pre-module and later assigned the different properties of the composites to the rectangular plate and performed random vibration analysis by applying PSD G acceleration load to the fixed supports. A total number of 9 critical points are selected on the plate for calculating the PSD responses [Acceleration, displacement and normal stresses] to the random vibrations which are given to the rectangular plate. Based on the results, critical point is located and best material and best layup sequence is selected for the composite plate.

Keywords: FEA, Random vibration analysis, PSD and PSD Response.

Nomenclature: ρ = Density, E = Young's Modulus, G = Shear Modulus & μ = Poisson's Ratio.

1. Introduction:

The structural loads can be classified in to two types based on the magnitude and direction with

IOP Conf. Series: Materials Science and Engineering 1185 (2021) 012035 doi:10.1088/1757-899X/1185/1/012035

respect to the time. If the load does not changes its magnitude and direction with respect to the given time, is called as static load. The failure comes under static load is called static failure. If the load changes its magnitude and direction with respect to the given time, is called as dynamic load. The failure comes under dynamic load is called fatigue failure. The structural loads can be classified in to several loads based on the time and frequency asillustrated in figure 1.

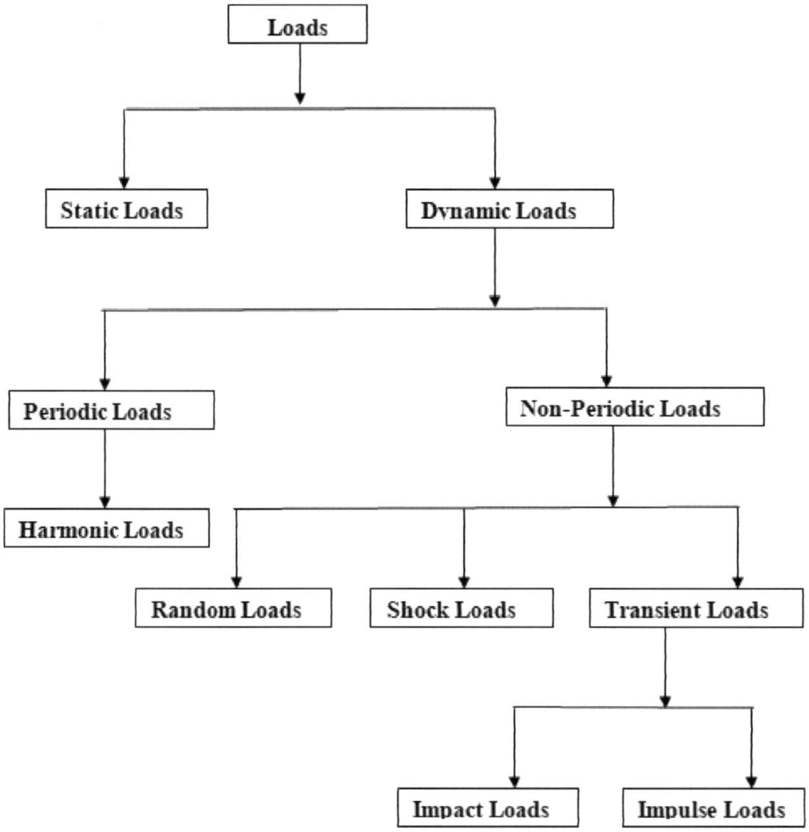

Figure 1 Type of Loads

Static Failure is gradual and fatigue failure is suddenly occur. The time and life of the static failure is more than the time and life of the fatigue failure. In reality pure cyclic loads very rare.In the lab we intentionally give the pure cyclic loads.

Note: In many real problems, it is found that exciting forces vary with time in a non-harmonic fashion that may be periodic or non-periodic.

Random vibrations are the loads having multiple amplitudes with multiple frequencies at the same time. These types of loads are frequency dependent. A seen pure sinusoidal vibration (harmonic) are having single amplitude with single frequency, but a random vibration loads are having multiple amplitudes with multiple frequencies. It is very difficult to define the random load on a structure. The random vibration loads are converted into bins and then converted in to Power spectral density (PSD) input. The typical type of random vibration load and psd load are shown in the below figures 2.0 and figure 3.

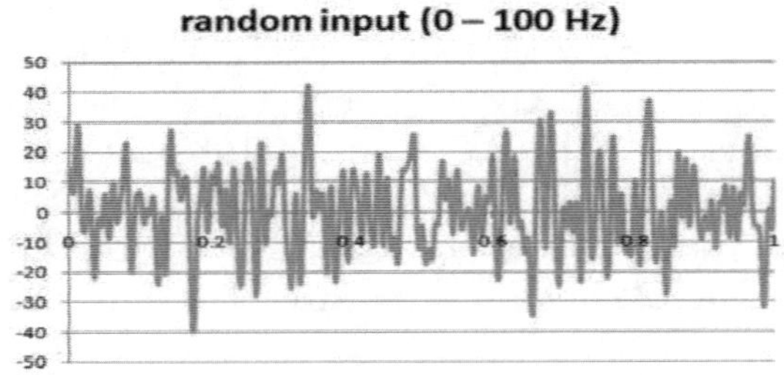

Figure 2 Random Vibration Load

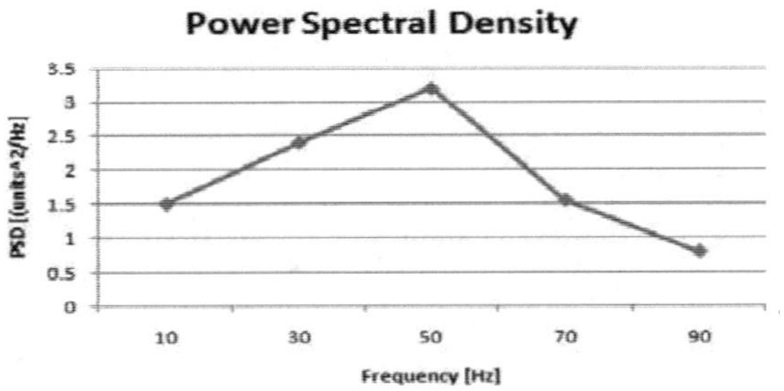

Figure 3 PSD Load

Karteek Navuri[1] performed random vibration analysis on the Flight data recorder (FDR) with ribs and without ribs, calculated PSD response at critical points using different types of isotropic and composite materials. Depend upon the PSD Response, best material for FDR is recommended. Wei Qu [2] calculated stress response to the random vibrations on the hydraulic composite pipe. Discrete type of analysis method for random vibration is used to solve the excitation of white noise. The results agreed with the simulation results, demonstrating that the analytical method could provide theoretical reference for the design, improved efficiency, and fatigue reduction of the composite pipe subjected to random vibration. Da Yu [3] concluded that the peak value of out-of-plane displacement of printed circuit board (PCB), when it is subjected to drop impact, is a major concern to electronic manufacturers as it relates to the maximum stress causing failure for the solder balls. M. A. Anuar [4] stated that, since the level of vibration always depends on the natural frequencies of the system, it is important to know the modal parameters of such system to control failure and provide prevention actions. The demand for structurally reliable Printed Circuit Boards (PCB) has increased as more functions are required from electronic products along with less weight and smaller size. Jayaraman S [5] performed shock response spectrum analysis of printed circuit boards subjected to a half-sine pulse excitation using finite element method. The objective of this paper is to predict the shock response spectrum of a printed circuit board due to launch environment. The analysis results are validated by conducting experimental tests of PCB. Soma Shekar [6] uses basic FEA tool to accurately investigate the dynamic characteristics of the PCB and avoid costly testing methods which require hardware. Here the normal modes & frequency response functions (FRF) of PCB are determined and validated using vibration test on PCB. The validated model is used to predict vibration response for random vibration input. It is shown here how the responses are accurately predicted for random vibration input for a design parameter variation of PCB. The results are also validated using vibration test on PCB. Hanna Fahlhren [7] performed Modal analysis on the electronic unit. The result of the analysis shows that the first natural frequency of the chassis body and the PCB are not divided enough. During resonance coupling of acceleration are occurred. A rib design is developed to eliminate the coupling and to increase the first natural frequency of the bodies.

2. Problem Description:

2.1 Problem Statement

The main aim of the problem is to study the dynamic response of the rectangular composite plate under random vibration analysis using FEA software ANSYS.

2.2 Problem Modelling

The geometry of the rectangular plate is modelled in the ANSYS Workbench and respective dimensions with views are shown and in the below figure 4.0

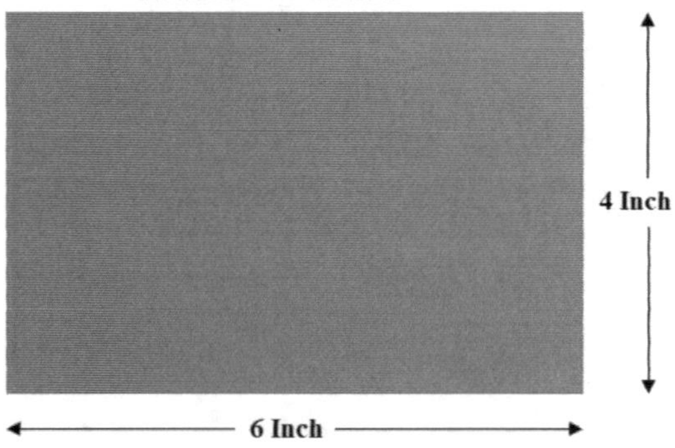

Figure 4 Dimensions of the Plate

2.3 Materials

Copper foil and Epoxy Glass UD are used for the random vibration analysis.

Copper Foil: $\rho = 8960$ kg/m^3, E = 128000, $\mu = 0.34$ and $\sigma_{yt} = 227$ MPa

Epoxy E Glass: $\rho = 8960$ kg/m^3, $E_{xy} = 45000$, $E_{yz} = 10000$, $E_{xz} = 10000$,
$\mu_{xy} = 0.3$, $\mu_{yz} = 0.4$ $\mu_{zx} = 0.3$
$G_{xy} = 5000$, $G_{xy} = 3846.2$ and $G_{xy} = 5000$

2.4 Mesh

Meshing is the process of converting geometry entities into the finite elements. Here quadrilateral elements i.e., Solid 186 Elements are used for the rectangular plate.

2.5. Loads & Boundary Conditions

The random vibration loads which are given to the rectangular plate are applied in the form base excitation i.e., PSD G acceleration is applied in the Y direction which is shown in the figure 5.0. The frequency range for the random vibrations is 0 to 2000 Hz. The boundary conditions are shown in the figure 6.0.

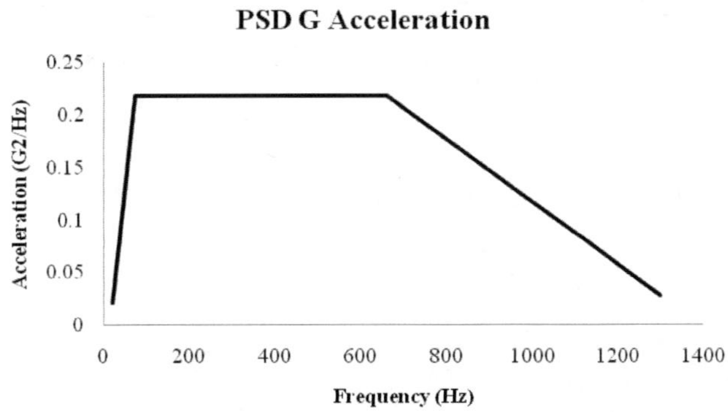

Figure 5 PSD G Acceleration

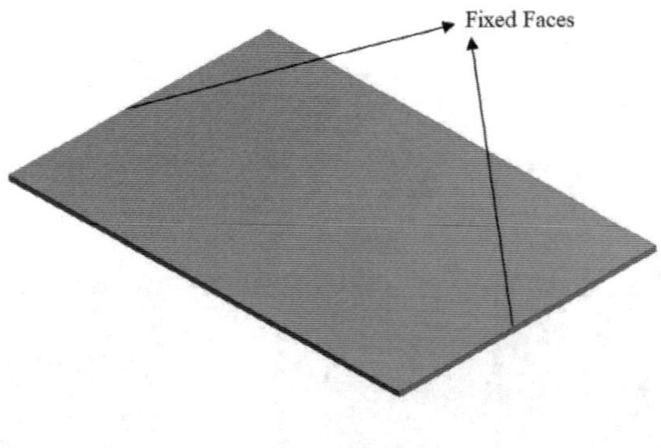

Figure 6 Boundary Conditions

2.6 Composite Plate [PCB Board]

Now day's three different types of PCB are used. Out of that FR4 PCB boards are used most commonly. In this research work, four types of layers are used i.e., FR4 single layer (1/1), FR4 multi layer (2/2), FR4 multi layer (2/4) and FR4 multi layer (4/4) with different fiber orientations.

IOP Conf. Series: Materials Science and Engineering 1185 (2021) 012035 doi:10.1088/1757-899X/1185/1/012035

FR4 MULTI LAYER (2/2) [$\mathbf{0^0/45^0\ /\text{-}45^0/0^0}$]

FR4 MULTI LAYER (2/4) [$\mathbf{0^0/45^0\ /\text{-}45^0\ /45^0\ /\text{-}45^0/0^0}$]

FR4 MULTI LAYER (4/4) [$\mathbf{0^0/45^0\ /\text{-}45^0\ /0^0/0^0/45^0\ /\text{-}45^0/0^0}$]

Figure 7 The FR4 Plate with different orientations

FR4 SINGLE LAYER (1/1) [$\mathbf{0^0/0^0}$]	→	Model 1
FR4 MULTI LAYER (2/2) [$\mathbf{0^0/45^0\ /\text{-}45^0/0^0}$]	→	Model 2
FR4 MULTI LAYER (2/4) [$\mathbf{0^0/45^0\ /\text{-}45^0\ /45^0\ /\text{-}45^0/0^0}$]	→	Model 3
FR4 MULTI LAYER (4/4) [$\mathbf{0^0/45^0\ /\text{-}45^0\ /0^0/0^0/45^0\ /\text{-}45^0/0^0}$]	→	Model 4

2.7 Fiber Orientations

The fibres of the Epoxy E-Glass UD are oriented in many angles. The orientation of the fibers are shown in the below figure 8.0.

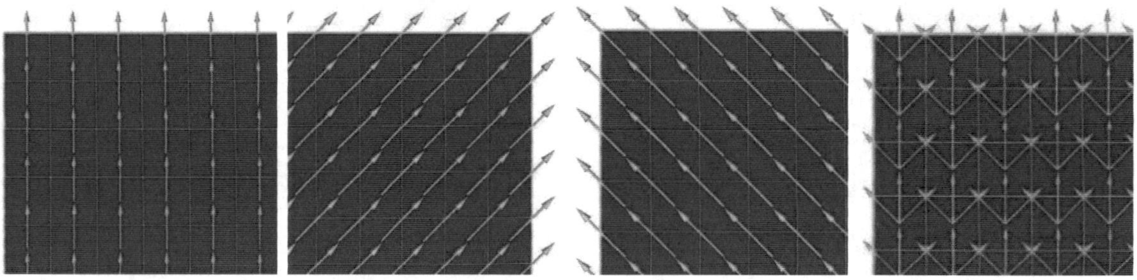

0^0 Orientation **45^0 Orientation** **-45^0 Orientation** **0^0 45^0 -45^0 Orientation**

Figure 8 Fiber orientation

IOP Conf. Series: Materials Science and Engineering 1185 (2021) 012035 doi:10.1088/1757-899X/1185/1/012035

3 Random Vibration Analysis

The random vibration analysis on the composite plate are performed using above fiber orientations, list of results are calculated on the critical points which are shown in the below figure 9.0.

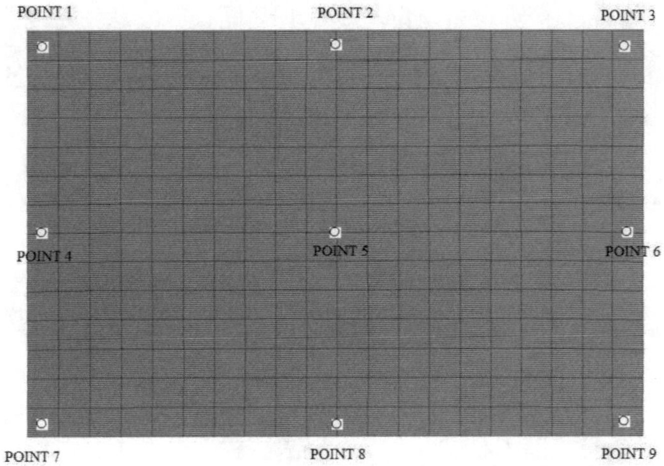

Figure 9 Critical Points on the composite plate

It is recommended to conduct the modal analysis of the composite plate in between the range of minimum 1.5 times of the random vibration load frequency. The modal natural frequency range is 2500 because the random vibration load range is 1300 Hz. The natural frequencies of the composite plate (model 1) are listed in the below table 1.0.

Table 1.0 Natural Frequencies of the composite plate (Model 1)

Mode No.	Frequency (Hz)
1	330.18
2	402.2
3	772.55
4	920.51
5	1027.1
6	1431.1
7	1627.4
8	1836.2
9	1961.1
10	2270.9
11	2409.3

From the above table, a total number of 11 natural frequencies are obtained in between the 0 to 2500 range. The first natural frequency is 330.18 Hz. The list of first six mode shapes of the composite plate are shown in the figure 10.0

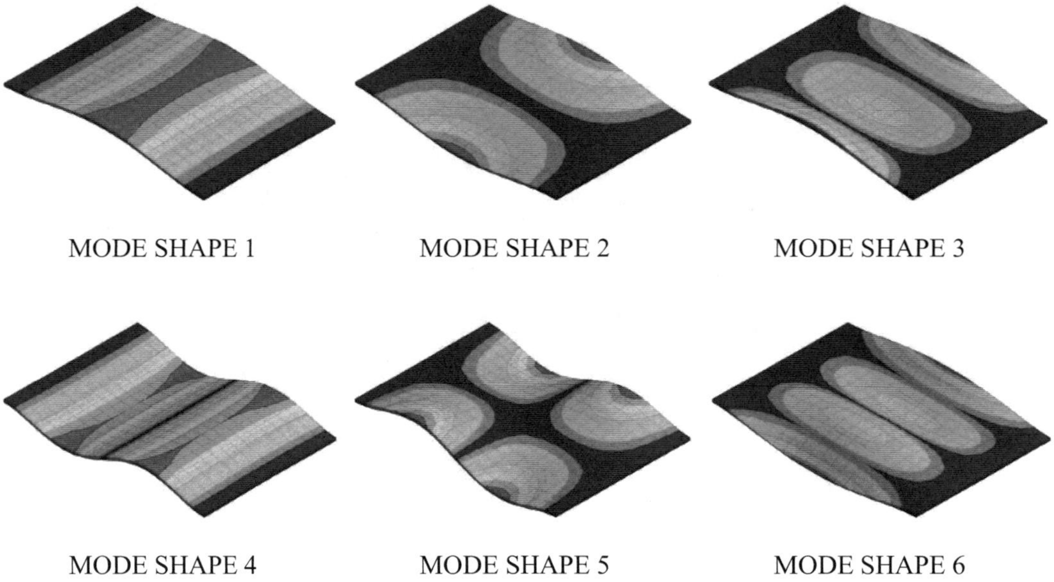

Figure 10 Mode Shapes of the Composite Plate

The response PSD (Acceleration, Displacement and Normal Stress in Y direction) of the rectangular plate (Model 1) to PSD G Acceleration load is listed in the table 2.0.

Table 2.0 Response PSD of the composite plate (Model 1)

Point	Acceleration (m/s^2)	Displacement (mm)	Normal Stress (MPa)
1	133.3	0.38177	0.27653
2	1059.1	0.46281	1.211e-002
3	133.47	0.38163	0.72142
4	133.35	0.38168	0.49258
5	956.88	0.45015	8.6226e-003
6	133.33	0.38174	0.30892
7	133.42	0.38181	0.15304
8	1054.9	0.46225	1.3076e-002
9	133.43	0.38181	0.14824

From the above table it is observed that, the accelerations and displacements are maximum at the middle point of the free ends, the stresses are maximum at the fixed supports. The natural frequencies of the composite plate (model 2) are listed in the below table 3.

Table 3 Natural Frequencies of the composite plate (Model 2)

Mode No.	Frequency (Hz)
1.	427.83
2.	584.96
3.	1184.1
4.	1347.
5.	1415.8
6.	2239.1
7.	2340.4
8.	2601.1
9.	2997.3

From the above table, a total number of 9 natural frequencies are obtained in between the 0 to 2500 range. The first natural frequency is 427.83 Hz. The response PSD (Acceleration, Displacement and Normal Stress in Y direction) of the rectangular plate (Model 2) to PSD G Acceleration load is listed in the table 4.0.

Table 4 Response PSD of the composite plate (Model 2)

Point	Acceleration (m/s^2)	Displacement (mm)	Normal Stress (MPa)
1	133.64	0.38428	2.6524
2	1219.6	0.42573	2.2442e-002
3	133.62	0.3842	6.7648
4	133.59	0.38423	4.8845
5	1078.5	0.41779	7.3861e-002
6	133.69	0.38427	3.0725
7	133.82	0.3843	1.522
8	1215.4	0.42548	2.0605e-002
9	133.81	0.3843	1.4608

IOP Conf. Series: Materials Science and Engineering 1185 (2021) 012035 doi:10.1088/1757-899X/1185/1/012035

From the above table it is observed that, the accelerations and displacements are maximum at the middle point of the free ends, the stresses are maximum at the fixed supports. The natural frequencies of the composite plate (Model 3) are listed in the below table 5.

Table 5 Natural Frequencies of the composite plate (Model 3)

Mode No.	Frequency (Hz)
1.	427.96
2.	585.21
3.	1184.4
4.	1347.6
5.	1416.3
6.	2240.2
7.	2341.1
8.	2602.
9.	2998.5

From the above table, a total number of 9 natural frequencies are obtained in between the 0 to 2500 range. The first natural frequency is 427.96 Hz. The response PSD (Acceleration, Displacement and Normal Stress in Y direction) of the rectangular plate (Model 3) to PSD G Acceleration load is listed in the table 6.

Table 6 Response PSD of the composite plate (Model 3)

Point	Acceleration (m/s^2)	Displacement (mm)	Normal Stress (MPa)
1	133.64	0.38429	2.6526
2	1219.9	0.42571	2.2389e-002
3	133.62	0.38421	6.7545
4	133.59	0.38424	4.8815
5	1078.6	0.41777	7.3795e-002
6	133.69	0.38428	3.0707
7	133.82	0.38431	1.5205
8	1215.7	0.42546	2.0597e-002
9	133.81	0.38431	1.4609

From the above table it is observed that, the accelerations and displacements are maximum at the middle point of the free ends, the stresses are maximum at the fixed supports. The natural frequencies of the composite plate (Model 4) are listed in the below table 7.

Table 7 Natural Frequencies of the composite plate (Model 4)

Mode No.	Frequency (Hz)
1.	366.87
2.	510.54
3.	1020.7
4.	1174.2
5.	1235.5
6.	1975.2
7.	2031.4
8.	2277.6
9.	2607.6

From the above table, a total number of 11 natural frequencies are obtained in between the 0 to 2500 range. The first natural frequency is 336.87 Hz. The response PSD (Acceleration, Displacement and Normal Stress in Y direction) of the rectangular plate (Model 4) to PSD G Acceleration load is listed in the table 8.

Table 8 Response PSD of the composite plate (Model 4)

Point	Acceleration (m/s^2)	Displacement (mm)	Normal Stress (MPa)
1	133.41	0.38542	3.9617
2	1131.8	0.44793	1.946e-002
3	133.59	0.38531	10.079
4	133.48	0.38535	7.2279
5	1000.1	0.43603	0.11256
6	133.47	0.3854	4.5454
7	133.52	0.38545	2.2616
8	1127.9	0.44755	1.3612e-002
9	133.52	0.38544	2.1809

IOP Conf. Series: Materials Science and Engineering 1185 (2021) 012035 doi:10.1088/1757-899X/1185/1/012035

From the above table it is observed that, the accelerations and displacements are maximum at the middle point of the free ends, the stresses are maximum at the fixed supports.

4. Conclusions:

Random vibration analysis is performed on the composite plate with different lamina sequence and orientations. PSD Response (Acceleration, Displacement and Normal Stress) are calculated using FEA Software. The list of observations and conclusions are made

1. The natural frequencies of the composite plate are higher with model 3.
2. Accelerations are maximum at the middle point of the free edges and less accelerations are obtained in the composite plate with model 1
3. Directional displacements are maximum at the middle point of the free edges and less deformations are obtained in the composite plate with model 3
4. Normal stresses are maximum at the free edges and less Norma stresses are obtained in the composite plate with model 1
5. Out of the three results (Accelerations, displacements and Stresses), Acceleration magnitudes are considerable and it is recommended to use model 1 type for PCB boards based upon the random vibrations.

References:

[1] Karteek Navuri, Eswara Kumar A., Beulah Mani P. and B. Satya Krishna "RANDOM VIBRATION ANALYSIS OF MECHANICAL HARDWARE OF FLIGHT DATA RECORDER", *ARPN Journal of Engineering and Applied Sciences, VOL. 11, NO. 18, SEPTEMBER 2016.*

*[2]*Wei Qu, Huailiang Zhang, Wenqian Sun and Wei Li, "Stress response of the hydraulic composite pipe subjected to random vibration*" Composite Structures 255 (2021) 112958*

[3] Da Yu, Jae B. Kwak, Seungbae Park and John Lee, "Dynamic responses of PCB under product-level free drop impact" *Microelectronics Reliability 50 (2010) 1028–1038.*

[4] M.A.Anuara, A.A.Mat Isaa, Ummi Z.A.R, "Modal Characteristics study of CEM-1 Single-Layer Printed Circuit Board Using Experimental Modal Analysis" *Procedia Engineering 41 (2012) 1360 – 1366.*

[5] Jayaraman S, Manish Trikha, Somashekar, Kamesh D, M.Ravindra, "Response Spectrum Analysis of Printed Circuit Boards subjected to Shock Loads, *Procedia Engineering 144 (2016) 1469 – 1476.*

[6] Somashekar V.N, Harikrishnan S, PSM Aejaz Ahmed, Kamesh D "Vibration Response Prediction of the Printed Circuit Boards using Experimentally Validated Finite Element Model", *Procedia Engineering 144 (2016) 576 – 583.*

[7] Hanna Fahlgren. "Vibration analysis of electronic unit". A thesis submits to the school of engineering in jonkoping in the subject area Mechanical Engineering.

[8] Banu aytekin. "Vibrational Analysis of PCBS and Electronic Components". A thesis submitted to the graduate school of natural and applied sciences of middle east technical university.

*[9]*Ren Guoquan, Li Ben, Li Dongwei, Jiao yingqi. "Modal Analysis of the Printed circuit board based on finite element method". *2014 International Conference on computer science and electronic technology(ICCSET 2014).*
[10] Arshad khan, Devashish Sarkar, Reshad Ahmar and Hitenkumar patel. "Random vibration analysis and fatigue life evaluation of auxiliary heater barcket. SIMULIA India Regional users Meeting 2011.
[11] Curtean Razvan, Lupea Iulian. "A car component subjected to multiple sources of random vibrations. *RJAV vol IX issue 2/2012.*
[12] G. Durga Prasad, Dr.K.Lalit Narayana, K.Manoj, B.V.Subramanyam. "Random vibration analysis of aerospace vehicle structure". *IPASJ International journal of mechanical engineering (IIJME), Vol 3, Issue 2, Feb 2015.*
[13] Jie Gu, Donald Barker and Michael Pecht. "Prognostics of electronics under vibration using Acceleration sensors". *Proceeding for 62nd Meeting of the Society for Machinery Failure Prevention Technology (MFPT), Virginia Beach, VA, May 2008. 131*
[14] G.S. Aglietti, and C. Schwingshackl. "Analysis of enclosures and anti vibration devices for Electronic Equipment for space Applications". School of Engineering Sciences, *Aeronautics and Astronautics, University of Southampton, UK.*
[15] Robin Alastair Amy, Guglielmo S. Aglietti and Guy Richardson. "Accuracy of simplified printed circuit board finite element models". *Elsevier 2009.*
[16] K.P. Subramanya, Jiwan Kumar Pandit, C.S. Prasad and M.R. Thyagaraj. "Vibration Analysis Study of Spacecraft Electronic Package". *International Journal of Science, Engineering and Technology Research (IJSETR), Volume 3, Issue 3, March 2014.*
[17] Hamid Reza Zandipour & Hasan Jalali. "Modeling & Vibration Analysis of Electronic Equipment in Aerospace Structures". *Indian Journal of Fundamental and Applied Life Sciences, 2014 Vol. 4.*
[18] Michael Nikkhoo. "Vibration Analysis for Electronic Equipment". *Heat sinks, Aug 2005.*
[19] Nilesh R. Bhavsar, H. P. Shinde, Mahesh Bhat. "Determination of Mechanical Properties of PCB". *ISSN (Print) : 2321-5747, Volume-2, Issue-4,2014.*
[20] J. Chung, H. H. Yoo. "Dynamic Analysis of a rotating cantilever beam by using the Finite Element Method". *Journal of Sound and Vibration (2002) 249(1), 147,164.*
[21] Usik Lee, Jinho Shin. "A frequency response function-based structural damage identification method". *Computers and Structures 80 (2002) 117–132 Elsevier science limited.*

Dynamic analysis of rocket motor case

Barla Kavya Sree[1*] and Ch Siva Rama Krishna[2]

[1*]P. G. Scholar, Department of Mechanical Engineering, Vignan's Institute of Information Technology (A), Visakhapatnam, India
[2]Associate Professor, Department of Mechanical Engineering, Vignan's Institute of Information Technology (A), Visakhapatnam, India

* Corresponding author e-mail: kavya4330@gmail.com

Abstract: Rocket Motor Case (RMC) is a major component of Rockets and other Satellite launching vehicles. So, these structures should be of low weight, i.e. Low weight to high strength ratio materials, i.e., composite and alloys materials are used for main components. In the present analysis, 3-Dimensional finite element based static structural analysis and modal analysis are performed on the rocket motor case using aluminum alloy, titanium alloy, alloy steel, carbon epoxy and e glass epoxy. The problem is modeled in student version of ANSYS software. Stresses under radial, hoop, longitudinal directions are calculated with cylindrical co-ordinate system and list of natural frequencies and mode shapes are noted with respect to pre-stressed constrained and free- free boundary conditions. Later rectangular, semicircular, semi hexagonal and triangular types of stiffeners are introduced to the rocket motor case to perform static structural and modal analysis. The effect of stiffeners to the rocket motor case has studied for different materials under static and dynamic loading conditions. Best stiffener for rocket motor case is recommended based on the results, i.e., rectangular section having higher natural frequencies and semi hexagonal shape having less stresses in their respective directions.

Keywords: FEM, Static Structural Analysis, Modal Analysis, Rocket Motor Case.

Nomenclature: σyt = Yield Strength, ρ = Density, E = Young's Modulus, G = Shear Modulus& μ = Poisson's ratio.

1 Introduction

Rocket Motor Case [RMC] provides the pressure shell for propellant as well as the structural shell for the missile. General Dynamics Ordnance and Tactical Systems produce both metal and composite rocket motor cases. They are the leaders in the industrial world in designing, analysis, testing and manufacturing consumer required solid propellant rocket motor cases. Today, as the advancement in carbon fiber technology and high temperature resin offer advantages in performance and cost effectiveness. General Dynamics Ordnance and Tactical Systems produce products based on customers in the design, development and production of rocket motor cases. Rocket motors don't require additional oxygen from atmosphere for combustion of the fuel. In a rocket motor, the chemical energy inside the fuel is converted to the thermal energy by a combustion process. The simple configuration of rocket motor consists of a dome, Igniter, motor case, insulation, propellant and nozzle [1] as shown in figure 1.

IOP Conf. Series: Materials Science and Engineering 1185 (2021) 012036 doi:10.1088/1757-899X/1185/1/012036

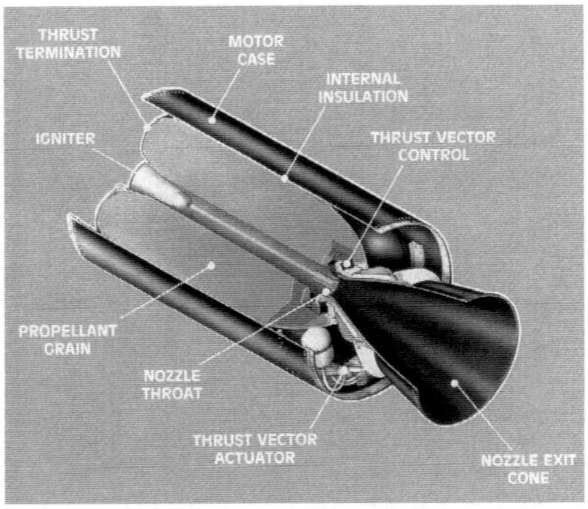

Figure 1. Solid rocket motor

A solid-fuel rocket is a rocket with a motor that uses solid propellants (fuel/oxidizer). Solid rockets are still used today in model rockets and on larger applications for their simplicity and reliability. V Ramanjaneyulu [1] performed static structural analysis on the composite rocket motor case (CRMC) under realistic conditions. He stated that there is a gradual increase of Hoop stress from outer layer to inner layer in all parts of the CRMC. Eswara Kumar A [2] performed static and dynamic analysis on cylinder with various stiffeners; he stated that the helical stiffener shaped design has the higher specific structural stiffness, lower von mises and hoop stress. Eswara Kumar A [3] performed buckling analysis on the cylinder with varying atmospheric temperatures.It was observed that there is no considerable change of buckling load with convective heat transfer co-efficient. But buckling load is decreasing with increase in temperature of inner surface of cylinder. Rakesh Potluri [4] Successful optimization using the response surface optimization procedure for a hybrid laminate has been performed. Bettiet al [5] has successfully performed the Design and Development of VEGA Solid Rocket Motors Composite Cases which has become major support for the development of the CRMC. Sayman [6] studied analysis of multi-layered composite cylinders under hygro thermal loading. In this study, he stated that under hygro thermal loadings condition stress analysis is developed for thick or thin multi-layered composite cylinders. Xia et al. [7] performed multi-layered filament-wound composite pipes under internal pressure.

2. Problem Description

2.1 Problem Statement

The main aim of the problem is to study the dynamic analysis of rocket motor case with different materials and effect of cylinder thickness on the stresses. The problem can be divided into two cases as discussed below.

2.2 Problem Modelling

The geometry of rocket motor case contains many components like igniter end dome, nozzle end dome, cylinder straight end dome, igniter end polar boss insulation, nozzle end polar boss insulation, igniter end bulk head, nozzle end bulk head, flexible rubbers at nozzle end and igniter end bulk heads, nozzle end closure and igniter end closure. The dimensions of the rocket motor case are taken from the Research Article [1] and it is shown in the figure 2.

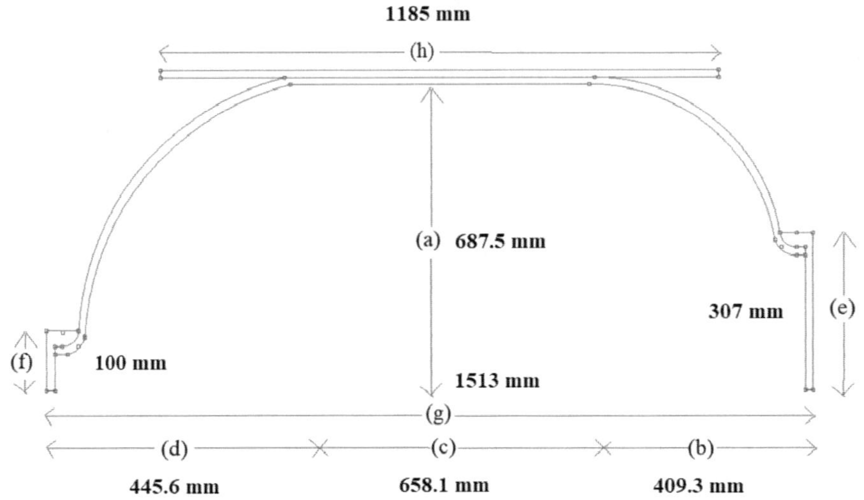

Figure 2. Cross-sectional Dimensions of the Rocket Motor Case.

The 3D geometry model of the rocket motor case with skirt is modeled in the FEA software i.e.,ANSYS 2020 R2 Academic version. The geometry of the model is shown in the figure 3.

Figure 3. Rocket Motor Case

2.3 Materials:

In this research work, 2 types of materials are used, i.e., metals and composite materials. Aluminum is assigned to the Igniter end bulk head, Nozzle end bulk head, igniter end polar boss and nozzle end polar boss. Rubber is assigned to the igniter end and nozzle end rubber portion. Steel is assigned to the igniter end and nozzle end closures. The main objective of the problem is to increase the stiffness and strength of the rocket motor case. The list of the materials used for the rocket motor case and skirt is listed in the below table 1.

IOP Conf. Series: Materials Science and Engineering 1185 (2021) 012036 doi:10.1088/1757-899X/1185/1/012036

Table 1. Isotropic Materials used for rocket motor case with skirt.

Material Name	ρ (g/cm3)	E (MPa)	μ	σyt
Aluminum Alloy 2024 (Heat Treated)	2.79	72000	0.33	455
Titanium Alloy	4.60	110000	0.36	1240
Alloy Steel (Heat Treated)	7.84	207000	0.3	1400 – 2000
Rubber	0	3.2	0.45	-
Aluminum	0.0013	70000	0.32	-
Steel	7.85	210000	0.3	-

The main parts of the assembly i.e., cylindrical motor case with skirt is too focused mainly. The orthotropic conditions of the composite materials which were assigned to the motor case and skirt for the static and Dynamic analysis are as shown below in the table 2.

Table 2. Orthotropic properties of the composite materials.

Material	ρ (g/cm3)	EX (MPa)	EY (MPa)	EZ (MPa)	μXY	μXY	μXY	GXY, (MPa)	GYZ, (MPa)	GXZ, (MPa)
Carbon Epoxy	1.6	36600	46800	30300	0.04	0.289	0.21	4900	4500	4900
E-Glass Epoxy	1.98	19980	19640	14210	0.204	0.312	0.288	5800	5600	5800

Note:
The longitudinal tensile and compressive strength of the carbon epoxy is 640 MPa
The longitudinal tensile and compressive strength of the E Glass epoxy is 428 MPa

2.4 Connections
Connections are given to the components of the rocket motor case to stick together. Bonded contacts are given between the parts of the rocket motor case assembly.

2.5 Mesh
Meshing is defined as conversion of geometric entities into the finite elements. Meshing is properly done on the rocket motor case for convergence of the applied load. Here two types of elements are used for the analysis i.e., solid 186 and solid 187. Solid 186 is a 20-node quadratic element which is applied for the central dome and skirt components. Solid 187 is a 10 node tetrahedron element which is applied for the igniter end and nozzle end domes.

2.6 Loads and Boundary Conditions
The structure of the rocket motor case is analyzed under static load of 9 MPa of internal pressure which is applied on the inside of the surface of the rocket motor case. Igniter end bulk head is fixed in all degrees of freedom i.e., fixed support is given to the igniter bulk head faces which are shown in the figure 4.

.

Figure 4. Loads and Boundary Conditions

3. Static Structural Analysis

The structure of the rocket motor case is analysed under static structural loads i.e., internal pressure load of 9 MPa and the maximum stresses in radial direction, hoop direction and longitudinal direction using aluminium, titanium, steel alloy, carbon epoxy and E Glass epoxy are calculated and listed in the table 3.0. The maximum radial stress of the motor case using aluminium alloy is shown in the below figure 5.

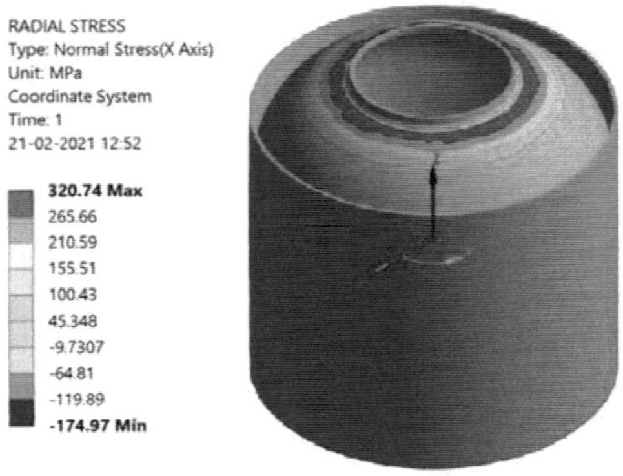

Figure 5. Radial stress of the Rocket Motor case using aluminium alloy

Table 3. Stresses of the Rocket Motor Case under Static Load.

Material	Radial Stress (MPa)	Circumferential Stress (MPa)	Longitudinal Stress (MPa)
Aluminum Alloy	320.74	456.91	336.16
Titanium Alloy	324.59	458.04	334.63
Alloy Steel	317.5	456.41	337.89
Carbon Epoxy	288.65	478.7	332.2
E Glass Epoxy	304.37	460.13	315.24

From the above table, it is noticed that the maximum stresses are coming in circumferential direction of the rocket motor case using titanium alloy in metals and carbon Epoxy in composites. The stresses which are obtained in the static analysis are within the limit of the respective material strength.

IOP Conf. Series: Materials Science and Engineering 1185 (2021) 012036 doi:10.1088/1757-899X/1185/1/012036

4. Modal Analysis

The modal analysis of the rocket motor case using Aluminium Alloy, Titanium Alloy, Alloy Steel, Carbon Epoxy and E-Glass Epoxy is performed under free-free and pre-stressed boundary conditions. The natural frequencies of the RMC using aluminium alloy and mode shapes of the rocket motor case are shown in the below figure 6.

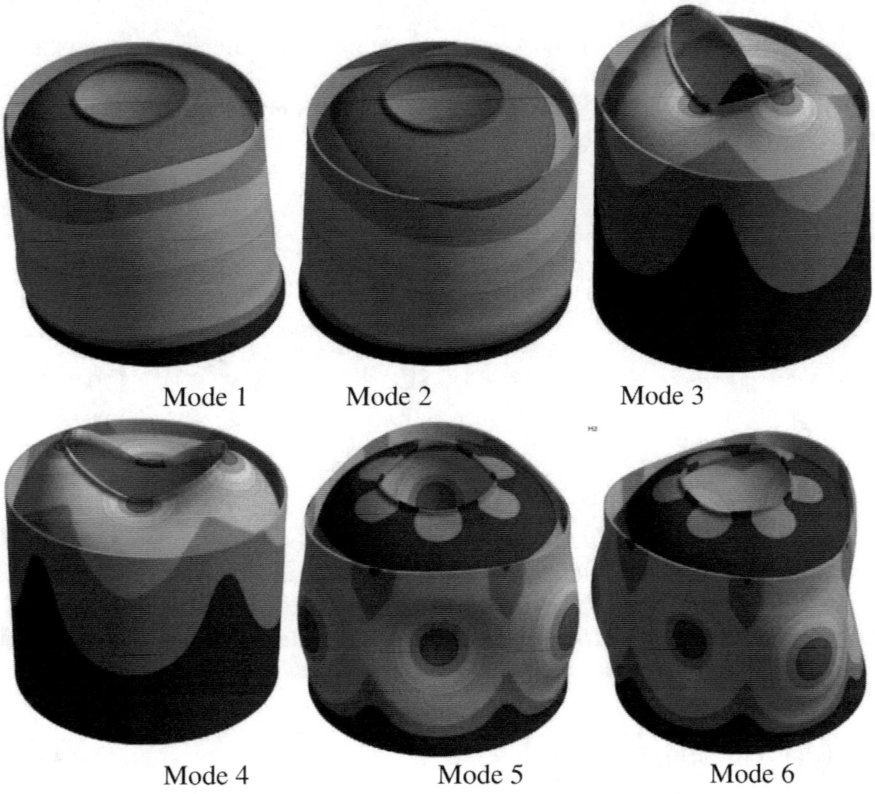

Figure 6. Mode Shapes of the Rocket Motor Case using Aluminium

The natural frequencies of the rocket motor case with different materials under free-free boundary conditions are listed in the below table 4.

Table 4. Natural frequencies of the structure under free - free boundary conditions.

Material Mode No	Aluminium Alloy (Hz)	Titanium Alloy (Hz)	Alloy Steel (Hz)	Carbon Epoxy (Hz)	E Glass Epoxy (Hz)
1	0.	0.	0.	0.	0.
2	0.	0.	0.	0.	0.
3	0.	0.	0.	0.	0.
4	6.953	6.7499	6.9859	4.9313	3.7338
5	7.2463	7.0326	7.2832	5.4714	3.9058
6	7.9235	7.5826	8.0699	5.6368	4.4128
7	226.6	218.73	228.69	184.7	132.19
8	226.96	219.07	229.06	185.04	132.4
9	429.2	412.72	434.73	298.52	234.79
10	429.4	412.92	434.92	298.63	234.88
11	448.89	433.89	452.46	316.02	247.95
12	448.93	433.93	452.5	316.24	247.97

From the above table, it is observed that the first six natural frequencies of the structure of the rocket motor case is nearly equal to zero implies that the structure is free to vibrate in linear X, Y, Z direction and rotate about X, Y, Z Axis. It is recommended to consider the natural frequencies from the seventh mode of the structure.
From the above table it is observed that the structure of the rocket motor case with alloy steel has having higher natural frequencies compared to the other materials. The natural frequency is mainly dependent on the mass and stiffness of the structure. Alloy steel better stiffness results for the higher natural frequencies under free – free boundary conditions.
The natural frequencies of the rocket motor case with different materials under pre-stressed boundary conditions are listed in the below table 5.

Table 5. Natural frequencies of the structure under pre-stressed boundary conditions.

Material Mode No	Aluminium Alloy (Hz)	Titanium Alloy (Hz)	Alloy Steel (Hz)	Carbon Epoxy (Hz)	E Glass Epoxy (Hz)
1	316.01	291.72	284.57	187.39	170.43
2	316.02	291.95	284.86	187.4	17043
3	323.60	301.82	322.41	286.95	245.73
4	323.82	301.83	322.42	287.01	245.76
5	492.23	462.92	477.67	298.46	299.39
6	492.47	463.16	477.9	385.92	323.3

From the above table it is observed that the structure of the rocket motor case with aluminium alloy is having higher natural frequencies compared to the other materials. The natural frequency is mainly dependent on the mass and stiffness of the structure. Aluminium with low mass and better stiffness results for the higher natural frequencies under pre stressed constrained boundary conditions.

5. Rocket Motor Case with Stiffeners

The design of the rocket motor case is modified by adding stiffeners for improving the strength and stiffness of the structure with respect to the static and dynamic loads. Here four types of stiffeners are used i.e., Rectangle, Semi Circle, Semi Hexagon and Triangular shapes. The volume of the structure is constant with different stiffeners. The type of stiffeners is discussed below and shown in the figure 7.

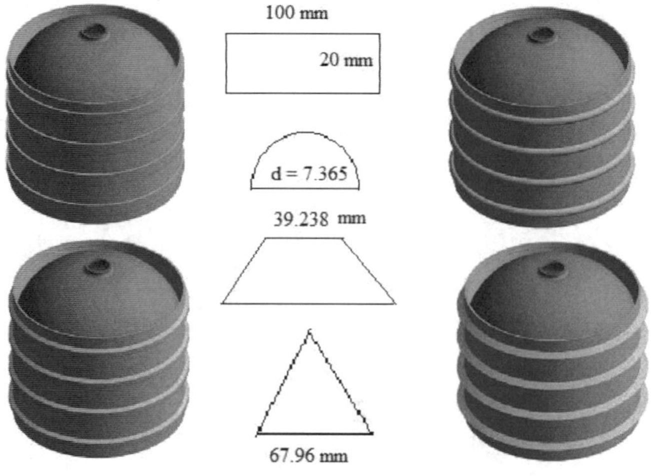

Figure 7. Rocket Motor Case with Stiffeners

Static Structural Analysis are performed and results with rectangular, semicircular, semi hexagonal and triangular sections using aluminium are noted in the below table 6.

Table 6. Static structural results with aluminium.

Stress\Shape	Rectangular	Semi Circular	Semi Hexagonal	Triangular
Radial Stress	297.73 MPa	297.47 MPa	297.64 MPa	297.29 MPa
Circumferential Stress	441.44 MPa	441.25 MPa	440.87 MPa	441.11 MPa
Longitudinal Stress	362.50 MPa	369.45 MPa	357.31 MPa	361.18 MPa

From the above table, the maximum stresses are located in the circumferential direction and minimum stresses are located in radial direction.
From the above table, it is observed that the minimum radial stress is produced with the triangular shape. The minimum circumferential stress is produced with the semi hexagonal shape. The minimum longitudinal stress is produced with the semi hexagonal shape.
The static structural analysis of the rocket motor case with stiffeners are performed with remaining materials (Titanium Alloy, Alloy Steel, Carbon Epoxy and E Glass Epoxy) and it is noticed that the obtained stresses are within the limits of respective material strength.
Dynamic Analysis (Modal Analysis) is performed on the rocket motor case with rectangular, Semi Circular, semi hexagonal and triangular sections using aluminium under pre-stressed constrained boundary conditions and respective natural frequencies with respect to the mode number are listed in the below table 7.

Table 7. Natural frequencies of the rocket motor case with Pre-stressed Constrained boundary conditions.

Shape Mode No	Rectangular (Hz)	Semi Circular (Hz)	Semi Hexagonal (Hz)	Triangular (Hz)
Mode 1	320.01	309.92	314.21	311.02
Mode 2	320.06	310.08	314.74	311.9
Mode 3	336.48	315.31	316.29	317.07
Mode 4	336.82	315.39	316.61	317.16
Mode 5	503.97	490.15	496.03	499.68
Mode 6	504.8	490.4	497.13	500.51

From the above table, it is observed that the higher natural frequencies are obtained with rectangular section, followed by the triangular, semi hexagonal and semi circular under pre-stressed constrained boundary conditions.
The modal analysis of the rocket motor case with stiffeners is performed with remaining materials (Titanium Alloy, Alloy Steel, Carbon Epoxy and E Glass Epoxy) and it is noticed that the maximum value of natural frequencies is obtained with the rectangular section.
The volume of the rocket motor case with and without stiffeners are listed in the below table 8.

Table 8. Volume of the rocket motor case.

	Rocket Motor Case	Rocket Motor Case with stiffener
Volume (mm3)	1.8222e+008 mm^3	2.01E+08 mm^3

The percentage increment of volume with stiffener is listed below.

$$\frac{2.01E+08-1.8222e+008}{1.8222e+008} X\ 100 = 10.12$$

From the above calculation, the total weight of the RMC with stiffeners is increased by 10.12% of its total weight without stiffener.

6 Conclusions

From the present work, the following observations and conclusions are drawn from the analysis of Rocket motor case.

1) Higher stresses are obtaining in circumferential direction compared to the radial and longitudinal directions.
2) The stresses which are produced from the rocket motor case using different materials are within the limits of respective material strength.
3) Rocket motor case using Aluminum having higher natural frequencies than the other materials.
4) The stresses of the rocket motor case with semi hexagonal shape stiffener having less value compared to the other stiffeners.
5) The natural frequencies of the rocket motor case with rectangular shape stiffener are higher compared to the other stiffeners.
6) It is recommended to use rectangular or semi hexagonal shape stiffeners to the rocket motor case for static and dynamic loads.

References

[1] Ramanjaneyulu V, Balakrishna Murthy V, Chandra Mohan R, Naga Raju Ch, 2018 "*Analysis of Composite Rocket Motor Case using Finite Element Method*"Materials Today: Proceedings **5** 4920–4929.

[2] Eswara Kumar A, Krishna Santosh R, Ravi Teja S, Abishek E, 2018 "*Static and Dynamic Analysis of Pressure Vessels with Various Stiffeners*" Materials Today: Proceedings **5** 5039–5048.

[3] Eswara Kumar A, Karteek Navuri, Manideep K, Priyanka M, 2018 "*Effect of Thermal Environment on Buckling of Thick Cylinder Subjected to Combined Axial Compressive and External Pressure loads*" Materials Today: Proceedings **5** 3298–3305.

[4] Rakesh Potluri, Eswara Kumar A, Karteek Navuri, Nagaraju M, Duduku Mojeswara Rao, 2016*"Buckling analysis of a ring stiffened hybrid composite cylinder"*,IOP Conf. Series: Materials Science and Engineering **149** 012148 doi:10.1088/1757-899X/149/1/012148.

[5] Betti F, Perugini P, Mataloni A, Tessitore N, Leone A, Di Cosmo A, 2007*"Design and Development of VEGA Solid Rocket Motors Composite Cases*", **43rd** AIAA/ASME/SAE/ASEE Joint Propulsion Conference & Exhibit, Cincinnati, *OH., 8-11, pp.117.*

[6] Sayman O, *2005"Analysis of Multi-Layered Composite Cylinders under Hygrothermal Loading",Composites Part A, pp.* **1-11**.

[7] Xia M, Takayanagi H, Kemmochi K, 2001 "*Analysis of Multi-Layered Filament-Wound Composite Pipes under Internal Pressure*", Composite Structures, **53**, 483-491.

[8] Xia M, Takayanagi H, Kemmochi K, 2002 "*Bending Behavior of Filament- Wound Fiber-Reinforced Sandwich Pipes*", Composite Structures, **56,** 201-210.

[9] Pawar D, Abhay Utpat, 2014"*Analysis of Composite Laminate for Maximum Stiffness*", International Journal of Recent Technology and Engineering (IJRTE) ISSN: 2277-3878, 3, 2.

[10] Madhavi Sinha, Pandit S N, 2012 "Design *and Burst Pressures Analysis of CFRP Composite Pressure Vessel for Various Fiber Orientations Angles''*, International Journal of Advances in Engineering Science and Technology, 13856236. .

[11] Vinay K. Goyal, Jacob I. Rome, Patrick M. Schubel, 2008"*structural analysis of solid rocket motor*", **49th** AIAA/ASME/ASCE/AHS/ASC Structures, Structural Dynamics, and Materials Conference, Schaumburg, IL.

[12] Aziz Onder, 2007 *"First failure pressure of composite Pressure vessels"*, Master's Thesis in Mechanical Engineering, Mechanics Program, Graduate School of Natural and Applied Sciences of DokuzEylül University.

[13] Tsouvalis N C, Zafeiratou A A, Papazoglou Gabrielides N C, Kaklis P. D, 2001 *"Numerical modeling of composite laminated cylinders in compression using a novel imperfections modeling method"*, Composites: Part B **32**, 387-399.

[14] Madhavi M, Rao K V J, Narayana Rao K, 2009 *"Design and Analysis of Filament Wound Composite Pressure Vessel with Integrated end Domes"*, Defence Science Journal, **59**, **1**, 73-81.

[15] David L. Gray, Flower Mound, Daniel J. Moser, *"Finite Element Analysis of a Composite Overwrapped Pressure Vessel"*, 40th AIAA/ASME/SAE/ASEE Joint Propulsion Conference and Exhibit, Fort Lauderdale, Florida. Composites Inc., Sandy, UT, 84092, 11-14July 2004.

A Study on Mechanical Properties of Hybrid FiberReinforced Polymer Composites

Badavatu Srikar [1], Lokireddi Venkata Venu Gopala Rao [2]

[1]P. G. Scholar, Department of Mechanical Engineering, Vignan's Institute of Information Technology (A), Visakhapatnam, India
[2]Professor, Department of Mechanical Engineering, Vignan's Institute of Information Technology (A), Visakhapatnam, India

* Corresponding author e-mail: badavatusrikar@gmail.com

Abstract:-These daysalmost all structures are being replaced with composite materials due to their high strength to weight ratio along with good mechanical properties. Thecomposite properties will change based on the factors like size, shape, length, weight, nature and resin used. In current scenario to achieve the particular properties almost all industries are using the hybrid composites. In present work, four types of woven fabrics are taken i.e., Carbon, E-Glass, Flax and Bamboo. From these four types, a total of 4 composites (Carbon, E-Glass, Flax & Bamboo) and total of 6 hybrid composites (Flax + E-Glass, Bamboo + E-Glass, Flax + Carbon, Bamboo + Carbon, Bamboo + Flax and Carbon + E-Glass) are fabricated with ASTM D3039 and tested under tensile machine. The effect of hybrid composite models on the density, young's modulus and yield strength are calculated. The best hybrid composite material is selected based upon the low density, high tensile strength and high young's modulus.

Keywords: Hybrid composite, Carbon, E-Glass, Bamboo, Flax, Epoxy and Hardener.

I. Introduction

Composite material is a material made from two or more constituent materials with significantly different physical or chemical properties, which are combined to produce a material with characteristics which are stronger and more durable than the previous components. The previous individual components will form a bond which indeed increases the desirable characteristics. The latest material can be preferred for many reasons for it's usage like stronger, harder, lighter, less expensive when compared to traditional materials.Composite material is generally classified into two differenttypes, the first type of classification is usually made with respect to the matrix constituent and the second type of classification refers to the reinforcement. S. Prabhakaran[1] et al analysed in detail the damping and sound absorption characteristics of flax and glass fiber reinforced composites and found that flax fiber reinforced composite has 21.42% and 25% higher sound absorption coefficient and at low frequencies it has 51.03% higher vibration damping than the glass fibre reinforced composite and the weight ,flexural strength and flexural modulus of flax fiber are between than glass fiber reinforced composites without sacrificing stiffness-to-weight ratios. Dakai Chen [2] et al analysed in detail the effect of ammonium polyphosphate and plasticizer poly (Butylene adipate-co-terephtlate) on the sound absorption property of ramie fibre, and found out composite with short ramie fiber reinforced PLLA composite. The SEM reveal micro-phase separation and porosity between single fibre bundles to short fibre makes more better acoustical absorptive.

Y.Hong et al [3] analyzed in detail the damping characterizes of several advanced fiber reinforced composite Finite Element analysis is carried out based on modal strain energy method to find out modal shapes, modal frequencies and modal loss factor. By using different layer angels and layup sequences, the composite structure were optimized. They observed increase of frequency trend with carbon/glass fiber composite in bending made with layup sequence 0,90 and reverse trend with layup sequence of 30,45,60. Jean Marie Berthelot [4] et al done experiments on unidirectional glass and Kevlarfiber composite materials by using a cantilever beam for damping values and compared these results with literature models such as Adams-bacon analysis, Ni – Adama analysis and complex stiffness model. A deviation of 10 to 35 of fibre orientation is seen in Kevlarfiber composites he obtained better results in modelling than experimental due to the specimen difference in both. In ritz method the specimen is rectangular so width is considered while AB analysis doesn't consider this width the inplane shear damping of shear energy to energy dissipated which is different for each analysis.

Jean Marie Berthelot [5] et al carried out experiments viz., orthotropic composites with viscoelastic layers on a rectangular laminates using ritz method. Natural frequencies and along with modes are obtained, the transverse shear energies can be obtained in each layer and damping is evaluated for different energies and different angles of laminate. The strains in each viscoelastic layers are observed. If we take multiple damping layers, laminates and complex structures the damping modelling can be used by finite element procedure. Ahmed Alb- El- hamid hamada [6] et al carried out experiments on a composite with different cores (steel and aluminium) as well as composite material with different fiber orientation, aligned longitudinally, transversely making 45^0 with each other and randomly oriented on which FFT is carried out and found that coatedlaminated beams have high damping ability than isotropic metal beams. Lamina coated has better than normal structure that too without change in natural frequency of structure.C.Kyriazoglou et al [7] carried out experiments of hybrid method which is applied to homogeneous material for damping by FEA and damping testing equipment. These both have good coordination while testing the specimen GFRP and CFRP which is done at 0^0 and 90^0 continuous GFRP lamina,woven (0^0,90^0),GFRP 8HSW, woven (0^0,90^0) and (+- 45^0)5HSWFRP lamina are found.The objective of the problem is to study the mechanical properties of Plane Composites (carbon Epoxy, E-Glass Epoxy, Bamboo Epoxy and Flax Epoxy) and hybrid composites (Flax + E-Glass Epoxy, Bamboo + E-Glass Epoxy, Flax + Carbon Epoxy, Bamboo + Carbon Epoxy, Bamboo + Flax Epoxy and Carbon + E-Glass Epoxy).

2. Experimentation

2.1 ASTM D3039

The specimen dimensions recommended by the ASTM D3039 for tensile test is shown in Figure 1.

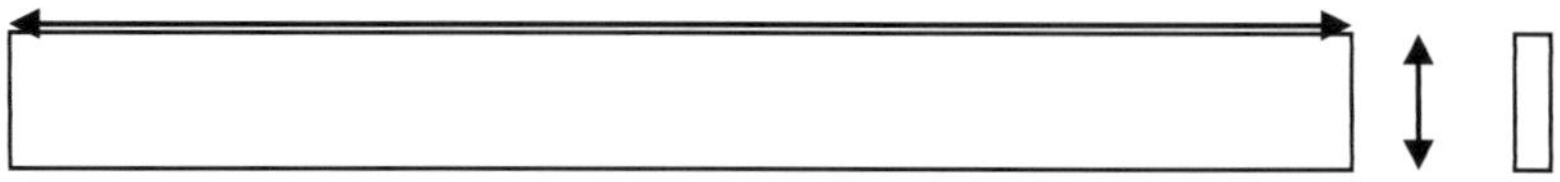

Figure 1 ASTM D3039 Specimen Dimensions

Length of the work piece	=	250 mm
Width of the work piece	=	25 mm
Thickness of the work piece	=	2.5 mm

2.2 Carbon Fibre Reinforced Polymer (CFRP)

Carbon fiber reinforced polymer (CFRP), or carbon fiber reinforced thermoplastic (CFRTP) is an very strong and light weight fiber reinforced plastic which contains carbon fibers. In CFRPs the reinforcement is the carbon fiber along with the matrix is usually a polymer resin. The material properties depends on those very two elements.

The main applications are guitar picks, laptop cases and audio Components such as loud speakers. The advantages are high tensile strength, low thermal expansion, light weight, corrosion resistant and electrical conductivity. The disadvantages are expensive in cost, Cracks propagation and holes are created while machining.

Two directional carbon woven fabricsare gathered from the market in the form of sheets of length 1.5 meters. The required length for the carbon lamina is 260 mm in length and 27 mm width and thickness of the sheet is 0.3 mm. The carbon woven sheet is shown in the below Figure 2.

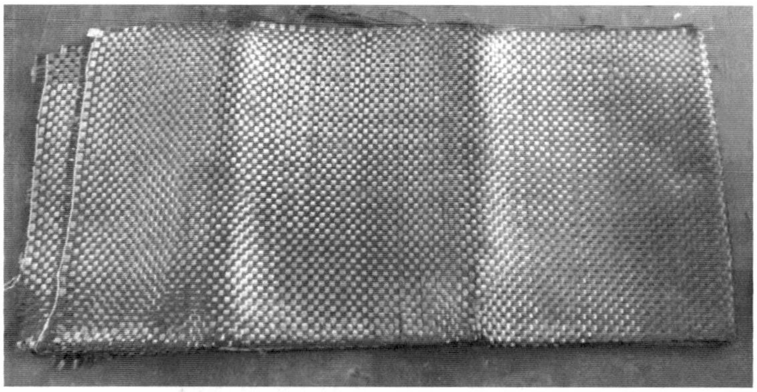

Figure 2 Carbon woven Fabric

Epoxy Resin:

The type of epoxy resin used in this work is Araldite LY556 which chemically belongs to epoxies family 25°C. The advantages of the epoxy resin are water resistant, low shrinkage, positive mechanical properties, excellent adhesive properties, chemical & Heat Resistance and very good electrical insulating properties.

The disadvantages of the epoxy resin are harmful to health and do not perform well in colder temperatures. Epoxy resin LY556 is used as the matrix for binding purpose to the carbon lamina which is shown in the below figure.Hardener HY951 is used for the epoxy resin as it is mixed with the 1:10 ratio of the epoxy resin which is shown in the below Figure 3.

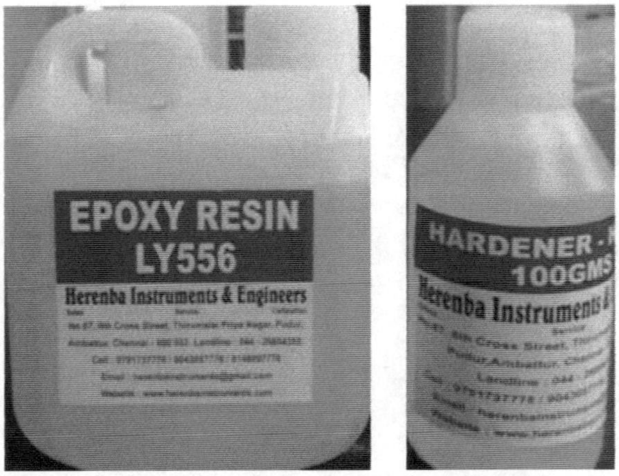

Figure 3 Epoxy Resin LY556 and Hardener

The laminate's of the Carbon Epoxy, E-Glass Epoxy, Flax Epoxy and Bamboo Epoxy are shown in the below Figure 4. Clear films are used for the preparation of the different lamina sheets for non sticky purpose.

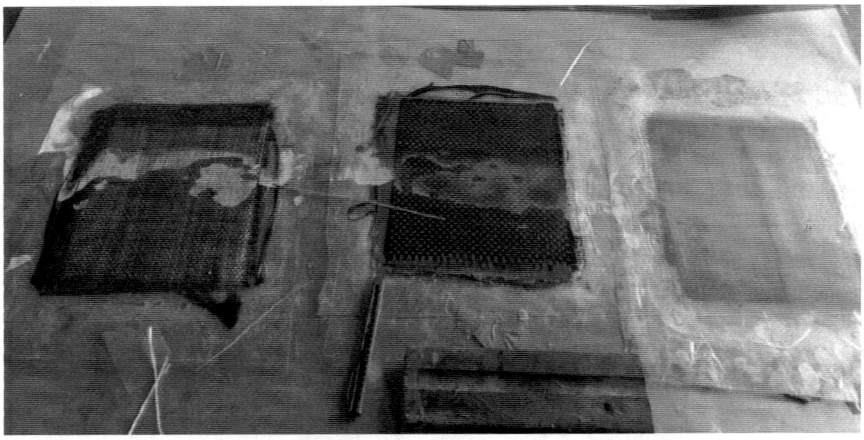

Figure 4 Laminate Preparation

2.3 Glass Fiber Reinforced Polymer (GFRP):

The reinforced polymers consisting of glass fibers are known as glass fiber reinforced polymers. Glass fibers have been around us for centuries,these are produced in bulk in strand form is done from 1932 when Games slayter a researcher accidentally discovered them at owens-illinois directed a jet of compressed air at a stream of molten glass and produced fibers.

The physical properties of GFRP are high corrosion resistance, low dielectric constant, higher electrical resistivity and high tensile strength. The applications of GFRP are automobile parts, home furniture, boats.

The advantages of GFRP are low modulus of elasticity, self abrasiveness, relatively low fatigue resistance, higher density compared to carbon fibers and organic fibers, high tensile strength and good insulation. The disadvantages are brittle and weak abrasive resistance.

Two directional E-Glass woven fabrics are gathered from the market in the form of sheets of length 1.5 meters. The required length for the E-Glass lamina is 260 mm in length and 27 mm width. The E Glass woven sheet is shown in the below Figure 5.

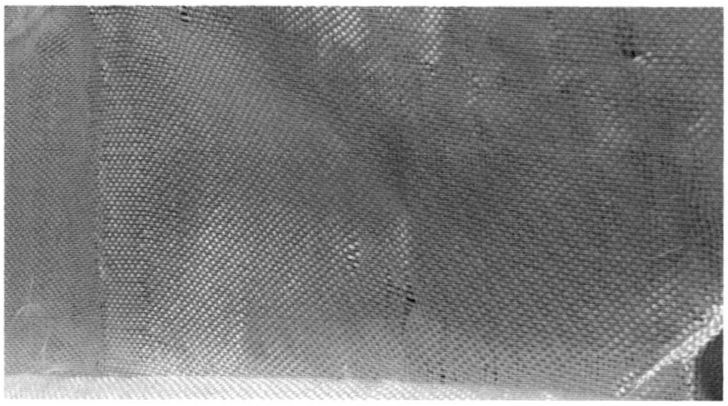

Figure 5 E Glass woven Fabric

2.4 Flax Fiber Reinforced Polymer

Flax fiber reinforced polymer is a natural polymer comes from the flowering plant which is of best family. It is comparatively has high strength than many natural fibers. The advantages of the FFRP are lower pollution level during production, CO2 Neutral, low cost, low Density, renewable resource and recyclable material.

The disadvantages of the FFRP composite are Stress – Strain response is nonlinear, highly inhomogeneous, degradation of properties and properties depend on growing and processing conditions.

Two directional flax woven fabrics are gathered from the market in the form of sheets of length 3 meters. The required length for the flax lamina is 260 mm in length and 27 mm width. The flax woven sheet is shown in the below Figure 6.

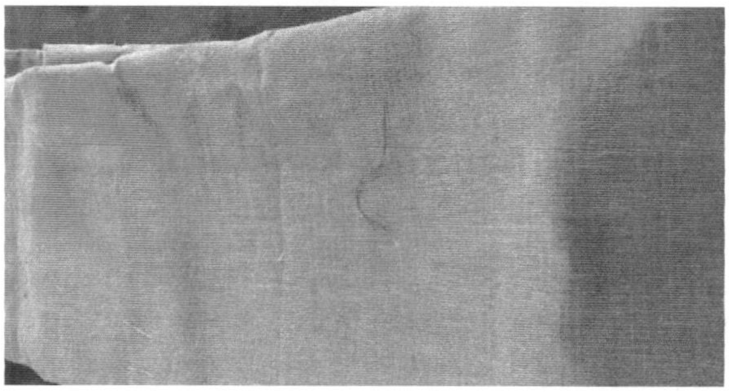

Figure 6 Flax woven Fabric

2.5 Bamboo Fiber Reinforced Polymer

Bamboo fiber is cellulose fiber plant in grasses family which is produced from natural bamboo plant which is grown within 6 to 8 months of tall 40 m. Now days, bamboo fibers are using frequently in the market due to their own properties. There are around 1000 species of bamboo recognized worldwide. Its fiber has many advantages over cotton fiber. Applications: Textile industries, used for making fabric as meso, antibacterial property, moisture absorption, elasticity, UV resistant, building materials, bathroom products, decorating items, hygiene products, etc.

Two directional bamboo woven fabrics are taken in the form of sheets of length 3 meters. The required length for the bamboo lamina is 260 mm in length and 27 mm width. The bamboo woven sheet is shown in the below Figure 7.

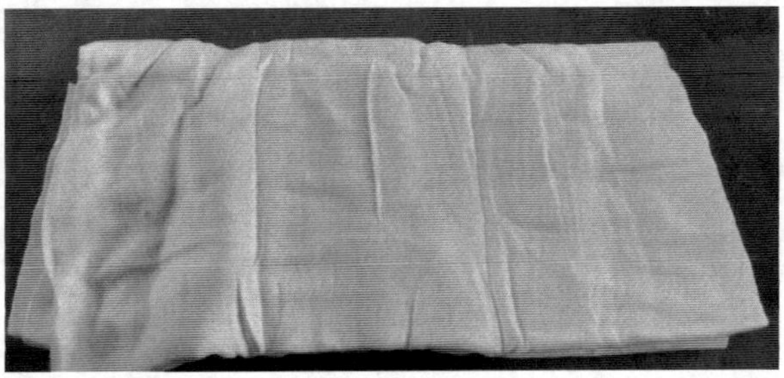

Figure 7 Bamboo woven Fabric

2.6 Laminates:

Bamboo epoxy laminate is made up of ten bamboo lamina (Fiber orientation of 0^0, 90^0) glued together by the epoxy resin and the final laminate figure is shown in the below Figure 8.

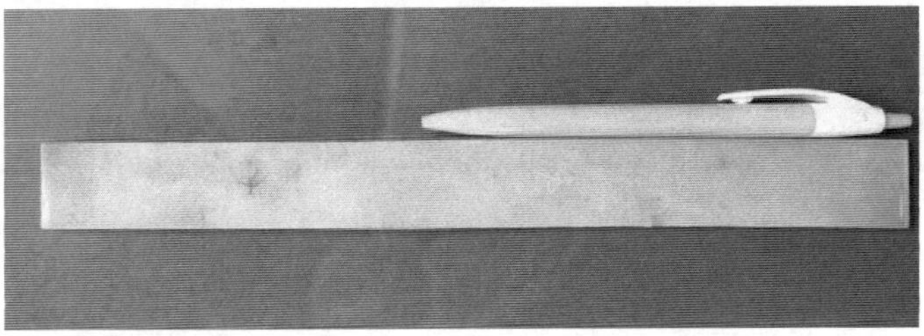

Figure 8 Composite laminate (Bamboo Epoxy)

Flax epoxy laminate is made up of three flax lamina (Fiber orientation of 0^0, 90^0) glued together by the epoxy resin and the final laminate figure is shown in the below figure 9.

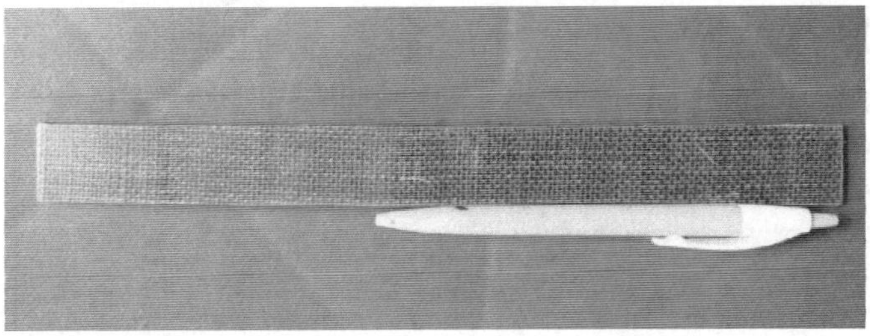

Figure 9 Composite laminate (Flax Epoxy)

E Glass epoxy laminate is made up of six E Glass lamina (Fiber orientation of 0^0, 90^0) glued together by the epoxy resin and the final laminate figure is shown in the below Figure 10.

IOP Conf. Series: Materials Science and Engineering 1185 (2021) 012037 doi:10.1088/1757-899X/1185/1/012037

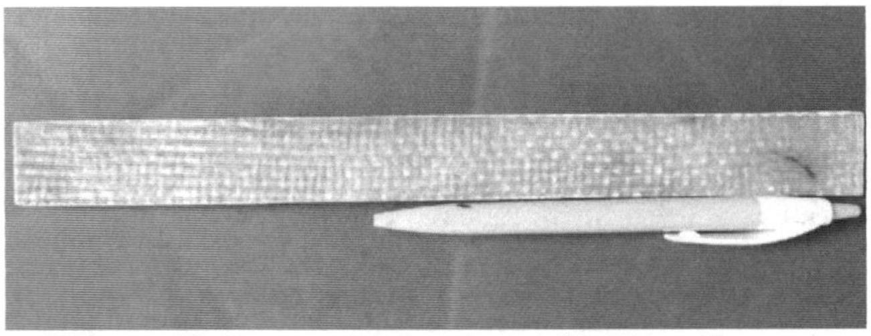

Figure 10 Composite laminate (E Glass Epoxy)

Carbon epoxy laminate is made up of six carbon lamina (Fiber orientation of 0^0, 90^0) glued together by the epoxy resin and the final laminate figure is shown in the below Figure 11

Figure 11 Composite laminate (Carbon Epoxy)

A Hybrid reinforced polymer Composite laminate having combination of Flax + E Glass is made up of two flax lamina's and three E Glass lamina (Fiber orientation of 0^0, 90^0) glued together by the epoxy resin and the final laminate figure is shown in the below Figure 12

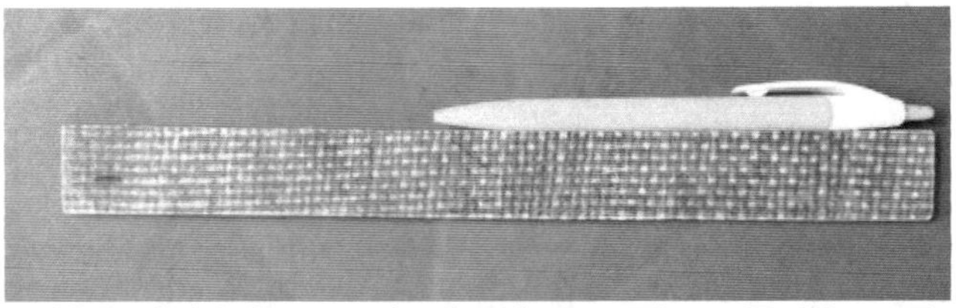

Figure 12 Hybrid Composite laminate (Flax +E-Glass Epoxy)

A Hybrid reinforced polymer Composite laminate having combination of Bamboo + E Glass is made up of three Bamboo lamina's and three E Glass lamina (Fiber orientation of 0^0, 90^0) glued together by the epoxy resin and the final laminate figure is shown in the below Figure 13.

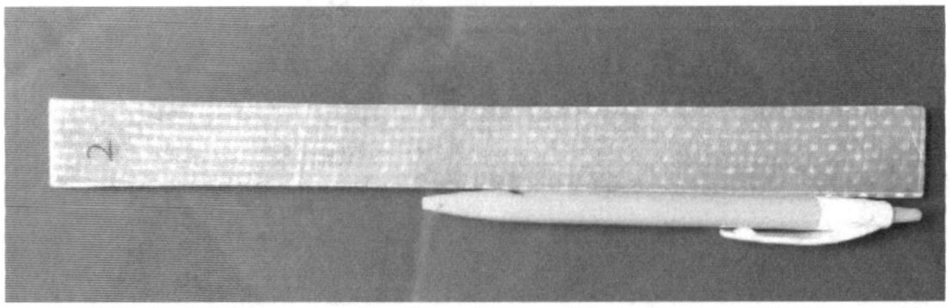

Figure 13 Hybrid Composite laminate (Bamboo + E-Glass Epoxy)

IOP Conf. Series: Materials Science and Engineering 1185 (2021) 012037 doi:10.1088/1757-899X/1185/1/012037

A Hybrid reinforced polymer Composite laminate having combination of Flax + Carbon is made up of two flax lamina's and two Carbon lamina (Fiber orientation of 0^0, 90^0) glued together by the epoxy resin and the final laminate figure is shown in the below Figure 14.

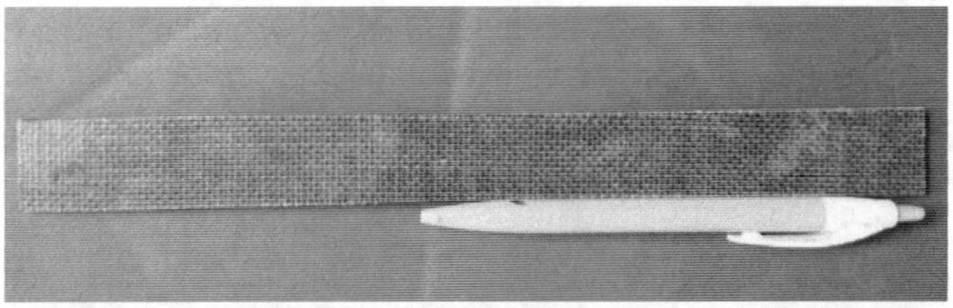

Figure 14 Hybrid Composite laminate (Flax + Carbon Epoxy)

A Hybrid reinforced polymer Composite laminate having combination of Bamboo + Carbon is made up of four bamboo lamina's and three Carbon lamina (Fiber orientation of 0^0, 90^0) glued together by the epoxy resin and the final laminate figure is shown in the below figure 15.

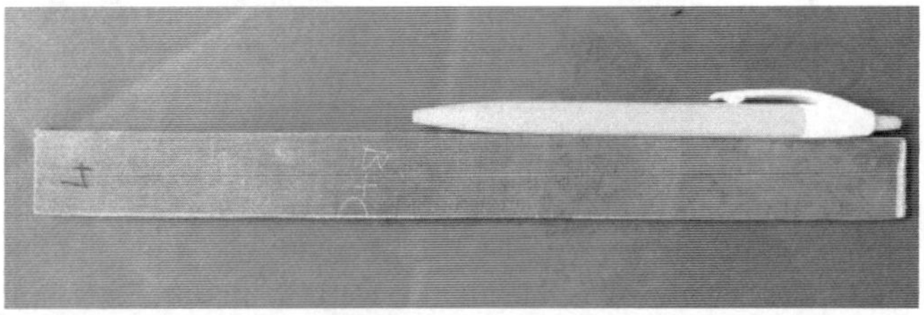

Figure 15 Hybrid Composite laminate (Bamboo + Carbon Epoxy)

A Hybrid reinforced polymer Composite laminate having combination of Bamboo + Flax is made up of three bamboo lamina's and two flax lamina (Fiber orientation of 0^0, 90^0) glued together by the epoxy resin and the final laminate figure is shown in the below Figure 16.

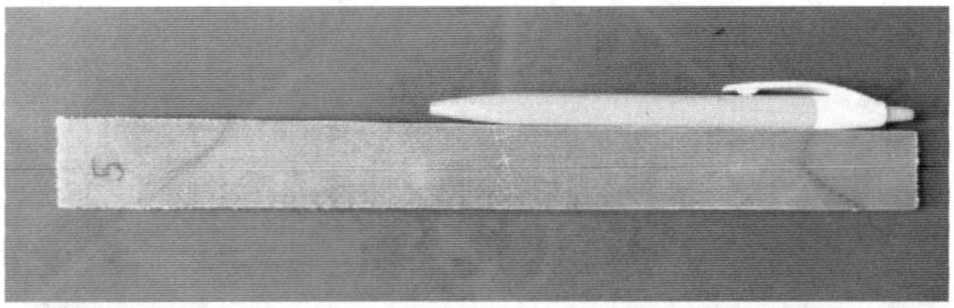

Figure 16 Hybrid Composite laminate (Bamboo + Flax Epoxy)

A Hybrid reinforced polymer Composite laminate having combination of Carbon + E Glass is made up of three carbon lamina's and three E Glass lamina (Fiber orientation of 0^0, 90^0) glued together by the epoxy resin and the final laminate figure is shown in the below Figure 17.

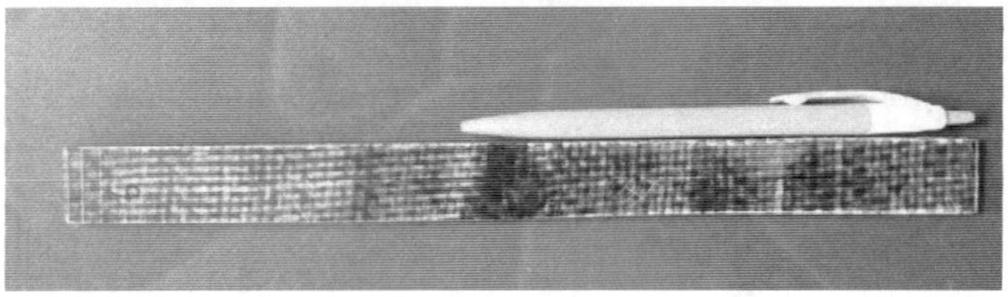

Figure 17 Hybrid Composite laminate (Carbon +E-Glass Epoxy)

Note: The above specified specimens are fabricated up to 3 sets, i.e., each variety is fabricated with 3 specimens.

3. Testing and Results

Tensile test on the specimens is performed by the universal tensile machine which uses the standards of ASTM D3039 and the calculated results and the density of the plane composites are shown in Figure 18.

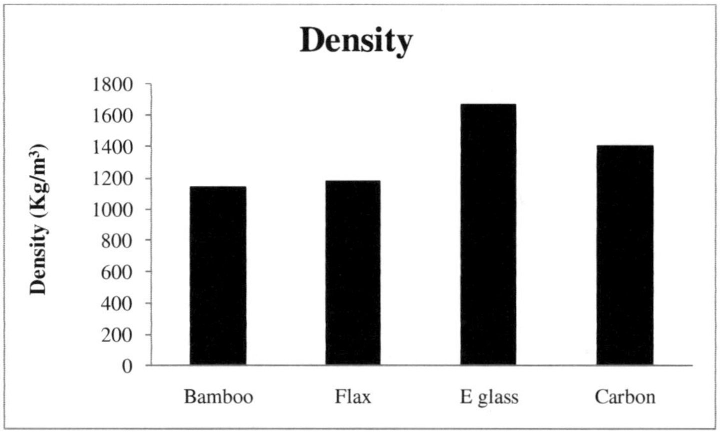

Figure 18 Densities of the Plane Composites

From the above figure, it is shown that the maximum value of the density takes place with the E Glass Epoxy and minimum value of the density takes place with bamboo Epoxy. The young's modulus of the plane composites are shown in the below Figure 19.

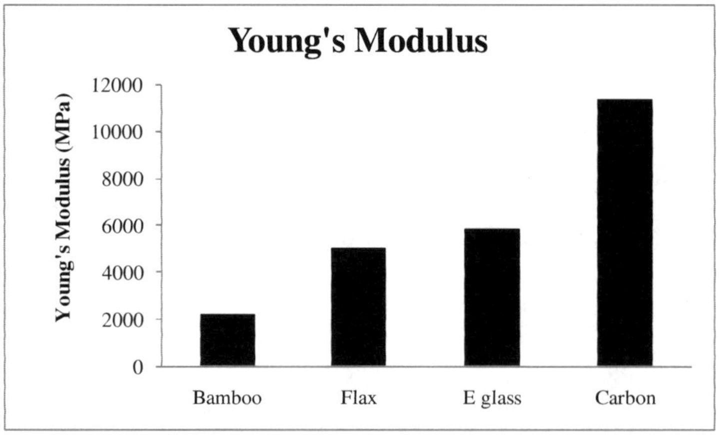

Figure 19 Young's Modulus of the Plane Composites

From the above figure, it is shown that the maximum value of the young's modulus takes place with the Carbon Epoxy and minimum value of the density takes place with

Bamboo Epoxy. The tensile strength of the plane composites are shown in the below Figure 20.

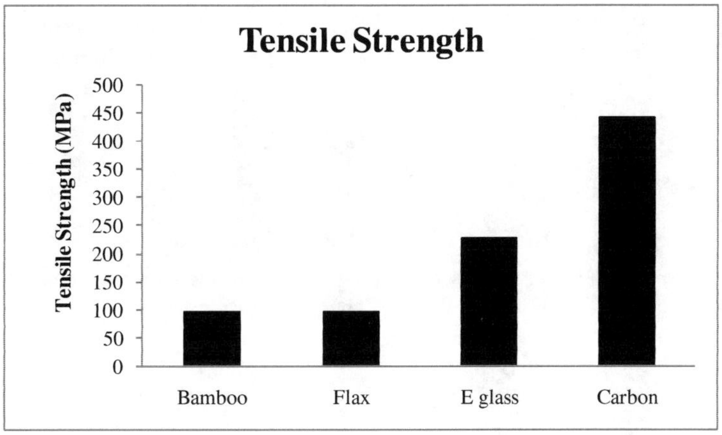

Figure 20 Tensile Strength of the Plane Composites

From the above figure, it is shown that the maximum value of the tensile strength takes place with the carbon epoxy and minimum value of the tensile strength takes place with bamboo epoxy. The densities of the hybrid composites are shown in the below Figure 21.

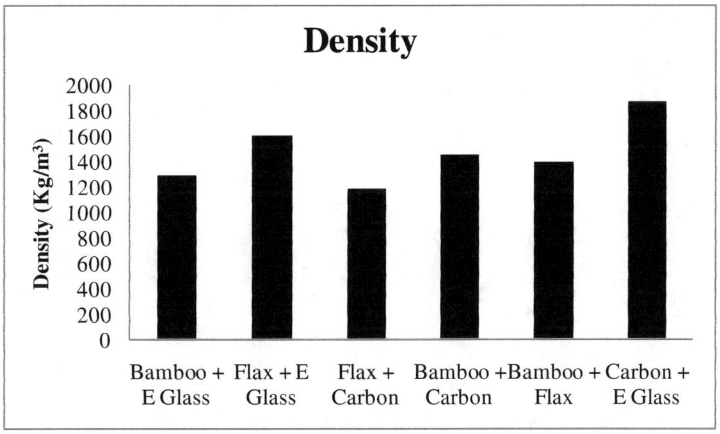

Figure 21 Densities of the Hybrid Composites

From the above figure, it is shown that the maximum value of the density takes place with the Carbon + E Glass Epoxy and minimum value of the density takes place with Flax + Carbon Epoxy. The young's modulus of the hybrid composites are shown in the below Figure 22.

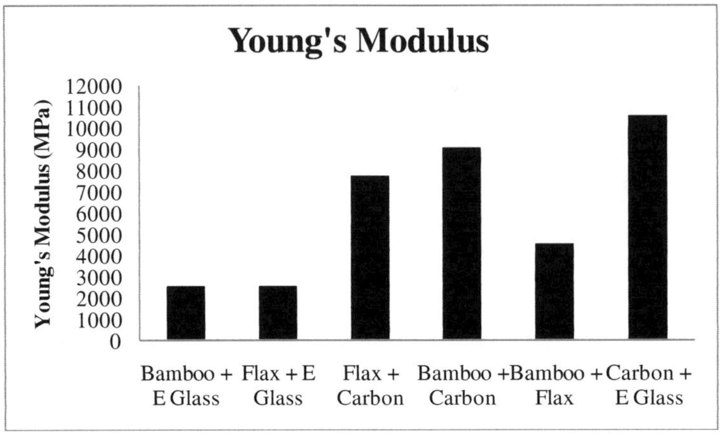

Figure 22 Young's Modulus of the Hybrid Composites

From the above figure, it is shown that the maximum value of the young's modulus takes place with the Carbon + E Glass Epoxy and minimum value of the density takes place with Flax + E Glass Epoxy. The tensile strength of the hybrid composites are shown in the below Figure 23.

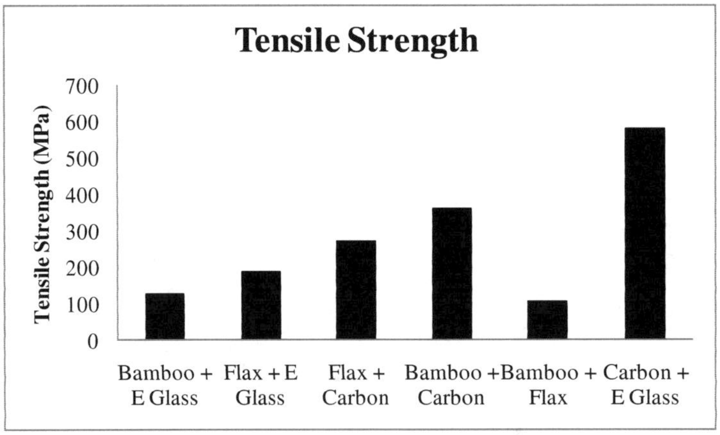

Figure 23 Tensile Strength of the Hybrid Composites

From the above figure, it is shown that the maximum value of the Tensile Strength takes place with the Carbon + E Glass Epoxy and minimum value of the Tensile Strength takes place with Bamboo + Flax Epoxy.

4. Conclusions

From the experimental results for the mechanical properties of the plane composites, carbon epoxy posse's higher young's modulus and higher tensile strength. Bamboo epoxy having less young's modulus and less yield strength values. From the hybrid composite test results, carbon + E Glass epoxy hybrid composite posses higher young's modulus and higher yield strength compared to other materials. So, it is recommended to use Carbon + E-Glass Epoxy hybrid composite for higher load applications.

IOP Conf. Series: Materials Science and Engineering 1185 (2021) 012037 doi:10.1088/1757-899X/1185/1/012037

References

[1] S Prabhakaran, V Krishnaraj, M Senthil kumar and R Zitoune, 2014 "Sound and Vibration Damping Properties of Flax Fiber Reinforced Composites", *Procedia Engineering 97 573 – 581.*

[2] Ke Jun, Wu Zhenyu, Chen Xiaoying Ying Zhiping, 2019 "A review on material selection, design method and performance investigation of composite leaf springs, *Composite Structures*, 226, 111277.

[3] Y. Hong, X.D. He, R.G. Wang, Y.B. Li, J.Z. Zhang, and H.M. Zhang, 2011"Damping Analysis of Fibre-Reinforced Composites and Structure, National Key Laboratory for Advanced Composites in Special Environment", *Harbin Institute of Technology, Harbin 150080, P.R. China.*

[4] Jean-Marie Berthelot,Youssef Sefrani,2004 "Damping analysis of unidirectional glass and Kevlar fibre composites", *Composites Science and Technology 64, 1261–78.*

[5] Muhammet Raci Aydın, Ömer Gündoğdu, Barbaros Kaya, Gürbüz Bayraktar, Okan Kaan Aksuoğlu and Osman Hotunlu, 2018 "Effect of Orientation Angles on Vibration Properties at Carbon Fiber reinforced Polymeric Composites, Engineering Sciences, *ISSN: 1308 7231.*

[6] A Treviso, B Van Genechten, D Mundo and M Tournour, 2015" Damping in composite materials: Properties and models", *Composites Part B 78, 144-152.*

[7] T Wollmann, N Modler, M Dannemann, A Langkamp, S Nitschke and A Filippatos, 2016 "Design and testing of composite compressor blades with focus on the vibration Behaviour", *Composites Part A: Applied Science and Manufacturing, (16) 30191-9.*

[8] Hui Lu, Xianping Wang, Tao Zhang, Zhijun Cheng, and Qianfeng Fang, 2009 "Design, Fabrication, and Properties of High Damping Metal Matrix Composites—A Review", *Materials (Basel). ; 2(3): 958–977.*

[9] S Mahmoudi, A Kervoelen, G Robin, L Duigou, EM Daya and JM Cadou, 2018 "Experimental and numerical investigation of the damping of flax-epoxy composite plates". *Composite Structures, 208, 426-433.*

[10] Atul Sharma, M L Aggarwal, Lakhwinder singh. 2017 "Experimental investigation into the effect of noise and damping using composite spur gear", *Materials Today: Proceedings 4 2777–82.*

[11] A S Hadi & J N Ashton, 1996 "Measurement and theoretical modelling of the damping properties of a uni-directional glass/epoxy composite", *Composite Structure 34 381-385.*

[12] Sam Ashworth, Jem Rongong, Peter Wilson, James Meredith," 2016 Mechanical and damping properties of resin transfer moulded jute-carbon hybrid composites", Composites Part B: Engineering, *105, 60-66.*

[13] Agnivesh Kumar Sinha, Harendra K Narang and Somnath Bhattacharya, 2017 "Mechanical properties of natural fibre polymer composites", *J Polym Eng ; 37(9): 879–895.*

[14] Ioana C Finegan, Ronald F Gibson b, 1999 "Recent research on enhancement of damping in polymer composites", *Composite Structures 44 89-98.*

[15] Y. Hong, X.D. He, R.G. Wang, 2012" Vibration and damping analysis of a composite blade", *Materials and Design 34 98–105.*

Transient Structural Analysis of Electric Bus Chassis Frame

A Hari Kishan, Pilla Kondalarao

Department of Mechanical Engineering, Vignan's Institute of Information Technology, Visakhapatnam 530049, India

* Corresponding author e-mail: kishanviit13@gmail.com, nani451018@gmail.com.

Abstract: From the few past years, usage of electric vehicles is increasing over the usage of the petrol or diesel vehicles. The main reason behind that is increasing cost of fuel and air pollution. In this vehicle chassis frame also play a major role for giving the support to the other parts acting as skeleton of the structure. During heavy load and dynamic loading conditions the structure of the chassis frame may fail and causes breakdown. It is very important to analyze the chassis frame under different loading conditions i.e., static and dynamic loading conditions. From this research paper work, the structure of the chassis frame with box section, channel section, channel section (with increased thickness) and channel section using stiffener are analyzed under static loading and transient loading (Time varying loads) in FEA Software (ANSYS 2020R2 Student Version). The material of the chassis frame is selected as Steel, Aluminum of type 6061-T6 , Aluminum of type 7075-T6 and Titanium alloy. Loads and boundary conditions are applied as per the research article [1] and proper mesh is given to the geometry. From the static and transient analysis it is concluded that channel section using stiffener is having less stresses and less deformations compared to the box section, channel section and channel section with increased thickness using with different materials. Chassis frame with steel having less stresses and low deformation compared to the remaining materials. By this result, it has concluded that the channel section with stiffener using steel material gives better performance compared to the other sections.

Keywords: FEM, Static Structural Analysis, Transient Structural Analysis.
Nomenclature: σyt = Yield Strength, ρ = Density, E = Young's Modulus &μ = Poisson's ratio.

1. Introduction:

"Chassis" a French term which means the complete automobiles without body and it includes all the systems like power plant, transmission, steering, suspension, wheels tyres, auto electric system etc. without body. There are 3 types of chassis frame sections which are shown in the Figure 1.

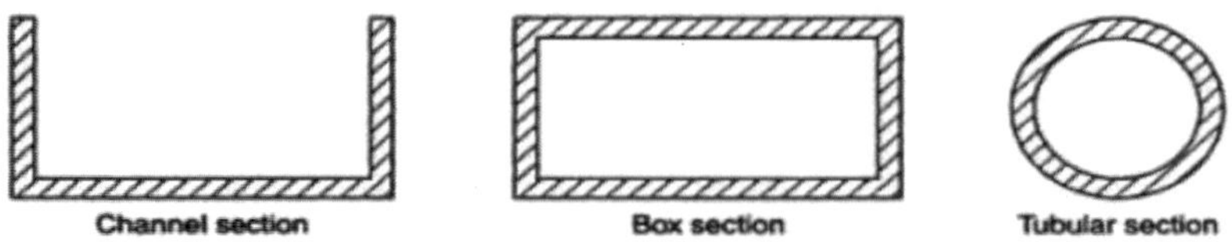

Figure 1. Different type of chassis frame sections

Note: Channel Section is good for bending; tubular section is good for twisting and box section is good for both bending and twisting.

S. Nanda Kumar [1] performed FEA on the chassis frame and calculated the stresses. He proposed that we can achieve weight reduction of 65.61% and 64.33% by using chassis frame with aluminum 6061-T6 and aluminum 7075-T6 instead of steelthrough which overall performance is improved. Patel Vijay Kumar V [2] performed FE analysis on chassis frame and proposed that weight of the chassis frame is reduced by 6.68 % and the results are also reduced respectively 12.14 %, 8.55 % and 11.20 %. It is clear that design is safe.Shaik Neelophar Begum [3] performed FE analysis on chassis frame and proposed that in C type Cross section,stresses anddeformations are more where as I type cross section, they are less and where as they are moderate in Box type cross section and also moderate total deformation and equivalent stresses. Rohan Y Garud [4] has carried out FEA on the chassis, it can be concluded that 8 mm thick Advanced High Strength Steel chassis shows better result as compared to original 5 mm thick steel chassis. Amar R [5] has concluded that, Eicher 11.10 chassis was subjected to finite element analysis and it says that thethe design is safeas the Von-mises stresses are less than the permissible value. And maximum equivalent stress and total deformation are obtained through the analysis done. Vishal Francis [6] conducted FE analysis on the ladder type chassis frame of jeep and concluded shear stresses are less than the permissiblevalue so the design is safe for all three materials. Stresses are minimum in aluminum alloy compared to the mild steel andaluminum alloy. Ramesh kumar [7] performed design and optimization analysis on the chassis frame. Fatigue analysis is carried out on the chassis using ANSYS software to identify the weak points and the life of the chassis.

2. Problem Description

2.1 Problem Statement

To perform Transient structural analysis on the Chassis Frame using 3D FEA and to verify the design of components in strength point of view under transient loads.

2.2 Geometry

The structure of the Chassis Frame [Box and Channel Section] is designed in the ANSYS WORKBENCH 2020 R2 Student Version (Figure 2). The dimensions of the Chassis Frame are taken from the research article [1]. The isometric view of the chassis frame is shown in the below Figure. 3.

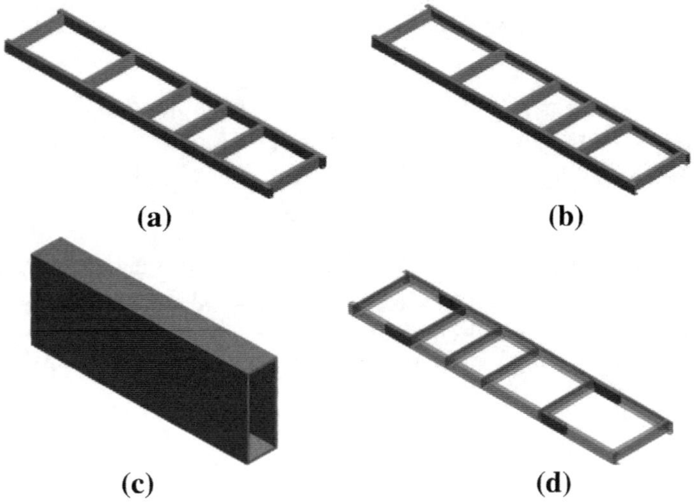

Figure 2. Chassis Frame (a) Box Section (b) Channel Section (c) Stiffener (d) Channel Section with Stiffener

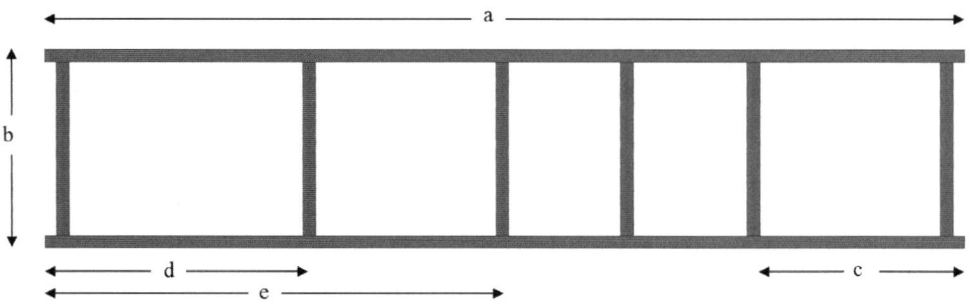

Figure 3. Dimensions of the Chassis Frame

The Dimensions of the Chassis frame is shown in Figure 4.

a = 11470 mm, b = 2420 mm, t = 10 mm, c = 2550 mm, d = 3250 mm, e = 5735 mm

Width of the Frame's Cross-section is 300 mm, Height of the Frame's Cross-section is 150 mm and thickness of the chassis frame is 10 mm.

2.3 Materials

In this research work, four different types of materials are used for the chassis frame i.e., Steel A709M, Aluminium 6061-T6, Aluminium Alloy 7075-T6 and Titanium Alloy. The properties of the materials are listed in the below Table 1

Table 1. Mechanical Properties of the materials.

Material Name	ρ (kg/m^3)	E (MPa)	μ	σ_{yt} (MPa)
Steel A709M Grade 345 W	7850	200000	0.3	260
Aluminium Alloy 6061-T6	2700	69000	0.33	275
Aluminium Alloy 7075-T6	2800	71700	0.33	505
Titanium Alloy	4620	96000	0.36	930

2.4 Mesh

Meshing is the process of converting geometry entities into the finite elements. The type of element used for the chassis frame with box and channel sections is solid 186. SOLID186 is a higher order 3-D 20-node solid element that exhibits quadratic displacement behavior.

2. 5 Loads & Boundary Conditions

The load acting on the frame is 214839 N and this particular load is acting on chassisframe mainly on two longitudinal member. So total load acting is distributed between the two members of the frame equally. Therefore the individual weight on the frame is 107419.5 N.The realistic boundary conditions are applied to the chassis frame by giving the fixed support to the horizontal channels of the frame as shown in the below figure.5.

Figure 4. Loads and boundary conditions on the Chassis Frame

2.5.1 Transient Load

The magnitude of the transient load which is applied on the chassis frame of the vehicle with different sections is shown in the below figure

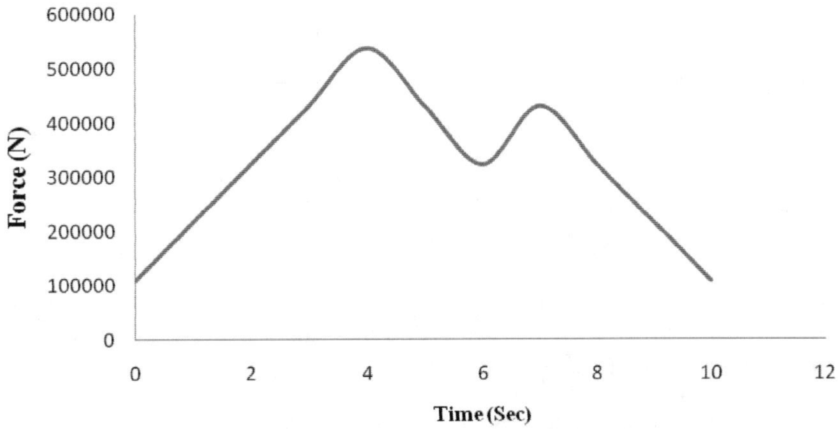

Figure 5. Transient Load

2.6 Connections

The realistic condition for connections (Contacts) is bolded joint connection between the frame members to from assembly frame. Here, the contacts are assumed to be bonded contacts.

2.7 Case Studies

The following case studies are studied on the chassis frame under static structural and transient structural loads. The list of case studies is described in the Table 2.

Table 2. Case Studies

S. No	Type	Thickness (mm)	Volume (mm^3)
1	Chassis with Box Section	10	4.373e+008
2	Chassis with Channel Section	10	2.7578e+008
3	Chassis with Channel Section with increased Thickness	17	4.373e+008
4	Chassis with Box Section using Stiffeners	10	4.373e+008

3. Static Structural Analysis of Chassis Frame

Static structural analysis is performed for initial static conditions to check whether the structure is with stand the loads in strength point of view on the chassis frame with box section, channel section, channel section with increased thickness and channel section with stiffeners using different materials and the calculated results are listed in the below tables.

The static structural analysis of chassis frame with box section using Steel A709M is performed and maximum von-misses stress of the structure is shown in the below figure.

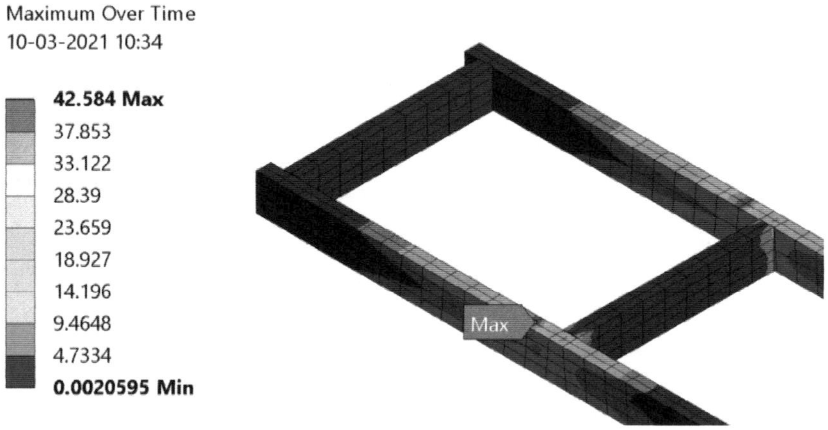

Figure 6. Von-misses Stress of Chassis Frame with Steel A709M

From the above figure, it is shown that the maximum von-misses stress of the chassis frame using steel A709M is 42.584 MPa, which is less than the material strength (260 MPa). From the above figure, it is clear thatthe maximum stress is located on the rear hang part of the chassis frame in bottom surface.

The static structural analysis for the chassis frame with cases [1, 2, 3 & 4] using Steel A709M Grade 345 W, Aluminium Alloy 6061-T6, Aluminium Alloy 7075-T6 and Titanium Alloy are performed and stress values are listed in the below Table 3.

Table 3. Von-misses Stress values under Static Structural Analysis

Material	Box Section	Channel Section	Channel Section with Increased Thickness	Channel Section with Stiffener
Steel A709M	42.584	59.970	43.007	39.729
Aluminum 6061-T6	42.850	59.593	42.901	37.551
Aluminum 7075-T6	42.850	59.593	42.901	37.551
Titanium Alloy	43.188	59.071	42.751	37.134

From the above table, it is noticed that the maximum stresses are obtained with the channel section with 10mm thickness and minimum stresses are obtained with the channel section with stiffeners. The graphical plot of the stress values of the leaf springs using different materials with different sections are plotted in the below Figure 7.

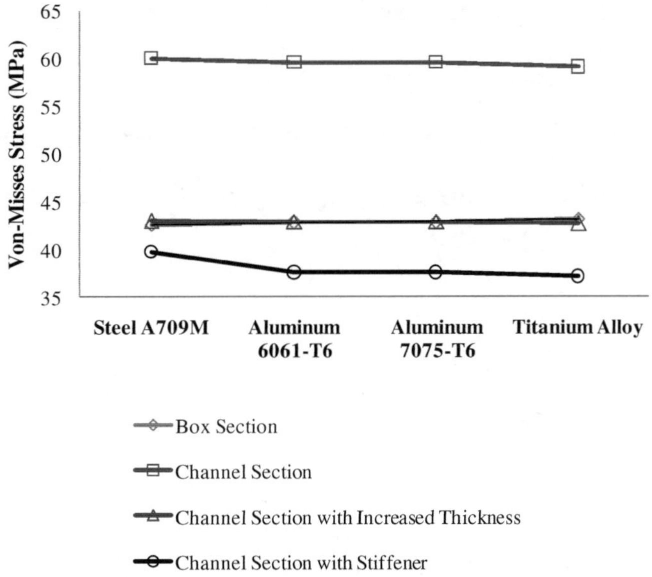

Figure 7. Von-misses Stresses of the Chassis Frame under Static Loads

From the above figure it is shown that, channel section with 10mm thickness has maximum stress values, channel section with stiffener has minimum stress values. The box section and channel section with increased thickness has similar value of stresses for every material. Steel A709M having higher stresses and Titanium has less stress values.

4. Transient Structural Analysis of Chassis Frame

The chassis frame is subjected to Transient structural analysis with box section with Steel A709M and maximum von-misses stress over the time is calculated using FEA software and it is shown in the below Figure 8.

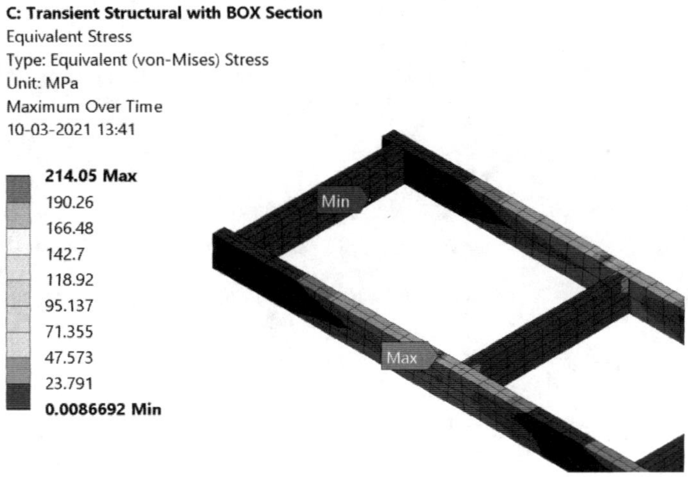

Figure 8. Maximum Von-misses stress of the chassis frame with box section

From the above figure, it is shown that the maximum von-misses stress of the chassis frame using steel A709M over the time range of 1 sec to 10 sec is 214.05 MPa, which is less than the material strength (260 MPa). From the above figure, it is clear that the maximum stress is located on the rear hang part of the chassis frame in bottom surface where constrains are given.The minimum von-misses stress is located on the extreme end of the rear hang part of the chassis frame.The minimum, maximum and average stress values of the chassis frame with box section using steel A709M material is listed in the below Table 3.

Table 4. Stress values of the Chassis Frame using Steel A709M

Time (Sec)	Maximum Stress (MPa)	Time (Sec)	Maximum Stress (MPa)
1	85.53	6	128.36
2	128.5	7	171.12
3	171.04	8	128.37
4	214.05	9	85.507
5	171.1	10	42.804

It is noted that the maximum stress value of the chassis frame using Steel A709M over the time range of the 1 sec to 10 sec of transient load are 214.05 MPa which is takes place at 4th second of the transient load.The maximum von-misses stresses of the chassis frame [case 1, 2, 3 & 4] using different materials at time of 4 sec are shown in the below Figure 9.

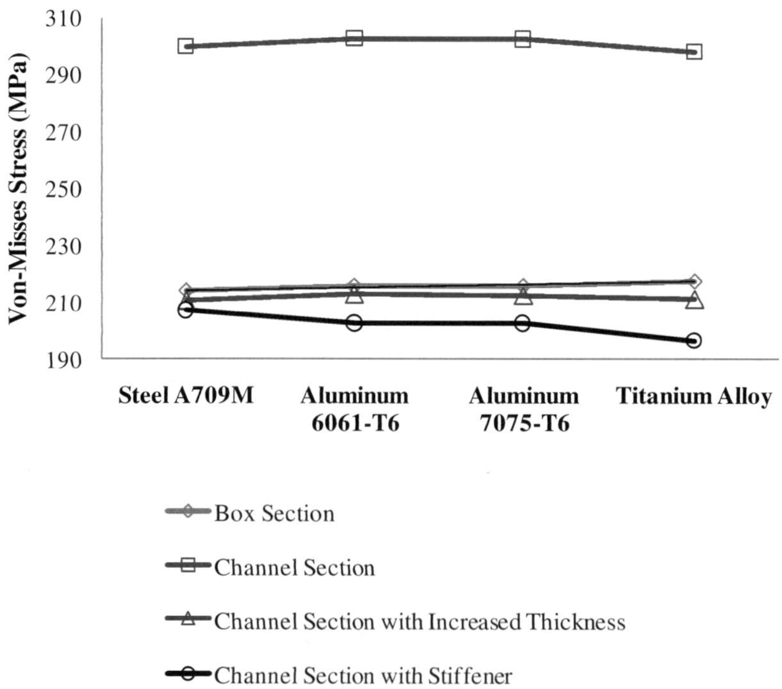

Figure 9. Von-misses Stress of the Chassis Frame under Transient Load

From the above figure, it is shown that the maximum stresses of the chassis frame using different materials at 4 sec of transient time are obtaining with the channel section. Minimum value of stress values are obtained with the channel section with the stiffeners.

The total deformation of the chassis frame with box section using Steel A709M for transient load range of 1 sec to 10 sec is shown in the below Figure 10.

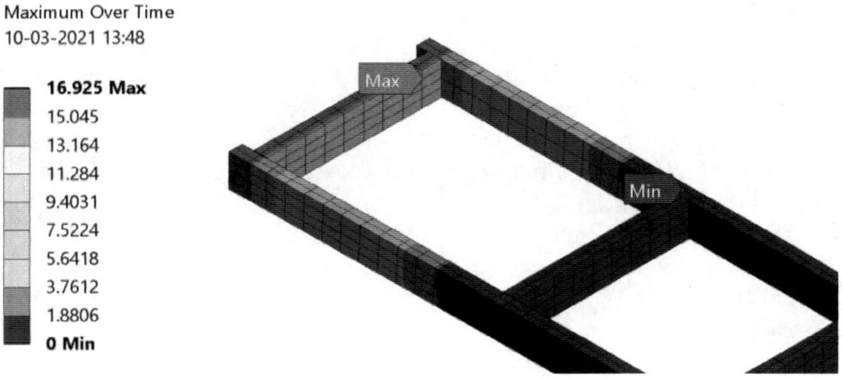

Figure 10. Total Deformation of the chassis frame with box section using Steel A709M

From the above figure, it is shown that the maximum total deformation of the chassis frame is 16.925 mm.The maximum total deformation is located on the extreme rear hang part of the chassis frame.The minimum total deformations are located at the fixed support locations of the chassis frame with box section. The total deformations of the chassis frame [case 1, 2, 3 & 4] using different materials are plotted in the below Figure 11.

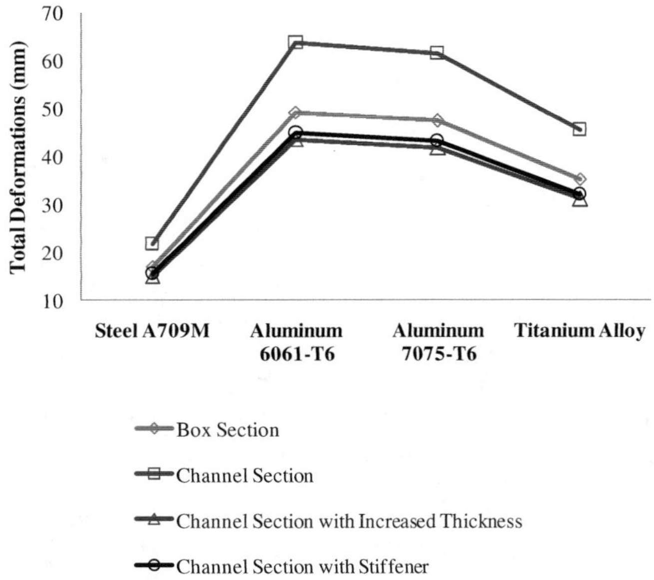

Figure 11. Total Deformations of the Chassis Frame under Transient Load

From the above figure, it is observed that the maximum deformations are obtained with the channel section with 10mm thickness and minimum deformations are obtained with the channel section with increased thickness.

5. Conclusions

The chassis frame of the electric bus is analysed under static and transient loads and list of conclusions are drawn from the work.

1. Von-misses stresses of the chassis frame with box section, channel section, channel section with increased thickness and channel section with stiffener are within the limits of the material strengths. That indicates that the structure withstands the static and transient loads.
2. The von-misses stresses of the chassis frame with channel section with 10mm thickness posses higher values due to the less cross sectional area of the section compared to the other sections.
3. The total deformations of the chassis frame with channel section with 10mm thickness posse's higher values due to the less cross sectional area of the section compared to the other sections.
4. The von-misses stresses of the chassis frame with channel section using stiffeners posses less values compared to the other sections.
5. The total deformations of the chassis frame with channel section using increased thickness posses less values compared to the other sections.
6. Structure using Steel A709M showing better results in stresses and deformations compared to the other materials.
7. The structure of the chassis frame with channel section using stiffener has low weight and higher strength.
8. It is recommended to study the structure under other dynamic conditions using super alloys.

References

1. S. Nanda Kumar, S.Seenivasan, A. Mahammed Salih and Muhammed Saifudheen, 2020 "Weight optimization and structural analysis of an electric bus chassis frame" *Materials Today proceedings* (In press)
2. Patel Vijay Kumar V and R. I. Patel, 2012."Structural Analysis of Automotive chassis and design Modification for weight reduction", *International Journal of Engineering Research and Technology,*, **1**(3) 1-6.
3. Shaik Neelophar Begum, S. P. Bhanu Murthy, 2016 "Modeling and Structural Analysis of Vehicle chassis frame made of polymeric composite material", *International Research Journal of Engineering and Technology,* **03**, 08.
4. Rohan Y Garud, Shahid C Tamboli, Dr. Anand Pandey, "Structural Analysis of Automotive Chassis, Design Modification and optimization", *International Journal of Applied Engineering Research,* **13**, 11.
5. Amar R. Contractor, Gaurav P. Rathod and Tushar M. Patel, Design and analysis of ladder Frame chassis considering support at contract region of leaf spring and chassis frame", *IOSR Journal of mechanical and civil Engineering,* **12**, 2.
6. Vishal Francis, Rajnish Kumar Rai, Anup Kumar Singh, Pratyush Kumar Singh and Himanshu Yadav, "Structural Analysis of Ladder Chassis Frame for Jeep Using Ansys", *International Journal of Modern Engineering Research,* **4**, 4.
7. Ramesh Kumar.S,Dhandapani. N.V, Parthiban.S, Kamalraj.D, Meganathan.S and Muthuraja.S, "Design and Analysis of a automotive chassis frame using Finite element Method", *International Journal of Pure and Applied Mathematics,* **118**, 20.
8. Naveen Kumar Chandramohan, Mohanraj Shanmugam, S. Sathiyamurthy, S. Tamil Prabakaran, S. Saravanakumar V.S. Shaisundaram, 2020 "Comparision of chassis frame

design of Go-Kart Vehicle Powered by internal combustion engine and electric motor". *Materials Today Proceeding.*(In press)

9. Nipun Sinha and Kaushik Kumar, 2020 "Efficacy of Vehicle Chassis of Polymetric Composite", *Materials Today proceedings,* **22,** 2683-2466.

Enhancing diesel engine performance by using nano-dispersing agents in fuel: A review

Manoj Kumar Elipey [1], *, Gopisetty Ravi Chandra [1], P. Sneha Latha [2], Anchupogu Praveen [3], B. Ratna Sunil [3], *

[1] Department of Mechanical Engineering,
Vardhaman College of Engineering, Shamshabad 501218, India
[2] Department of Mechanical Engineering, School of Engineering and Technology,
Sri Padmavati Mahila Visvavidyalayam,Tirupati 517502, India
[3] Department of Mechanical Engineering,
Bapatla Engineering College, Bapatla 522101, India

* Corresponding authors e-mail: emanoj2000@gmail.com, bratnasunil@gmail.com

Abstract. Biofuel is a renewable, non-toxic and ecofriendly alternative fuel becoming promising for compression ignition (CI) engines. After reviewing various biofuels used in the CI engines, it was noticed that biofuels show a considerable impact on the efficiency and controlling the pollutants released from the CI engines. Fuel modification is the best technique to enhance the performance of the engine and to control the exhaust pollutants compared to other strategies. Recently, the usage of nanoparticles dispersed fuels is found to be a viable strategy to improve the properties of fuel, engine efficiency and to reduce the engine exhaust emissions. In this paper, a brief review on the effect of nano-additives in fuel on engine efficiency is presented. Preparation methods of nanoparticles dispersed fuels and future perspectives are also discussed.

Keywords: Biofuel, nanoparticles, diesel engine, alternate fuels, engine efficiency.

1. Introduction

Diminishing fossil fuel reserves and improved global warming situations in the world lead to many researchers to search for alternative resources of fuel. Diesel engine is used in many applications such as power generation, automobile sector and agriculture equipment due to its higher efficiency compared to the gasoline engine. But, the diesel engine also emits the major pollutants which include carbon monoxide (CO), carbondioxide (CO2), nitrogen oxides (NOx), hydrocarbons (HC) and smoke [1]. These pollutants are a serious concern in the context of global warming and associated ill effects on the environment. Biodiesel is an alternative fuel for diesel engine and can be used without modification of the engine. Biofuels are nontoxic, renewable, non-toxic and environmental friendly. Biodiesel used in the diesel engines leads to controle the amount of emitting pollutants such as hydrocarbon, carbon monoxide, smoke and particulate matter, but improves the nitrogen oxides. By using biodiesel, the engine performance can be improved [2]. On the other side, adding different nanoparticles in the fuels has been observed as a promising strategy to increase the engine efficiency and also to decrease the level of pollutants emitted from the engine significantly.

2. Effect of biodiesel and its blends used in the diesel engine

The usage of biodiesel and its blends used CI engines was elaborately investigated by several research groups and suggested that different methods to reduce the viscosity of the fluid with the help of various production techniques. H. G. How et.al [3] reported that a variety of biodiesel blends (B10 to B30 and B50) and demonstrated a lower heat release rate compared with diesel. They observed that BSEC was diminished with an increment in BTE for biodiesel blends. Further, increasing the biodiesel percentage in biodiesel blends reduces the CO emissions and slightly enhances the NOx emissions. M. Vijayakumar et. al. [4] produced the various blends of Mahua methyl ester (MME) and used to operate the CI engine. From the studies, they demonstrated an increase in the fuel consumption than the diesel for B20 and B40 blends and BTE was also observed as slightly decreased. The reduced trend was noted for exhaust emissions, smoke opacity and also CO when biodiesel was used. A. Praveen et. Al [5] compared the engine performance of a diesel engine using canola and sesame blended biodiesel and observed that the CBD20 test fuel shows the higher brake thermal efficiency than the other fuels. It is also noticed that the emission of NOx in the case of biodiesel were enhanced by 7.4-9.8% than the diesel fuel. Banapurmath et. al. [6] used cotton seed methyl esters and the performance of CI engine was tested. The results show that a decrement in BTHE when biodiesel was used than the diesel fuel. This can be claimed to the poor fuel spray behaviour, increased viscosity and decreased calorific value exhibited by methyl esters. The BSFC of CSME biodiesel blends was reported to be slightly greater than the diesel fuel. In addition, it is evident that for all the biodiesel blends, the emission of CO_2, CO and HC were considerably reduced along with increased NOx emissions. Dinesh et. al. [7] also investigated the influence of Tamanu added biodiesel on the performance of CI engine and better efficiency and lowered exhaust emissions was observed with B10 fuel. Karthik et. al [8] have done various tests using rubber seed oil (RSO20) as the fuel for a CI engine and compared the performance with diesel. It is observed that BTHE measurement with RSO20 at 30° was higher and BSFC was also increased with the increased injection time. Because of the higher cylinder temperature, CO and UHC emissions were lowered for the increased injection time. Further, the NOx emissions were increased with higher injection timing and decreased by the retarded injection timing. The reported findings establish the potential of biodiesel blends on enhancing the engine efficiency and also controlling the pollutants which are inevitable in the emissions of a CI engine.

3. Influence of nano-additives used in biodiesel blends of a diesel engine

Basically, nanoparticals are the powders or colloids in the range of few tens of nanometers and exhibit a higher surface area. Therefore the properties associated with nanoparticles are significantly different compared with microparticles. The recent developments in producing nanomaterials have expanded the applications of nanomaterials in energy sector. Quick oxidation, higher stability, lower melting point, high generation of heat during combustion, decreased heat of fusion and higher heat transfer nature make nanoparticles as promising candidates for various energy engineering applications. Particularly, by using as dispersing agents in fuel, the engine performance and the characteristics of the exhaust gases can be significantly altered [9, 10]. The promising role of nanoadditives to alter several properties which play very important role in energy engineering has been well established. Several applications which include thermal conductivity fluids in the cooling systems of engines and fluids used in propulsion systems are a few examples. As suggested by Ribeiro N M et al. [11], the nanoadditives that are added to the diesel must have a few specific functions as summarized here. (1) the exhaust emissions must be reduced, (2) the oxygen concentration must be increased in the filter as well as in the, (3) the fuel stability should be increased, (4) the viscosity will be improved, (5) the reduced flash point and ignition delay period, (6) improved mode of contact from chemical-chemical to metal-metal; when subjected to higher loads. From the literature, it can be understood that several group of nanomaterials were used as dispersing agents in the fuels which include metallic nanopowders, polymers, ceramic nanomaterials, non-organic materials, nano metal oxides and carbon based nanomaterials such as carbon nanotubes (CNTs) [12]. Fig 1 shows different steps involved in blending nanoparticles into diesel-biodiesel and conducting the engine performance tests.

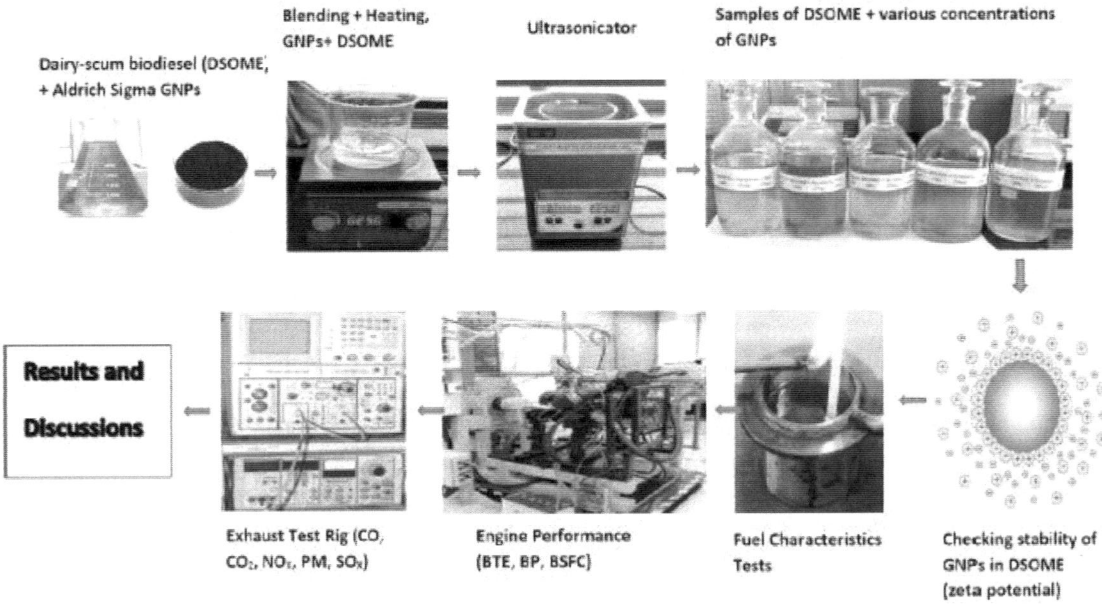

Fig 1 Sequence of steps in studying the nanoparticle dispersed fuel on engine perofrmance [12].

Metallic oxide nanoparticles are the most common dispersants used to blend in diesel – biodiesel. In addition to decreasing the fuel consumption and controlling the exhaust emissions, metallic nanoparticles such as manganese (Mn), aluminium (Al), titanium (Ti), cerium (Ce) zirconium (Zr), zinc (Zn) and copper behave as catalysts and complete the process of fuel combustion. From the works of Nanthagopal et. al [13], the effect of nano-ZnO and nano-TiO_2 added to Calophyllum Inophyllum biodiesel and promising effect on improving the kinematic viscosity was observed. Anchupogu et. al. [14] also conducted experiments by using MWCNTs dispersed Calophyllum Inophyllum biodiesel. They concluded that the pollutants in the exhaust gases were decreased and BTE was increased due to the added MWCNTs. Similar studies were also reported by Ranjan et. al [15] Chinnasamy et.al [16] and Khadijeh et.al [17] with the added nanodispersing agents into the fuel. Anchupogu praveen et. al [18] studied the effect of nano-TiO_2 in diesel fuel for operating the diesel engine. They reported that the BTHE was increased with the addition of nanoparticles and reduction of tail pipe emissions was considerable. Paramashivaiah et.al [19] added Graphene to diesel blend of Simarouba methyl ester in a diesel engine and improved BTE (9.14%) was observed in addition to decreased unburnt hydrocarbons, CO, oxides of nitrogen by 15.38%, 42.85%, 12.71% respectively with SME2040 fuel. Rashmi et. al [20] investigated the nano-fuel used in the diesel engine. The obtained results demonstrate increased efficiency (7%) by adding nano- CuO to the fuel and reduced BSFC (6%) respectively, compared to diesel fuel. Hence, it can be concluded that variety of nano-dispering agents are added to diesel and biodiesels, which can be an effective way to enhance the efficiency of the engine and to control the polluting emissions. Fig 2 shows the comparison of different nano particles dispersed in diesel to improve the engine performance as compared by M.E.M. Soudagar et al. [12].

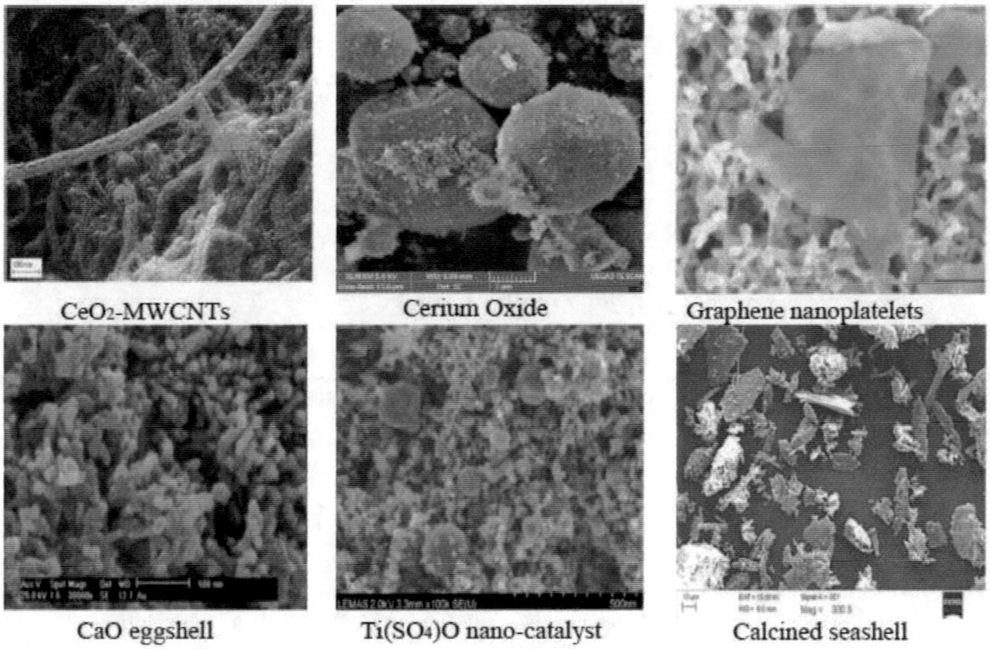

Fig 2 SEM images of different nanoparticles used to disperse in fuel to assess the performance of diesel engine [12].

4. Preparation of nanoparticle dispersed fuels

Preparing the nanofluids involves proper blending of the nanoparticles into the dispersing solvent. In producing nanoparticles dispersed fuels, the size, shape and chemical nature of the nanoparticle are the important factors must be considered. Usually, nanofluids can be prepared by one step method or two step method. An idle procedure should results homogeneous mixing of nanoparticles throughout the fuel and also must reduce the agglomeration. The prepared solution with a stable nanodispersing particles are highly preferred to yield better results.

In one step preparation method, the preparation and dispersion of nanoparticles can be completed in a single process. This was introduced by Akoh et al. [21]. One step method is also known as Vacuum Evaporation onto a Running Oil Substrat (VEROS) route. Lower level of agglomeration, better stability and decreases cost of preparation are the advantages with one step method. Furthermore, drying the nanoparticles and dispersing in to another fluid can be completely eliminated with the one step process [22-28]. Two step preparation method is most commonly used route where the nanoparticles are separately synthesized and then mixed in the fuel. Commercially available nanoparticles are also used to blend in the fuel. Surfactants are also used to decrease the agglomeration of the nanoparticles Usually, nanoparticles tend to form clusters by agglomeration due to the availability of more surface energy. Several reports present the preparation of nanoparticles through different routes and then mixing along with the surfactants into the fuel [29, 30, 31, 32]. Table 1 list the important surfactants used along with nanoparticles and their significance in adding to fuel.

Table 1 different surfactants used for different nanoparticles to disperse in the fuel

Surfactant	Purpose	Authors
Sodium dodecyl sulfate (SDS)	Modifies the hydrophobic nature of the surfaces of MWCNTs	33
Cetyl Trimethyl Ammonium Bromide (CTAB)	It creates an envelope on the surface of the nanoparticle and controls the sedimentation	34
Sodium dodecyl sulfate (SDS)	This surfactant decreases the agglomeration significantly	35
Tetramethylammonium hydroxide (TMAOH)	It stabilize the nanoparticles and avoids agglomeration	36
Sorbitan Olieate	It decreases the agglomeration and improves the stability of metal nanopowders.	37
Dodecenyl succinic anhydride (DDSA)	Exhibits adhesive nature and stabilizes the nanoparticles in diesel,	38
Polyethylene sorbitol ester (Tween80) and Sorbitan monostearate (Span80)	It is soluble in organic solvents and insoluble in water	39
Oleic acid	With the increased concentration of oleic acid, the tendency to form nano-agglomeration is decreased	40
Sodium dodecyl sulfate (SDS)	Homogenizes the suspension of the nanoparticles and decreases the clustering and agglomeration.	41
Sodium dodecylbenzenesulfonate (SDBS)	This surfactant acts as an anti-hygroscopic agent	42
Sorbitan monooleate	It improves the suspension stability of the nanoparticles	43

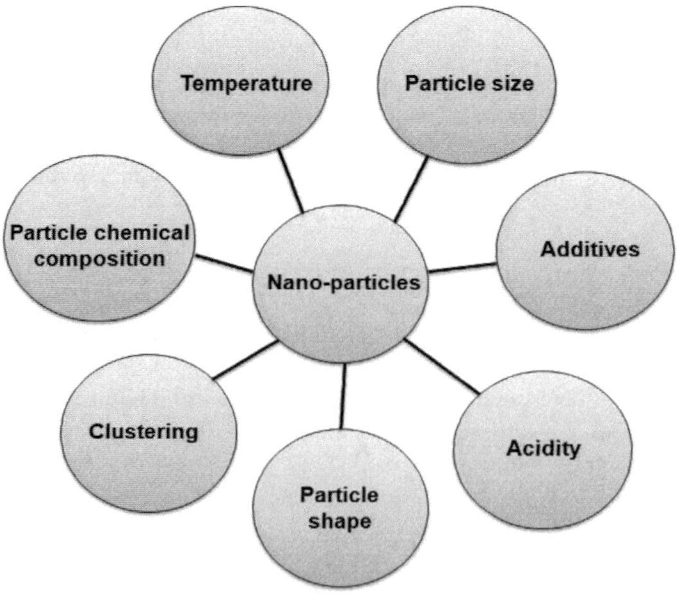

Fig 3 Different factors which affect the performance of nanoparticles as dispersing agents in fuel.

5. Challenges and perspectives

The available literature strongly demonstrates the promising role of nanoadditives to decrease exhaust emissions, carbon monoxide and particulate matter. This was claimed to the promising role of nanoparticles on improving the ignition behavior and more oxygen content to promote better combustion. Majority of the findings are advocating the several advantages of nanoparticle addition, however, a few studies demonstrate increased carbon monoxide concentration due to the addition of $FeCl_3$ nanoparticles. Another limitation with these nanoparticles was the emission of more NO_X when Al-Mg and Co_3O_4 additives were used due to the presence of more oxygen. When applying at higher engine loads, the nano-oxygenated additives decrease the CO emission by increasing the combustion chamber temperature. Additionally, oxygenated additives reduce the cylinder temperature. Other factor such as temperature, particle size, type of additives, acidity, particle shape, resistance to agglomeration and particles chemical composition are a few other factors essentially must be understood while selecting any nanoparticles to mix with the fuel to improve the engine performance as schematically shown in Fig 3.

The recent developments really define a promising scope by using nanoparticles to enhance the engine performance and for using in the biodiesel also reduce the demand for the fossil based fuels. However, there exists a need for comparative studies of different nanoparticles with various biodiesel combinations. Furthermore, combination of metallic and organic nanoparticles as the dispersing material in fuels is still in need to investigate elaborately. Studies on stabilization of nanoparticles in the fuels for longer periods is not completely explored which is essentially required to bring the advantages of using nanoparticles as dispersants in fuels. On the other hand, understanding the reactive nature of added nanoparticles on the surfaces of engine parts and exhaust piping systems is crucial. Furthermore, the emissions from the diesel engines which use nanoadditives must be investigated to find the level of nanoparticles in the exhaust gases and their effect on the environment and human health. Economical issues associated with the production of nanoparticles, blending and then storing for continuous usage also must be considered in order to assess the promising future of nanoadditive fuels for diesel engines.

6. Conclusions

The developments in using nanoparticles as promising additives in fuels for diesel engines are quiet interesting and potential to address the day to day increasing demand for fossil based fuels in the field of energy engineering. The available literature demonstrate that wide variety of nanoparticles including metllic powders, polymers, carbon based nanomaterials and organic particles as the promising dispersing to increase the efficiency of the engine. However, the overall effect of these nano-additives on the engine components, exhaust systems, environment and human heath are the other aspects of the technology must be essentially assessed to bring the advantages of nanoparticles as dispersing agents in fuels of diesel engines.

References

[1] Agarwal D, Agarwal A K 2007 *Applied Thermal Engineering* **27** 2314.
[2] AtabaniA E, dasilva C A. Calophyllum Inophyllum L. 2014 *Renewable and Sustainable Energy Reviews* **37** 644.
[3] How H G, Masjuki H H, Kalam M A, Teoh Y H, Chuah H G. 2018 *Fuel* **227** 154.
[4] Vijay Kumar M, Veeresh Babu A, Ravi Kumar P. 2019 *International Journal of Ambient Energy* **40(3)** 304.
[5] A. Praveen, G. Lakshmi Narayana Rao, B. Balakrishna and K. Seshagirirao 2019 AIP Conference Proceedings **2161** 020029.

[6] Banapurmath N R, Tewari P G, Hosmath R S 2008 *Renewable Energy* **33** 1982.
[7] Dinesh K, Tamilvanan A, Vaishnavi S, Gopinath M, Raj MKS. 2019 *Biofuels* **10(3)** 347.
[8] KarthikN, Rajasekar R, Siva R, Mathiselvan G. 2019 International Journal of Ambient Energy **40(3)** 292.
[9] Saxena V, Kumar N, Saxena V K. 2017 *Renew Sustain Energy Rev* **70** 563.
[10] Venu H, Madhavan V. 2016 *J Mech Sci Technol* **30(5)** 2361.
[11] Ribeiro N M, et al.. 2007 *Energy Fuels* **21(4)** 2433.
[12] Manzoore Elahi M. Soudagar, Nik-Nazri Nik-Ghazali, Md. Abul Kalam, I.A. Badruddin, N.R. Banapurmath, Naveed Akram, 2018 *Energy Conversion and Management* **178** 146.
[13] Nanthagopal K, Thundil K R, Ashok B. 2019 *Waste Biomass Valor* **10** 2001.
[14] Anchupogu Praveen G. Lakshmi Narayana Rao & B. Balakrishna 2018 *International Journal of Ambient Energy*, **40(8)** 791.
[15] Ranjan A, DawnaS S,Jayaprabakar J, Nirmala N, Saikiran K, Sriram S S. 2018 *Fuel* **220** 780.
[16] Chinnasamy C, TamilselvamP, Ranjith R. 2018 *Energy Sources, Part A: Recovery, Utilization, and Environmental Effects* **40** (22) 2735.
[17] Khadijeh HM, Ahmad TA, Barat G,Ahmad AM. 2017 *Fuel* **196** 110.
[18] Anchupogu, Praveen ; Rao, G. Lakshmi Narayana ; Balakrishna, B ; Sankar, B. Ravi ; Umamaheswarrao, P. 2018 *Advanced Science Letters*, 24(8), 5712.
[19] Paramashivaiah B M, Rajashekhar C R. 2017 *I J E Transcations B: Applications* **30 (8)** 1205.
[20] Rashmi RS, Animesh J.. Fuel 2019; 236: 365–372.
[21] Akoh H, Tsukasaki Y, Yatsuya S, Tasaki A. 1978 J Cryst Growth **45** 495.
[22] Tran P, Soong Y. *Preparation of nanofluids using laser ablation in liquid technique*. Pittsburgh, PA, and Morgantown, WV: National Energy Technology Laboratory (NETL); 2007.
[23] Zhu H-T, Lin Y-S, Yin Y-S. 2004 *J Colloid Interface Sci* **277(1)** 100.
[24] Wang X-Q, Mujumdar A S. 2007 *Int J Therm Sci* **46(1)** 1.
[25] Li Y, Tung S, Schneider E, Xi S. 2009 Powder Technol **196(2)** 89.
[26] Das S K, Choi S U, Yu W, Pradeep T. *Nanofluids: science and technology*. John Wiley & Sons; 2007.
[27] Lo C-H, Tsung T-T, Chen L-C, Su C-H, Lin H-M. 2005 *J Nanopart Res* **7(2–3)** 313.
[28] Chang H, Kao M-J. 2007 J Chin Soc Mech Eng **28(2)** 187.
[29] Lee J-H, et al. 2008 *Int J Heat Mass Transf* **51(11–12)** 2651.
[30] Das S K, Choi S U, Patel H E. 2006 *Heat Transfer Eng* **27(10)** 3.
[31] Praveen A, Lakshmi Narayana Rao G, Balakrishna B. 2018 *Egypt J Pet* **27(4)** 731.
[32] Zafarani-Moattar M T, Majdan-Cegincara R. 2012 *J Chem Thermodyn* **54** 55.
[33] Hwang Y, Park H, Lee J, Jung W. 2006 Curr Appl Phys **6** 67.
[34] Ganesh D, Gowrishankar G. 2011 Electrical and Control Engineering. IEEE. 3453.
[35] Vadivel M, Babu R R, Arivanandhan M, Ramamurthi K, Hayakawa Y. 2015 *RSC Adv* **5(34)** 27060.
[36] Shafii M, Daneshvar F, Jahani N A, Mobini K. 2011 *Adv Mech Eng* **3** 529049.
[37] Gan Y, Qiao L. 2011 *Combust Flame* **158(2)** 354.
[38] Sajeevan A C, Sajith V.2013 *J Eng*. 589382.
[39] Rao D S, Dash R K.. 2015 *Adv Mater Lett* **6(3)** 247.
[40] Balamurugan S, Sajith V. 2013 *Adv Mater Res* **685** 128.
[41] Ghadimi A, Metselaar I H. 2013 *Exp Therm Fluid Sci* **51** 1.
[42] Paramashivaiah B, Rajashekhar C 2016 *IOP Conf Ser: Mater Sci Eng* **149(1)** 012083.
[43] Saraee H S, Jafarmadar S, Taghavifar H, Ashrafi S. 2015 *Int J Environ Sci Technol* **12(7)** 2245.

AUTHOR INDEX

Institute of Physics
Dirac House, Temple Back
Bristol BS1 6BE UK

ISSN: 1757-8981
ISBN 978-1-7138-4645-1

9 781713 846451

CURRAN ASSOCIATES INC.
proceedings.com